Springer Series in
Computational
Mathematics

W0049950

13

Eugene L. Allgower   Kurt Georg

Numerical Continuation Methods

An Introduction

With 37 Figures

Springer-Verlag Berlin Heidelberg New York
London Paris Tokyo Hong Kong

Eugene L. Allgower
Kurt Georg

Department of Mathematics
Colorado State University
Fort Collins, CO 80523, USA

Mathematics Subject Classification (1980): 65 H 10, 65 K 05, 90 C 30

ISBN-13: 978-3-642-64764-2   e-ISBN-13: 978-3-642-61257-2
DOI: 10.1007/978-3-642-61257-2

© Springer-Verlag Berlin Heidelberg 1990
Softcover reprint of the hardcover 1st edition 1990

2141/3140 - 5 4 3 2 1 0

Foreword

Over the past ten to fifteen years two new techniques have yielded extremely important contributions toward the numerical solution of nonlinear systems of equations. These two methods have been called by various names. One of the methods has been called the predictor-corrector or pseudo arc-length continuation method. This method has its historical roots in the imbedding and incremental loading methods which have been successfully used for several decades by engineers and scientists to improve convergence properties when an adequate starting value for an iterative method is not available. The second method is often referred to as the simplicial or piecewise linear method. This method has its historical roots in the Lemke-Howson algorithm for solving nonlinear complementarity problems. The idea of complementary pivoting has been adapted and applied to the calculation of fixed points of continuous maps and of semi-continuous set valued maps. In this book we endeavor to provide an easy access for scientific workers and students to the numerical aspects of both of these methods.

As a by-product of our discussions we hope that it will become evident to the reader that these two seemingly very distinct methods are actually rather closely related in a number of ways. The two numerical methods have many common features and are based on similar general principles. This holds even for the numerical implementations. Hence we have elected to refer to both of these methods as continuation methods. The techniques based on predictor and corrector steps and exploiting differentiability are referred to as "predictor-corrector continuation methods". The techniques based on piecewise linear approximations are referred to as "piecewise linear continuation methods". Chapters 3–10 treat the predictor-corrector methods primarily, and chapters 12–16 treat the piecewise linear methods. Chapter 11 bridges the two approaches since it deals with a number of applications were either or both of these numerical methods may be considered. On the other hand, it will also become evident that when the two methods are regarded as numerical tools, there are tasks for which each tool is more appropriate. The reader who has a particular class of problems in mind should be able to determine from our discussions which of the two tools is preferable for his task.

This brings us to the point of discussing some of the philosophy of our book. It is our aim to present the basic aspects of these two rather general numerical techniques, and to study their properties. In presenting formulations for algorithms we have employed pseudo codes using a PASCAL syntax, since it seems to us to be the clearest way of giving a description from which the reader can directly proceed to write a program. We offer some FORTRAN programs and numerical examples in the appendix, but these are primarily to be regarded as illustrations. We hope that the reader will experiment with our illustration programs and be led to make improvements and adaptations suited to his particular applications. Our experience with students and scientists on the American and European continents indicates that FORTRAN remains to date the more widely spread programming language. We emphasize that our programs should not be regarded as perfected library programs.

The FORTRAN code for program 5 was developed by S. Gnutzmann. It represents a simplified version of a more complex PASCAL program which he developed in his Ph.D. thesis. The appendix also contains a detailed description (program 6) of a sophisticated PL program SCOUT which has been developed by H. Jürgens and D. Saupe. This program is not listed due to limitations of space. However, an application of the program on an interesting numerical example is described. We wish to thank our colleagues for these contributions.

The codes of the above programs will also be available for a limited time via electronic mail. Readers wishing to obtain them are requested to provide a valid E-mail address to:

Kurt Georg
Department of Mathematics
Colorado State University
Ft. Collins, CO 80523, USA   **for programs 1–5**

Hartmut Jürgens or Dietmar Saupe
Institut für Dynamische Systeme
Fachbereich Mathematik und Informatik
Universität Bremen
2800 Bremen 33
Federal Republic of Germany   **for program 6 (SCOUT)**

Due to their temporary nature, we refrain from providing the current E-mail addresses of the authors.

We include a bibliography which is rather extensive as of this date. Nevertheless, we are certain that some works will inevitably have slipped by on us. Our aim in presenting this bibliography is to present a fairly complete catalogue of publications on the topic of numerical continuation methods. Hence it has not been possible for us to cite or comment upon all of these articles in the text of the book.

We have in recent years lectured on the material in this book to students of mathematics, engineering and sciences at American and European universities. Our experience suggests that a background in elementary analysis e.g. the implicit function theorem, Taylor's formula, etc. and elementary linear algebra are adequate prerequisites for reading this book. Some knowledge from a first course in numerical analysis may be helpful. Occasionally we need some slightly more sophisticated tools e.g. parts of chapters 8 and 11. Such passages can be skimmed over by the mathematically inexperienced reader without greatly damaging the comprehension of the other parts of the book.

At this point it gives us pleasure to acknowledge the help and support which we received during the preparation of this book. First of all to our wives who had to tolerate much tardiness and absence during the writing. To Anna Georg we owe special thanks for typing, learning TeX and preparing much of the bibliography.

We received a great deal of encouragement over the years and invitations for visits to the University of Bonn from H. Unger and to Colorado State University from R. E. Gaines respectively. During the writing of parts of this book both authors received support from the National Science Foundation under grant # DMS - 8805682 and from the Deutsche Forschungsgemeinschaft under Sonderforschungsbereich 72 at the University of Bonn. E. Allgower also received support from the Alexander von Humboldt Foundation.

A number of our friends have been kind enough to critically read parts of our manuscript while it was in preparation. We are grateful for the helpful comments and corrections given by S. Gnutzmann, D. Saupe, P. Schmidt, Y. Yamamoto. Many typos and mistakes in preliminary versions were caught by students in our courses at Colorado State University and in the seminar of K. Böhmer at the University of Marburg. For this we would like to make a well deserved acknowledgment.

Fort Collins, Colorado

January, 1990   Eugene L. Allgower and Kurt Georg

Table of Contents

Table of Pseudo Codes

Chapter 1. Introduction

Continuation, embedding or homotopy methods have long served as useful tools in modern mathematics. Their use can be traced back at least to such venerated works as those of Poincaré (1881–1886), Klein (1882–1883) and Bernstein (1910). Leray & Schauder (1934) refined the tool and presented it as a global result in topology viz. the homotopy invariance of degree. The use of deformations to solve nonlinear systems of equations may be traced back at least to Lahaye (1934). The classical embedding methods may be regarded as a forerunner of the predictor-corrector methods which we will treat extensively in this book.

Stated briefly, a homotopy method consists of the following. Suppose one wishes to obtain a solution to a system of N nonlinear equations in N variables, say

$$(1.1) \qquad\qquad F(x) = 0\,,$$

where $F : \mathbf{R}^N \to \mathbf{R}^N$ is a mapping which, for purposes of beginning our discussion we will assume is smooth. When we say a map is smooth, we shall mean that it has as many continuous derivatives as the subsequent discussion requires. We do this to make our statements less cumbersome. Let us consider the situation in which very little a priori knowledge concerning zero points of F is available. Certainly, if on the contrary a good approximation x_0 of a zero point $\bar{x}$ of F is available, it is advisable to calculate $\bar{x}$ via a Newton-type algorithm defined by an iteration formula such as

$$(1.2) \qquad\qquad x_{i+1} := x_i - A_i^{-1} F(x_i), \quad i = 0, 1, \ldots$$

where A_i is some reasonable approximation of the Jacobian $F'(x_i)$.

Since we assume that such a priori knowledge is not available, the iteration (1.2) will often fail, because poor starting values are likely to be chosen. As a possible remedy, one defines a homotopy or deformation $H : \mathbf{R}^N \times \mathbf{R} \to \mathbf{R}^N$ such that

$$(1.3) \qquad\qquad H(x,1) = G(x)\,, \quad H(x,0) = F(x)\,,$$

where $G : \mathbf{R}^N \to \mathbf{R}^N$ is a (trivial) smooth map having known zero points and H is also smooth. Typically, one may choose a **convex homotopy** such as

$$(1.4) \qquad\qquad H(x, \lambda) := \lambda G(x) + (1 - \lambda)F(x),$$

and attempt to trace an implicitly defined curve $c(s) \in H^{-1}(0)$ from a starting point $(x_1, 1)$ to a solution point $(\bar{x}, 0)$. If this succeeds, then a zero point $\bar{x}$ of F is obtained. Another standard deformation which is often used is the **global homotopy**

$$(1.5) \qquad\qquad H(x, \lambda) := F(x) - \lambda F(x_1).$$

The reader will have realized that several questions immediately arise:

1. When is it assured that a curve $c(s) \in H^{-1}(0)$ with $(x_1, 1) \in$ rangec exists and is smooth?
2. If such a curve exists, when is it assured that it will intersect the target homotopy level $\lambda = 0$ in a finite length?
3. How can we numerically trace such a curve?

The first question is answered by the implicit function theorem, namely if $(x_1, 1)$ is a **regular zero point** of H i.e. if the Jacobian $H'(x_1, 1)$ has full rank N, then a curve $c(s) \in H^{-1}(0)$ with initial value $c(0) = (x_1, 1)$ and tangent $\dot{c}(0) \neq 0$ will exist at least locally i.e. on some open interval around zero. Furthermore, if zero is a **regular value** of H i.e. if all zero points of H are regular points, then this curve is diffeomorphic to a circle or the real line. This can be seen by a more sophisticated application of the implicit function theorem as given by Milnor (1969).

The second question is linked with existence theorems in nonlinear analysis, in particular with solution techniques using deformation or degree. Generally, it is sufficient to require some boundary condition which essentially prevents the curve from running to infinity before intersecting the homotopy level $\lambda = 0$, or from returning back to level $\lambda = 1$, see figure 1.a.

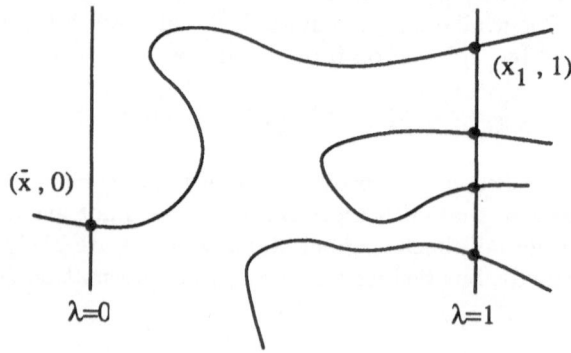

Figure 1.a Different possibilities for the curve $c(s)$

Such boundary conditions must be investigated in the context of a particular problem. In chapter 11, several such cases are discussed.

It is evident that the map $H : \mathbf{R}^N \times \mathbf{R} \to \mathbf{R}^N$ which we first introduced as a homotopy, could equally well have arisen as an arbitrary underdetermined system of equations. Typical sources of such problems are discretizations of operator equations involving a natural parameter λ, e.g. boundary value problems, integral equations, delay-differential equations etc. In such cases, the general theory of such problems often provides some qualitative a priori information about the curve $c(s)$. The numerical tracing of $c(s)$ will yield more qualitative and quantitative data.

The essential theme of this book is to deal extensively with the third question. If the curve c can be parametrized with respect to the parameter λ, then the classical embedding methods can be applied. These have been extensively surveyed by Ficken (1951), Wasserstrom (1973) and Wacker (1978). The basic idea in these methods is explained in the following algorithm for tracing the curve from, say $\lambda = 1$ to $\lambda = 0$.

(1.6) Embedding Algorithm.   *comment:*

    input

        begin

        $x_1 \in \mathbf{R}^N$ such that $H(x_1, 1) = 0$;   *starting point*

        $m > 0$ integer;   *number of increments*

        end;

    $x := x_1$; $\lambda := (m - 1)/m$; $\Delta\lambda := 1/m$;

    for $i = 1, \dots, m$ **do**

    begin

    solve $H(y, \lambda) = 0$ iteratively for y

        using x as starting value;   *e.g. use a Newton-type iteration (1.2)*

    $x := y$; $\lambda := \lambda - \Delta\lambda$;

    end;

    output x.   *solution obtained*

The idea behind the embedding algorithm is quite clear: if the increment $\Delta\lambda$ is chosen sufficiently small, then the iterative process will generally converge since the starting value x will be close to the solution of $H(y, \lambda) = 0$. The drawback of this method is clearly that it will fail when turning points of the curve with respect to the λ parameter are encountered, see figure 1.b.

In some instances, even if the curve is parametrizable with respect to λ, it may be necessary to choose an extremely small increment $\Delta\lambda$ in order for the imbedding algorithm to succeed. The failure or poor performance of the above embedding method can be attributed to the fact that the parameter λ may be ill suited as a parametrization for the curve. One remedy is to consider that

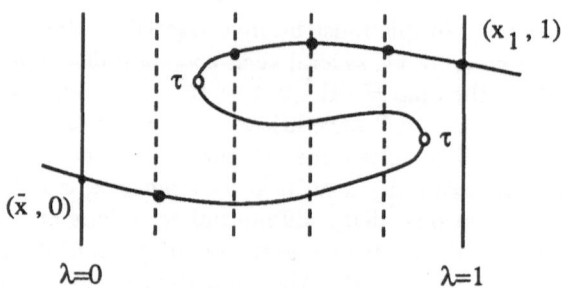

Figure 1.b Failure at turning points τ

the arclength is a natural parameter for the curve. For numerical purposes, it is unnecessary to explicitly use the arclength parameter. It is sufficient to merely approximate this idea e.g. using pseudo-arclength methods or merely suitable parameter switching. However, for purposes of exposition, we use the arclength parameter as an expedient device.

The curve c, now parametrized with respect to arclength s, may be regarded as the solution of an initial value problem which is obtained by differentiating the equation

(1.7) $H\big(c(s)\big) = 0$

with respect to s:

(1.8) $H'(c)\dot{c} = 0, \quad \|\dot{c}\| = 1, \quad c(0) = (x_1, 1).$

The idea of using a differential equation such as (1.8) may be traced at least to Davidenko (1953). Now it is clear that methods for numerically solving initial value problems may be applied to (1.8). However, the reader will suspect that this is an unnatural approach, since (1.8) seems to be a more complicated problem than to solve (1.7). In fact, we should not lose sight of the fact that the solution curve c consists of zero points of H, and as such it enjoys powerful local contractive properties with respect to iterative methods such as those of Newton-type. Hence, one is led to numerically integrate (1.8) very coarsely and then locally use an iterative method for solving (1.7) as a stabilizer. This will be the general idea in our descriptions of predictor-corrector methods. Chapters 3–10 extensively describe the numerical aspects of predictor-corrector continuation methods.

Since the late 1970's, numerous authors have contributed to a still growing body of literature concerning the class of methods which we designate here as predictor-corrector continuation methods. Meanwhile, Scarf (1967) gave a numerically implementable proof of the Brouwer fixed point theorem, based upon a complementary pivoting algorithm of Lemke & Howson (1964) and Lemke (1965). Eaves (1972) observed that a related class of algorithms can be obtained by considering piecewise linear (PL) approximations of homotopy maps. Thus the PL continuation methods began to emerge as a parallel to the predictor-corrector methods. Although PL methods can be viewed in the more general context of comlementary pivoting algorithms, we have elected to concentrate on viewing them as a special class of continuation methods.

The PL methods require no smoothness of the underlying equations and hence have, at least in theory, a more general range of applicability. In fact, they can be used to calculate fixed points of set-valued maps. They are more combinatorial in nature and are closely related to the topological degree, see Peitgen & Prüfer (1979), Peitgen (1982) and Peitgen & Siegberg (1981). PL continuation methods are usually considered to be less efficient than the predictor-corrector methods when the latter are applicable, especially in higher dimensions. The reasons for this lie in the facts that steplength adaptation and exploitation of special structure are more difficult to implement in the PL methods. Chapters 12–14 extensively describe the numerical aspects of PL methods. Chapter 15 deals with the approximation of implicitly defined surfaces and makes use of methods involving both predictor-corrector and PL methods. Chapter 16 presents some update methods which may be useful for interpreting and implementing both predictor-corrector and PL methods, and some questions regarding numerical stability are discussed.

In the text of the book, specific ideas and continuation techniques are often described and illustrated by means of pseudo-codes using the PASCAL syntax. Actual implementations and FORTRAN programs are left to the appendix. There we present several implementations of continuation methods as illustrations of the discussions of the text, and also as examples that the methods presented can be customized to solve a variety of significant problems. To illustrate the characteristics of the algorithms, each one is applied to a simple but meaningful example. In these codes, some efficiency is sacrificed for the sake of simplicity and ease of understanding. In particular, the numerical linear algebra solvers are only given to make the presentation selfcontained. We strongly recommend that they ought to be replaced by efficient library routines. In doing so, some parts of the codes may need to be adapted. Actually, one of our reasons for presenting these codes is to get the reader started with implementing such methods, to experiment with them, make them more efficient and adapt them to his particular needs.

To date a number of program packages for different purposes and applications in numerical continuation have appeared in the literature. We make no attempt to compare them. Those of which we are aware are listed under the following entries in our bibliography. Predictor-corrector codes are found in Bank & Chan (1986), Doedel (1981), Holodniok & Kubíček (1984), Kubíček (1976), Mejia (1986), Morgan (1987), Rheinboldt (1986), Rheinboldt & Burkardt (1983), Seydel (1988), Watson & Fenner (1980). Piecewise linear codes are found in Gnutzmann (1988), Jürgens & Saupe, Todd (1981).

An extensive bibliography is given at the end of the book. It has not been possible for us to discuss or even cite all of the listed works in the text. The bibliography has been given in an attempt to provide an up to date collection of the literature concerning numerical continuation methods.

Chapter 2. The Basic Principles of Continuation Methods

2.1 Implicitly Defined Curves

In the introduction some contexts were described in which underdetermined systems of nonlinear equations $H(x, \lambda) = 0$ arose. We saw that in general, such a system implicitly defines a curve or one-manifold of solution points. The theme of this book is to describe methods for numerically tracing such curves. In this chapter, we begin by describing some basic ideas. To make the context of our discussion precise, let us make the following

(2.1.1) Assumption. $H : \mathbf{R}^{N+1} \to \mathbf{R}^N$ *is a smooth map.*

When we say that a map is smooth we shall mean that it has as many continuous derivatives as the subsequent discussion requires. We do this merely to eliminate complicated hypotheses which are not intrinsically important. Under smoothness, the reader may even assume C^∞ i.e. a map has continuous partial derivatives of all orders.

(2.1.2) Assumption. *There is a point $u_0 \in \mathbf{R}^{N+1}$ such that:*

(1) $H(u_0) = 0$;
(2) *the* **Jacobian matrix** $H'(u_0)$ *has maximum rank i.e.*
$\operatorname{rank}(H'(u_0)) = N$.

Given assumptions (2.1.1) and (2.1.2), we can choose an index i, $1 \leq i \leq N+1$, such that the submatrix of the Jacobian $H'(u_0)$ obtained by deleting the i th column is non-singular. It follows from the Implicit Function Theorem that the solution set $H^{-1}(0)$ can be locally parametrized with respect to the i th co-ordinate. By a re-parametrization, we obtain the following

(2.1.3) Lemma. *Under the assumptions (2.1.1) and (2.1.2), there exists a smooth curve $\alpha \in J \mapsto c(\alpha) \in \mathbf{R}^{N+1}$ for some open interval J containing zero such that for all $\alpha \in J$:*

(1) $c(0) = u_0$;

(2) $H\big(c(\alpha)\big) = 0$;

(3) $\mathrm{rank}\Big(H'\big(c(\alpha)\big)\Big) = N$;

(4) $c'(\alpha) \neq 0$.

By differentiating equation (2.1.3)(2) it follows that the tangent $c'(\alpha)$ satisfies the equation

(2.1.4) $H'\big(c(\alpha)\big)c'(\alpha) = 0$

and hence the tangent spans the one-dimensional kernel $\ker\Big(H'\big(c(\alpha)\big)\Big)$, or equivalently, $c'(\alpha)$ is orthogonal to all rows of $H'\big(c(\alpha)\big)$. There still remains a freedom of choice in parametrizing the curve. For our subsequent discussions, it is convenient (but not essential) to parametrize the curve with respect to the arclength parameter s such that

$$ds = \left[\sum_{j=1}^{N+1} \left(\frac{dc_j(\alpha)}{d\alpha} \right)^2 \right]^{\frac{1}{2}} d\alpha,$$

where c_j denotes the j th co-ordinate of c.

Upon replacing α by s we obtain

$$\|\dot{c}(s)\| = 1, \, s \in J$$

for some new interval J. Here we have adopted the conventions

$$\dot{c} = \frac{dc}{ds};$$

$$\|x\| = \text{the Euclidean norm of } x;$$

which we will use in the remainder of the book. The kernel of the Jacobian $H'\big(c(s)\big)$ has exactly two vectors of unit norm which correspond to the two possible directions of traversing the curve. In general, one will wish to traverse the solution curve in a consistent direction. In order to specify the orientation of traversing, we introduce the $(N+1)\times(N+1)$ **augmented Jacobian** matrix defined by

(2.1.5) $\begin{pmatrix} H'\big(c(s)\big) \\ \dot{c}(s)^* \end{pmatrix}.$

Hereafter we use the notation $A^* = $ transpose of the matrix, column or row A. Since the tangent $\dot{c}(s)$ is orthogonal to the N linearly independent rows of the Jacobian $H'\big(c(s)\big)$, it follows that the augmented Jacobian (2.1.5) is non-singular for all $s \in J$. Hence the sign of its determinant stays constant on J and it can be used to specify the direction in which the curve is traversed. Let us adopt the convention to call the **orientation of the curve** positive if this determinant is positive. We note in passing that this is the convention usually adopted in differential geometry. We summarize the above discussion in the following

(2.1.6) Lemma. *Let $c(s)$ be the positively oriented solution curve parametrized with respect to arclength s which satisfies $c(0) = u_0$ and $H(c(s)) = 0$ for s in some open interval J containing zero. Then for all $s \in J$, the tangent $\dot{c}(s)$ satisfies the following three conditions:*

(1) $H'(c(s))\dot{c}(s) = 0$;

(2) $\|\dot{c}(s)\| = 1$;

(3) $\det \begin{pmatrix} H'(c(s)) \\ \dot{c}(s)^* \end{pmatrix} > 0.$

The above three conditions uniquely determine the tangent $\dot{c}(s)$. More generally, the preceding discussion motivates the following

(2.1.7) Definition. *Let A be an $N \times (N+1)$-matrix with rank$(A) = N$. The unique vector $t(A) \in \mathbf{R}^{N+1}$ satisfying the three conditions*

(1) $At = 0$;

(2) $\|t\| = 1$;

(3) $\det \begin{pmatrix} A \\ t^* \end{pmatrix} > 0$;

*is called the **tangent vector induced by** A.*

It can be seen from the Implicit Function Theorem that the tangent vector $t(A)$ depends smoothly on A:

(2.1.8) Lemma. *The set $\mathcal{M}$ of all $N \times (N+1)$-matrices A having maximal rank N is an open subset of $\mathbf{R}^{N \times (N+1)}$, and the map $A \in \mathcal{M} \mapsto t(A)$ is smooth.*

Proof. $\mathcal{M}$ is the set of all $N \times (N+1)$-matrices A such that $\det(AA^*) \neq 0$, and this set is open since the determinant is a continuous function. The tangent vector $t(A)$ is locally defined by the equations

$$\begin{pmatrix} At \\ \frac{1}{2}t^*t - \frac{1}{2} \end{pmatrix} = 0.$$

The derivative of the left hand side with respect to t is the square matrix

$$\begin{pmatrix} A \\ t^* \end{pmatrix}$$

which is invertible for $A \in \mathcal{M}$ and $t = t(A)$. The conclusion now follows from the Implicit Function Theorem.   ∐

In the context of definition (2.1.7), lemma (2.1.6) states that the solution curve c has a derivative $\dot{c}(s)$ which is the tangent vector induced by the Jacobian matrix $H'(c(s))$. Another way of stating this is that $c(s)$ is the local solution of the

(2.1.9) Defining Initial Value Problem.

$$(1) \quad \dot{u} = t\big(H'(u)\big);$$
$$(2) \quad u(0) = u_0.$$

In the above equation (2.1.9)(1) the right hand side is of course only defined for points u such that the Jacobian $H'(u)$ has maximal rank. Let us therefore give the following standard

(2.1.10) Definition. *Let $f : \mathbf{R}^p \to \mathbf{R}^q$ be a smooth map. A point $x \in \mathbf{R}^p$ is called a* **regular point** *of f if the Jacobian $f'(x)$ has maximal rank $\min\{p, q\}$. A value $y \in \mathbf{R}^q$ is called a* **regular value** *of f if x is a regular point of f for all $x \in f^{-1}(y)$. Points and values are called* **singular** *if they are not regular.*

Note that y is vacuously a regular value of f if $y \notin \operatorname{range}(f)$. The celebrated theorem of Sard (1942) states that almost all $y \in \mathbf{R}^q$ are regular values of f, see (11.2.2)–(11.2.3) for the exact statements and Abraham & Robbin (1967), Milnor (1969) or Hirsch (1976) for proofs and further details. On the other hand, it can be easily seen that the set of regular points is open:

(2.1.11) Lemma. *Let $f : \mathbf{R}^p \to \mathbf{R}^q$ be a smooth map. Then the set*

$$\{x \in \mathbf{R}^p \mid x \text{ is a regular point of } f \}$$

is open.

Proof. Consider the case $p \geq q$. Then x is regular if and only if

$$\det\big(f'(x)f'(x)^*\big) \neq 0,$$

and the set of such x is open since the map $x \mapsto f'(x)$ is continuous. The case $p < q$ is treated analogously by considering the determinant of $f'(x)^* f'(x)$. □

In view of lemmas (2.1.8) and (2.1.11), it is now clear that the right hand side of the defining initial value problem (2.1.9) is a smooth vector field defined on the open set of regular points of H. As a partial converse of the discussion leading to (2.1.9), it is easily seen that the equation $H \equiv \text{const.}$ solves the differential equation:

(2.1.12) Lemma. *If $u(s)$ is a solution of the differential equation $\dot{u} = t\big(H'(u)\big)$, then $H(u(s))$ is constant.*

Proof. Since the derivative of $H\big(u(s)\big)$ with respect to s is $H'\big(u(s)\big)\dot{u}(s)$, and since the vector field $t\big(H'(u)\big)$ represents $\ker\big(H'(u)\big)$, we obtain $\frac{d}{ds}H\big(u(s)\big) = 0$ which yields the assertion. □

Since we have assumed that u_0 is a regular point of H, it follows from classical existence and uniqueness results for initial value problems [see e.g. Hartmann (1964)] that there is a maximal interval (a, b) of existence for the solution $c(s)$ of (2.1.9). Hereafter, $c(s)$ will denote this maximal solution. It is of course possible that $a = -\infty$ or $b = \infty$. Since we assume that $H(u_0) = 0$, the initial value condition (2.1.9)(2) implies $H(c(0)) = 0$, and the preceding lemma shows that all points of the solution curve are regular zero points of H.

(2.1.13) Lemma. *If* $-\infty < a$ *then the curve* $c(s)$ *converges to a limit point* $\tilde{u}$ *as* $s \to a$, $s > a$ *which is a singular zero point of* H. *An analogous statement holds if* $b < \infty$.

Proof. Since $c(s)$ satisfies the defining initial value problem (2.1.9), we have

$$c(s_1) - c(s_2) = \int_{s_2}^{s_1} t\Big(H'(c(\xi))\Big)\, d\xi \ \text{ for } \ s_1, s_2 \in (a, b).$$

Because the integrand has unit norm, it follows that

$$\|c(s_1) - c(s_2)\| \le |s_1 - s_2| \ \text{ for } \ s_1, s_2 \in (a, b).$$

If $\{s_n\}_{n=1}^{\infty} \subset (a, b)$ is a sequence such that $s_n \to a$ as $n \to \infty$, then the above inequality shows that the sequence $\{c(s_n)\}_{n=1}^{\infty}$ is Cauchy. Hence it converges to a point $\tilde{u}$. By continuity it follows that $H(\tilde{u}) = 0$. The remaining assertion will be shown by contradiction. Suppose that $\tilde{u}$ is a regular point of H. Then using the initial point $u(0) = \tilde{u}$ in the defining initial value problem (2.1.9), we obtain a local solution $\tilde{c}(s)$. Since $c(a + \xi) = \tilde{c}(\xi)$ for $\xi > 0$ holds by the uniqueness of solutions, it follows that c can be extended beyond a by setting $c(a + \xi) := \tilde{c}(\xi)$ for $\xi \le 0$, contradicting the maximality of the interval (a, b). $\square$

We can now state the main result of this section.

(2.1.14) Theorem. *Let zero be a regular value of* H. *Then the curve* c *is defined on all of* **R**, *and satisfies one of the following two conditions:*

(1) *The curve* c *is diffeomorphic to a circle. More precisely, there is a period* $T > 0$ *such that* $c(s_1) = c(s_2)$ *if and only if* $s_1 - s_2$ *is an integer multiple of* T;

(2) *The curve* c *is diffeomorphic to the real line. More precisely, c is injective, and* $c(s)$ *has no accumulation point for* $s \to \pm\infty$.

Proof. Since zero is a regular value, no zero point of H is singular, and by lemma (2.1.13), c is defined on all of **R**. Furthermore, since the defining differential equation (2.1.9)(1) is autonomous, its solutions are invariant under translations, i.e. for all $s_0 \in$ **R**, the curve $s \mapsto c(s_0 + s)$ is also a solution of (2.1.9)(1). Let us now consider the two possibilities:

(i) c is not injective. We define $T := \min\{s > 0 \mid c(s) = c(0)\}$. By the uniqueness of the solutions of initial value problems and by the above mentioned translation invariance, the assertion (1) follows.

(ii) c is injective. We show assertion (2) by contradiction. Let us assume without loss of generality that $\tilde{u}$ is an accumulation point of $c(s)$ as $s \to \infty$. By continuity, $H(\tilde{u}) = 0$. Since $\tilde{u}$ is a regular point of H, we can use the initial point $u(0) = \tilde{u}$ in the defining initial value problem (2.1.9) to obtain a local solution $\tilde{c}$. By uniqueness, the two curves c and $\tilde{c}$ must coincide locally, and hence there exists an $s_1 > 0$ such that $c(s_1) = \tilde{u}$. Since $\tilde{u}$ is also an accumulation point of $c(s_1 + s)$ as $s \to \infty$, and since the curve $s \mapsto c(s_1 + s)$ is also a solution curve, the above argument can be repeated to obtain an $s_2 > 0$ such that $c(s_1 + s_2) = \tilde{u}$. This contradicts the injectivity of c. $\square$

A more topological and global treatment of the Implicit Function Theorem can be found in the books of Hirsch or Milnor. For a discussion of the Implicit Function Theorem in a Banach space context see, for example, the book by Berger (1977). The discussion of this section can also be given in a Banach space context, note however, that the orientation concept would need to be otherwise formulated.

Among the main applications of tracing c we mention the numerical tasks of solving nonlinear eigenvalue problems in finite dimensions, and the solving of nonlinear systems of equations via homotopy deformation methods. Some of these applications will be dealt with in detail in later chapters.

Since the solution curve c is characterized by the defining initial value problem (2.1.9), it is evident that the numerical methods for solving initial value problems can immediately be used to numerically trace the curve c. This is not, however, in general an efficient approach. As our discussions of the continuation methods in subsequent chapters will show, such approaches ignore the contractive properties which the curve c has relative to Newton-type iterative methods, because it is a set of zero points of H. There are essentially two different methods for numerically tracing c which will be considered in this book:

- Predictor-Corrector (PC) methods,
- Piecewise-Linear (PL) methods.

In the next two sections we briefly sketch the basic ideas of both methods. In subsequent chapters, many explicit details of these methods will be discussed.

2.2 The Basic Concepts of PC Methods

The idea in PC methods is to numerically trace the curve c of section 2.1 by generating a sequence of points u_i, $i = 1, 2, \ldots$ along the curve satisfying a chosen tolerance criterion, say $\|H(u_i)\| \leq \varepsilon$ for some $\varepsilon > 0$. We assume here that a regular starting point $u_0 \in \mathbf{R}^{N+1}$ is given such that $H(u_0) = 0$.

It seems intuitively plain, and we shall indicate a proof in chapter 3, that for $\varepsilon > 0$ sufficiently small, there is a unique parameter value s_i such that the point $c(s_i)$ on the curve is nearest to u_i in Euclidean norm. Figure 2.2.a portrays this assertion. We caution the reader that the figures throughout this book must be regarded as portraying curves in $\mathbf{R}^{N+1}$, thus our points u_i cannot be regarded as lying "above" or "below" the curve c.

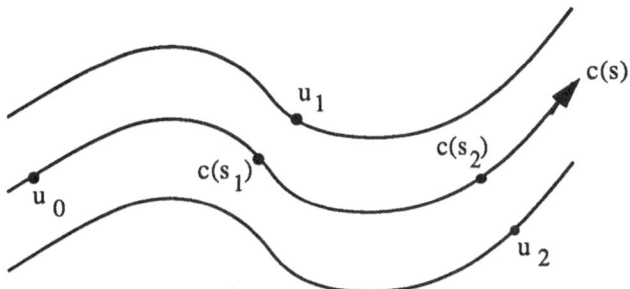

Figure 2.2.a The point $c(s_i)$ is the best approximation to u_i on the curve c

To describe how points u_i along the curve c are generated, suppose that a point $u_i \in \mathbf{R}^{N+1}$ has been accepted such that $\|H(u_i)\| \leq \varepsilon$. If u_i is a regular point of H, then the results of section 2.1 can be applied. Hence, there exists a unique solution curve $c_i : J \to \mathbf{R}^{N+1}$ defined on its maximal interval of existence J, which satisfies the initial value problem

$$(2.2.1) \qquad \begin{aligned} \dot{u} &= t\big(H'(u)\big); \\ u(0) &= u_i. \end{aligned}$$

To obtain a new point u_{i+1} along c, we first make a **predictor step**. Typically, a predictor step is obtained as a simple numerical integration step for the initial value problem (2.2.1). Very commonly, an **Euler predictor** is used:

$$(2.2.2) \qquad v_{i+1} = u_i + ht\big(H'(u_i)\big),$$

where $h > 0$ represents a "stepsize". The manner in which h is to be chosen will be discussed in detail in chapter 6. As has already been mentioned, a powerful corrector step is available due to the fact that the solution curve

c satisfies the equation $H(u) = 0$. Consequently, even for a poor predictor point v_{i+1}, an iterative corrector process will exhibit rapid convergence to the solution curve c. To illustrate this, let w_{i+1} denote the point on c which is nearest to v_{i+1}, see figure 2.2.b.

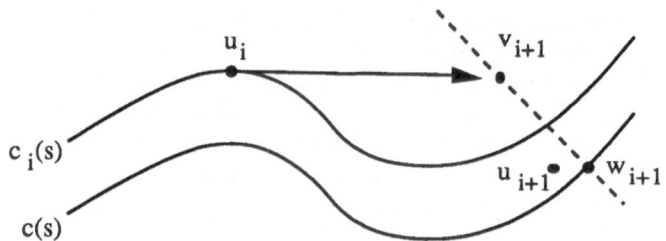

Figure 2.2.b Predictor point v_{i+1} and corrector point u_{i+1}

The point w_{i+1} solves the following optimization problem:

$$(2.2.3) \qquad ||w_{i+1} - v_{i+1}|| = \min_{H(w)=0} || w - v_{i+1} ||.$$

If u_i is sufficiently close to the curve c and the stepsize h is sufficiently small, then the predictor point v_{i+1} will be sufficiently close to the curve c so that the minimization problem has a unique solution w_{i+1}. An obvious way to numerically approximate w_{i+1} in (2.2.3) is to utilize a Newton-like method. Suppose that by one or two iterations of such a method we obtain a point u_{i+1} approximating w_{i+1} within a given tolerance, say $||H(u_{i+1})|| \leq \varepsilon$. Then u_{i+1} is taken as our next point along the curve. The PC continuation method for approximating c thus consists of repeatedly performing predictor and corrector steps such as those described above. To construct an efficient and robust PC method which can successfully approximate complicated or difficult curves, a number of important items remain to be carefully developed e.g.

(1) an effective step size adaptation;
(2) an efficient implementation of the corrector step;
(3) an efficient incorporation of higher order predictors;
(4) handling or approximating special points on the curve such as turning points, bifurcation points or other points of special interest.

These problems will be dealt with in our later chapters. We again emphasize that the PC continuation methods are considerably different than the well known methods for the numerical integration of initial value problems which are also called predictor-corrector methods. Although the predictor steps in both methods are similar in nature, the corrector process in the continuation methods thrives upon the powerful contractive properties of the solution set $H^{-1}(0)$ for iterative methods such as Newton's method. This is a property which solution curves of general initial value problems do not enjoy, in fact their corrector processes converge in the limit only to an approximating point, the approximating quality of which depends on the stepsize h.

2.3 The Basic Concepts of PL Methods

Whereas a PC method involves approximately following the exact solution curve c of section 2.1, in a PL method, one follows exactly a piecewise-linear curve c_T which approximates c. In particular, the curve c_T is a polygonal path relative to underlying triangulation T of $\mathbf{R}^{N+1}$. To describe how this is done, it is necessary to introduce a definition of a triangulation of $\mathbf{R}^{N+1}$ which is adequate for PL algorithms.

(2.3.1) Definition. Let $v_1, v_2, \ldots, v_{j+1} \in \mathbf{R}^{N+1}$, $j \leq N+1$, be affinely independent points (i.e. $v_k - v_1$, $k = 2, \ldots, j+1$ are linearly independent). The convex hull

$$[v_1, v_2, \ldots, v_{j+1}] := \mathrm{co}\{v_1, v_2, \ldots, v_{j+1}\}$$

is the j-**simplex** in $\mathbf{R}^{N+1}$ having vertices $v_1, v_2, \ldots, v_{j+1}$. The convex hull $[w_1, \ldots, w_{r+1}]$ of any subset $\{w_1, \ldots, w_{r+1}\} \subset \{v_1, v_2, \ldots, v_{j+1}\}$ is an r-**face** of $[v_1, v_2, \ldots, v_{j+1}]$.

(2.3.2) Definition. A triangulation T of $\mathbf{R}^{N+1}$ is a subdivision of $\mathbf{R}^{N+1}$ into $(N+1)$-simplices such that

(1) any two simplices in T intersect in a common face, or not at all;

(2) any bounded set in $\mathbf{R}^{N+1}$ intersects only finitely many simplices in T.

Since our aim in this section is merely to give the basic ideas of a PL algorithm, we shall defer giving constructions of triangulations until later. More details will be given in chapter 12.

(2.3.3) Definition. For any map $H : \mathbf{R}^{N+1} \to \mathbf{R}^N$, the **piecewise linear approximation** H_T to H relative to the triangulation T of $\mathbf{R}^{N+1}$ is the map which is uniquely defined by

(1) $H_T(v) = H(v)$ for all vertices of T;

(2) for any $(N+1)$-simplex $\sigma = [v_1, v_2, \ldots, v_{N+2}] \in T$, the restriction $H_T|_\sigma$ of H_T to σ is an affine map;

As a consequence, if $u = \sum_{i=1}^{N+2} \alpha_i v_i$ is a point in σ, then its barycentric coordinates α_i satisfy $\sum_{i=1}^{N+2} \alpha_i = 1$ and $\alpha_i \geq 0$ for $i = 1, \ldots, N+2$, and since H_T is affine, we have

$$H_T(u) = H\left(\sum_{i=1}^{N+2} \alpha_i v_i\right) = \sum_{i=1}^{N+2} \alpha_i H(v_i).$$

The set $H_T^{-1}(0)$ contains a polygonal path $c_T : \mathbf{R} \rightarrow \mathbf{R}^{N+1}$ which approximates c. Error estimates for the truncation error of such approximations will be given in chapter 15. Tracing the path is carried out via PL-steps similar to the steps used in linear programming methods such as the Simplex Method. Figure 2.3.a portrays the basic idea of a PL method.

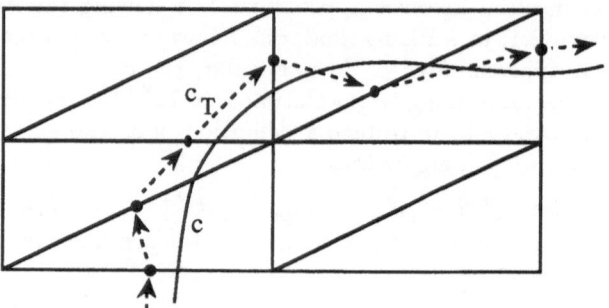

Figure 2.3.a PL path following

It is also possible to blend the two basic continuation techniques. For example, the PL curve c_T approximating c may be used as a "global" predictor for obtaining points on c when c_T lies sufficiently near c. That is, for every point $v \in \text{range}(c_T)$ there is a nearest point $w \in \text{range}(c)$ as in (2.2.3). Here again a Newton-type iterative method may be used to perform the corrector steps. Such a corrector process is described in a general context in section 15.2.

Let us also point out that for PL methods, a smoothness assumption concerning H is not necessary. For certain applications the piecewise linear path c_T may still be of interest even if it does not closely approximate c. This holds in particular in the case of homotopy methods for computing fixed points under the hypotheses of the Kakutani fixed point theorem.

In subsequent chapters we will expand upon the ideas for implementing both the PC and the PL methods. We will first deal with the PC methods in chapters 3–10, and then with the PL methods in chapters 12–15.

Chapter 3. Newton's Method as Corrector

3.1 Motivation

Let zero be a regular value of the smooth map $H : \mathbf{R}^{N+1} \to \mathbf{R}^N$. We again consider the solution curve c contained in $H^{-1}(0)$ defined by the initial value problem (2.1.9), where the initial point $c(0) = u_0$ such that $H(u_0) = 0$ is assumed to be given. The PC methods which were generally outlined in section 2.2 motivate the following

(3.1.1) Generic Predictor-Corrector Method. *comment:*

   input

      begin

         $u \in \mathbf{R}^{N+1}$ such that $H(u) = 0$; *initial point*

         $h > 0$; *initial steplength*

      end;

   repeat

      predict a point v such that *predictor step*

         $H(v) \approx 0$ and $\|u - v\| \approx h$;

      let $w \in \mathbf{R}^{N+1}$ approximately solve *corrector step*

         $\min_{w} \{ \|v - w\| \mid H(w) = 0 \}$;

      $u := w$; *new point along $H^{-1}(0)$*

      choose a new steplength $h > 0$; *steplength adaptation*

      until traversing is stopped.

It was suggested that a straightforward way of solving the minimization problem

(3.1.2) $$\min_{w} \{ \|v - w\| \mid H(w) = 0 \}$$

in the corrector step might be a Newton-type method. In this chapter the
basic ideas for doing this will be outlined and analyzed. In chapter 6 details
concerning steplength adaptation are discussed.

As is well known, Newton's method for solving the equation $f(x) = 0$
generally takes the form

$$x_{i+1} = x_i - f'(x_i)^{-1} f(x_i)$$

where $f : \mathbf{R}^N \to \mathbf{R}^N$ is a smooth map. In the present context, the Jacobian
H' is not a square matrix and therefore cannot be inverted. Hence Newton's
method has to be accordingly modified. This can be done by introducing
a certain right inverse H'^+ of H' which is motivated by the minimization
problem (3.1.2). Such a suitable right inverse is provided by the Moore-
Penrose inverse, which we need to introduce only for the special case of $N \times$
$(N + 1)$-matrices of maximal rank. For general discussions of the Moore-
Penrose inverse see the textbook of Golub & Van Loan (1983).

3.2 The Moore-Penrose Inverse in a Special Case

Let us consider the simplest example of an implicitly defined curve namely a
line in $\mathbf{R}^{N+1}$. More precisely, let us consider the special case of an affine map

$$H(u) := Au - b$$

where A is an $N \times (N + 1)$ matrix with maximal rank and $b \in \mathbf{R}^N$. Then the
curve c implicitly defined by the equation $H(u) = 0$ is a straight line. Figure
3.2.a portrays this situation.

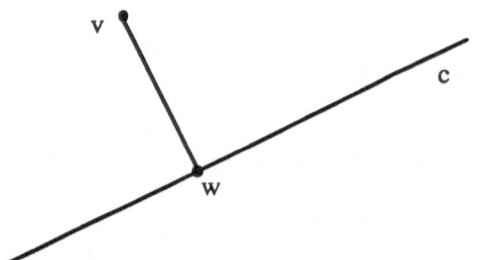

Figure 3.2.a Minimal solution for the linear
case (3.2.1)

Since a straightforward calculation shows that the condition $H(w) = 0$
is equivalent to $A(w - v) = -H(v)$, the minimization problem (3.1.2) takes
the particular form

(3.2.1) $\min_{w} \{ \, \|w - v\| \mid A(w - v) = -H(v) \, \}$

From lemma (3.2.3) below it follows that the minimal solution $w - v$ can be written as $w - v = -A^{+}H(v)$ where A^{+} is the Moore-Penrose inverse of A. Since in this special case $A = H'(v)$, we obtain the "Newton step" $w = v - H'(v)^{+}H(v)$. In our present context, the Moore-Penrose inverse is given by the following

(3.2.2) Definition. *Let A be an $N \times (N+1)$ matrix with maximal rank. Then the **Moore-Penrose inverse** of A is defined by $A^{+} = A^{*}(AA^{*})^{-1}$.*

The next lemma describes the familiar normal equations for linear least squares problems:

(3.2.3) Lemma. *Let A be an $N \times (N+1)$ matrix with maximal rank, and let $t(A)$ be its tangent vector, cf. (2.1.7). Then the following statements are equivalent for all $b \in \mathbf{R}^{N}$ and $x \in \mathbf{R}^{N+1}$:*

(1) $Ax = b$ *and* $t(A)^{*}x = 0$;
(2) $x = A^{+}b$;
(3) x *solves the problem:* $\min_{w} \{ \|w\| \mid Aw = b \}$.

Proof. We first observe that $At(A) = 0$ implies $t(A)^{*}A^{*} = 0$. Using definition (3.2.2), it can be seen by multiplying that

(3.2.4)
$$\begin{pmatrix} A \\ t(A)^{*} \end{pmatrix} \left(A^{+}, t(A) \right) = \mathrm{Id}$$

holds. Statement (1) is equivalent to

$$\begin{pmatrix} A \\ t(A)^{*} \end{pmatrix} x = \begin{pmatrix} b \\ 0 \end{pmatrix},$$

which by (3.2.4) is equivalent to (2). This shows (1) $\Leftrightarrow$ (2). To show (2) $\Leftrightarrow$ (3), let $x = A^{+}b$. The general solution of the equation $Aw = b$ is given by $w = x + \alpha t(A)$, $\alpha \in \mathbf{R}$. Since x and $t(A)$ are orthogonal by (1), we have $\|w\|^{2} = \|x\|^{2} + \alpha^{2}$, and it is clear that w has minimal norm if and only if $w = x$. $\square$

In our subsequent discussion we make use of the following properties of the Moore-Penrose inverse. Here and in the following we denote by $y \perp x$ orthogonality, i.e. $y^{*}x = 0$, and $X^{\perp} := \{y \mid y \perp x \text{ for all } x \in X\}$ denotes the orthogonal complement of X.

(3.2.5) Lemma. *If A is an $N \times (N+1)$ matrix with maximal rank, then*

(1) $A^{+}A$ *is the orthogonal projection from $\mathbf{R}^{N+1}$ onto $\{t(A)\}^{\perp} = \mathrm{range}(A^{*})$, i.e. $A^{+}A = \mathrm{Id} - t(A)t(A)^{*}$.*
(2) $AA^{+} = \mathrm{Id}$.

(3) If B is any right inverse of A, then $A^+ = \big(\mathrm{Id} - t(A)t(A)^*\big)B$.

Proof. Assertion (2) follows immediately from (3.2.4). To prove (1), let us first recall the familiar fact of linear algebra that $\mathrm{range}(A^*) = \ker(A)^\perp$ holds, and hence it is clear that $\{t(A)\}^\perp = \mathrm{range}(A^*)$. If $x \in \{t(A)\}^\perp$, then $x = A^+Ax$ by (3.2.3)(1). Since $A^+At(A) = 0$, assertion (1) follows. To prove assertion (3), we set $B_0 := \big(\mathrm{Id} - t(A)t(A)^*\big)B$. For $b \in \mathbf{R}^N$, let $x := B_0 b$. From the definition of B_0 it follows immediately that $Ax = b$ and $t(A)^*x = 0$. Now assertion (3) is obtained from lemma (3.2.3). $\qquad\square$

3.3 A Newton's Step for Underdetermined Nonlinear Systems

Let us now consider the general nonlinear map $H : \mathbf{R}^{N+1} \to \mathbf{R}^N$. A necessary condition for a solution to (3.1.2) is obtained via the method of Lagrange multipliers. Hence, if w is a solution of (3.1.2), then it satisfies the Lagrangian equations

$$
\begin{aligned}
H(w) &= 0; \\
w - v &= H'(w)^* \lambda
\end{aligned}
\tag{3.3.1}
$$

for some vector of multipliers $\lambda \in \mathbf{R}^N$. The second condition is equivalent to $w - v \in \mathrm{range}\big(H'(w)^*\big) = \big\{t(H'(w))\big\}^\perp$. Thus a necessary condition for w to solve (3.1.2) is that w satisfies the equation

$$
\begin{aligned}
H(w) &= 0; \\
t\big(H'(w)\big)^*(w - v) &= 0.
\end{aligned}
\tag{3.3.2}
$$

In Newton's method, the nonlinear system (3.3.2) is solved approximately via a linearization about v. To illustrate this, let us consider the Taylor expansion about v:

$$
\begin{aligned}
H(w) &= H(v) + H'(v)(w - v) + O\big(\|w - v\|^2\big) \\
t\big(H'(w)\big)^*(w - v) &= t\big(H'(v)\big)^*(w - v) + O\big(\|w - v\|^2\big)
\end{aligned}
\tag{3.3.3}
$$

For the reader who is unfamiliar with the **Landau symbol** O, let us briefly say that for our purposes it suffices to know that

$$
f(h) = O(h^m)
$$

means that

$$
\|f(h)\| \le C|h|^m
\tag{3.3.4}
$$

for small h and a constant $C > 0$. If the function f in (3.3.4) depends on some additional parameter, e.g. α, then we say that $f(h) = O(h^m)$ holds uniformly in α if the constant C in (3.3.4) can be chosen independently of α. Local uniformity is defined in an analogous manner. For example, the asymptotic relationship in (3.3.3) is locally uniform in v. Our subsequent discussions involving asymptotic arguments are to be understood in this locally uniform sense. The reader who wishes to see more about the Landau notation may consult a general book about it, such as de Bruijn's book on asymptotic analysis.

To return to our discussion, a linearization of (3.3.2) consists of neglecting the higher order terms $O(\|w-v\|^2)$ of (3.3.3). As is usual in Newton's method, we obtain an approximation $\mathcal{N}(v)$ to the solution w of (3.3.2), which has a truncation error of second order. Hence, the Newton point $\mathcal{N}(v)$ satisfies the following equations:

(3.3.5)
$$H(v) + H'(v)\big(\mathcal{N}(v) - v\big) = 0;$$
$$t\big(H'(v)\big)^*\big(\mathcal{N}(v) - v\big) = 0.$$

Using (3.2.3)(1), we are therefore led to the following

(3.3.6) Definition. Let $v \in \mathbf{R}^{N+1}$ be a regular point of H. Then the **Newton point** $\mathcal{N}(v)$ for approximating the solution of (3.1.2) is given by

$$\mathcal{N}(v) := v - H'(v)^+ H(v).$$

The map $\mathcal{N}$ defined on the regular points of H will also be called the **Newton map**.

Note that this Newton step is analogous to the classical Newton's method, with the only formal difference being that the Moore-Penrose inverse $H'(v)^+$ replaces the classical inverse.

The following algorithm sketches a particular version of the predictor-corrector method (3.1.1) incorporating an Euler predictor step, cf. (2.2.2), and the above described Newton's method as a corrector.

(3.3.7) Euler-Newton Method.   *comment:*

   input

      begin

        $u \in \mathbf{R}^{N+1}$ such that $H(u) = 0$;   *initial point*

        $h > 0$;   *initial steplength*

      end;

   repeat

     $v := u + ht\big(H'(u)\big)$;   *predictor step*

     repeat

       $w := v - H'(v)^+ H(v)$;   *corrector loop*

       $v := w$;

     until convergence;

     $u := w$;   *new point along $H^{-1}(0)$*

     choose a new steplength $h > 0$;   *steplength adaptation*

     until traversing is stopped.

3.4 Convergence Properties of Newton's Method

Full theoretical discussions of Newton's method (3.2.6) using the Moore-Penrose inverse can be found in several text books e.g. Ortega & Rheinboldt (1970) or Ben-Israel & Greville (1974). Since our context deals with the curve-following problem, we confine our discussion to the case $H : \mathbf{R}^{N+1} \to \mathbf{R}^N$.

(3.4.1) Theorem. *Let $H : \mathbf{R}^{N+1} \to \mathbf{R}^N$ be a smooth map having zero as a regular value. Then there exists an open neighborhood $U \supset H^{-1}(0)$ such that the following assertions hold.*

(1) *The* **solution map** *$v \in U \mapsto S(v) \in H^{-1}(0)$ such that $S(v)$ solves the minimization problem problem (3.1.2) is uniquely defined and smooth.*

(2) *For each $v \in U$, the Newton sequence $\{\mathcal{N}^i(v)\}_{i=1}^{\infty}$ converges to a point $\mathcal{N}^{\infty}(v)$ in $H^{-1}(0)$.*

(3) *The following estimates hold locally uniformly for $v \in U$:*

(a) $$\|\mathcal{N}^2(v) - \mathcal{N}(v)\| = O\big(\|\mathcal{N}(v) - v\|^2\big);$$

(b) $$\|\mathcal{N}^{\infty}(v) - \mathcal{N}(v)\| = O\big(\|\mathcal{N}^{\infty}(v) - v\|^2\big);$$

(c) $$\|\mathcal{N}(v) - S(v)\| = O\big(\|v - S(v)\|^2\big);$$

(d) $$\|\mathcal{N}^{\infty}(v) - S(v)\| = O\big(\|v - S(v)\|^2\big).$$

(4) *The relation $\mathcal{N}(U) \subset U$ holds.*

Proof. We shall only sketch the main points. Furnishing additional details may be an exercise for the reader. By a standard continuity argument, the minimization problem in (3.1.2) has at least one solution for every v, and the map $\mathcal{S}$ can be defined by selecting one such solution. By (3.3.2) which was a consequence of the method of Lagrange multipliers, the pair $(\mathcal{S}(v), v))$ is a zero point of the map $R : \mathbf{R}^{N+1} \times \mathbf{R}^{N+1} \to \mathbf{R}^{N+1}$ defined by

$$R(w, v) := \begin{pmatrix} H(w) \\ t(H'(w))^*(w - v) \end{pmatrix}.$$

For a zero point $v_0 \in H^{-1}(0)$, the following partial derivative is easily calculated:

$$R_w(v_0, v_0) = \begin{pmatrix} H'(v_0) \\ t(H'(v_0))^* \end{pmatrix}.$$

The latter matrix is the augmented Jacobian, see (2.1.7), which is nonsingular. Since $R(v_0, v_0) = 0$, we can apply the Implicit Function Theorem and obtain that the map $v \mapsto \mathcal{S}(v)$ is uniquely defined and smooth on a neighborhood $U(v_0)$ of v_0 containing only regular points of H. Assertion (1) now follows by using

$$U_1 := \bigcup_{v_0 \in H^{-1}(0)} U(v_0)$$

as the neighborhood of $H^{-1}(0)$.

To obtain the convergence and estimates, let us consider a fixed $\tilde{v} \in H^{-1}(0)$. Furthermore, we choose an $\varepsilon > 0$ such that the closed ball

$$B_\varepsilon := \{v \in \mathbf{R}^{N+1} \mid ||v - \tilde{v}|| \le \varepsilon\}$$

is contained in U_1. We define the following constants for B_ε:

(3.4.2)
$$\begin{aligned}
\alpha &:= \max\{||H'(v)|| \mid v \in B_\varepsilon\}; \\
\beta &:= \max\{||H'(v)^+|| \mid v \in B_\varepsilon\}; \\
\gamma &:= \max\{||H''(v)|| \mid v \in B_\varepsilon\}; \\
\rho &:= \max\{||(H'(v)^+)'|| \mid v \in B_\varepsilon\}.
\end{aligned}$$

To simplify our estimates below, we consider a $\delta > 0$ such that

(3.4.3)
$$\delta + \beta\alpha\delta + (\beta\gamma)(\beta\alpha\delta)^2 \le \frac{\varepsilon}{2};$$
$$(\beta\gamma)(\beta\alpha\delta) \le \frac{1}{2}.$$

Let $v \in B_\delta$. The following estimates will show that all for all iterates $\mathcal{N}^i(v) \in B_e$ holds, and hence the bounds in (3.4.2) can be applied. From Taylor's formula it follows that

$$H(v) = H(\tilde{v}) + \int_0^1 H'(\tilde{v} + \xi(v - \tilde{v}))d\xi\,(v - \tilde{v})$$

and hence by (3.4.2)

(3.4.4) $\|H(v)\| \leq \alpha\|v - \tilde{v}\| \leq \alpha\delta.$

Also from Taylor's formula, see e.g. Berger (1977), p.75, we have

$$H(\mathcal{N}(v)) = H(v) + H'(v)(\mathcal{N}(v) - v)$$
$$+ \frac{1}{2}\int_0^1 H''\big(v + \xi(\mathcal{N}(v) - v)\big)2(1 - \xi)d\xi\,[\mathcal{N}(v) - v, \mathcal{N}(v) - v].$$

Using the fact that $H(v) + H'(v)(\mathcal{N}(v) - v) = 0$ and taking norms, it follows from (3.4.2),(3.4.4) that

$$\|H(\mathcal{N}(v))\| \leq \frac{1}{2}\gamma\|\mathcal{N}(v) - v\|^2$$

(3.4.5) $$\leq \frac{1}{2}\gamma\|\beta H(v)\|^2$$

$$\leq \frac{1}{2}\gamma(\beta\alpha\delta)^2.$$

This immediately implies

(3.4.6) $\|\mathcal{N}^2(v) - \mathcal{N}(v)\| = \|H'(\mathcal{N}(v))^+ H(\mathcal{N}(v))\| \leq \frac{1}{2}\beta\gamma\|\mathcal{N}(v) - v\|^2$

$$\leq \frac{1}{2}\beta\gamma(\beta\alpha\delta)^2.$$

Proceeding recursively, we obtain

(3.4.7) $$\|H(\mathcal{N}^i(v))\| \leq \beta^{-1}\big(\frac{1}{2}\beta\gamma\big)^{2^i-1}(\beta\alpha\delta)^{2^i};$$

$$\|\mathcal{N}^{i+1}(v) - \mathcal{N}^i(v)\| \leq \big(\frac{1}{2}\beta\gamma\big)^{2^i-1}(\beta\alpha\delta)^{2^i}.$$

Summing the right hand side over i and comparing with the corresponding geometric series yields the estimates

$$\sum_{i=1}^\infty \big(\frac{1}{2}\beta\gamma\big)^{2^i-1}(\beta\alpha\delta)^{2^i} = \big(\frac{1}{2}\beta\gamma\big)^{-1}\sum_{i=1}^\infty \big(\frac{1}{2}\beta\gamma\big)^{2^i}(\beta\alpha\delta)^{2^i}$$

$$\leq \big(\frac{1}{2}\beta\gamma\big)^{-1}\sum_{i=1}^\infty \big(\frac{1}{2}\beta\gamma\big)^{2i}(\beta\alpha\delta)^{2i}$$

$$= \big(\frac{1}{2}\beta\gamma\big)^{-1}\frac{\big(\frac{1}{2}\beta\gamma\big)^2(\beta\alpha\delta)^2}{1 - \big(\frac{1}{2}\beta\gamma\big)^2(\beta\alpha\delta)^2}$$

$$\leq (\beta\gamma)(\beta\alpha\delta)^2$$

since $\left(\frac{1}{2}\beta\gamma\right)^2(\beta\alpha\delta)^2 < \frac{1}{2}$ follows from (3.4.3). This implies that the Newton iterates $\mathcal{N}^i(v)$ form a Cauchy sequence which converges to the limit point $\mathcal{N}^\infty(v)$, and from the first inequality in (3.4.7) it follows that $H\left(\mathcal{N}^\infty(v)\right) = 0$. This proves assertion (2). Since the bounds we are using in (3.4.3) are locally uniform, (3.4.6) already establishes the estimate (3)(a). We now proceed to obtain (3)(b). The last estimates above yield

$$\|\mathcal{N}(v) - \mathcal{N}^\infty(v)\| \leq \sum_{i=1}^\infty \|\mathcal{N}^{i+1}(v) - \mathcal{N}^i(v)\|,$$

$$\|\tilde{v} - \mathcal{N}^\infty(v)\| \leq \|\tilde{v} - v\| + \|v - \mathcal{N}(v)\| + \|\mathcal{N}(v) - \mathcal{N}^\infty(v)\|$$
$$\leq \delta + \beta\alpha\delta + (\beta\gamma)(\beta\alpha\delta)^2,$$

and (3.4.3) now implies that

(3.4.8) $$\mathcal{N}^\infty(v) \in B_{\frac{\varepsilon}{2}}.$$

It is easy to see, cf. (3.4.5) that we may replace $\|\mathcal{N}(v) - v\|$ by $\beta\alpha\delta$ in the above estimates. This yields

$$\|\mathcal{N}^\infty(v) - \mathcal{N}(v)\| \leq \beta\gamma\|\mathcal{N}(v) - v\|^2.$$

From this inequality and (3.4.3) it follows that

$$\|\mathcal{N}^\infty(v) - v\| \geq \|\mathcal{N}(v) - v\| - \|\mathcal{N}^\infty(v) - \mathcal{N}(v)\|$$
$$\geq \|\mathcal{N}(v) - v\| - \beta\gamma\|\mathcal{N}(v) - v\|^2$$
$$\geq \|\mathcal{N}(v) - v\| - \beta\gamma(\beta\alpha\delta)\|\mathcal{N}(v) - v\|$$
$$\geq \frac{1}{2}\|\mathcal{N}(v) - v\|$$

and consequently

(3.4.9) $$\|\mathcal{N}^\infty(v) - \mathcal{N}(v)\| \leq 4\beta\gamma\|\mathcal{N}^\infty(v) - v\|^2$$

establishes (3)(b). To obtain (3)(c), we first note that $\|v - \mathcal{S}(v)\| \leq \|v - \tilde{v}\| \leq \delta \leq \frac{\varepsilon}{2}$ yields $\mathcal{S}(v) \in B_\varepsilon$, and hence the bound (3.4.2) apply to the estimates below. Once again applying Taylor's formula, we have

$$H(v) = H\left(\mathcal{S}(v)\right) + H'\left(\mathcal{S}(v)\right)\left(v - \mathcal{S}(v)\right)$$
$$+ \frac{1}{2}\int_0^1 H''\left(\mathcal{S}(v) + \xi\left(v - \mathcal{S}(v)\right)\right)2(1 - \xi)d\xi\left[v - \mathcal{S}(v), v - \mathcal{S}(v)\right].$$

Since $w = \mathcal{S}(v)$ satisfies (3.3.2), the difference $v - \mathcal{S}(v)$ is orthogonal to $\ker\left(H'(\mathcal{S}(v))\right)$, and therefore by (3.2.5)(1) we have

$$v - \mathcal{S}(v) = H'\left(\mathcal{S}(v)\right)^+ H(v)$$
$$- \frac{1}{2}H'\left(\mathcal{S}(v)\right)^+ \int_0^1 H''\left(\mathcal{S}(v) + \xi\left(v - \mathcal{S}(v)\right)\right)2(1 - \xi)d\xi\left[v - \mathcal{S}(v), v - \mathcal{S}(v)\right].$$

Now subtracting the equation $v - \mathcal{N}(v) = H'(v)^+ H(v)$ yields

$$\mathcal{N}(v) - \mathcal{S}(v) = \left(H'(\mathcal{S}(v))^+ - H'(v)^+ \right) H(v)$$

$$- \frac{1}{2} H'(\mathcal{S}(v))^+ \int_0^1 H'' \Big(\mathcal{S}(v) + \xi(v - \mathcal{S}(v)) \Big) 2(1 - \xi) d\xi \left[v - \mathcal{S}(v), v - \mathcal{S}(v) \right].$$

By an argument as that in establishing (3.4.4) we obtain $\|H(v)\| \leq \alpha \|v - \mathcal{S}(v)\|$, and the bounds in (3.4.3) now yield

$$\|\mathcal{N}(v) - \mathcal{S}(v)\| \leq \rho \|v - \mathcal{S}(v)\| \, \alpha \|v - \mathcal{S}(v)\| + \frac{1}{2} \beta \gamma \|v - \mathcal{S}(v)\|^2,$$

which establishes (3)(c). To prove the last estimate (3)(d), we note that

$$\|\mathcal{N}^\infty(v) - \mathcal{S}(v)\| \leq \|\mathcal{N}^\infty(v) - \mathcal{N}(v)\| + \|\mathcal{N}(v) - \mathcal{S}(v)\|$$

and (3)(b–c) imply

$$\|\mathcal{N}^\infty(v) - \mathcal{S}(v)\| = O\big(\|\mathcal{N}^\infty(v) - v\|^2\big) + O\big(\|v - \mathcal{S}(v)\|^2\big),$$

and by using

$$\|\mathcal{N}^\infty(v) - v\|^2 \leq \|\mathcal{N}^\infty(v) - \mathcal{S}(v)\|^2 + \|\mathcal{S}(v) - v\|^2 + 2\|\mathcal{N}^\infty(v) - v\| \, \|\mathcal{S}(v) - v\|$$

we obtain

$$\|\mathcal{N}^\infty(v) - \mathcal{S}(v)\| = O\Big(\|\mathcal{N}^\infty(v) - \mathcal{S}(v)\| + \|\mathcal{S}(v) - v\| \Big) \|\mathcal{N}^\infty(v) - \mathcal{S}(v)\|$$
$$+ O\big(\|v - \mathcal{S}(v)\|^2 \big),$$

which implies

$$\|\mathcal{N}^\infty(v) - \mathcal{S}(v)\| = \frac{O\big(\|v - \mathcal{S}(v)\|^2\big)}{1 - O\Big(\|\mathcal{N}^\infty(v) - \mathcal{S}(v)\| + \|\mathcal{S}(v) - v\| \Big)}$$
$$= O\big(\|v - \mathcal{S}(v)\|^2\big),$$

and this proves (3)(d).

Let us finally show, that we can find an open neighborhood U of $H^{-1}(0)$ which is stable under the Newton map $\mathcal{N}$ and such that all the convergence properties and estimates hold for all iterates of $v \in U$. In the above discussion, for a given $\tilde{v} \in H^{-1}(0)$ we found $\varepsilon = \varepsilon(\tilde{v})$ and $\delta = \delta(\tilde{v})$ so that all the estimates are satisfied on the corresponding neighborhoods $B_{\delta(\tilde{v})}(\tilde{v})$ and $B_{\varepsilon(\tilde{v})}(\tilde{v})$. We first define the neighborhood

$$U_2 := \bigcup_{\tilde{v} \in H^{-1}(0)} \text{int}\big(B_{\delta(\tilde{v})}(\tilde{v}) \big)$$

where "int" denotes the interior. Now we set

$$U := \{u \mid u, \mathcal{N}(u), \mathcal{N}^2(u), \ldots \in U_2\}.$$

It is clear that $H^{-1}(0)$ is contained in U, and that the asserted convergence properties and estimates hold for all Newton iterates of $v \in U$. So it only remains to show that U is open. Let $v \in U$. We will show that some open neighborhood of v is also contained in U. We choose $\tilde{v} := \mathcal{N}^{\infty}(v)$. It is possible to find an open neighborhood V of $\tilde{v}$ so that

$$\mathcal{N}^i(V) \subset B_{\frac{\delta(\tilde{v})}{2}}(\tilde{v}) \quad \text{for} \quad i = 0, 1, 2, \ldots$$

For example, if $\eta > 0$ satisfies

$$\eta + \beta\alpha\eta + (\beta\gamma)(\beta\alpha\eta)^2 \leq \frac{\delta(\tilde{v})}{2},$$

where the constants correspond to our choice $\tilde{v} := \mathcal{N}^{\infty}(v)$, then the open set $V = \text{int}(B_{\eta}(\tilde{v}))$ is such a possible neighborhood, see the first inequality in (3.4.3), which was used to obtain (3.4.8). Let $k > 0$ be an index such that $\mathcal{N}^i(v) \in V$ for $i \geq k$. Then

$$\{u \mid u, \mathcal{N}(u), \ldots, \mathcal{N}^k(u) \in U_2, ,\mathcal{N}^k(u) \in V\}$$

is an asserted open neighborhood of v, since it is a finite intersection of open sets which contain v and which are contained in U by our choice of V. □

Chapter 4. Solving the Linear Systems

As has been seen in the preceding chapters, the numerical tracing of $c(s)$ will generally involve the frequent calculation of both the tangent vectors and the execution of the corrector steps. This will require a sufficient amount of linear equation solving to warrant that it be done in an efficient and carefully considered manner. Here too, we shall treat the details of numerical linear algebra only in the context which concerns us viz. the calculation of tangent vectors $t(A)$, and performing the operations $w = A^+ b$ where A is an $N \times (N+1)$ matrix with $\mathrm{rank}(A) = N$ which arise in the corrector steps. Readers interested in further background concerning numerical linear algebra may consult such textbooks on the subject as that of Golub & Van Loan.

In the discussion which follows, we mainly concentrate upon the QR decomposition of A^*. Our reasons for this are that by using the QR decomposition, scaling of the dependent variable becomes unnecessary, also the method is numerically stable without any pivoting, and it is easy to describe. The QR decomposition of A^* might be accomplished in different ways e.g. Householder transformations or Givens rotations. We have elected to describe in detail the latter, because we will use them also later when we describe updating methods for approximating the Jacobian matrix. In the last section of the chapter we will outline the numerical steps when using an LU decomposition of A^*. In chapter 10 we discuss how any general linear equation solver can be incorporated in the continuation method. This is of particular interest for large sparse systems where a user may wish to apply a particular solver.

4.1 Using a QR Decomposition

Let us indicate briefly how $t(A)$ and A^+ can be easily obtained once a QR factorization of A^* is available. We assume that A is an $N \times (N+1)$ matrix with rank$(A) = N$, and that a decomposition

$$A^* = Q \begin{pmatrix} R \\ 0^* \end{pmatrix}$$

is given, where Q is an $(N+1) \times (N+1)$ orthogonal matrix i.e. $Q^*Q = $ Id, and R is a nonsingular $N \times N$ upper triangular matrix, i.e. $R[i,j] = 0$ for $i > j$ and $R[i,i] \neq 0$. Hence if z denotes the last column of Q, then $Az = 0$ and $||z|| = 1$. The question which remains, is how to choose the sign of z so that

$$\det \begin{pmatrix} A \\ z^* \end{pmatrix} > 0,$$

in order to satisfy the orientation condition (2.1.7)(3). To answer this, note that

$$(A^*, z) = Q \begin{pmatrix} R & 0 \\ 0^* & 1 \end{pmatrix}$$

implies

(4.1.1) $$\det \begin{pmatrix} A \\ z^* \end{pmatrix} = \det(A^*, z) = \det Q \det R.$$

Hence, $t(A) = \pm z$ according as the determinant in (4.1.1) is positive or negative. Now, $\det R$ is the product of the diagonal elements of R, and its sign is easily determined. Also sign $\det Q$ is usually easily obtained. For example, if Givens rotations are used, it is equal to unity. If Householder reflections are used, each reflection changes the sign, and so sign $\det Q = (-1)^p$ where p is the number of reflections which are involved in the factorization of A^* by Householder's method. In any event, the question of determining $t(A)$ is now easily resolved. Note that the selection of the appropriate sign does not cost any additional computational effort.

Let us now turn to the problem of determining the Moore-Penrose inverse. From (3.2.2), we have $A^+ = A^*(AA^*)^{-1}$, and from

$$A^* = Q \begin{pmatrix} R \\ 0^* \end{pmatrix} \quad \text{and} \quad A = (R^*, 0)Q^*$$

we obtain

$$A^+ = Q \begin{pmatrix} (R^*)^{-1} \\ 0^* \end{pmatrix}.$$

Of course, as is usual in solving linear systems of equations, we do not invert R^*, but rather we calculate $w = A^+ b$ by a forward solving $R^* y = b$ i.e.

$$\text{for } i := 1, \dots, N$$

$$y[i] := \left(b[i] - \sum_{k=1}^{i-1} R[k,i] y[k] \right) / R[i,i]$$

and a matrix multiplication

$$w = Q \begin{pmatrix} y \\ 0 \end{pmatrix}.$$

4.2 Givens Rotations for Obtaining a QR Decomposition

At this point it may be helpful to some readers if we carry out an example of a QR decomposition for A^*. As we have already indicated, we choose to illustrate the use of Givens rotations, since this is convenient for our later description of updating approximations to the Jacobian. The reader who is interested in utilizing other methods such as the fast Givens or Householder methods, can see how to do this in section 6.3 of the book by Golub & Van Loan. Givens rotations act only on two co-ordinates and may hence be described by a matrix of the form

$$G = \begin{pmatrix} s_1 & s_2 \\ -s_2 & s_1 \end{pmatrix}$$

such that $s_1^2 + s_2^2 = 1$, for then $GG^* = I$ and $\det G = 1$. For any vector $x \in \mathbf{R}^2$,

$$Gx = \begin{pmatrix} ||x|| \\ 0 \end{pmatrix} \quad \text{if} \quad s_1 := \frac{x[1]}{||x||}, \quad s_2 := \frac{x[2]}{||x||}.$$

The reduction of A^* to upper triangular form R is accomplished via a succession of Givens rotations acting on varying pairs of co-ordinates. We illustrate this by the following pseudo code:

(4.2.1) QR Decomposition.   *comment:*

$Q := \text{Id}; \; R := A^*;$ *initialization*

for $i = 1$ **to** N **do**

    for $k = i + 1$ **to** $N + 1$ **do**

      begin

      $(s_1, s_2) := (R[i, i], R[k, i]);$ *calculate Givens rotation*

      if $s_2 \neq 0$ **then**   *else: no rotation is necessary*

        begin

        $s := \sqrt{s_1^2 + s_2^2}; \; (s_1, s_2) := s^{-1}(s_1, s_2);$

$$\begin{pmatrix} e_i^* R \\ e_k^* R \end{pmatrix} := \begin{pmatrix} s_1 & s_2 \\ -s_2 & s_1 \end{pmatrix} \begin{pmatrix} e_i^* R \\ e_k^* R \end{pmatrix}; \qquad \text{\textit{Givens rotation on rows } } i, k$$

$$\begin{pmatrix} e_i^* Q \\ e_k^* Q \end{pmatrix} := \begin{pmatrix} s_1 & s_2 \\ -s_2 & s_1 \end{pmatrix} \begin{pmatrix} e_i^* Q \\ e_k^* Q \end{pmatrix}; \qquad \text{\textit{Givens rotation on rows } } i, k$$

      end;

    end;

$Q := Q^*.$ *to make Q consistent with the above discussion*

The above illustration is given only to make our discussion complete and self-contained. In the interest of simplicity and brevity, we have formulated it in a slightly inefficient way. A number of improvements in efficiency could be made, see the standard literature and library routines.

4.3 Error Analysis

In the process of performing the numerical calculations of linear algebra, roundoff errors arise from machine arithmetic. We assume in this section that the reader is familiar with the standard error analysis of numerical linear algebra, see e.g. Golub & Van Loan (1983) or Stoer & Bulirsch (1980). The standard analysis shows that the relative error of the solution to the equation $Bx = b$ for the square matrix B is estimated by

$$\frac{||\Delta x||}{||x||} \leq \text{cond}(B) \left(\frac{||\Delta B||}{||B||} + \frac{||\Delta b||}{||b||} \right) + O(||\Delta B||^2 + ||\Delta B|| \; ||\Delta b||),$$

where $\text{cond}(B) := ||B|| \; ||B^{-1}||$ is the **condition number** of B and $||\Delta x||$ represents the error in x etc. In this section we briefly show that an essentially analogous result holds for underdetermined systems of equations. We note that this is not true for overdetermined systems of equations (least squares solutions), see the above mentioned references.

(4.3.1) Definition. Let A be an $N \times (N+1)$-matrix with maximal rank N. Then the **condition number** of A is defined by $\operatorname{cond}(A) := ||A|| \, ||A^+||$.

As always, $|| \; ||$ denotes the Euclidean norm. It can be seen that $||A||^2$ is the largest eigenvalue of AA^* and $||A^+||^{-2}$ the smallest eigenvalue. Let us now investigate the sensitivity of the solution $x = A^+ b$ with respect to perturbations in the entries of A and b. For the following lemma we use the fact that the estimate

(4.3.2) $$(B + \Delta B)^{-1} = B^{-1} - B^{-1}(\Delta B)B^{-1} + O(||\Delta B||^2)$$

holds for nonsingular square matrices B and perturbations ΔB having sufficiently small norm. Note that the existence of $(B + \Delta B)^{-1}$ is also implied.

(4.3.3) Lemma. *Let A be an $N \times (N+1)$-matrix with maximal rank N, and let $t := t(A)$ be the induced tangent. If ΔA is an $N \times (N+1)$-matrix with sufficiently small norm, then $(A + \Delta A)$ also has maximal rank, and the following estimate holds:*

$$(A + \Delta A)^+ = A^+ - A^+(\Delta A)A^+ + tt^*(\Delta A)^*(A^+)^* A^+ + O(||\Delta A||^2).$$

Proof. By definition (3.2.2) we have

$$(A + \Delta A)^+$$
$$= (A + \Delta A)^* \left[(A + \Delta A)(A + \Delta A)^* \right]^{-1}$$
$$= (A + \Delta A)^* \left[AA^* + A(\Delta A)^* + (\Delta A)A^* + O(||\Delta A||^2) \right]^{-1}$$
$$= (A + \Delta A)^* \left[(AA^*)^{-1} - (AA^*)^{-1}\left(A(\Delta A)^* + (\Delta A)A^* \right)(AA^*)^{-1} \right]$$
$$\quad + O(||\Delta A||^2)$$
$$= A^+ - A^+ \left(A(\Delta A)^* + (\Delta A)A^* \right)(AA^*)^{-1} + (\Delta A)^*(AA^*)^{-1} + O(||\Delta A||^2)$$
$$= A^+ - A^+(\Delta A)A^+ + (\operatorname{Id} - A^+ A)(\Delta A)^*(AA^*)^{-1} + O(||\Delta A||^2).$$

Since $(A^+)^* A^+ = (AA^*)^{-1}$ by (3.2.2) and $\operatorname{Id} - A^+ A = tt^*$ by (3.2.5)(1), the assertion follows.   □

(4.3.4) Lemma. *Let A be an $N \times (N+1)$-matrix with maximal rank N, and let ΔA be an $N \times (N+1)$-matrix with sufficiently small norm. For $b, \Delta b \in \mathbf{R}^N$, let $x := A^+ b$ and $x + \Delta x := (A + \Delta A)^+(b + \Delta b)$. Then the following estimate holds for the relative error in x:*

$$\frac{||\Delta x||}{||x||} \leq \operatorname{cond}(A) \left(2\frac{||\Delta A||}{||A||} + \frac{||\Delta b||}{||b||} \right) + O(||\Delta A||^2) + ||\Delta A|| \, ||\Delta b||.$$

Proof. From the preceding lemma we obtain

$$x + \Delta x = A^+ b - A^+(\Delta A) A^+ b + tt^*(\Delta A)^*(A^+)^* A^+ b + A^+(\Delta b)$$
$$+ O(||\Delta A||^2 + ||\Delta A|| \, ||\Delta b||).$$

Now using $x = A^+ b$, $||t|| = 1$ and taking norms yields

$$||\Delta x|| \le ||A^+|| \, ||\Delta A)|| \, ||x|| + ||\Delta A|| \, ||A^+|| \, ||x|| + ||A^+|| \, ||\Delta b||$$
$$+ O(||\Delta A||^2 + ||\Delta A|| \, ||\Delta b||).$$

Now we divide by $||x||$ and use the estimate $||x|| \ge ||A||^{-1}||b||$ to obtain the assertion. $\square$

We may now refer to standard results on the roundoff errors (backward error analysis in the sense of Wilkinson) of decomposition methods and conclude: if we use a QR factorization method such as the one described in the previous section, then we can expect a relative roundoff error in the Newton step $w = u - H'(u)^+ H(u)$ of the order of magnitude $\text{cond}(H'(u))\varepsilon$ where ε represents the relative machine error.

Let us now give an analogous error estimate discussion for the calculation of the tangent vector.

(4.3.5) Lemma. *Let A be an $N \times (N + 1)$-matrix with maximal rank N, and let $t := t(A)$ be the induced tangent. If ΔA is an $N \times (N + 1)$-matrix with sufficiently small norm, then the following estimate holds:*

$$t(A + \Delta A) = t(A) - A^+(\Delta A) t(A) + O(||\Delta A||^2).$$

Proof. We make the ansatz $t(A + \Delta A) = \rho(t(A) - A^+ y)$. Since A^+ is a bijection from $\mathbf{R}^N$ onto $\{t(A)\}^\perp$, the equation has a unique solution $y \in \mathbf{R}^N$ and $0 < \rho \le 1$. By the definition of the induced tangent, y must satisfy the equation $(A + \Delta A)(t(A) - A^+ y) = 0$ which implies $(\text{Id} + (\Delta A)A^+)y = (\Delta A)t(A)$. Hence $y = (\Delta A)t(A) + O(||\Delta A||^2)$. From the orthogonality $t(A) \perp A^+ y$, we obtain $\rho^{-2} = 1 + ||A^+ y||^2 = 1 + O(||\Delta A||^2)$ and hence $\rho = 1 + O(||\Delta A||^2)$. The assertion now follows from the above estimates for y and ρ.
                                                                $\square$

(4.3.6) Lemma. *Let A be an $N \times (N + 1)$-matrix with maximal rank N, and let ΔA be an $N \times (N + 1)$-matrix with sufficiently small norm. Then the following estimate holds for the relative error of the induced tangent vectors:*

$$\frac{||t(A + \Delta A) - t(A)||}{||t(A)||} \le \text{cond}(A)\frac{||\Delta A||}{||A||} + O(||\Delta A||^2).$$

Proof. From the previous lemma we have $t(A + \Delta A) - t(A) = -A^+(\Delta A)t(A) + O(||\Delta A||^2)$. Now by taking norms and regarding that $||t(A)|| = 1$, the assertion follows immediately. $\square$

Analogously to the remark following lemma (4.3.4) we conclude: if we use a QR factorization method such as the one described in the previous section, then we can expect a relative roundoff error in the evaluation of the tangent $t(H'(u))$ of the order of magnitude $\text{cond}(H'(u))\varepsilon$ where ε represents the relative machine error.

4.4 Scaling of the Dependent Variables

In the numerical solution of systems of equations it is sometimes advisable for reasons of stability, to perform a scaling of the dependent variables. Then instead of solving $H(u) = 0$, one solves an equivalent system

$$\tilde{H}(u) = DH(u)$$

where D is a chosen $N \times N$ diagonal matrix with positive diagonal entries. This scaling induces a row scaling of the corresponding Jacobian

$$\tilde{H}'(u) := DH'(u).$$

The aim of the row scaling is to decrease $\text{cond}(DH'(u))$.

It turns out that the QR decomposition as described in the section 4.2, is invariant under such scaling. Indeed, if $H'(u)Q = (R^*, 0)$ as in section 4.1, then also $DH'(u)Q = D(R^*, 0)$. We note that in performing the QR decomposition by e.g. Givens rotations as described in section 4.2, only elements of the same row are compared and transformed. Thus the relative precision with which Q and $D(R^*, 0)$ are calculated, is actually independent of the choice of D. Furthermore, the Newton steps are easily seen to satisfy $w = u - \tilde{H}'(u)^+\tilde{H}(u) = u - H'(u)^+H(u)$ and hence are invariant under such scalings. The above remarks serve to show that also from the point of view of stability with respect to roundoff errors, there is no advantage to be gained from different choices of scalings. Consequently, if we employ a QR decomposition of $H'(u)^*$ as described in section 4.2, then for numerical stability considerations we can conclude that the tracing of a curve in $H^{-1}(0)$ is automatically performed with a scaling which is optimal with respect to

(4.4.1) $\inf_D \text{cond}(DH'(u)).$

Stated in other terms, this means that if the QR implementation for predictor and corrector steps is used, then scaling of the dependent variables is unnecessary. Of course, when a curve in $H^{-1}(0)$ is being traversed, it may be advisable to monitor the condition number, and to do this with a minimum of computational cost. There are some reasonably fast and efficient algorithms for estimating the condition of a triangular matrix. Since it is not our main concern to estimate the condition very exactly, but merely to detect places on $H^{-1}(0)$ where bad conditioning occurs, we suggest using the following measure if a QR decomposition in the above sense is used:

Let us first note that $\text{cond}(DH'(u)) = \text{cond}(DR^*)$, since $\text{cond}(Q) = 1$. If we set $D[i,i] := R[i,i]^{-1}$ for $i = 1, \ldots, N$, then the diagonal elements of DR^* are all unity. If for all off-diagonal elements of DR^* the absolute value can be estimated by $O(1)$, then we do not expect bad conditioning.

4.5 Using LU Decompositions

We conclude this chapter with a brief discussion of the analogous steps which must be made when LU decompositions of the Jacobians are made instead of the steps described in section 4.1 for QR decompositions. See also a more recent forward error analysis given by Stummel & Hainer (1982, chapter 6) for this case. Let us again assume that A is an $N \times (N + 1)$-matrix with maximal rank N. We consider a decomposition of the form

$$(4.5.1) \qquad PA^* = L \begin{pmatrix} U \\ 0^* \end{pmatrix},$$

where L is a lower triangular $(N + 1) \times (N + 1)$-matrix, U is an $N \times N$ upper triangular matrix, and P an $(N + 1) \times (N + 1)$ permutation matrix corresponding to partial pivoting which is in general necessary to improve the numerical stability.

Let us first consider the calculation of the tangent vector $t(A)$. From (4.5.1) it follows that

$$(4.5.2) \qquad A = (U^*, 0)L^*P.$$

Hence, if we set

$$y := P^*(L^*)^{-1}e_{N+1}$$

then it is readily seen from (4.5.2) that $Ay = 0$. Of course $y \neq 0$, and can be calculated by one backsolving and a permutation of its co-ordinates. Hence $t(A) = \pm y/\|y\|$, where the sign is determined by evaluating the sign of the determinant of

$$(A^*, y) = \left(P^* L \begin{pmatrix} U \\ 0^* \end{pmatrix}, \ P^*(L^*)^{-1}e_{N+1} \right)$$

$$= P^* L \left(\begin{pmatrix} U \\ 0^* \end{pmatrix}, \ L^{-1}(L^*)^{-1}e_{N+1} \right).$$

Since $L^{-1}(L^*)^{-1}$ is positive definite, the last entry of $L^{-1}(L^*)^{-1}e_{N+1}$ must be positive, and hence

$$(4.5.3) \qquad \text{sign det}(A^*, y) = \text{sign det}(P)\det(L)\det(U).$$

The right hand side is easily determined. Hence $t(A) = \pm y/\|y\|$ according as the above determinant is positive or negative.

Let us now turn to the problem of determining the Moore-Penrose inverse. From (4.5.2) it follows that

$$B := P^*(L^*)^{-1} \begin{pmatrix} (U^*)^{-1} \\ 0^* \end{pmatrix}$$

is a right inverse of A, and hence $A^+ = (\mathrm{Id} - t(A)t(A)^*)B$ by (3.2.5)(3). Finally, let us note that a calculation of $w = A^+b$ amounts to essentially one forward solving with U^*, one backsolving with L^*, and one scalar product for the orthogonal projection with $(\mathrm{Id} - t(A)t(A)^*)$.

Chapter 5. Convergence of Euler-Newton-Like Methods

In this chapter we analyze the convergence properties of an Euler-Newton method under the simplifying assumption that a sufficiently small uniform steplength is maintained.

5.1 An Approximate Euler-Newton Method

Let $H : R^{N+1} \to R^N$ be a smooth map having zero as a regular value and let $H(u_0) = 0$. An Euler-Newton method for numerically tracing the path c given by the defining initial value problem (2.1.9) was outlined in (3.3.7).

Often it may be preferred to save computational effort in the corrector process by replacing the current Jacobian matrix $H'(v)$ or $H'(w)$ by an approximation, say A. First of all, because of rounding errors, we cannot in general expect to represent $H'(v)$ precisely. Furthermore, we may not even want to calculate $H'(v)$ at all, but we may rather prefer to approximate $H'(v)$ by e.g. a difference approximation, or updating method. For this reason, we incorporate into our illustration algorithm (5.1.1) below an approximation A to $H'(u)$ satisfying $\|H'(v) - A\| \leq ch$, for a step length $h > 0$. In addition to this, for reasons of numerical stability, we also find it occasionally desirable to incorporate a perturbation of the equation $H(u) = 0$, so that actually $H(u) = p$ is solved for some $p \in R^N$ such that $\|p\|$ is small. The following illustration algorithm is stated in an artificial form since it is meant to show that an Euler-Newton PC method will succeed in tracing the curve c if the uniform step size $h > 0$ is sufficiently small.

(5.1.1) Illustration Algorithm.   *comment:*

    input

      begin

        $u_0 \in \mathbf{R}^{N+1}$ such that $H(u_0) = 0$;   *initial point*

        $h > 0$;   *fixed steplength*

        $C > 0$; $\varepsilon > 0$;   *constants for characterizing the approximations below*

      end;

    $u := u_0$; print u;   *points generated along $H^{-1}(0)$*

    repeat

      choose any $N \times (N+1)$-matrix A such that

        $\|H'(u) - A\| \leq Ch$ and $\mathrm{rank}(A) = N$;   *approximate Jacobian*

      $v := u + ht(A)$;   *predictor step*

      choose any $p \in \mathbf{R}^N$ such that $\|p\| \leq \varepsilon h^2$;   *perturbation*

      $w := v - A^+\big(H(v) - p\big)$;   *corrector step*

      $u := w$; print u;   *points generated along $H^{-1}(0)$*

      until traversing is stopped.

5.2 A Convergence Theorem for PC Methods

The following theorem shows that a PC-method indeed approximates a solution curve if the steplength h is sufficiently small. For simplicity, we consider the situation of algorithm (5.1.1). Analogous proofs can be given for other versions of PC algorithms. We shall only sketch the proof by giving the main arguments and omitting tedious technical details.

(5.2.1) Theorem. *Let $H : R^{N+1} \to R^N$ be a smooth map having zero as a regular value and let $H(u_0) = 0$. Denote by $c_h(s)$ the polygonal path, starting at u_0, going through all points u generated by the algorithm (5.1.1). Denote by $c(s)$ the corresponding curve in $H^{-1}(0)$ given by the defining initial value problem (2.1.9). For definiteness, we assume that $c_h(0) = c(0) = u_0$, and that both curves are parametrized with respect to arclength. Then, for a given maximal arclength $s_0 > 0$, and for given constants C, $\varepsilon > 0$ as in the algorithm, there exist constants $C_0, \gamma > 0$ and a maximal steplength $h_0 > 0$ such that*

(1) $\|H(u)\| \leq 2\varepsilon h^2$ *for all nodes u of c_h,*

(2) $\|H\big(c_h(s)\big)\| \leq \big(3\varepsilon + \frac{1}{2}\gamma\big)h^2$ *for $0 < h \leq h_0$,*

(3) $\|c_h(s) - c(s)\| \leq C_0 h^2$ *for $0 < h \leq h_0$*

holds for all $s \in [0, s_0]$.

The last statement means that the arclength of the solution curve is approximated by the polygonal path with order $O(h^2)$.

Proof. We only give a sketch of the proof. Let U be a compact neighborhood of $c[0, s_0]$ which consists only of regular points of H. Adopting the notation of (3.4.2), we define the following constants for U:

(5.2.2)
$$\alpha := \max\{\|H'(v)\| \mid v \in U\};$$
$$\beta := \max\{\|H'(v)^+\| \mid v \in U\};$$
$$\gamma := \max\{\|H''(v)\| \mid v \in U\}.$$

From the estimates below it is evident that the algorithm (5.1.1) generates only predictor and corrector points in U for sufficiently small steplength h and so long as the maximal arclength s_0 is not exceeded.

The proof for assertion (1) proceeds by induction. Let us assume that the estimate (1) is true for a current corrector point u. The next predictor and corrector points are respectively

$$v = u + ht(A) \quad \text{and} \quad w = v - A^+\big(H(v) - p\big).$$

We need to show that w also satisfies the estimate (1). Defining the constant bilinear form M_1 by the mean value

$$M_1 := \int_0^1 H''(u + \xi ht(A))2(1 - \xi)d\xi,$$

Taylor's formula expanded about u takes the form

$$H(v) = H(u) + hH'(u)t(A) + \frac{h^2}{2}M_1\big[t(A), t(A)\big].$$

Now from

$$H'(u)t(A) = \big(H'(u) - A\big)t(A),$$

(5.2.2) and the induction hypothesis we obtain the estimate

(5.2.3)
$$\|H(v)\| \leq \varepsilon h^2 + Ch^2 + \frac{\gamma}{2}h^2.$$

Defining the mean value

$$M_2 := \int_0^1 H''(v + \xi(w - v))2(1 - \xi)d\xi,$$

Taylor's formula expanded about v takes the form

$$
\begin{aligned}
H(w) &= H(v) + H'(v)(w - v) + \frac{1}{2} M_2[w - v, w - v] \\
&= H(v) + A(w - v) + (H'(v) - A)(w - v) + \frac{1}{2} M_2[w - v, w - v] \\
&= p + (H'(v) - A)(w - v) + \frac{1}{2} M_2[w - v, w - v].
\end{aligned}
$$

Taking into account the estimates

$$
(5.2.4) \quad
\begin{cases}
\|p\| \le \varepsilon h^2; \\
\|H'(v) - A\| \le \|H'(v) - H'(u)\| + \|H'(u) - A\| \\
\qquad\qquad \le (\gamma + C)h; \\
\|A^+\| \le \dfrac{\|H'(v)^+\|}{1 - \|H'(v) - A\|\,\|H'(v)^+\|} \\
\qquad \le \dfrac{\beta}{1 - (C + \gamma)\beta h}; \quad \text{cf. lemma } (5.2.8) \text{ below} \\
\|w - v\| \le \|A^+\|(\|H(v)\| + \|p\|) \\
\qquad \le \|A^+\| \left(1 + C + \dfrac{\gamma}{2} + 2\varepsilon\right) h^2; \quad \text{cf. } (5.2.3) \\
\|M_2[w - v, w - v]\| \le \gamma \|w - v\|^2;
\end{cases}
$$

we obtain constants C_1 and C_2 such that

$$
\|H(w)\| \le \varepsilon h^2 + C_1 h^3 + C_2 h^4 \le 2\varepsilon h^2
$$

for h sufficiently small. This completes the inductive step for proving assertion (1).

To prove assertion (2), we use the Taylor formulae

$$
H(u) = H(u_\tau) + H'(u_\tau)(u - u_\tau) + \frac{1}{2} A_1[u - u_\tau, u - u_\tau],
$$

$$
H(w) = H(u_\tau) + H'(u_\tau)(w - u_\tau) + \frac{1}{2} A_2[w - u_\tau, w - u_\tau],
$$

where $u_\tau := \tau u + (1 - \tau)w$ for $0 \le \tau \le 1$, and A_1, A_2 are the mean values

$$
\int_0^1 H''(u_\tau + \xi(u - u_\tau))2(1 - \xi)d\xi \quad \text{and} \quad \int_0^1 H''(u_\tau + \xi(w - u_\tau))2(1 - \xi)d\xi
$$

of H'' on the segments $[u, u_\tau]$ and $[w, u_\tau]$ respectively. Multiplying the first equation by τ and the second by $1 - \tau$, summing and taking norms yields the estimate

$$\|\tau H(u) + (1 - \tau)H(w) - H(\tau u + (1 - \tau)w)\|$$
$$\leq \frac{1}{2}\|\tau A_1[u - u_\tau, u - u_\tau] + (1 - \tau)A_2[w - u_\tau, w - u_\tau]\|$$
$$\leq \frac{1}{2}\gamma(h^2 + 4\varepsilon^2 h^4).$$

The last inequality follows from (5.2.2), assertion (1) and the Pythagoras formula. Assertion (2) now follows from the above estimate for sufficiently small h. This part of the proof is analogous to a more general result in (15.5.2).

To prove assertion (3), let u be a point which the algorithm (5.1.1) has generated. Consider the respective predictor and corrector points

$$v = u + ht(A), \qquad w = v - A^+(H(v) - p).$$

Since $v - w$ is orthogonal to $t(A)$, we have $\|w - u\|^2 = \|w - v\|^2 + h^2$, and from the estimates (5.2.4) we obtain

(5.2.5) $\|w - u\| = h(1 + O(h^2))\,,$

see figure 5.2.a. The quantity $\|w - u\|$ represents the arclength of the polygonal path c_h between two nodes u and w.

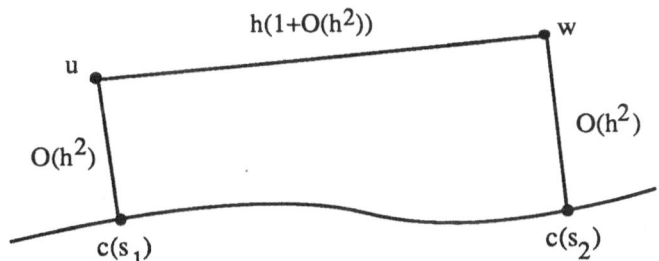

Figure 5.2.a Diagram for the estimates

In theorem (3.4.1) we considered the map $u \mapsto S(u)$ where $S(u)$ solves the problem

$$\min_x \{ \|u - x\| \mid H(x) = 0\}$$

for u sufficiently near $H^{-1}(0)$. Hence, by assertion (1), the points

$$c(s_1) := S(u) \quad \text{and} \quad c(s_2) := S(w)$$

are well defined for $h > 0$ sufficiently small, and we obtain the estimates

(5.2.6) $\|c(s_1) - u\| = O(h^2) \quad \text{and} \quad \|c(s_2) - w\| = O(h^2).$

Furthermore, the Lagrange equations (3.3.2) show that

$$\big(u - c(s_1)\big) \perp \dot{c}(s_1) \quad \text{and} \quad \big(w - c(s_2)\big) \perp \dot{c}(s_2).$$

Setting $\Delta s := s_2 - s_1$, and using the fact that $\|\dot{c}(s)\|^2 \equiv 1$ implies $\dot{c}(s) \perp \ddot{c}(s)$, Taylor's formula yields

$$\|c(s_2) - c(s_1)\|^2 = \left\| \int_0^1 \dot{c}(s_1 + \xi \Delta s) d\xi \right\|^2$$
$$= \|\dot{c}(s_1)\Delta s + \tfrac{1}{2}\ddot{c}(s_1)(\Delta s)^2 + O((\Delta s)^3)\|^2$$
$$= (\Delta s)^2 + O((\Delta s)^4)$$

and consequently

(5.2.7) $$\|c(s_2) - c(s_1)\| = \Delta s \big(1 + O((\Delta s)^2)\big).$$

Let us observe that the above relations immediately lead to the coarse estimates

$$h = \Delta s + O((\Delta s)^2) = \Delta s + O(h^2).$$

Thus it is justified to replace $O(h)$ by $O(\Delta s)$ etc. in the estimates below. From the orthogonality relation $(w - c(s_2)) \perp \dot{c}(s_2)$ and Taylor's formula we obtain

$$\big(w - c(s_2)\big)^* \big(c(s_2) - c(s_1)\big) = \big(w - c(s_2)\big)^* \big(\dot{c}(s_2)\Delta s + O((\Delta s)^2)\big)$$
$$= O(h^2)O((\Delta s)^2).$$

Similarly,

$$\big(u - c(s_1)\big)^* \big(c(s_2) - c(s_1)\big) = O(h^2)O((\Delta s)^2).$$

From these estimates we obtain

$$\|w - u\|^2 = \|(w - c(s_2)) + (c(s_2) - c(s_1)) + (c(s_1) - u)\|^2$$
$$= \|c(s_2) - c(s_1)\|^2 + O(h^4).$$

Taking square roots and using (5.2.7) yields

$$\|w - u\| = \|c(s_2) - c(s_1)\| + \frac{1}{2}\frac{O(h^4)}{\|c(s_2) - c(s_1)\|}$$
$$= \Delta s \big(1 + O((\Delta s)^2)\big) + O(h^3)$$
$$= \Delta s + O(h^3).$$

Summing up all arclengths between the nodes of c_h and using assertion (1) yields assertion (3). $\qquad \square$

The above proof made use of the following

(5.2.8) Lemma. *Let B, E be $N \times (N+1)$-matrices such that B has maximal rank and $\|B^+\| \, \|E\| < 1$. Then $B - E$ has maximal rank, and*

$$\|(B - E)^+\| \leq \frac{\|B^+\|}{1 - \|B^+\| \, \|E\|}.$$

Proof. By the hypothesis, the Neumann series

$$C := \sum_{i=0}^{\infty} B^+ (EB^+)^i$$

is well defined and generates a right inverse C of $B - E$. Furthermore,

$$\|C\| \leq \frac{\|B^+\|}{1 - \|B^+\| \, \|E\|}.$$

Let $t := t(B - E)$ denote the induced tangent. From $(3.2.5)(3)$ we have $(B - E)^+ = (\mathrm{Id} - tt^*)C$, and since $\|(\mathrm{Id} - tt^*)\| = 1$, the assertion follows. $\square$

Chapter 6. Steplength Adaptations for the Predictor

The convergence considerations of section 5.2 were carried out under the assumption that the steplength of the algorithm (5.1.1) was uniformly constant throughout. This is of course not efficient for any practical implementation. The discussion in chapter 5 did not indicate any means of choosing the steplength $h > 0$. To some extent of course, the steplength strategy depends upon the accuracy with which it is desired to numerically trace a solution curve. In any case, an efficient algorithm for this task needs to incorporate an automatic strategy for controlling the steplength. In this respect the PC methods are similar to the methods for numerically integrating initial value problems in ordinary differential equations. Indeed, one expedient way of tracing an implicitly defined curve c is to merely use a numerical initial value problem solver on the defining initial value problem (2.1.9). Although such an approach has been successfully used by some authors to solve a large variety of practical problems in science and engineering (for some examples, see the bibliography), the general opinion is that it is preferable to exploit the contractive properties of the zero set $H^{-1}(0)$ relative to such iterative methods as those of Newton type.

In this chapter we shall outline in detail some possible steplength strategies. One of them, based upon asymptotic estimates in the mentality of initial value solvers, is due to Georg (1983). It is simple to implement and our experience with it has been successful. Another method, due to Den Heijer & Rheinboldt (1981) is based upon an error model for the corrector iteration. This error model is justified by considerations from the Newton-Kantorovitch theory. Den Heijer & Rheinboldt report results of numerical experiments utilizing the error model steplength strategy. An advantage of this method is that it has a wider range of applicability. A third class of steplength strategies can be obtained by adapting analogues of the variable order steplength strategies used in multistep methods for initial value problems see e.g. Shampine & Gordon (1975). The method proposed by Schwetlick & Cleve (1987) is close in spirit to this approach. Finally, Kearfott (1989–90) proposes interval arithmetic techniques to determine a first order predictor which stresses secure path following. As usual, we assume throughout this chapter that $H : \mathbf{R}^{N+1} \to \mathbf{R}^N$ is a smooth map having zero as a regular value, and for some initial value $u_0 \in H^{-1}(0)$, we consider the smooth curve c in $H^{-1}(0)$ given by the defining initial value problem (2.1.9).

6.1 Steplength Adaptation by Asymptotic Expansion

The basic idea in this approach is to observe the performance of the corrector procedure and then to adapt the steplength $h > 0$ accordingly. More precisely, suppose that a point $u \in H^{-1}(0)$ has been obtained. Suppose further that a steplength $h > 0$ is given. The Euler predictor point is $v(h) = u + ht\big(H'(u)\big)$. Then a Newton type iterative corrector process is performed to generate a next point $z(h) \in H^{-1}(0)$.

The steplength strategies which we discuss in this section are based upon a posteriori estimates of the performance of the corrector process in order to answer the following question: Given the manner in which the corrector process starting at $v(h)$ performed, which steplength $\tilde{h}$ would have been "best" for obtaining $z(\tilde{h})$ from u? This "ideal" steplength $\tilde{h}$ is determined via asymptotic estimates, and it is then taken as the steplength for the next predictor step. This strategy depends primarily upon two factors: the particular predictor-corrector method being utilized, and the criteria used in deciding what performance is considered "best".

Let us begin with a simple example in which we describe how the rate of contraction of the Newton process can be used to govern the steplength of the straightforward Euler-Newton method (3.3.7). If u and $v(h)$ are given as above, then the first corrector point is given by

$$w(h) := v(h) - H'\big(v(h)\big)^+ H\big(v(h)\big).$$

Let us call the quotient of the first two successive Newton steps

(6.1.1) $$\kappa(u,h) := \frac{\|H'\big(v(h)\big)^+ H\big(w(h)\big)\|}{\|H'\big(v(h)\big)^+ H\big(v(h)\big)\|}$$

the **contraction rate** of the corrector process. Since Newton's method is locally quadratically convergent, it is plain that $\kappa(u,h)$ will decrease (and hence Newton's method will become faster) if h decreases and hence $v(h)$ approaches $H^{-1}(0)$. The following lemma characterizes the asymptotic behavior of $\kappa(u,h)$ with respect to h.

(6.1.2) Lemma. *Suppose that*

$$H''(u)\big[t\big(H'(u)\big), t\big(H'(u)\big)\big] \neq 0$$

(i.e. c has non-zero curvature at u), then

(6.1.3) $$\kappa(u,h) = \kappa_2(u)h^2 + O(h^3)$$

for some constant $\kappa_2(u) \geq 0$ which is independent of h and depends smoothly on u.

Proof. From Taylor's formula we have

$$H(v(h)) = C_1(u)h^2 + O(h^3),$$
$$\text{where } \ C_1(u) := \frac{1}{2}H''(u)\Big[t(H'(u)), t(H'(u))\Big],$$

since $H(u) = 0$ and $H'(u)t(H'(u)) = 0$.
Because the map $u \to H'(u)^+$ is smooth, we have

$$H'(v(h))^+ = H'(u)^+ + O(h)$$

and hence

$$H'(v(h))^+ H(v(h)) = C_2(u)h^2 + O(h^3),$$
$$\text{where } \ C_2(u) := H'(u)^+ C_1(u).$$

Now

$$\begin{aligned}
H(w(h)) &= H(v(h)) - H'(v(h))H'(v(h))^+ H(v(h)) \\
&\quad + \frac{1}{2}H''(v(h))\Big[H'(v(h))^+ H(v(h)), H'(v(h))^+ H(v(h))\Big] \\
&\quad + O(h^5) \\
&= C_3(u)h^4 + O(h^5),
\end{aligned}$$
$$\text{where } \ C_3(u) := \frac{1}{2}H''(u)[C_2(u), C_2(u)].$$

Furthermore we have

$$H'(v(h))^+ H(w(h)) = C_4(u)h^4 + O(h^5),$$
$$\text{where } \ C_4(u) := H'(u)^+ C_3(u).$$

Finally we obtain

$$\kappa(u, h) = \kappa_2(u)h^2 + O(h^3),$$
$$\text{where } \ \kappa_2(u) := \frac{\|C_4(u)\|}{\|C_2(u)\|}.$$

Note that the hypothesis implies that $C_1(u) \neq 0$ and hence $C_2(u) \neq 0$. The smoothness of $\kappa_2(u)$ follows from the smoothness of the vectors $C_2(u)$ and $C_4(u)$. $\qquad\square$

In view of the asymptotic relation (6.1.3) the steplength modification $h \rightarrow \tilde{h}$ is now easy to explain. Assume that an Euler-Newton step has been performed with steplength h. Then $H'(v(h))^+ H(v(h))$ and $H'(v(h))^+ H(w(h))$ will have been calculated and thus $\kappa(u, h)$ can be obtained without any significant additional cost. Now an a-posteriori estimate

$$(6.1.4) \qquad \kappa_2(u) = \frac{\kappa(u, h)}{h^2} + O(h)$$

is available.

In order to have a robust and efficient PC method we want to continually adapt the steplength h so that a nominal prescribed contraction rate $\tilde{\kappa}$ is maintained. The choice of $\tilde{\kappa}$ will generally depend upon the nature of the problem at hand, and on the desired security with which we want to traverse $H^{-1}(0)$. That is, the smaller $\tilde{\kappa}$ is chosen, the greater will be the security with which the PC method will follow the curve. Once $\tilde{\kappa}$ has been chosen, we will consider a steplength $\tilde{h}$ to be adequate if $\kappa(u, \tilde{h}) \approx \tilde{\kappa}$. By using (6.1.1) and (6.1.4) and neglecting higher order terms we obtain the formula

$$(6.1.5) \qquad \tilde{h} = h \sqrt{\frac{\tilde{\kappa}}{\kappa(u, h)}}$$

as the steplength for the next predictor step.

There are several factors which ought to be regarded in governing the steplength. In addition to contraction, other important factors are:

• the first corrector steplength

$$(6.1.6) \qquad \delta(u, h) := \|H'(v(h))^+ H(v(h))\|,$$

which approximates the distance to the curve;
• and the **angle** between two consecutive steps

$$(6.1.7) \qquad \alpha(u, h) := \arccos\left(t(H'(u))^* t(H'(v(h)))\right),$$

which is a measure of the curvature.

The following lemma may be proven in a similar fashion to (6.1.2). The proof is in fact simpler and will be omitted.

(6.1.8) Lemma. *The following asymptotic expansions hold:*

$$\delta(u, h) = \delta_2(u)h^2 + O(h^3),$$
$$\alpha(u, h) = \alpha_1(u)h + O(h^2),$$

where the asymptotic constants $\delta_2(u)$ and $\alpha_1(u)$ are independent of h and depend smoothly on u.

By fixing a nominal distance $\tilde{\delta}$ and a nominal angle $\tilde{\alpha}$, we obtain further steplength adaptations

$$(6.1.9) \qquad \tilde{h} = h \sqrt{\frac{\tilde{\delta}}{\delta(u, h)}} \qquad \text{and} \qquad \tilde{h} = h \frac{\tilde{\alpha}}{\alpha(u, h)}$$

where $\delta(u, h)$, $\alpha(u, h)$ respectively, are the distance and angle which have been measured after performing the current Euler-Newton step.

In order to formulate a safeguarded curve tracing algorithm, we recommend monitoring several factors such as those outlined above. Thus, at any point several adequate steplengths $\tilde{h}$ are calculated and the smallest of these possibilities is then chosen as the steplength for the next predictor step. The following algorithm sketches how the above described steplength adaptations can be incorporated into an Euler-Newton method such as (3.3.7).

(6.1.10) Steplength Adaptation
Via Asymptotic Estimates. *comment:*

  input
    begin
    $u \in \mathbf{R}^{N+1}$ such that $H(u) = 0$; *initial point*
    $h > 0$; *initial steplength*
    $\tilde{\kappa}$; *nominal contraction rate*
    $\tilde{\delta}$; *nominal distance to curve*
    $\tilde{\alpha}$; *nominal angle*
    end;
  repeat
    $v := u + ht\big(H'(u)\big)$; *predictor step*
    calculate the quantities
      $\alpha(u, h)$, $\delta(u, h)$, $\kappa(u, h)$; *cf. (6.1.7), (6.1.6), (6.1.1)*
    repeat
      $w := v - H'(v)^+ H(v)$; *corrector loop*
      $v := w$;
    until convergence;

    $f := \max \left\{ \sqrt{\dfrac{\kappa(u, h)}{\tilde{\kappa}}}, \sqrt{\dfrac{\delta(u, h)}{\tilde{\delta}}}, \dfrac{\alpha(u, h)}{\tilde{\alpha}} \right\}$; *cf. (6.1.5), (6.1.9)*

$$f := \max\{\min\{f, 2\}, \tfrac{1}{2}\}; \qquad\qquad \textit{deceleration factor } f \in [\tfrac{1}{2}, 2]$$

$$h := h/f; \qquad \textit{steplength adaptation using deceleration factor } f$$

if $f < 2$ **then** $u := w;$ \qquad *new point along $H^{-1}(0)$ is accepted*

until traversing is stopped.

The evaluation of $\delta(u, h)$ and $\kappa(u, h)$ is available at essentially no additional cost after performing the first and second steps respectively in the corrector loop. If $f = 2$ i.e. if the observed quantities $\alpha(u, h)$, $\delta(u, h)$ or $\kappa(u, h)$ exceed a certain tolerance above the nominal values, the predictor step is repeated with a reduced steplength. The test for $f = 2$ should actually be performed within the corrector loop in order to avoid unnecessary corrector steps which would in any event be discarded later.

We emphasize that the steplength adaptation described above make use of the existence of an underlying asymptotic expansion such as (6.1.3) and (6.1.8). Such asymptotic expansions vary according to the ingredients of the particular predictor-corrector process. Let us illustrate this with an example: If we change the corrector loop in (3.3.7) to the cord iteration

$$w := v - H'(u)^+ H(v)$$

then the analogue of formula (6.1.1) would be

$$\kappa(u, h) = \frac{\|H'(u)^+ H(w(h))\|}{\|H'(u)^+ H(v(h))\|}$$

where $w(h) := v(h) - H'(u)^+ H(v(h))$. It is easy to see that the asymptotic expansion analogous to (6.1.3) would be

$$\kappa(u, h) = \kappa_1 h + O(h^2)$$

and the analogue of the steplength adaptation (6.1.5) would be

$$\tilde{h} = h \, \frac{\tilde{\kappa}}{\kappa(u, h)}.$$

6.2 The Steplength Adaptation of Den Heijer & Rheinboldt

The steplength adaptation in section 6.1 was based on the consideration of asymptotic expansions. The reader familiar with steplength adaptation in numerical integration of initial value problems will have perceived that this philosophy of steplength adaptation proceeds along similar lines. A totally different approach has been given by Den Heijer & Rheinboldt (1981). Their approach is based on an error model for the corrector iteration, and they control the steplength by the number of steps which are taken in the corrector iteration until a given stopping criterion is fulfilled. This approach is a useful alternative to the previous steplength adaptation whenever the corrector steps may be viewed as an iterative process in which some variable number of iterations may be performed at each point along the curve. Let us also note that for this approach, it is not necessary to assume that an asymptotic expansion holds as was done in section 6.1. We shall sketch a somewhat modified and simplified version of the Den Heijer & Rheinboldt steplength strategy. The modification also reflects that we view a corrector step as always being orthogonal to the current tangent. We begin with a description of a general PC method in which the modified Den Heijer & Rheinboldt steplength strategy may be used.

For a predictor point v near $H^{-1}(0)$, let $T(v)$ be one step of an iterative process for determining an approximation of the nearest point to v on $H^{-1}(0)$. Then the steplength strategy is guided by the observed performance of the corrector iterates v, $T(v)$, $T^2(v)$, ... The following algorithm illustrates the basic steps of a PC method involving the adaptation of the steplength via an error model. The precise determination of the deceleration factor f is the main content of the remainder of this section.

(6.2.1) Steplength Adaptation Via Error Models.   *comment:*

   input

      begin

      $\tilde{k} \in \mathbf{N}$;   *desired number of corrector iterations*

      $u \in \mathbf{R}^{N+1}$ such that $H(u) = 0$;   *initial point*

      $h > 0$;   *initial steplength*

      end;

   repeat

      $v_0 := u + ht\big(H'(u)\big)$;   *predictor step*

      repeat for $k = 0, 1, 2, \ldots$

         $v_{k+1} := T(v_k)$;   *corrector loop*

      until convergence;

if the corrector loop was successful **then**
  begin
    determine f
      as a function of $k, \tilde{k}, v_0, v_1, \ldots, v_k$;   *see discussion below*
    $f := \max\{\min\{f, 2\}, \frac{1}{2}\}$;   *deceleration factor $f \in [\frac{1}{2}, 2]$*
    $h := h/f$;   *steplength adaptation using deceleration factor f*
    $u := v_k$;   *new point along $H^{-1}(0)$*
  end
  else $h := \dfrac{h}{2}$;   *repeat predictor with reduced h*
until traversing is stopped.

Since in the discussion to follow, the dependence of the various points upon the steplength h will be of importance, let us denote the predictor point from the current point $u \in H^{-1}(0)$ by

$$v_0(h) = u + ht(H'(u))$$

and the subsequent corrector steps by

(6.2.2) $v_{i+1}(h) = T(v_i(h))$.

Let us further suppose that for $h > 0$ sufficiently small, the following limit exists and

$$v_\infty(h) := \lim_{i \to \infty} v_i(h) \in H^{-1}(0).$$

We make the following assumption concerning the iteration process T: the angle between $v_0(h) - v_\infty(h)$ and $t(H'(u))$ is $\pi/2 + O(h)$. In other words, we assume that the iterative corrector method T operates essentially orthogonally to the curve c initiating at u i.e. given by the defining initial value problem

$$\dot{c} = t(H'(c)),$$
$$c(0) = u.$$

It is immediately seen that also the angle between $v_0(h) - c(h)$ and $t(H'(u))$ is $\pi/2 + O(h)$. From this it can easily be seen that $\|c(h) - v_\infty(h)\| = O(h^3)$. Since Taylor's formula implies $c(h) - v_0(h) = \frac{1}{2}\ddot{c}(0)h^2 + O(h^3)$, we finally have

(6.2.3) $\|v_\infty(h) - v_0(h)\| = \dfrac{1}{2}\|\ddot{c}(0)\|h^2 + O(h^3)$.

A crucial assumption in the steplength strategy of Den Heijer & Rheinboldt is that an error model is available for the iterative corrector process

(6.2.2). This means that it is assumed there exists a constant $\gamma > 0$ (which is independent of h) such that for the **modified error**

(6.2.4) $$\varepsilon_i(h) := \gamma \|v_\infty(h) - v_i(h)\|$$

inequalities of the following type

(6.2.5) $$\varepsilon_{i+1}(h) \leq \psi(\varepsilon_i(h))$$

hold, where $\psi : \mathbf{R} \to \mathbf{R}$ is a known monotone function such that $\psi(0) = 0$. Of course, the **error model function** ψ will depend in an essential way upon the special form of the iterative process T.

For example, when T is the unmodified Newton's method

$$T(v_i) = v_i - H'(v_i)^+ H(v_i),$$

Den Heijer & Rheinboldt suggest two models which are derived by estimates arising from the Newton-Kantorovitch theory:

$$\psi(\varepsilon) = \frac{\varepsilon^2}{3 - 2\varepsilon}, \quad 0 \leq \varepsilon \leq 1,$$

$$\psi(\varepsilon) = \frac{\varepsilon + \sqrt{10 - \varepsilon^2}}{5 - \varepsilon^2} \varepsilon^2, \quad 0 \leq \varepsilon \leq 1.$$

The idea behind such models is that more realistic error models for a particular corrector process T lead to more realistic steplength strategies.

Suppose that for a given steplength h, a predictor step and a subsequent iterative corrector procedure (6.2.2) stopped after k iterations yielding the point $v_k(h) \approx v_\infty(h)$. That is, we assume that a certain stopping criterion used in Algorithm (6.2.1) is fulfilled for the first time after $k > 1$ iterations. The particular form of the stopping criterion plays no role in the discussion to follow. We may a posteriori evaluate the quotient

$$\omega(h) := \frac{\|v_k(h) - v_{k-1}(h)\|}{\|v_k(h) - v_0(h)\|} \approx \frac{\|v_\infty(h) - v_{k-1}(h)\|}{\|v_\infty(h) - v_0(h)\|} = \frac{\varepsilon_{k-1}(h)}{\varepsilon_0(h)}.$$

Using the estimate $\varepsilon_{k-1}(h) \leq \psi^{k-1}(\varepsilon_0(h))$, we obtain

$$\omega(h) \leq \frac{\psi^{k-1}(\varepsilon_0(h))}{\varepsilon_0(h)}.$$

This motivates taking the solution ε of the equation

(6.2.6) $$\omega(h) = \frac{\psi^{k-1}(\varepsilon)}{\varepsilon}$$

as an estimate for $\varepsilon_0(h)$.

Similarly to the strategy of the previous section, we now try to choose the steplength $\tilde{h}$ so that the corrector process satisfies the stopping criterion after a chosen number $\tilde{k}$ of iterations. Such a steplength leads to the modified error $\varepsilon_0(\tilde{h})$. Now we want the modified error $\varepsilon_{\tilde{k}}(\tilde{h})$ after $\tilde{k}$ iterations to be so small that the stopping criterion is satisfied. Using the inequality $\varepsilon_{\tilde{k}}(\tilde{h}) \leq \psi^{\tilde{k}}\big(\varepsilon_0(\tilde{h})\big)$, we accept the solution ε of the equation

$$(6.2.7) \qquad\qquad \psi^{\tilde{k}}(\varepsilon) = \psi^{k}\big(\varepsilon_0(h)\big)$$

as an estimate for $\varepsilon_0(\tilde{h})$. Now we use the asymptotic expansion (6.2.3) to obtain the approximation

$$(6.2.8) \qquad\qquad f^2 = \left(\frac{h}{\tilde{h}}\right)^2 \approx \frac{\varepsilon_0(h)}{\varepsilon_0(\tilde{h})}$$

which we use to determine f and in turn $\tilde{h}$.

We summarize the above discussion by indicating how the line

"determine f as a function of k, $\tilde{k}$, v_0, v_1, $\ldots$, v_k"

in Algorithm (6.2.1) can be specified:

$$(6.2.9) \qquad \begin{cases} \omega := \dfrac{\|v_k - v_{k-1}\|}{\|v_k - v_0\|}; \\[2ex] \text{determine } \varepsilon \text{ so that } \dfrac{\psi^{k-1}(\varepsilon)}{\varepsilon} = \omega; \\[2ex] \text{determine } \tilde{\varepsilon} \text{ so that } \psi^{\tilde{k}}(\tilde{\varepsilon}) = \psi^{k}(\varepsilon); \\[2ex] f := \sqrt{\dfrac{\varepsilon}{\tilde{\varepsilon}}}\,; \end{cases}$$

As an example of the above, let us consider a simple model of superlinear convergence:

$$(6.2.10) \qquad\qquad \psi(\varepsilon) = \varepsilon^p, \qquad \text{where } p > 1.$$

Then (6.2.9) assumes the form

$$(6.2.11) \qquad \begin{cases} \omega := \dfrac{\|v_k - v_{k-1}\|}{\|v_k - v_0\|}; \\[2ex] \varepsilon = \omega^{\frac{1}{p^{k-1}-1}}; \\[2ex] \tilde{\varepsilon} = \varepsilon^{p^{k-\tilde{k}}}; \\[2ex] f := \omega^{\frac{1-p^{k-\tilde{k}}}{2(p^{k-1}-1)}}; \end{cases}$$

An interesting case to consider in (6.2.11) is the limit $p \to 1$, since this may be viewed as the value for f in the case of a linear error model. By l'Hopital's rule

(6.2.12)
$$\lim_{p \to 1} f = \omega^{\frac{\tilde{k}-k}{2(k-1)}}.$$

On the other hand, we can also consider a linear model directly:

(6.2.13)
$$\varepsilon_{i+1}(h) \le \lambda \varepsilon_i(h)$$

for some $\lambda \in (0,1)$. A discussion similar to that above gives estimates:

$$\omega \approx \frac{\varepsilon_{k-1}(h)}{\varepsilon_0(h)} \approx \frac{\lambda^{k-1} \varepsilon_0(h)}{\varepsilon_0(h)} \quad \text{implying} \quad \lambda \approx \omega^{\frac{1}{k-1}}.$$

Now (6.2.7) assumes the form

$$\lambda^{\tilde{k}} \varepsilon_0(\tilde{h}) \approx \lambda^k \varepsilon_0(h),$$

implying

(6.2.14)
$$f^2 = \left(\frac{h}{\tilde{h}} \right)^2 \approx \frac{\varepsilon_0(h)}{\varepsilon_0(\tilde{h})} \approx \lambda^{\tilde{k}-k} \approx \omega^{\frac{\tilde{k}-k}{k-1}}$$

which again yields (6.2.12).

In practical cases, very often the contraction rate λ of the linear convergence model (6.2.13) will also depend on the steplength h. For example, the two chord methods

$$v_{i+1}(h) = T(v_i(h)) = v_i(h) - H'(u)^+ H(v_i(h)),$$
$$v_{i+1}(h) = T(v_i(h)) = v_i(h) - H'(v_0(h))^+ H(v_i(h)),$$

could be described by the error model

(6.2.15)
$$\varepsilon_{i+1}(h) \le \lambda h^q \varepsilon_i(h)$$

with $q = 1$ and $q = 2$ respectively. If we use this error model in the context of the above discussion, we obtain the following estimates:

$$\omega \approx \frac{\varepsilon_{k-1}(h)}{\varepsilon_0(h)} \approx \frac{(\lambda h^q)^{k-1} \varepsilon_0(h)}{\varepsilon_0(h)} \quad \text{implies} \quad \lambda h^q \approx \omega^{\frac{1}{k-1}}.$$

Now (6.2.7) assumes the form

$$(\lambda \tilde{h}^q)^{\tilde{k}} \varepsilon_0(\tilde{h}) \approx (\lambda h^q)^k \varepsilon_0(h).$$

Substituting

$$\lambda \approx \frac{\omega^{\frac{1}{k-1}}}{h^q}, \qquad \left(\frac{h}{\tilde{h}} \right)^2 \approx \frac{\varepsilon_0(h)}{\varepsilon_0(\tilde{h})}, \qquad f = \frac{h}{\tilde{h}}$$

into the previous relation yields

(6.2.16)
$$f^{2+q\tilde{k}} \approx \omega^{\frac{\tilde{k}-k}{k-1}}$$

from which the deceleration factor f is determined. Note that (6.2.14) and (6.2.16) agree if $q = 0$.

6.3 Steplength Strategies Involving Variable Order Predictors

In the previous sections the steplength strategies were based upon the Euler predictor, which is only of order one. This crude predictor is very often satisfactory since it is usually used in conjunction with very strongly contracting correctors such as Newton type correctors. Nevertheless, one may expect to obtain improved efficiency in traversing by using variable order predictors and formulating corresponding steplength strategies. The strategy which we present here is motivated by similar strategies used in the numerical solution of initial value problems, see e.g. Shampine & Gordon (1975). Recently Lundberg & Poore (1989) have made an implementation of such an approach. Their numerical results show that there is often a definite benefit to be derived by using higher order predictors. The main difference from the approach given here is that they make crucial use of an accurate approximation of the arc length as opposed to our approach using local parametrizations.

The high order predictors considered by us and also by Lundberg & Poore (1989) are based on polynomial interpolation. In view of the stability of Newton's method as a corrector, it may be advantageous to use more stable predictors. Mackens (1989) has proposed such predictors which are based on Taylor's formula and which are obtained by successive numerical differentiation in a clever way. However, the gain in stability has to be paid for by additional evaluations of the map H and additional applications of the Moore-Penrose inverse of the Jacobian H' (where it may be assumed that H' has already been decomposed). It would be interesting to compare these two essentially different predicting methods numerically.

The strategy presented here is based upon the requirement that we wish to generate predictor points v such that $\text{dist}(v, H^{-1}(0)) \leq \varepsilon_{\text{tol}}$ for a steplength h which is as large as possible. Here our tolerance ε_{tol} should be sufficiently small so that the corrector procedure converges reasonably well. In many cases, a user may have difficulty forcasting a suitable tolerance a priori. On the other hand, there is no need for this tolerance to be fixed throughout the continuation process. We will show at the end of this section how the ideas of the previous section can be used for adapting ε_{tol}.

Let us begin the discussion by considering a typical situation in which the points u_0, u_1, ..., u_n along the solution curve c in $H^{-1}(0)$ have already been generated. In certain versions of the continuation method, also the tangents $t_0 := t(H'(u_0))$, ..., $t_n := t(H'(u_n))$ are evaluated. The idea is to use an interpolating polynomial P_q of degree q with coefficients in $\mathbf{R}^{N+1}$ in order to obtain a predictor point $P_q(h)$ from the current point u_n which satisfies the above tolerance requirement. The expectation is that by increasing the order q a larger steplength h will be permitted. On the other hand, due to the instability of polynomial extrapolation, there is clearly a limit for the order q which can be permitted, and this limit may depend on the local properties of the solution curve. The choice of the steplength and the order of the next predictor point $P_q(h)$ will be based on the a priori estimates

(6.3.1) $$\|P_{q+1}(h) - P_q(h)\| \approx \text{dist}\left(P_q(h), H^{-1}(0)\right).$$

In order to use the interpolating polynomial P_q, we need to express it in terms of a suitable parameter ξ. Naturally, the arclength parameter s which we always consider for theoretical discussions would be ideal to use, see Lundberg & Poore (1989). However, for purposes of exposition, we shall avoid the additional complexity of obtaining precise numerical approximations of the arclength s_i such that $c(s_i) = u_i$. We therefore propose to use a local parametrization ξ induced by the current approximation $t \approx t(H'(u_n))$, which does not need to be very accurate. We assume however, that the normalization $\|t\| = 1$ holds. This local parametrization $c(\xi)$ is defined as the locally unique solution of the system

(6.3.2)
$$\begin{aligned} H(u) &= 0, \\ t^*(u_n + \xi t - u) &= 0, \end{aligned}$$

for ξ in some open interval containing zero. It follows immediately that

(6.3.3) $$c(\xi_i) = u_i \quad \text{where} \quad \xi_i = t^*(u_i - u_n).$$

Differentiating (6.3.2) with respect to ξ yields

(6.3.4) $$\frac{dc(\xi)}{d\xi} = \frac{\dot{c}(s)}{t^*\dot{c}(s)}.$$

If the tangents t_i at the points u_i are available for use, we may form a Hermite interpolating polynomial P_q. Otherwise, a standard interpolating polynomial using Newton's formula is generated. For details concerning interpolating polynomials, we refer the reader to standard textbooks e.g. Stoer & Bulirsch (1980).

(6.3.5) Interpolating Formulae.
The interpolating polynomial which does not use the tangents t_i is given by the **Newton formula**

$$\begin{aligned} P_{n,q}(h) = c[\xi_n] &+ c[\xi_n, \xi_{n-1}](h - \xi_n) + c[\xi_n, \xi_{n-1}, \xi_{n-2}](h - \xi_n)(h - \xi_{n-1}) \\ &+ \ldots + c[\xi_n, \ldots, \xi_{n-q}](h - \xi_n) \cdots (h - \xi_{n-q+1}), \end{aligned}$$

where the coefficients are the divided differences which are obtained from the table:

ξ_n	$c[\xi_n]$			
ξ_{n-1}	$c[\xi_{n-1}]$	$c[\xi_n, \xi_{n-1}]$		
ξ_{n-2}	$c[\xi_{n-2}]$	$c[\xi_{n-1}, \xi_{n-2}]$	$c[\xi_n, \xi_{n-1}, \xi_{n-2}]$	
$\vdots$	$\vdots$	$\vdots$	$\vdots$	$\ddots$

The interpolating polynomial which uses the tangents t_i is given by the **Hermite formula**

$$P_{n,2i+2}(h) = c[\xi_n] + c[\xi_n, \xi_n](h - \xi_n) + c[\xi_n, \xi_n, \xi_{n-1}](h - \xi_n)^2 + \cdots$$
$$+ c[\xi_n, \xi_n, \ldots, \xi_{n-i}, \xi_{n-i}, \xi_{n-i-1}](h - \xi_n)^2 \cdots (h - \xi_{n-i})^2,$$
$$P_{n,2i+1}(h) = c[\xi_n] + c[\xi_n, \xi_n](h - \xi_n) + c[\xi_n, \xi_n, \xi_{n-1}](h - \xi_n)^2 + \cdots$$
$$+ c[\xi_n, \xi_n, \ldots, \xi_{n-i}, \xi_{n-i}](h - \xi_n)^2 \cdots (h - \xi_{n-i-1})^2(h - \xi_{n-i}),$$

where the divided differences are obtained from the table:

$$
\begin{array}{llllll}
\xi_n & c[\xi_n] \\
\xi_n & c[\xi_n] & c[\xi_n, \xi_n] \\
\xi_{n-1} & c[\xi_{n-1}] & c[\xi_n, \xi_{n-1}] & c[\xi_n, \xi_n, \xi_{n-1}] \\
\xi_{n-1} & c[\xi_{n-1}] & c[\xi_{n-1}, \xi_{n-1}] & c[\xi_n, \xi_{n-1}, \xi_{n-1}] & c[\xi_n, \xi_n, \xi_{n-1}, \xi_{n-1}] \\
\vdots & \vdots & \vdots & \vdots & \vdots & \ddots
\end{array}
$$

The entries of the tables are given by

$$\xi_i := t^*(u_i - u_n)$$
$$c[\xi] := u_i$$
$$c[\xi_i, \xi_i] = \frac{dc(\xi_i)}{d\xi} := \frac{t_i}{t^* t_i}$$
$$c[\xi_i, \ldots, \xi_j] := \frac{c[\xi_i, \ldots, \xi_{j+1}] - c[\xi_{i-1}, \ldots, \xi_j]}{\xi_i - \xi_j} \quad \text{for } i > j, \ \xi_i \neq \xi_j.$$

For simplicity, we will now confine the presentation of the steplength strategy to the case of the Newton formula. The discussion for the Hermite formula proceeds analogously. Motivated by the remarks leading to (6.3.1) we take the term

(6.3.6) $\qquad e(h, n, q) := \|c[\xi_n, \ldots, \xi_{n-q-1}]\| (h - \xi_n) \cdots (h - \xi_{n-q})$

as an estimate for $\text{dist}\big(P_{n,q}(h), H^{-1}(0)\big)$. The error term $e(h, n, q)$ is a polynomial of degree $q + 1$ which is strictly increasing for $h > 0$. Since $e(0, n, q) = 0$, the equation

(6.3.7) $\qquad\qquad\qquad e(h_{n,q}, n, q) = \varepsilon_{\text{tol}}$

has exactly one solution $h_{n,q} > 0$. This solution can be easily approximated e.g. via the secant method using the starting points

$$h = 0 \quad \text{and} \quad h = \left(\frac{\varepsilon_{\text{tol}}}{\|c[\xi_n, \ldots, \xi_{n-q-1}]\|} \right)^{\frac{1}{q+1}}.$$

Due to instability of high order polynomial interpolation, the typical behavior
of the solutions (6.3.7) will be that

$$(6.3.8) \qquad h_{n,1} < h_{n,2} < \cdots < h_{n,q} \geq h_{n,q+1}$$

holds for some $q \geq 1$. The basic idea of the present steplength and order adaptation is to choose the order q and the steplength $\tilde{h} = h_{n,q}$ in the
next predictor step according to (6.3.8). However, some stabilizing safeguards
should also be employed. We suggest a safeguard such as allowing at most a
doubling of the previous steplength h i.e.

$$(6.3.9) \qquad \tilde{h} := \min\{2h, h_{n,q}\}.$$

Furthermore, to increase the stability of the predictors, we will in fact take
the lowest order i such that $e(\tilde{h}, n, i) \leq \varepsilon_{tol}$. If a predictor-corrector step is
rejected, we repeat it with a reduced steplength $\tilde{h} := h/2$. Again we will take
the lowest possible order as above.

Let us describe how, in analogy to the steplength adaptation of Den Heijer
& Rheinboldt of the previous section, the tolerance ε_{tol} may be adapted at
each step in order to obtain a desired uniform number $\tilde{k}$ of corrector iterates
during the traversing. Let us again adopt the notation of section 6.2. Upon
rereading the discussion leading to (6.2.9) it is clear that the ratio $\varepsilon/\tilde{\varepsilon}$ in (6.2.9)
estimates the ratio between an observed distance and a desired distance to
the solution curve. Consequently, we forecast the tolerance via the formula

$$(6.3.10) \qquad \varepsilon_{tol} := \varepsilon_{tol}\frac{\tilde{\varepsilon}}{\varepsilon}.$$

We finally summarize the discussion of this section in the following algorithm.
As in section 6.2, we assume that the iterative corrector procedure is given by
the corrector operator T, and that the function ψ describes an error model
for T.

(6.3.11) Steplength and Order Adaptation
Via Interpolation.

$\qquad\qquad\qquad\qquad\qquad\qquad\qquad$ *comment:*

input

$\quad$ **begin**

$\quad$ $\tilde{k} \in \mathbf{N};$ $\qquad\qquad\qquad$ *desired number of corrector iterations*

$\quad$ $\varepsilon_{\text{tol}};$ $\qquad\qquad\qquad\qquad$ *initial tolerance for predictor*

$\quad$ $q_{\max};$ $\qquad\qquad\qquad\qquad$ *maximal order for predictor*

$\quad$ $u_0, u_1 \in H^{-1}(0);$ $\qquad\qquad\qquad$ *initial points*

$\quad$ **end**;

$h := \|u_1 - u_0\|;$ $\qquad\qquad\qquad$ *initial steplength*

for $n = 1, 2, 3, \ldots$ **do** $\qquad\qquad\qquad$ *begin of PC loop*

$\quad$ **begin**

$\quad$ $t := \dfrac{u_n - u_{n-1}}{\|u_n - u_{n-1}\|};$ $\qquad\qquad\qquad$ *approximate tangent*

$\quad$ $q := 1;$ $\qquad\qquad\qquad\qquad$ *initial order*

$\quad$ **if** $n = 1$ **then go to** 1;

$\quad$ **while** $q < q_{\max}$ **and** $q < n - 1$ **do**

$\quad\quad$ **begin**

$\quad\quad$ **if** $e(2h, n, q) < \varepsilon_{\text{tol}}$ **then** $\qquad\qquad$ *see (6.3.5) – (6.3.6)*

$\quad\quad\quad$ **begin** $h := 2h;$ **go to** 1; **end**;

$\quad\quad$ solve $e(h_{n,q}, n, q) = \varepsilon_{\text{tol}}$ for $h_{n,q};$

$\quad\quad$ **if** $q > 1$ **and** $h_{n,q-1} \geq h_{n,q}$ **then**

$\quad\quad\quad$ **begin** $q := q - 1;$ $h := h_{n,q-1};$ **go to** 1; **end**;

$\quad\quad$ $q := q + 1;$

$\quad\quad$ **end**;

$\quad$ $h := h_{n,q};$

$\quad$ 1: $v_0 := P_{n,q}(h);$ $\qquad\qquad\qquad$ *predictor step, see (6.3.5)*

$\quad$ **repeat for** $k := 0, 1, 2, \ldots$

$\quad\quad$ $v_{k+1} := T(v_k);$ $\qquad\qquad\qquad$ *corrector loop*

$\quad$ **until** convergence;

$\quad$ **if** the corrector loop was successful **then**

$\quad\quad$ **begin**

$\quad\quad$ **if** $k = 1$ **then** $f := \frac{1}{2}$

$\quad\quad$ **else**

$\quad\quad\quad$ **begin**

$\quad\quad\quad$ $\omega := \dfrac{\|v_k - v_{k-1}\|}{\|v_k - v_0\|};$

determine ε so that $\dfrac{\psi^{k-1}(\varepsilon)}{\varepsilon} = \omega$;

determine $\tilde{\varepsilon}$ so that $\psi^{\tilde{k}}(\tilde{\varepsilon}) = \psi^k(\varepsilon)$;

$f := \dfrac{\varepsilon}{\tilde{\varepsilon}}$;

end;

$f := \max\{\min\{f, 2\}, \frac{1}{2}\}$;   *factor $f \in [\frac{1}{2}, 2]$*

$\varepsilon_{\text{tol}} := \dfrac{\varepsilon_{\text{tol}}}{f}$;   *adaptation of ε_{tol}*

$u := v_k$;   *new point along $H^{-1}(0)$*

end

else $h := \dfrac{h}{2}$;   *repeat predictor with reduced h*

end:

Chapter 7. Predictor-Corrector Methods Using Updating

In iterative methods for numerically finding zero-points of a smooth map $F : \mathbf{R}^N \to \mathbf{R}^N$ it is often preferable to avoid the costly recalculation and decomposition of the Jacobian F' at each iteration by using an approximation to F'. This results in sacrificing quadratic convergence in exchange for a superlinear convergence which is nearly as good, or a rather fast rate of linear convergence. When N is of at least moderate size, or F' is otherwise cumbersome to calculate, this trade-off is usually to be preferred. It is reasonable to expect that the same situation should also hold in the corrector step of the predictor-corrector methods for numerically tracing $H^{-1}(0)$ where $H : \mathbf{R}^{N+1} \to \mathbf{R}^N$ is a smooth map. Indeed, since the corrector process needs to be performed essentially at every predicted point, the possibility of saving computational effort in the corrector process becomes even more attractive for the predictor-corrector curve tracing methods.

In this chapter we will describe how to incorporate an analogue of Broyden's update method into a predictor-corrector algorithm. In preparation for this, we shall first recall the Broyden update method for solving the zero-point problem $F(x) = 0$, and examine some of its aspects. For a general reference on update methods we suggest the book of Dennis & Schnabel (1983) or the review article by Dennis & Moré (1977).

7.1 Broyden's "Good " Update Formula

Suppose that $F : \mathbf{R}^N \to \mathbf{R}^N$ is a smooth map and that $F(\bar{x}) = 0$ with $F'(\bar{x})$ having maximal rank N. It is well-known (also see section 3.3) that Newton's method

$$x_{n+1} = x_n - F'(x_n)^{-1}F(x_n), \qquad n = 0, 1, \dots$$

is locally quadratically convergent i.e.

$$\|x_{n+1} - \bar{x}\| = O(\|x_n - \bar{x}\|^2)$$

for x_0 sufficiently near $\bar{x}$. Even when an adequate x_0 has been chosen, there remains the drawback that after every iteration the Jacobian matrix $F'(x_n)$

needs to be calculated and a new matrix decomposition has to be obtained in order to solve the linear system

$$F'(x_n)s_n = -F(x_n)$$

for $s_n = x_{n+1} - x_n$. On the other hand, if an approximate Jacobian is held fixed, say for example $A := F'(x_0)$, a familiar Newton-Chord method is obtained:

$$x_{n+1} = x_n - A^{-1}F(x_n).$$

This method offers the advantage that A stays fixed. Thus, once the matrix decomposition for A has been obtained, further iterations may be cheaply carried out. The drawback of the Newton-Chord method is that the local convergence is only linear. That is,

$$||x_{n+1} - \bar{x}|| \leq \kappa ||x_n - \bar{x}|| \quad \text{for some} \quad \kappa \in (0,1)$$

when x_0 is sufficiently near $\bar{x}$ and A is sufficiently near $F'(\bar{x})$. If $A = F'(x_0)$ is taken, it is easy to show that $\kappa = O(||\bar{x} - x_0||)$. Thus the contraction rate κ will become much better when x_0 is chosen near the solution point $\bar{x}$.

The method of **Broyden** (1965) involves the use of previously calculated data to iteratively improve the quality of the approximation $A \approx F'(x_n)$ via successive rank-one updates. A more general class of update methods usually called Quasi-Newton methods, have since been developed, which take into account possible special structure of the Jacobian F' such as positive definiteness, symmetry or sparseness. For surveys of these methods we refer the reader to the above mentioned literature. It is possible to prove local superlinear convergence of a large class of these methods under the standard hypotheses for Newton's method, see Dennis & Walker (1981).

For general purpose updates, i.e. when no special structure is present, the consensus of opinion appears to be that the so-called "good formula" of Broyden remains the favorite rank-one update available. For this reason we will confine our discussion of update methods for curve-following to an analogue of Broyden's "good formula". Similar but more complicated extensions of the discussion below can be given for the case that special structure is present, see also Bourji & Walker (1987), Walker (1990).

Let us motivate our discussion by reviewing the Broyden update formula for solving $F(x) = 0$ via a Newton-type method where $F : \mathbf{R}^N \to \mathbf{R}^N$ is a smooth map. From Taylor's formula, we have

$$(7.1.1) \qquad F'(x_n)(x_{n+1} - x_n) = F(x_{n+1}) - F(x_n) + O(||x_{n+1} - x_n||^2).$$

By neglecting the higher order term in (7.1.1), and setting

$$(7.1.2) \qquad s_n := x_{n+1} - x_n, \qquad y_n := F(x_{n+1}) - F(x_n),$$

we obtain the **secant equation**

(7.1.3)
$$As_n = y_n \,,$$

which should be satisfied (at least to first order) by an approximate Jacobian $A \approx F'(x_n)$.

When Newton-type steps

(7.1.4)
$$x_{n+1} = x_n - A_n^{-1} F(x_n)$$

are performed using some approximate Jacobian $A_n \approx F'(x_n)$, it is natural to require that the next approximate Jacobian A_{n+1} should satisfy the secant equation

(7.1.5)
$$A_{n+1} s_n = y_n \,,$$

since the data s_n and y_n are already available. Clearly, the equation (7.1.5) does not uniquely determine A_{n+1}, since it involves N equations in N^2 unknowns. An additional natural consideration is that if A_n was a "good" approximation to $F'(x_n)$, then this quality ought to be incorporated in formulating subsequent approximations. This leads to the idea of obtaining A_{n+1} from A_n by the **least change principle** i.e. among all matrices A satisfying the secant equation, we choose the one with the smallest "distance" from A_n. Thus we are led to the following

(7.1.6) Definition of Broyden's "good update formula". *We define the updated approximate Jacobian A_{n+1} as the solution of the problem*

$$\min_A \{\|A - A_n\|_F \mid As_n = y_n\}.$$

The norm $\| \cdot \|_F$ is the **Frobenius norm**:

$$\|A\|_F = \left(\sum_{i,j=1}^{N} (A[i,j])^2 \right)^{\frac{1}{2}}.$$

This is a simple matrix norm which enables the minimization for (7.1.6) to be done row-wise. A straightforward calculation (using orthogonal projections) shows that the solution to (7.1.6) is given explicitly by

(7.1.7)
$$A_{n+1} = A_n + \frac{y_n - A_n s_n}{\|s_n\|^2} s_n^* \,,$$

which is generally referred to as Broyden's "good" update formula.

The following theorem shows the superlinear convergence of Broyden's modification of Newton's method. Our proof will essentially follow that of Dennis & Moré (1974). We give the following proof in some detail in order to emphasize the fact that the superlinear convergence of Broyden's method is a surprising result, since $\|A_n - F'(x_n)\| \to 0$ as $n \to \infty$ need not necessarily hold. Similarly, in the context of numerically following a curve $H^{-1}(0)$ with the exclusive use of analogous Broyden updates, there is little hope for maintaining a good approximation of the Jacobian H' without taking further measures. We will discuss this point again in section 7.2.

(7.1.8) Theorem (Broyden). *Let* $F : \mathbf{R}^N \to \mathbf{R}^N$ *be a smooth map and let* $\bar{x} \in \mathbf{R}^N$ *be a regular zero point of* F. *Suppose*

$$x_{n+1} = x_n - A_n^{-1}F(x_n)$$

is the Newton-type method where A_n *is updated according to Broyden's formula (7.1.7). If* x_0 *is sufficiently near* $\bar{x}$, *and if* A_0 *is sufficiently near* $F'(x_0)$, *then the sequence* $\{x_n\}$ *converges* **superlinearly** *to* $\bar{x}$ *i.e.*

$$\frac{\|x_{n+1} - \bar{x}\|}{\|x_n - \bar{x}\|} \to 0 \qquad as \quad n \to \infty.$$

Proof. Let us begin with the obvious remark that if $x_n = x_{n+1}$ for some n, then clearly $x_n = \bar{x} = x_{n+1}$, and hence the assertion concerning the order of convergence is meaningless.Consequently, we assume hereafter

$$s_n = x_{n+1} - x_n \neq 0 \quad \text{for} \quad n = 0, 1, 2, \ldots$$

Thus the Broyden update (7.1.7) is well defined for all n.

Let us adopt the following notation:

$$\text{(7.1.9)} \quad \begin{cases} x_{k+1} := x_k - A_k^{-1}F(x_k) \\[2mm] s_k := x_{k+1} - x_k \\[2mm] y_k := F(x_{k+1}) - F(x_k) \\[2mm] A_{k+1} := A_k + \dfrac{(y_k - A_k s_k)s_k^*}{\|s_k\|^2} \\[2mm] A := F'(\bar{x}) \\[2mm] \varepsilon_k := x_k - \bar{x} \\[2mm] E_k := A_k - A \end{cases}$$

From (7.1.9) we have

$$\text{(7.1.10)} \quad \begin{aligned} E_{k+1} &= A_{k+1} - A \\ &= A_k - A + \frac{(y_k - A_k s_k)s_k^*}{\|s_k\|^2} \\ &= A_k - A + \frac{(A - A_k)s_k s_k^*}{\|s_k\|^2} + \frac{(y_k - A s_k)s_k^*}{\|s_k\|^2} \\ &= E_k\left(I - \frac{s_k s_k^*}{\|s_k\|^2}\right) + \frac{(y_k - A s_k)s_k^*}{\|s_k\|^2} \end{aligned}$$

Furthermore, from (7.1.9) and the mean value theorem we obtain

$$\text{(7.1.11)} \quad \begin{aligned} y_k - A s_k &= F(x_{k+1}) - F(x_k) - F'(\bar{x})(x_{k+1} - x_k) \\ &= \int_0^1 \left[F'(x_k + \xi s_k) - F'(\bar{x})\right] s_k \, d\xi \\ &= O(\|\varepsilon_k\| + \|\varepsilon_{k+1}\|)s_k . \end{aligned}$$

The last equation follows from the smoothness of F' and the inequality

(7.1.12) $$||x_k + \xi s_k - \bar{x}|| \le ||\varepsilon_k|| + ||\varepsilon_{k+1}|| \quad \text{for} \quad \xi \in [0, 1],$$

which is easy to establish. From (7.1.10) and (7.1.11), we have

(7.1.13) $$E_{k+1} = E_k \left(I - \frac{s_k s_k^*}{||s_k||^2} \right) + O(||\varepsilon_k|| + ||\varepsilon_{k+1}||).$$

Our next step is to prove the following

(7.1.14) **Claim.** *For every $C > 0$, there exist $\varepsilon, \delta > 0$ such that*

(7.1.15) $$||\bar{x} - (x - B^{-1}F(x))|| \le \frac{1}{2}||\bar{x} - x||$$

whenever

(7.1.16) $$||B - A|| \le \delta + 3C\varepsilon \quad \text{and} \quad ||x - \bar{x}|| \le \varepsilon.$$

The above claim is intuitively clear, since it describes the local linear convergence of Newton-type methods. We prove (7.1.14) using the Taylor formula

$$F(x) = F(x) - F(\bar{x}) = F'(\bar{x})(x - \bar{x}) + O(||x - \bar{x}||^2).$$

Hence

$$B^{-1}F(x) = B^{-1}F'(\bar{x})(x - \bar{x}) + O(||x - \bar{x}||^2)$$
$$= B^{-1}B(x - \bar{x}) + O(||B - F'(\bar{x})|| \cdot ||x - \bar{x}||) + O(||x - \bar{x}||^2).$$

Since the constants implied by the Landau symbols are all locally uniform, the claim (7.1.14) follows easily.

We choose $C > 0$, $\varepsilon > 0$, $\delta > 0$ such that (7.1.14–16) holds and also (7.1.13) holds in the specific form

(7.1.17) $$\left\| E_{k+1} - E_k \left(I - \frac{s_k s_k^*}{||s_k||^2} \right) \right\| \le C(||\varepsilon_k|| + ||\varepsilon_{k+1}||)$$

for $||\bar{x} - x_k|| \le \varepsilon$ and $||\bar{x} - x_{k+1}|| \le \varepsilon$. Now we are able to give a precise meaning to the hypothesis that x_0 and A_0 are "sufficiently near" $\bar{x}$ and A respectively. Let us show that under the assumptions

(7.1.18) $$||A_0 - A|| \le \delta \quad \text{and} \quad ||\bar{x} - x_0|| \le \varepsilon$$

the hypothesis (7.1.16) of claim (7.1.14) holds. This is accomplished by showing the following estimates via induction:

(7.1.19)
$$\begin{cases} ||E_i|| = ||A_i - A|| \le \delta + 3C\varepsilon\left(\frac{1}{2} + \frac{1}{4} + \cdots + \frac{1}{2^i}\right), \\ ||\varepsilon_i|| = ||x_i - \bar{x}|| \le \dfrac{\varepsilon}{2^i}. \end{cases}$$

By (7.1.18), this assertion holds for $i = 0$. Then by the inductive hypothesis
(7.1.19), the conditions (7.1.16) are verified for $x = x_i$ and $B = A_i$, and hence
(7.1.15) implies

$$||x_{i+1} - \bar{x}|| \le \frac{1}{2}||x_i - \bar{x}|| \le \frac{\varepsilon}{2^{i+1}}.$$

Furthermore, by (7.1.17) we have

$$\begin{aligned} ||A_{i+1} - A|| &\le ||A_i - A|| + C\big(||\varepsilon_i|| + ||\varepsilon_{i+1}||\big) \\ &\le \delta + 3C\varepsilon\left(\frac{1}{2} + \frac{1}{4} + \frac{1}{2^i}\right) + C\left(\frac{\varepsilon}{2^i} + \cdots + \frac{\varepsilon}{2^{i+1}}\right) \\ &= \delta + 3C\varepsilon\left(\frac{1}{2} + \frac{1}{4} + \cdots + \frac{1}{2^{i+1}}\right). \end{aligned}$$

This completes the inductive proof of (7.1.19). Next we note the fact that the
corresponding rows of the two matrices

$$E_k\left(I - \frac{s_k s_k^*}{||s_k||^2}\right) \quad \text{and} \quad E_k\frac{s_k s_k^*}{||s_k||^2}$$

are orthogonal to each other. Hence, applying the Pythagoras equation on
(7.1.13), we obtain for the Frobenius norms

$$||E_{k+1}||^2_F = ||E_k||^2_F - \frac{||E_k s_k||^2}{||s_k||^2} + O(||\varepsilon_k|| + ||\varepsilon_{k+1}||).$$

Now by summing up both sides with respect to k and using (7.1.19), we have

(7.1.20)
$$\sum_{k=0}^{\infty}\left(\frac{||E_k s_k||}{||s_k||}\right)^2 < \infty,$$

since $||E_{k+1}||_F$ and $||E_k||_F$ remain bounded and the asymptotic term is sum-
mable. From (7.1.20) we have

(7.1.21)
$$\frac{||E_k s_k||}{||s_k||} \to 0 \quad \text{as} \quad k \to \infty.$$

Since $E_k s_k = (A_k - A)s_k = (A_k - A)(x_{k+1} - x_k)$ and $A_k(x_{k+1} - x_k) = -F(x_k)$ from the Newton step, we have

$$E_k s_k + F(x_k) + As_k = 0$$

and hence

(7.1.22) $$E_k s_k + F(x_{k+1}) + As_k - y_k = 0.$$

From (7.1.11), we obtain

$$\frac{\|As_k - y_k\|}{\|s_k\|} \to 0 \quad \text{as} \quad k \to \infty.$$

Hence by (7.1.21–22), we have

(7.1.23) $$\frac{\|F(x_{k+1})\|}{\|s_k\|} \to 0 \quad \text{as} \quad k \to \infty.$$

On the other hand, from

$$F(x_k) = F(x_k) - F(\bar{x}) = F'(\bar{x})\varepsilon_k$$

(7.1.24)
$$+ \frac{1}{2} \int_0^1 F''(\bar{x} + \xi\varepsilon_k) 2(1 - \xi)\, d\xi\, [\varepsilon_k, \varepsilon_k]$$

we obtain the estimate

$$\|F(x_k)\| \geq \|F'(\bar{x})^{-1}\|^{-1} \|\varepsilon_k\| + O(\|\varepsilon_k\|^2).$$

Now for ε sufficiently small, (7.1.23) implies

(7.1.25) $$\frac{\|\varepsilon_{k+1}\|}{\|s_k\|} \to 0 \quad \text{as} \quad k \to \infty.$$

On the other hand,

$$\|s_k\| = \|x_{k+1} - x_k\| \leq \|\varepsilon_{k+1}\| + \|\varepsilon_k\|$$

implies

$$\frac{\|\varepsilon_{k+1}\|}{\|s_k\|} \geq \frac{\|\varepsilon_{k+1}\|}{\|\varepsilon_{k+1}\| + \|\varepsilon_k\|} = \frac{1}{1 + \|\varepsilon_k\|/\|\varepsilon_{k+1}\|}.$$

Hence (7.1.25) implies

$$\frac{\|\varepsilon_k\|}{\|\varepsilon_{k+1}\|} \to \infty \quad \text{as} \quad k \to \infty,$$

or equivalently,

$$\frac{\|\varepsilon_{k+1}\|}{\|\varepsilon_k\|} \to 0 \quad \text{as} \quad k \to \infty,$$

which concludes the proof of (7.1.8).   □

7.2 Broyden Updates Along a Curve

In this section we develop an update method for approximating a curve of zero points in $H^{-1}(0)$, which is analogous to the Broyden method for approximating an isolated zero point discussed in the previous section. We assume that $H : \mathbf{R}^{N+1} \to \mathbf{R}^N$ is a smooth map and zero is a regular value of H. To begin our discussion, let us suppose we are given two approximate zero-points $u_n, u_{n+1} \in \mathbf{R}^{N+1}$ and the corresponding values $H(u_n), H(u_{n+1})$. Analogously to our discussion in section 7.1, we set

(7.2.1)
$$s_n := u_{n+1} - u_n$$
$$y_n := H(u_{n+1}) - H(u_n)$$

and we consider an analogous **secant equation**

(7.2.2)
$$As_n = y_n$$

where $A \approx H'(u_n)$ is an approximate Jacobian. The corresponding **Broyden update on the points** u_n, u_{n+1} is given by

(7.2.3)
$$A_{n+1} := A_n + \frac{(y_n - A_n s_n)s_n^*}{||s_n||^2}.$$

Let us begin by introducing a generic predictor-corrector algorithm employing Broyden's update method.

(7.2.4) Generic Euler-Newton Method Using Updates.   comment:
  input
    begin
    $u \in \mathbf{R}^{N+1}$ such that $H(u) = 0$;   initial point
    $A \approx H'(u)$;   initial approximate Jacobian
    $h > 0$;   initial stepsize
    end;
  repeat
    $v := u + ht(A)$;   predictor step
    update A on u, v;   see (7.2.3)
    $w := v - A^+ H(v)$;   corrector step
    update A on v, w;   see (7.2.3)
    $u := w$;   new approximate along $H^{-1}(0)$
    choose a new stepsize $h > 0$;   stepsize adaptation
  until traversing is stopped.

The first remark we have to make concerning algorithm (7.2.4) is that it does not generate a reliable approximation of the Jacobian $H'(u)$ i.e. a relation such as

$$\|A - H'(u)\| = O(h),$$

which we assumed in our previous convergence analysis, see (5.2.1), does not hold in general. In our discussion of section 7.1 it was already seen that the Jacobian is not necessarily well approximated by the update formula. The reason behind this is that we are in general not assured that the update data spans the whole space sufficiently well. To put this more precisely, let S denote the $(N + 1) \times k$-matrix, whose columns have unit norm and indicate the last k directions in which the update formula was used in algorithm (7.2.4). Then the condition

$$\text{cond}(SS^*) < C$$

for some $C > 0$ independent of h and of whichever last k directions S are taken, is sufficient to ensure the convergence of algorithm (7.2.4) for h sufficiently small, in the sense of (5.2.1), see chapter 4 of Georg (1982) for a sketch of a proof. However, it is unrealistic to expect that this condition will in general be satisfied. Rather than to present the very technical proof of the above mentioned convergence, let us instead describe how this difficulty may be circumvented. We begin our discussion with a study of some of the properties of the Broyden update defined in (7.2.3). Let us recall a well known fact concerning the determinant of a special rank one update matrix.

(7.2.5) Lemma. *Let* $u, v \in \mathbf{R}^N$. *Then*

$$\det(\text{Id} + uv^*) = 1 + v^* u.$$

Proof. For $u = 0$ or $v = 0$, the result is trivial. Let us first assume that $v^* u \neq 0$. If $w_1, \ldots, w_{N-1}$ is any linearly independent set of vectors which are orthogonal to $v \neq 0$ then $(\text{Id} + uv^*)w_i = w_i$ for $i = 1, \ldots, N - 1$, and so $\lambda = 1$ is an eigenvalue of $\text{Id} + uv^*$ of multiplicity $N - 1$. Furthermore,

$$(I + uv^*)u = u + uv^* u = (1 + v^* u)u$$

shows that $\lambda = 1 + v^* u$ is also an eigenvalue of $\text{Id} + uv^*$. The result now follows from the fact that the determinant is equal to the product of the eigenvalues. The exceptional case $v^* u = 0$ can be treated by using the fact that the determinant is a continuous function of u. □

The following theorem furnishes a number of properties of the Broyden update which will be used for curve tracing.

(7.2.6) Theorem. *Suppose* $w_n \in \mathbf{R}^N$, $v_n \in \mathbf{R}^{N+1}$ *and* A_n *is an* $N \times (N+1)$-*matrix with* $\mathrm{rank} A_n = N$. *Define:*

$$A_{n+1} := A_n + w_n v_n^*,$$
$$D_n := 1 + v_n^* A_n^+ w_n,$$
$$t_n := t(A_n),$$

and assume $D_n \neq 0$. Then:

(1) $\mathrm{rank} A_{n+1} = N$,

(2) $t_{n+1} = \rho_n(t_n - \dfrac{v_n^* t_n}{D_n} A_n^+ w_n)$ for some $\rho_n \in \mathbf{R}$ with $|\rho_n| \in (0,1]$,

(3) $A_{n+1}^+ = (\mathrm{Id} - t_{n+1} t_{n+1}^*)\left(\mathrm{Id} - \dfrac{A_n^+ w_n v_n^*}{D_n}\right) A_n^+$,

(4) $\det \begin{pmatrix} A_{n+1} \\ t_{n+1}^* \end{pmatrix} = \dfrac{D_n}{\rho_n} \det \begin{pmatrix} A_n \\ t_n^* \end{pmatrix}$.

Proof. Ad (1): As in the proof of the well-known Sherman-Morrison formula, we have

$$(A_n + w_n v_n^*)\left(A_n^+ - \frac{A_n^+ w_n v_n^* A_n^+}{1 + v_n^* A_n^+ w_n}\right)$$

$$= \mathrm{Id} + w_n v_n^* A_n^+ - \frac{w_n v_n^* A_n^+}{1 + v_n^* A_n^+ w_n} - \frac{w_n(v_n^* A_n^+ w_n) v_n^* A_n^+}{1 + v_n^* A_n^+ w_n}$$

$$= \mathrm{Id} + w_n v_n^* A_n^+ - \left(\frac{1 + v_n^* A_n^+ w_n}{1 + v_n^* A_n^+ w_n}\right)(w_n v_n^* A_n^+)$$

$$= \mathrm{Id}.$$

From this it follows that A_{n+1} has a right inverse, namely

(7.2.7) $A_{n+1}\left(\mathrm{Id} - \dfrac{A_n^+ w_n v_n^*}{D_n}\right) A_n^+ = \mathrm{Id}$,

and assertion (1) is proven.
Ad(2): Analogously to the previous calculation we have

$$A_{n+1}\left(t_n - \frac{v_n^* t_n A_n^+ w_n}{D_n}\right)$$

$$= A_n(\mathrm{Id} + A_n^+ w_n v_n^*)\left(t_n - \frac{v_n^* t_n A_n^+ w_n}{1 + v_n^* A_n^+ w_n}\right)$$

$$= A_n\left(t_n + v_n^* t_n A_n^+ w_n - \frac{v_n^* t_n A_n^+ w_n}{1 + v_n^* A_n^+ w_n} - \frac{A_n^+ w_n(v_n^* A_n^+ w_n)(v_n^* t_n)}{1 + v_n^* A_n^+ w_n}\right)$$

$$= A_n\left(t_n + v_n^* t_n A_n^+ w_n - v_n^* t_n \left(\frac{1 + v_n^* A_n^+ w_n}{1 + v_n^* A_n^+ w_n}\right) A_n^+ w_n\right)$$

$$= A_n t_n = 0.$$

Now by normalizing with

$$\rho_n = \pm \left\| t_n - \frac{v_n^* t_n A_n^+ w_n}{D_n} \right\|^{-1}$$

we have

(7.2.8)
$$t_{n+1} = \rho_n \left(t_n - \frac{v_n^* t_n A_n^+ w_n}{D_n} \right).$$

Since by (3.2.4), $t_n^* A_n^+ = 0$, the right hand side of (7.2.8) is the sum of two orthogonal vectors, one of which viz. t_n has norm equal to unity, we have $|\rho_n| \le 1$. On the other hand, $\rho_n \ne 0$, since $D_n \ne 0$.

Ad (3): Multiplying (7.2.7) by A_{n+1}^+ and applying (3.2.5)(1) yields

$$A_{n+1}^+ = A_{n+1}^+ A_{n+1} \left(\mathrm{Id} - \frac{A_n^+ w_n v_n^*}{D_n} \right) A_n^+$$
$$= (\mathrm{Id} - t_{n+1} t_{n+1}^*) \left(\mathrm{Id} - \frac{A_n^+ w_n v_n^*}{D_n} \right) A_n^+ .$$

Ad (4): It is easy to calculate from (2) and (3) that

$$\begin{pmatrix} A_{n+1} \\ t_{n+1}^* \end{pmatrix} \left(\mathrm{Id} - \frac{A_n^+ w_n v_n^*}{D_n} \right) (A_n^+, t_n) = \begin{pmatrix} \mathrm{Id} & 0 \\ * & \rho_n^{-1} \end{pmatrix}$$

holds. Taking determinants of both sides and using (7.2.5) we have

$$\det \left[\begin{pmatrix} A_{n+1} \\ t_{n+1}^* \end{pmatrix} \left(\mathrm{Id} - \frac{A_n^+ w_n v_n^*}{D_n} \right) (A_n^+, t_n) \right]$$
$$= \det \begin{pmatrix} A_{n+1} \\ t_{n+1}^* \end{pmatrix} \left(1 - \frac{D_n - 1}{D_n} \right) \det(A_n^+, t_n)$$
$$= \rho_n^{-1} .$$

Hence

$$\det \begin{pmatrix} A_{n+1} \\ t_{n+1}^* \end{pmatrix} \det(A_n^+, t_n) = \frac{D_n}{\rho_n}.$$

Assertion (4) now follows from (3.2.4). $\square$

The preceding discussion shows that the orientation of the new tangent vector is obtained by setting $\mathrm{sign}\rho_n = \mathrm{sign}D_n$. The above formulae could be implemented in an Euler-Newton method based on updates such as (7.2.4). However, we usually prefer to update a decomposition of A_n, see chapter 16 for details. Using the above notation, the Broyden update formula on the points u_n, u_{n+1} uses the following data:

(7.2.9)
$$w_n = \frac{H(u_{n+1}) - H(u_n) - A_n(u_{n+1} - u_n)}{\|u_{n+1} - u_n\|},$$
$$v_n = \frac{u_{n+1} - u_n}{\|u_{n+1} - u_n\|}.$$

In (7.2.4), two types of updates arise, namely predictor updates and corrector updates. For the predictor update we have $u_{n+1} = u_n + ht_n$ and consequently

(7.2.10)
$$w_n = \frac{H(u_{n+1}) - H(u_n)}{h},$$
$$v_n = t_n.$$

Then $v_n^* A_n^+ = 0$ implies $D_n = 1$ and

$$A_{n+1}^+ = (\mathrm{Id} - t_{n+1} t_{n+1}^*) A_n^+.$$

From this it is clear that updates based only upon predictor steps cannot in general provide reasonable approximations of the Jacobian H'. For the corrector update we have $u_{n+1} = u_n - A_n^+ H(u_n)$ and consequently (7.2.6) implies

(7.2.11)
$$w_n = \frac{H(u_{n+1})}{\|A_n^+ H(u_n)\|},$$
$$v_n = -\frac{A_n^+ H(u_n)}{\|A_n^+ H(u_n)\|}.$$

In this case, $v_n^* t_n = 0$ and hence $t_{n+1} = \pm t_n$. From this it is again clear that updates based only upon corrector steps cannot provide reasonable approximations of the Jacobian H'. As a consequence of (7.2.11) we have

(7.2.12)
$$\|A_n^+ w_n\| = \frac{\|A_n^+ H(u_{n+1})\|}{\|A_n^+ H(u_n)\|}.$$

The vector $A_n^+ w_n$ arises naturally in the corrector update formula, and its norm gives a reasonable measure for the contraction rate of the corrector step. If this rate is large, then this may be attributed to one of two factors: either the predictor point was too far from $H^{-1}(0)$ or the Jacobian H' is poorly approximated by A_n. Often the first of these two possibilities is easy to check. Hence (7.2.12) affords us an empirical measure of the quality of the approximation of the Jacobian.

Let us return to the basic issue of algorithm (7.2.4), namely to assure that a good approximation of the Jacobian is maintained. To do this, we may be guided by a device proposed by Powell (1970) in which he suggests monitoring the directions of the differences contributing to the updates. In our context, this amounts to monitoring whether the condition number cond(SS^*) discussed prior to (7.2.5) is sufficiently small. Instead of following Powell's approach of introducing additional directions, we combine a stepsize adaptation with several tests, the most important being the measuring of the above contraction rate (7.2.12). If the criteria of the tests are satisfied, the algorithm performs a successive predictor-corrector step with increased stepsize. If however, the criteria are not satisfied, the predictor-corrector step is repeated with reduced stepsize. In both cases, the predictor and corrector update formulae are applied. This enables the method to generally update an approximation of the Jacobian in directions which are most needed. The following specific version of algorithm (7.2.4) incorporates these considerations.

(7.2.13) Euler-Newton Method
Using Updating And Steplength Adaptation.   *comment:*

> **input**
>> **begin**
>> $u \in \mathbf{R}^{N+1}$ such that $H(u) = 0$;   *initial point*
>> $A \approx H'(u)$;   *initial approximate Jacobian*
>> $h > 0$;   *initial stepsize*
>> $\delta_0 > 0$;   *minimal residual*
>> $\delta_1 > \delta_0$;   *maximal residual*
>> $\kappa \in (0,1)$;   *maximal contraction*
>> $\gamma \in (0, \frac{\pi}{2})$;   *maximal angle*
>> **end**;
>
> **repeat**
>> **1:** $s := t(A)$;   *save tangent*
>> $v := u + ht(A)$;   *predictor step*
>> $A := A + h^{-1}\big(H(v) - H(u)\big)t(A)^*$;   *predictor update*
>> **if** $\cos^{-1}\big(s^*t(A)\big) < \gamma$ **or** $\|A^+H(v)\| \geq \delta_1$ **then**   *angle, residual test*
>>> **begin** $h := \dfrac{h}{2}$; **go to 1**; **end**;
>>
>> **if** $\|A^+H(v)\| \leq \delta_0$ **then**   *perturbation*
>>> choose p such that $\|A^+\big(H(v) - p\big)\| = \delta_0$
>>
>> **else** $p := 0$;
>> $w := v - A^+\big(H(v) - p\big)$;   *corrector step*

$$A := A + ||w - v||^{-2}(H(w) - p)(w - v)^*; \qquad \text{\textit{corrector update}}$$

if $||A^+(H(v) - p)||^{-1}||A^+(H(w) - p)|| > \kappa$ **then** $\qquad$ *contraction test*

$\quad$**begin** $h := \dfrac{h}{2}$; **go to 1; end;**

$u := w$; $h := 2h$; $\qquad\qquad\qquad\qquad\qquad\qquad$ *PC step accepted*

until traversing is stopped.

Let us again emphasize that usually an implementation of the above algorithm updates some decomposition of the matrix A at each step. For details on such update methods we refer to chapter 16. The perturbation term p in the above algorithm serves several purposes. The main purpose is to prevent numerical instability in the corrector update formula due to cancellation of digits. As a general rule for choosing δ_0, one may require that $v - w$ should carry at least half as many significant digits as the computer arithmetic carries. Another purpose of the perturbation p is to safeguard the algorithm against intrinsic instabilities such as those arising from singular points on the curve e.g. bifurcation points. In fact, the above algorithm will usually bifurcate off at simple bifurcation points. This is discussed further in chapter 8.

Several versions of the above algorithm have been implemented and tested on a variety of problems, see e.g. Georg (1981–82). At least for non-sparse problems, they turned out to be very efficient. It should be noted that an iteration of the corrector step has not been incorporated. This needs only to be done if it is wished to follow the curve closely. However, one additional corrector step could easily be performed at low cost since $A^+(H(w) - p)$ is calculated in the "contraction test". Sometimes it may be desired to calculate some special points on a curve in $H^{-1}(0)$ exactly. The discussion for doing this is taken up in chapter 9.

Let us again emphasize that it is important that although a predictor or corrector point may not be accepted, because it may not satisfy the test criteria, the update information which it contributes to approximating the Jacobian is nevertheless utilized. Finally, let us point out that the essential feature of the above algorithm is that it updates **along** the curve as well as in the corrector steps. This is to be distinguished from Euler-Newton methods where the Jacobian is precisely calculated at the predictor point, and updates are only used to accelerate the Newton-type corrector iteration. In this case, proofs of superlinear convergence have been given by Bourji & Walker (1987), see also Walker (1990).

Chapter 8. Detection of Bifurcation Points Along a Curve

8.1 Simple Bifurcation Points

Up to this point we have always assumed that zero is a regular value of the smooth mapping $H : \mathbf{R}^{N+1} \to \mathbf{R}^N$. In the case that H represents a mapping arising from a discretization of an operator of the form $\mathcal{H} : E_1 \times \mathbf{R} \to E_2$ where E_1 and E_2 represent appropriate Banach spaces, it is often of interest to approximate bifurcation points of the equation $\mathcal{H} = 0$. It is often possible to choose the discretization H in such a way that also the resulting discretized equation $H = 0$ has a corresponding bifurcation point. Under reasonable non-degeneracy assumptions it is possible to obtain error estimates for the bifurcation point of the original problem $\mathcal{H} = 0$. We shall not pursue such estimates here and refer the reader to the papers of Brezzi & Rappaz & Raviart and Beyn.

Our aim in this chapter is to investigate bifurcation points of the discretized equation $H = 0$. We will show how certain types of bifurcation points along a solution curve $c(s)$ can be detected, and having detected a bifurcation point, how one can numerically switch from $c(s)$ onto a bifurcating branch. Usually, bifurcation points are defined in a Banach space context, see for example the book of Chow & Hale. Higher order bifurcations often arise from symmetries with respect to certain group operations, see the books of Golubitsky & Schaeffer (1985) and Golubitsky & Stewart & Schaeffer (1988). These symmetries can also be exploited numerically, see e.g. Cliffe & Jepson & Spence (1985), Cliffe & Spence (1985), Cliffe & Winters (1986), Healey (1988), Dellnitz & Werner (1989). Since we are primarily concerned with bifurcation in the numerical curve following context, we confine our discussion to the case of the equation $H = 0$ where $H : \mathbf{R}^{N+1} \to \mathbf{R}^N$ is sufficiently smooth. However, we note that the theoretical discussion below will essentially extend to the Banach space context if we assume that H is also a Fredholm operator of index one. This holds automatically in the case $H : \mathbf{R}^{N+1} \to \mathbf{R}^N$. Discretization errors of bifurcation problems have been analyzed by Beyn (1980), Brezzi & Rappaz & Raviart (1980–1981), Crouzeix & Rappaz (1989).

Some of the fundamental results on the constructive aspects of bifurcation theory and the numerical solution of bifurcation problems are due to Keller (1970), see also Keener & Keller (1974) and Keller (1977). The recent literature on the numerical treatment of bifurcation is very extensive. For an introduction into the field we suggest the lecture notes of Keller (1987). For surveys and bibliography we suggest Mittelmann & Weber (1980), Kubíček & Marek (1983), Küpper & Mittelmann & Weber (1984), Küpper & Seydel & Troger (1987). Some scientists related to the Theoretical Physics Division at the Harwell Laboratory in England have published extensive applications of numerical bifurcation techniques to classical problems in fluid dynamics, see the bibliography under Cliffe, Riley, Winters. Computer programs for numerical bifurcation currently available are for example given in Doedel (1986), Kubíček & Marek (1983), Rheinboldt (1986), Seydel (1988). Some authors, see e.g. Deuflhard (1979), propose basing the calculation of bifurcation or turning points on the failure of the corrector procedure. We cannot recommend such an approach.

In view of the extensive literature we will only touch upon the problem here, and we will confine our discussion to the task of detecting a simple bifurcation point along $c(s)$ and effecting a branch switching numerically. We will see that the detection of simple bifurcation points requires only minor modifications of algorithms such as (3.3.7). Let us begin by defining a bifurcation point.

(8.1.1) Definition. *Let* $H : \mathbf{R}^{N+1} \to \mathbf{R}^N$ *be sufficiently smooth. Suppose that* $c : J \to \mathbf{R}^{N+1}$ *is a smooth curve, defined on an open interval J containing zero, and parametrized (for reasons of simplicity) with respect to arc length such that* $H\big(c(s)\big) = 0$ *for* $s \in J$. *The point $c(0)$ is called a* **bifurcation point** *of the equation $H = 0$ if there exists an $\varepsilon > 0$ such that every neighborhood of $c(0)$ contains zero-points z of H which are not on $c(-\varepsilon, \varepsilon)$.*

An immediate consequence of this definition is that a bifurcation point of $H = 0$ must be a singular point of H. Hence the Jacobian $H'\big(c(0)\big)$ must have a kernel of dimension at least two. We consider the simplest case:

(8.1.2) Assumption. *$\bar{u}$ is a zero point of the smooth map $H : \mathbf{R}^{N+1} \to \mathbf{R}^N$ such that* $\dim \ker H'(\bar{u}) = 2$.

We now describe the Liapunov-Schmidt reduction in the above finite dimensional context. Let us introduce the decompositions

(8.1.3)
$$\mathbf{R}^{N+1} = E_1 \oplus E_2 \quad \text{and} \quad \mathbf{R}^N = F_1 \oplus F_2, \quad \text{where}$$
$$E_1 := \ker H'(\bar{u}), \quad E_2 := E_1^{\perp},$$
$$F_2 := \operatorname{range} H'(\bar{u}), \quad F_1 := F_2^{\perp}, \quad \text{and}$$
$$\dim E_1 = 2, \ \dim E_2 = N - 1, \ \dim F_1 = 1, \ \dim F_2 = N - 1.$$

Thus we may decompose H into the form

$$(8.1.4) \qquad H(u) = H(u_1, u_2) = \begin{pmatrix} H_1(u_1, u_2) \\ H_2(u_1, u_2) \end{pmatrix}$$

where $u_i \in E_i$ and $H_i : E_i \to F_i$, $i = 1, 2$. From the above choice of decompositions, we have

$$H'(u) = \begin{pmatrix} \partial_1 H_1(u_1, u_2) & \partial_2 H_1(u_1, u_2) \\ \partial_1 H_2(u_1, u_2) & \partial_2 H_2(u_1, u_2) \end{pmatrix}$$

and in particular,

$$(8.1.5) \qquad H'(\bar{u}) = \begin{pmatrix} 0 & 0 \\ 0 & \partial_2 H_2(\bar{u}_1, \bar{u}_2) \end{pmatrix}.$$

Here ∂_1, ∂_2 denote the partial derivative operators with respect to the parameters of E_1, E_2 respectively.

Note that $\partial_2 H_2(\bar{u})$ is a nonsingular $(N-1) \times (N-1)$-matrix. Since the equation $H_2(\bar{u}_1, \bar{u}_2) = 0$ has the solution point $(u_1, u_2) = (\bar{u}_1, \bar{u}_2)$, by the implicit function theorem, there exist neighborhoods U_1 of $\bar{u}_1$ in E_1 and U_2 of $\bar{u}_2$ in E_2 and a smooth map $\varphi : U_1 \to U_2$ such that

$$(8.1.6) \qquad H_2(u_1, u_2) = 0 \quad \text{if and only if} \quad u_2 = \varphi(u_1).$$

holds for all $u_1 \in U_1$, $u_2 \in U_2$. Thus, we have a local parametrization of the equation $H_2(u_1, u_2) = 0$ in terms of the variable u_1 in the 2-dimensional space E_1. Consequently, for all $u_1 \in U_1$, $u_2 \in U_2$, the equation $H(u) = 0$ is equivalent to $u_2 = \varphi(u_1)$ and $H_1(u_1, u_2) = 0$, or

$$(8.1.7) \qquad b(u_1) := H_1(u_1, \varphi(u_1)) = 0.$$

This is called the **bifurcation equation** for $H(u) = 0$ at the singular point $\bar{u}$.

We now want to obtain for the present context a characterization of simple bifurcation points, which is analogous to that of Crandall & Rabinowitz (1971). By differentiating the equation $H_2(u_1, \varphi(u_1)) = 0$ arising from (8.1.6) we have by the chain rule

$$\partial_1 H_2(\bar{u}) + \partial_2 H_2(\bar{u})\varphi'(\bar{u}_1) = 0.$$

Since $\partial_1 H_2(\bar{u}) = 0$ and $\partial_2 H_2(\bar{u})$ is nonsingular, it follows that

$$(8.1.8) \qquad \varphi'(\bar{u}_1) = 0.$$

Differentiating $b(u_1) = H_1(u_1, \varphi(u_1))$ twice we obtain for $u = (u_1, \varphi(u_1))$,

$$b'(u_1) = \partial_1 H_1(u) + \partial_2 H_1(u)\varphi'(u_1),$$
$$b''(u_1) = \partial_1^2 H_1(u) + 2\partial_1 \partial_2 H_1(u)\varphi'(u_1)$$
$$+ \partial_2^2 H_1(u)[\varphi'(u_1), \varphi'(u_1)] + \partial_2 H_1(u)\varphi''(u_1).$$

Setting $u_1 := \bar{u}_1$ and taking into account that $\varphi'(\bar{u}_1) = 0$, $\partial_1 H_1(\bar{u}) = 0$, $\partial_2 H_1(\bar{u}) = 0$, we obtain

$$(8.1.9) \qquad b(\bar{u}_1) = 0, \qquad b'(\bar{u}_1) = 0, \qquad b''(\bar{u}_1) = \partial_1^2 H_1(\bar{u}).$$

The simplest (generic) case is that the 2×2 Hessian matrix $b''(\bar{u}_1)$ is non-singular i.e. both eigenvalues are different from zero. We use the following 2-dimensional version of a celebrated theorem of Morse, see e.g. the book Hirsch (1976), p.145., in order to characterize the local structure of the solution set $b^{-1}(0)$.

(8.1.10) Lemma. *Let $\bar{u}_1 \in \mathbf{R}^2$, and let $b : \mathbf{R}^2 \to \mathbf{R}$ be a smooth function such that $b(\bar{u}_1) = 0$, $b'(\bar{u}_1) = 0$ and the Hessian $b''(\bar{u}_1)$ has nonzero eigenvalues λ_1, λ_2. Then there are open neighborhoods U of $0 \in \mathbf{R}^2$ and V of $\bar{u}_1 \in \mathbf{R}^2$ and a diffeomorphism $\psi : U \to V$ such that $\psi(0) = \bar{u}_1$ and $b(\psi(\xi_1, \xi_2)) = \lambda_1 \xi_1^2 + \lambda_2 \xi_2^2$ where $(\xi_1, \xi_2) \in U$.*

If both eigenvalues have the same sign, then $\bar{u}_1$ is an isolated zero point of b and consequently $\bar{u}$ is an isolated zero point of H. Such points are of no interest to us, since they cannot be obtained by traversing a solution curve of the equation $H = 0$. If the eigenvalues are of opposite sign, then the local structure of $b^{-1}(0)$ near $\bar{u}_1$ and consequently the local structure of $H^{-1}(0)$ near $\bar{u}$ are described by two curves, intersecting transversely at $\bar{u}_1$ and $\bar{u}$ respectively. By a transverse intersection we mean that the two corresponding tangents are linearly independent. Let $e \in \mathbf{R}^N$, $e \neq 0$ be a vector which spans $F_1 = \left(\text{range } H'(\bar{u})\right)^{\perp} = \ker H'(\bar{u})^*$. Then the component map H_1 corresponds to $e^* H$, and the preceding discussion motivates the following

(8.1.11) Definition. *Let $H : \mathbf{R}^{N+1} \to \mathbf{R}^N$ be sufficiently smooth. A point $\bar{u} \in \mathbf{R}^{N+1}$ is called a **simple bifurcation point** of the equation $H = 0$ if the following conditions hold:*

(1) $H(\bar{u}) = 0$,

(2) $\dim \ker H'(\bar{u}) = 2$,

(3) $e^* H''(\bar{u})\big|_{\left(\ker H'(\bar{u})\right)^2}$ *has one positive and one negative eigenvalue,*

where e spans $\ker H'(\bar{u})^*$.

The result of the preceding discussion can now be summarized in the following

(8.1.12) Theorem. *Let* $\bar{u} \in \mathbf{R}^{N+1}$ *be a simple bifurcation point of the equation* $H = 0$. *Then there exist two smooth curves* $c_1(s), c_2(s) \in \mathbf{R}^{N+1}$, *parametrized with respect to arclength* s, *defined for* $s \in (-\varepsilon, \varepsilon)$ *and* ε *sufficiently small, such that the following holds:*

(1) $H\big(c_i(s)\big) = 0$, $i \in \{1,2\}$, $s \in (-\varepsilon, \varepsilon)$,

(2) $c_i(0) = \bar{u}$, $i \in \{1,2\}$,

(3) $\dot{c}_1(0), \dot{c}_2(0)$ *are linearly independent,*

(4) $H^{-1}(0)$ *coincides locally with* $\mathrm{range}(c_1) \cup \mathrm{range}(c_2)$, *more precisely:* $\bar{u}$ *is not in the closure of* $H^{-1}(0) \setminus \big(\mathrm{range}(c_1) \cup \mathrm{range}(c_2)\big)$.

Using the notation of (8.1.11) and (8.1.12), let us differentiate the equation $e^* H\big(c_i(s)\big) = 0$ for $i \in \{1,2\}$ twice and evaluate the result at $s = 0$. We then obtain

$$e^* H''(\bar{u})\big[\dot{c}_i(0), \dot{c}_i(0)\big] + e^* H'(\bar{u})\ddot{c}_i(0) = 0.$$

Since e spans $\ker H'(\bar{u})^*$, the second term vanishes, and we are led to the following lemma which characterizes the tangents of the two curves at the bifurcation point $\bar{u}$ up to an obvious freedom of orientation.

(8.1.13) Lemma. *Let* $\bar{u} \in \mathbf{R}^{N+1}$ *be a simple bifurcation point of the equation* $H = 0$. *Under the notation of (8.1.11) and (8.1.12), we obtain*

(1) $\ker H'(\bar{u}) = \mathrm{span}\{\dot{c}_1(0), \dot{c}_2(0)\}$,

(2) $e^* H''(\bar{u})\big[\dot{c}_i(0), \dot{c}_i(0)\big] = 0$ *for* $i \in \{1,2\}$.

The following theorem furnishes a criterion for detecting a simple bifurcation point when traversing one of the curves c_i.

(8.1.14) Theorem. *Let* $\bar{u} \in \mathbf{R}^{N+1}$ *be a simple bifurcation point of the equation* $H = 0$. *Under the notation of (8.1.11) and (8.1.12), the determinant of the following augmented Jacobian*

$$\det \begin{pmatrix} H'\big(c_i(s)\big) \\ \dot{c}_i(s)^* \end{pmatrix}$$

changes sign at $s = 0$ *for* $i \in \{1,2\}$.

Proof. We treat the case $i = 1$. It is more convenient for the proof to use the matrix:

$$A(s) := \begin{pmatrix} \dot{c}_1(s)^* \\ H'\big(c_1(s)\big) \end{pmatrix}.$$

Consider an orthogonal $(N+1) \times (N+1)$-matrix $V = (v_1, \ldots, v_{N+1})$ where $v_1 := \dot{c}_1(0)$, $\mathrm{span}\{v_1, v_2\} = \ker H'(u)$, and an orthogonal $N \times N$-matrix $W = (w_1, \ldots, w_N)$ where $w_1 := e$ spans $\ker H'(\bar{u})^*$ as in (8.1.11). Since

$$\dot{c}_1(s)^* v_j = \dot{c}_1(0)^* v_j + O(s),$$
$$w_k^* H'\big(c_1(s)\big) v_j = w_k^* H'(\bar{u}) v_j + w_k^* H''(\bar{u})\big[\dot{c}_1(0), v_j\big] s + O(s^2),$$

we obtain:

$$
\textbf{(8.1.15)} \qquad \begin{pmatrix} 1 & 0^* \\ 0 & W^* \end{pmatrix} A(s) V = \begin{pmatrix} 1 + O(s) & O(s) & O(s) \\ O(s^2) & \rho s + O(s^2) & O(s) \\ O(s) & O(s) & B + O(s) \end{pmatrix}.
$$

The $(N-1) \times (N-1)$ block matrix B in (8.1.15) is nonsingular, see (8.1.5) and the remarks thereafter. The scalar ρ in (8.1.15) is given as the off-diagonal entry of the following symmetric 2×2-matrix

$$
\begin{pmatrix} e^* H''(\bar{u})[v_1, v_1] & e^* H''(\bar{u})[v_1, v_2] \\ e^* H''(\bar{u})[v_2, v_1] & e^* H''(\bar{u})[v_2, v_2] \end{pmatrix}.
$$

Since this matrix is nonsingular, cf. (8.1.11)(3), and since the diagonal entry $e^* H''(\bar{u})[v_1, v_1]$ vanishes, cf. (8.1.13), it follows that $\rho \neq 0$. Now by performing Gaussian elimination upon the first two columns of (8.1.15), we obtain a reduced form

$$
\begin{pmatrix} 1 + O(s) & O(1) & O(s) \\ 0 & \rho s + O(s^2) & O(s) \\ 0 & 0 & B + O(s) \end{pmatrix}
$$

which clearly has a determinant of the form

$$
\rho \det(B) s + O(s^2).
$$

It follows that the determinant of $A(s)$ changes sign at $s = 0$. $\qquad \Box$

Theorem (8.1.14) implies that when traversing a solution curve $c(s) \in H^{-1}(0)$, a simple bifurcation point is detected by a change in the orientation. Figure 8.1.a illustrates this. The arrows in the figure show the orientation.

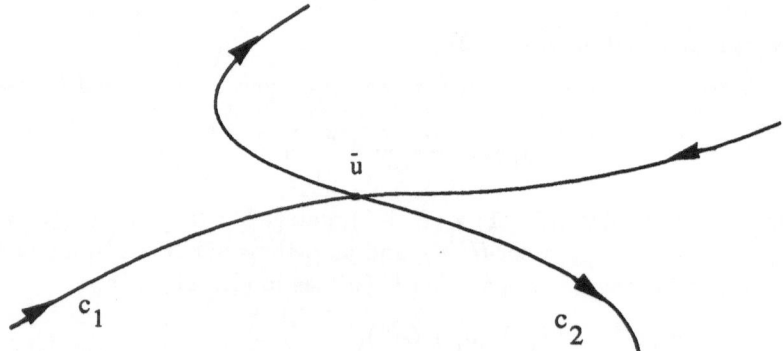

Figure 8.1.a Encountering a simple bifurcation point $\bar{u}$

Conversely, suppose that a smooth curve $c(s) \in H^{-1}(0)$ is traversed and that $c(0)$ is an isolated singular point of H such that

$$\det \begin{pmatrix} H'(c(s)) \\ \dot{c}(s)^* \end{pmatrix}$$

changes sign at $s = 0$. Using a standard argument in degree theory, see Krasnosel'skiĭ(1964) or Rabinowitz (1971), it can be shown that $c(0)$ is a bifurcation point of $H = 0$. However $c(0)$ is not necessarily a simple bifurcation point.

We next present an argument similar to that of Keller (1977) to show that an Euler-Newton type continuation method "jumps over" a simple bifurcation point. To this end, we give the following lemma which basically states that a small truncated cone with vertex $\bar{u}$, axis $\dot{c}_1(0)$ and aperture δ is contained in the region of attraction of the Newton corrector method, cf. figure (8.1.b). We shall only give the main ideas of the proof and omit some of the tedious technical details.

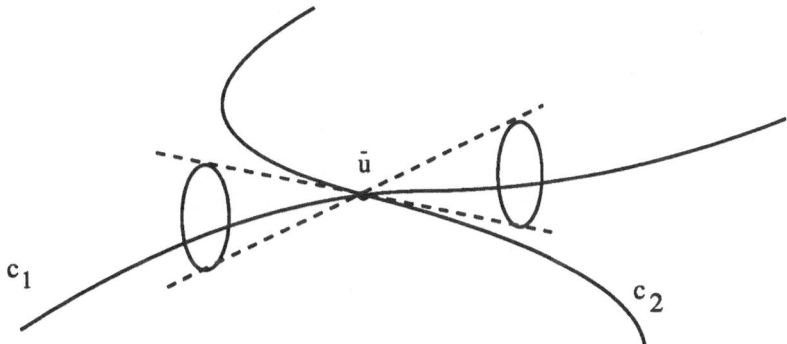

Figure 8.1.b The cone of attraction of the Newton corrector

(8.1.16) Lemma. *Let $\bar{u} \in \mathbf{R}^{N+1}$ be a simple bifurcation point of the equation $H = 0$. Under the notation of (8.1.11) and (8.1.12), there exist an open neighborhood U of $\{u \in H^{-1}(0) \mid u$ is a regular point of $H\}$, and positive numbers ε, δ such that*

(1) *the conclusion of theorem (3.4.1) holds;*

(2) $\bar{u} + (\dot{c}_1(0) + z)s \in U$ *for $0 < |s| < \varepsilon$ and $z \in \mathbf{R}^{N+1}$ with $\|z\| < \delta$.*

Outline of Proof. The proof of the first part is analogous to that of theorem (3.4.1) but it is even more technical and will be omitted. The main ideas of the proof of the second part involve asymptotic estimates of H, H', ... at

$$u := \bar{u} + (\dot{c}_1(0) + z)s.$$

We use the same notation as in the proof of (8.1.14). The first two estimates are immediate consequences of the Taylor formula and the results in (8.1.12–15):

$$W^* H(u) = \begin{pmatrix} O(\|z\|s^2) + O(s^3) \\ O(\|z\|s) + O(s^2) \end{pmatrix} ;$$

$$W^* H'(u)V = \begin{pmatrix} O(s^2) + O(\|z\|s) & \rho s + O(\|z\|s + s^2) & O(s) \\ O(s) & O(s) & B + O(s) \end{pmatrix} .$$

The following estimate for the tangent vector $t(H'(u))$ is obtained by directly solving $H'(u)\tau = 0$, $\tau[1] = 1$ for τ in the previous equation and normalizing $t(H'(u)) = \pm\tau/\|\tau\|$:

$$V^* t(H'(u)) = \begin{pmatrix} \pm 1 + O(s^2) + O(\|z\|s) + O(\|z\|^2) \\ O(s) + O(\|z\|) \\ O(s) \end{pmatrix} .$$

The next estimate is obtained from the previous estimates by a Gauss-Jordan reduction:

$$\begin{pmatrix} t(H'(u))^* V \\ W^* H'(u)V \end{pmatrix}^{-1} =$$

$$\begin{pmatrix} 1 + O(s^2) + O(\|z\|s) + O(\|z\|^2) & O(1) + O(\|z\|)s^{-1} & O(s) + O(\|z\|) \\ O(s) + O(\|z\|) & \rho^{-1}s^{-1} + O(1) & O(1) \\ O(s) & O(1) & B^{-1} + O(s) \end{pmatrix} .$$

In the above inverse, the first column again represents the tangent vector, and the remaining submatrix yields an estimate for the Moore-Penrose inverse:

$$V^* H'(u)^+ W = \begin{pmatrix} O(1) + O(\|z\|)s^{-1} & O(s) + O(\|z\|) \\ \rho^{-1}s^{-1} + O(1) & O(1) \\ O(1) & B^{-1} + O(s) \end{pmatrix} .$$

Combining the above results, we obtain the following estimate for a Newton step:

$$V^* H'(u)^+ H(u) = \begin{pmatrix} O(\|z\|s_i^2) + O(s^3) + O(\|z\|^2 s) \\ O(\|z\|s) + O(s^2) \\ O(\|z\|s) + O(s^2) \end{pmatrix} .$$

From all the above formulae, the following crucial norm estimates are obtained:

$$\|H'(u)^+\| = (|\rho|^{-1} + \|z\|)|s|^{-1} + O(1),$$

$$\|H'(u)^+ H(u)\| = O(\|z\|\|s|) + O(s^2),$$

$$\sup_{\|z\|\leq\delta} \|H''(u)\| \; \sup_{\|z\|\leq\delta} \|H'(u)^+\| \; \sup_{\|z\|\leq\delta} \|H'(u)^+ H(u)\| \to 0 \quad \text{as} \quad \delta, |s| \to 0.$$

The asserted convergence of Newton's method now can be shown by using a standard Newton-Kantorovich type argument, see e.g. Ortega & Rheinboldt (1970), p. 421.   □

To summarize the discussion of this section, we have seen that an Euler-Newton type continuation method as sketched in (3.3.7) detects simple bifurcation points on the curve $c(s)$ which is being traversed, when a change in

$$\operatorname{sign} \det \begin{pmatrix} H'(c(s)) \\ \dot{c}(s)^* \end{pmatrix}$$

occurs. Depending upon the method used to perform the decomposition of the Jacobian, the above orientation can often be calculated at very small additional cost. The Euler-Newton algorithm generally has no difficulty in "jumping over" i.e. proceeding beyond the bifurcation point $\bar{u}$. That is, for sufficiently small steplength h, the predictor point will fall into the "cone of attraction" of the Newton corrector. If it is wished, the simple bifurcation point $\bar{u}$ can be approximated precisely. We leave this discussion for Chapter 9, where the general question of calculating special points along a curve is addressed.

The following algorithm indicates the adaptations which must be made in order to proceed beyond a bifurcation point on a curve which is currently being traversed. Steplength adaptations and strategies for accepting or rejecting predictor-corrector steps are omitted since they have been treated previously.

(8.1.17) Jumping Over A Bifurcation Point. *comment:*

> **input**
>> **begin**
>> $u \in \mathbf{R}^{N+1}$ such that $H(u) = 0$;   *initial point*
>> $h > 0$;   *initial steplength*
>> $\omega \in \{+1, -1\}$;   *initial orientation of traversing*
>> **end**;
> **repeat**
>> $v := u + h\omega t\big(H'(u)\big)$;   *predictor step*
>> **repeat**
>>> $v := v - H'(v)^{+} H(v)$;   *corrector loop*
>> **until** convergence;
>> adapt steplength h;
>> **if** $t\big(H'(u)\big)^{*} t\big(H'(v)\big) < 0$ **then**   *test for bifurcation point*
>>> **begin**
>>> $\omega := -\omega$;   *reverses orientation of curve*
>>> print "bifurcation point between", u, "and", v, "encountered";
>>> **end**;
>> $u := v$;   *new point along $H^{-1}(0)$*
> **until** traversing is stopped.

In the corrector procedure, the Jacobian $H'(v)$ can be replaced by an approximation. Thus a chord method may be implemented. We emphasize however, that it is necessary to obtain a good approximation of the Jacobian at least once **at the predictor point** since otherwise the local convergence of the Newton corrector iterations cannot be guaranteed when jumping over a simple bifurcation point. As an example, let us consider a predictor step

$$v_0 := u + h\omega t(A_0) \qquad \text{with} \qquad A_0 \approx H'(u)$$

and a successive Broyden corrector iteration

$$v_{i+1} := v_i - A_i^+ H(v_i) \qquad \text{where} \qquad A_{i+1} := A_i - \frac{H(v_{i+1})\left(A_i^+ H(v_i)\right)^*}{\|A_i^+ H(v_i)\|^2}.$$

Since $\left(A_i^+ H(v_i)\right)^* t(A_i) = 0$, it follows from (7.2.6)(4) that

$$\det \begin{pmatrix} A_{i+1} \\ t(A_{i+1})^* \end{pmatrix} = \frac{D_i}{\rho_i} \det \begin{pmatrix} A_i \\ t(A_i)^* \end{pmatrix}$$

where $t(A_{i+1}) = \rho_i t(A_i)$ with $\rho_i = \pm 1$ and

$$D_i = 1 - \frac{\left(A_i^+ H(v_{i+1})\right)^* \left(A_i^+ H(v_i)\right)}{\|A_i^+ H(v_i)\|^2}.$$

However, if the step has jumped over a simple bifurcation point, we expect a change of orientation i.e. $\rho_i = -1$, and this can only occur if $D_i < 0$, which implies

$$\frac{\|A_i^+ H(v_{i+1})\|}{\|A_i^+ H(v_i)\|} > 1.$$

Hence the Newton step is not contractive. Thus we cannot expect that the corrector process will converge when we have jumped over a bifurcation point if a chord method is employed which uses a Jacobian approximation from the "wrong side" of the bifurcation point.

8.2 Switching Branches Via Perturbation

In the previous section we have seen that it is possible to detect and jump over simple bifurcation points while numerically tracing a solution curve $c_1(s) \in H^{-1}(0)$ via an Euler-Newton method. The more difficult task is to numerically branch off onto the second solution curve $c_2(s)$ at the detected bifurcation point $\bar{u}$. The simplest device for branching off numerically rests upon the Sard theorem (11.2.3). If a small perturbation vector $p \in \mathbf{R}^N$ is chosen at random, then the probability that p is a regular value of H is unity. Of course, in this case $H^{-1}(p)$ has no bifurcation point. This situation is illustrated in figure 8.2.a.

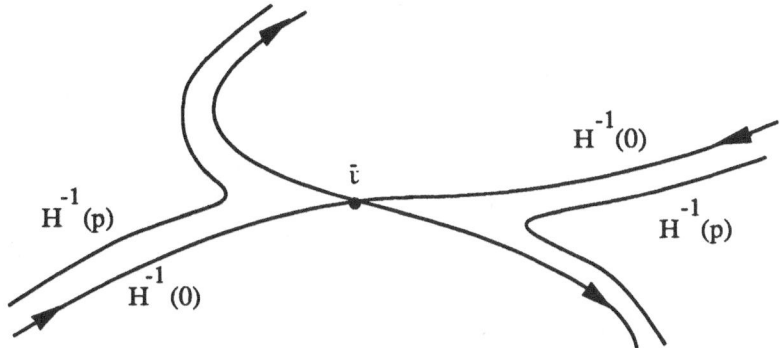

Figure 8.2.a Global perturbation of a simple bifurcation point

Since $p \in \mathbf{R}^N$ is chosen so that $||p||$ is small, the solution sets $H^{-1}(0)$ and $H^{-1}(p)$ are close together. On $H^{-1}(p)$, no change of orientation can occur. Therefore, corresponding solution curves in $H^{-1}(p)$ must branch off near the bifurcation point $\bar{u}$. There are essentially two approaches which can be taken with the use of perturbations.

(8.2.1) Global perturbation. One can incorporate a perturbation p of the curve following problem $H(u) = 0$ at the outset and follow the curve $H(u) = p$ throughout. By the Sard theorem (11.2.3), with probability unity, the corresponding solution curves $c_p(s)$ will contain no bifurcation points at all. This approach has recently been used by Glowinski & Keller & Reinhart (1984). As we shall see in our later discussion of piecewise linear methods, the global perturbation approach has analogous qualitative properties with respect to bifurcations as the piecewise linear methods for tracing implicitly defined curves. In the piecewise linear methods, this is achieved by considering lexicographically positive matrices, see (12.3.2). Hence, branching off is automatic at simple bifurcation points, however detection of the presence of a bifurcation point is more difficult.

(8.2.2) Monitoring the solution branches via local perturbations. In contrast to the global perturbation approach, one may traverse along a curve $c_1(s) \in H^{-1}(0)$ and monitor for changes in orientation as in (8.1.17). When a bifurcation point has been detected between two corrector points u and w, a local perturbation vector p can be chosen and the nearby curve $c_p(s) \in H^{-1}(p)$ may be traversed in the reversed direction starting at the solution w_p of

(8.2.3)
$$\min_{w_p} \{||w_p - w|| \mid H(w_p) = p\}.$$

After the local perturbation has served its purpose for branching off, it can be shut off. A similar technique was described by Georg (1981). Such a process of handling bifurcation is best done interactively. The following is an algorithm illustrating this technique.

(8.2.4) Interactive Tracing Of Bifurcation Branches.   *comment:*

> **input**
>> **begin**
>> $u \in \mathbf{R}^{N+1}$ such that $H(u) \approx 0$;   *initial point*
>> $\omega \in \{+1, -1\}$;   *initial orientation of traversing*
>> **end**;
>
> **1: enter**   *interactive input*
>> **begin**
>> initial steplength h;
>> arclength α to be traversed;
>> perturbation vector p;
>> $\sigma \in \{\text{true}, \text{false}\}$;   *logical variable for stopping*
>> $\rho \in \{\text{true}, \text{false}\}$;   *logical variable —*
>> **end**;   *for reversing the direction of traversing*
>
> **if** $\sigma = $ true **then** quit;
> **if** $\rho = $ true **then** $\omega := -\omega$;
> **repeat**
>> $u := u - H'(u)^+ \big(H(u) - p\big)$;   *initial corrector steps, cf. (8.2.3)*
>
> **until** convergence;
> **repeat**   *start of predictor-corrector loop*
>> $v := u + h\omega t\big(H'(u)\big)$;   *predictor step*
>> **repeat**
>>> $v := v - H'(v)^+ \big(H(v) - p\big)$;   *corrector steps*
>>
>> **until** convergence;
>> **if** $t\big(H'(u)\big)^* t\big(H'(v)\big) < 0$ **then**   *test for bifurcation point*
>>> **begin**
>>> $\omega := -\omega$;   *reverses orientation of curve*
>>> print "bifurcation point encountered";
>>> $u := v$;
>>> **go to** 1;
>>> **end**;
>>
>> $u := v$;
>> $\alpha := \alpha - h$;   *countdown of arclength*
>> adapt steplength h;
>
> **until** $\alpha < 0$;
> **go to** 1.

8.3 Branching Off Via the Bifurcation Equation

Although the branching off from $c_1(s)$ onto $c_2(s)$ via perturbation techniques works effectively, this approach can have some shortcommings. In general, it cannot be decided in advance which of the two possible directions along $c_2(s)$ will be taken. Furthermore, if the perturbation vector p is not chosen correctly (and it is not always clear how this is to be done), one may still have some difficulty in tracing $H^{-1}(p)$. The solution set $H^{-1}(0)$ can be approximated near the bifurcation point $\bar{u}$ only after $c_2(s)$ has been found by a perturbation technique. That is, one may follow $c_2(s)$ backwards and jump over $\bar{u}$ again along $c_2(s)$.

To obtain on approximation of $H^{-1}(0)$ near a simple bifurcation point $\bar{u}$, we now show how the theoretical results of section 8.1 can be numerically realized. Most of the papers which deal with the numerical treatment of simple bifurcation points involve such a numerical realization. Papers of such a nature are those of Keller (1977) and Rheinboldt (1978), see also the lecture notes of Keller (1987). In this section, we give a sketch of how a numerical analogue of the characterization of simple bifurcation points may be implemented in a way which is numerically stable, and does not require much computational effort. We will do this by outlining the essential steps, namely:

1. Approximation of a bifurcation point $\bar{u}$;
2. Approximation of the kernels $\ker H'(\bar{u})$ and $\ker H'(\bar{u})^*$;
3. Approximation of the bifurcation equation;
4. Approximation of the tangents $\dot{c}_1$ and $\dot{c}_2$ at $\bar{u}$.

Step 1: Approximation of a bifurcation point $\bar{u}$. Assume that a solution curve $c_1(s) \in H^{-1}(0)$ is currently being traversed. As has been noted in (8.1.14), the presence of a simple bifurcation point $c_1(\bar{s})$ will be signalled by a change of sign of the functional

$$(8.3.1) \qquad\qquad f(u) := \det \begin{pmatrix} H'(u) \\ t(H'(u))^* \end{pmatrix}$$

when traversing along $c_1(s)$ past $\bar{s}$. Of course in any implementation, the step size needs to be sufficiently small that not more than one bifurcation point is jumped over. The bifurcation point $c_1(\bar{s}) = \bar{u}$ can be approximated by calculating a solution of the scalar equation $f(c_1(s)) = 0$. As has been noted previously, the determinant (8.3.1) can be calculated with very little computational effort once a decomposition of $H'(u)$ has been obtained, see e.g. (4.1.1) or (4.5.3). We may use the technique described in chapter 9 to calculate the solution of $f(c_1(s)) = 0$. Some care has to be taken in doing this, since $\bar{u}$ is actually a singular point of H. Nonetheless, a (superlinearly convergent) secant method for approximating $\bar{s}$ and thereby $c_1(\bar{s}) = \bar{u}$, gives accurate results.

Step 2: Approximation of the kernels $\ker H'(\bar{u})$ **and** $\ker H'(\bar{u})^*$. We now assume that an approximation $\tilde{u}$ of a simple bifurcation point $\bar{u}$ has been obtained. To approximate $\ker H'(\bar{u})$ and $\ker H'(\bar{u})^*$, we need to calculate two vectors $\tau_1, \tau_2 \in \mathbf{R}^{N+1}$ such that $\ker H'(\bar{u}) = \mathrm{span}\{\tau_1, \tau_2\}$ and a vector $e \in \mathbf{R}^N$ such that $\ker H'(\bar{u})^* = \mathrm{span}\{e\}$. The task is complicated by the fact that we should not assume that the approximation $\tilde{u}$ of the bifurcation point $\bar{u}$ is precise. In addition, we should not assume that the Jacobian $H'(\tilde{u})$ is precisely approximated.

The approximation of τ_1 is easy, since by step 1, an approximation of the tangent $\dot{c}_1(\bar{s})$ for $c_1(\bar{s}) = \tilde{u}$ is already known. So we may set $\tau_1 = \dot{c}_1(\bar{s})$. Ideally, assuming $\tau_1 \in \ker H'(\bar{u})$, a τ_2 could be obtained by solving

$$(8.3.2) \qquad \begin{pmatrix} H'(\bar{u}) \\ \tau_1^* \end{pmatrix} \tau_2 = 0, \qquad \|\tau_2\| = 1.$$

A solution of (8.3.2) exists since $H'(\bar{u})$ has a 2-dimensional kernel. However, since everything is only approximated, this approach has to be modified. So instead we calculate τ_2 by solving the minimization problem

$$(8.3.3) \qquad \min_{\tau_2} \left\{ \|H'(\tilde{u})\tau_2\|^2 + (\tau_1^*\tau_2)^2 \ \middle| \ \|\tau_2\| = 1 \right\}.$$

To solve (8.3.3), let us set

$$(8.3.4) \qquad A := \begin{pmatrix} H'(\tilde{u}) \\ \tau_1^* \end{pmatrix}.$$

It can be seen that the solution of (8.3.3) is a unit eigenvector τ_2 of A^*A which corresponds to the smallest eigenvalue λ. Solving for τ_2 can be performed by the classical inverse iteration method, cf. Stoer & Bulirsch (1980), p. 356. Of course the matrix A has a bad condition number, since it approximates a singular matrix. However, since $\ker H'(\bar{u})$ is two-dimensional, the matrix A^*A should have only one eigenvalue which is close to zero. Thus, the inverse iteration method is numerically stable, see Peters & Wilkinson (1979). In fact, the smaller λ is, the faster the method converges.

In a similar fashion, the vector e spanning $\ker H'(\bar{u})^*$ can be approximated by the eigenvector of $H'(\tilde{u})H'(\tilde{u})^*$ corresponding to the smallest eigenvalue λ, i.e. by the solution of

$$(8.3.5) \qquad \min_{e} \left\{ \|H'(\tilde{u})^*e\| \ \middle| \ \|e\| = 1 \right\}.$$

Here too, the inverse iteration method may be utilized.

Let us sketch how τ_1, τ_2 and e may be calculated if a QR decomposition

$$H'(\tilde{u})^* = Q \begin{pmatrix} R \\ 0^* \end{pmatrix}$$

is given. In this case, one basis vector of $\ker H'(\tilde{u})$ is given by the last column of Q i.e. $\tau_1 := Q e_{N+1}$. It is now easily seen that τ_2 is a solution of (8.3.3) if and only if

$$\tau_2 = Q \begin{pmatrix} y \\ 0 \end{pmatrix}$$

and $y \in \mathbf{R}^N$ solves

$$\min_y \{ \|R^* y\| \, \big| \, \|y\| = 1 \}.$$

On the other hand, $\|H'(\tilde{u})^* e\| = \|Re\|$, and hence both problems (8.3.3) and (8.3.5) can be solved by utilizing the same triangular matrix R. We sketch how to do this in the following algorithm.

(8.3.6) Inverse Iteration
For Approximating $\ker H'(\tilde{u})$ **And** $\ker H'(\tilde{u})^*$.   *comment:*

  input $H'(\tilde{u})^* = Q \begin{pmatrix} R \\ 0^* \end{pmatrix}$;   *QR decomposition*

  $\tau_1 := Q e_{N+1}$;
  find solution k of $\min_k |R[k, k]|$;
  $y := e_k$;   *starting vector*

  repeat

   $x := R^{-1} y$; $\dfrac{x := x}{\|x\|}$;   *inverse iteration w.r.t.* RR^*

   $y := (R^*)^{-1} x$; $\dfrac{y := y}{\|y\|}$;   *inverse iteration w.r.t.* $R^* R$

  until convergence;

  $\tau_2 := Q \begin{pmatrix} y \\ 0 \end{pmatrix}$; $e := x$;

  print τ_1, τ_2, e.   *output*

Step 3: Approximation of the bifurcation equation. Assume that an approximation $\tilde{u}$ of the bifurcation point $\bar{u}$ is given, that τ_1, τ_2 approximately span $\ker H'(\tilde{u})$, and that e approximately spans $\ker H'(\tilde{u})^*$. The crucial equation to be solved is the bifurcation equation (8.1.7). To this end, we need to approximate the symmetric 2×2-matrix with entries

$$(8.3.7) \qquad \alpha[i, j] := \left(e^* H''(\tilde{u})[\tau_i, \tau_j] \right) \qquad i, j = 1, 2.$$

For this purpose we use difference formulae for approximating the second order partial derivatives

$$(8.3.8) \qquad \alpha[i, j] = \partial_i \partial_j g(0, 0)$$

where the function $g : \mathbf{R}^2 \to \mathbf{R}$ is given by

(8.3.9)
$$g(\xi_1, \xi_2) := e^* H(\tilde{u} + \xi_1 \tau_1 + \xi_2 \tau_2).$$

We use the following difference formulae:

$$\partial_1^2 g(0,0) := \varepsilon^{-2} \Big[g(\varepsilon, 0) - 2g(0,0) + g(-\varepsilon, 0) \Big] + O(\varepsilon^2);$$

$$\partial_2^2 g(0,0) := \varepsilon^{-2} \Big[g(0, \varepsilon) - 2g(0,0) + g(0, -\varepsilon) \Big] + O(\varepsilon^2);$$

$$\partial_1 \partial_2 g(0,0) := \frac{1}{4} \varepsilon^{-2} \Big[g(\varepsilon, \varepsilon) + g(-\varepsilon, -\varepsilon) - g(\varepsilon, -\varepsilon) - g(-\varepsilon, \varepsilon) \Big] + O(\varepsilon^2);$$

$$\partial_2 \partial_1 g(0,0) := \partial_1 \partial_2 g(0,0).$$

As usual, the meshsize ε needs to be chosen so as to counterbalance between the truncation error and the cancellation error. In many cases,

$$\varepsilon \approx \sqrt[3]{\text{relative machine error}}$$

is a suitable choice, see (10.3.24–27) for a justification. Hence, the bifurcation equation (8.1.7) can be approximated by a quadratic equation arising from the Hessian (8.3.7) obtained via 8 evaluations of H.

Step 4: Approximation of the tangents $\dot{c}_1$ and $\dot{c}_2$ at $\bar{u}$. According to (8.1.13)(2), a tangent vector t for a bifurcating branch of solutions $c_i(\bar{s})$ satisfies the equation

(8.3.10)
$$e^* H''(\bar{u})[t, t] = 0.$$

Hence by setting $t = \xi_1 \tau_1 + \xi_2 \tau_2$, we solve the approximation to (8.3.10) obtained via step 3:

(8.3.11)
$$\alpha[1,1]\xi^2 + 2\alpha[1,2]\xi_1\xi_2 + \alpha[2,2]\xi_2^2 = 0.$$

If the symmetric 2×2 α-matrix has one positive and one negative eigenvalue (which is to be expected in the case that $\bar{u}$ is a simple bifurcation point), then we obtain two linearly independent solutions of (8.3.11) which approximate the two tangents $\dot{c}_1(0)$ and $\dot{c}_2(0)$ (up to a scalar multiple). One tangent direction will readily be identified with the tangent $\dot{c}_1(\bar{s})$ of the currently traversed curve c_1 (if the approximations are any good), and the other tangent gives us a predictor direction in order to traverse the branch of solutions c_2 which at $\bar{u}$ bifurcates from our currently traversed solution curve c_1. Note that the computational cost of the above technique consists mainly of a decomposition of $H'(\tilde{u})$, several solvings of linear systems using this decomposition, and several computations of the map $H(u)$ at points u near $\bar{u}$. The above four steps illustrate that the theory of simple bifurcation points in the sense of Crandall & Rabinowitz is also numerically implementable. For an up to date survey on numerical treatment of bifurcation problems, we refer the reader to literature cited at the beginning of this chapter.

Chapter 9. Calculating Special Points of the Solution Curve

9.1 Introduction

One of the main purposes of numerical continuation methods concerns the accurate determination of certain points on a smooth curve $c(s)$ in $H^{-1}(0)$, which are of special interest. The following are some examples.

In the applications dealing with **homotopy methods**, the equation $H(x, \lambda) = 0$ for $x \in \mathbf{R}^N$ and $\lambda \in \mathbf{R}$ generally has a known starting point (x_0, λ_0). The homotopy path $c(s)$ passes through this point, and we seek a point $(\bar{x}, \bar{\lambda})$ on $c(s)$ such that $H(\bar{x}, \bar{\lambda}) = 0$ for a certain value $\bar{\lambda}$ of the homotopy parameter λ. Examples of applications of homotopy methods are given in chapter 11.

(9.1.1) Turning points in $H^{-1}(0)$ may be of interest when the equation represents a branch of solutions for a nonlinear eigenvalue problem involving the eigenvalue parameter λ. Such points are characterized by the fact that λ has a local extremum on $H^{-1}(0)$. In physics and engineering applications, a turning point can signify a change in the stability of the solutions. A vast literature exists for calculating turning points, the following papers are a sample: Chan (1984), Griewank & Reddien (1984), Kikuchi (1979), Mehlem & Rheinboldt (1982), Moore & Spence (1980), Pönisch & Schwetlick (1981–82), Schwetlick (1984), Seydel (1979), Ushida & Chua (1984).

(9.1.2) Simple bifurcation points have already been discussed in detail in chapter 8. There we showed how to detect the presence of such points along the curve c. It may also be of interest to accurately approximate a bifurcation point. They may also arise in nonlinear eigenvalue problems and are of great interest since they usually represent points at which the stability of the solutions changes.

To unify our discussion, let $f : \text{range } c \to \mathbf{R}$ be a smooth functional. There are two general types of special points on the curve c which we shall consider:

(9.1.3) Zero points. In this case we seek points $c(s)$ such that $f\big(c(s)\big) = 0$. The homotopy method is such a case if we set $f(x, \lambda) := \lambda - \bar{\lambda}$. Simple bifurcation points are another such case if we set e.g.

$$f\big(c(s)\big) := \det \begin{pmatrix} H'\big(c(s)\big) \\ \dot{c}(s)^* \end{pmatrix}.$$

(9.1.4) Extremal points. In this case we seek extreme points (usually maxima or minima) of $f\big(c(s)\big)$. Turning points are such a case if we set $f(x, \lambda) := \lambda$. Certain regularization methods may also be formulated as determining a turning point on an implicitly defined curve. For general references on regularization methods see Tikhonov & Arsenin (1977) or Groetsch (1984). We now treat these two general cases in greater detail.

9.2 Calculating Zero Points f(c(s))=0

Let $H : \mathbf{R}^{N+1} \rightarrow \mathbf{R}^N$ be a smooth map, let $c(s) \in H^{-1}(0)$ be a smooth solution curve parametrized with respect to arclength (for the sake of convenience), and let $f : \text{range}\,c \rightarrow \mathbf{R}$ be a smooth functional. Suppose that some point $c(s_n)$ has been found which is an approximate zero point of f. For example, it would be reasonable to take $c(s_n)$ as an approximate zero point if a predictor-corrector method produced two successive points $c(s_{n-1})$ and $c(s_n)$ such that $f\big(c(s_{n-1})\big) f\big(c(s_n)\big) < 0$. Then it is reasonable to replace the usual steplength adaptation used to traverse the curve c by a **Newton-type steplength adaptation** which is motivated by the following one-dimensional Newton method for solving the equation $f\big(c(s)\big) = 0$:

(9.2.1)
$$s_{n+1} = s_n - \frac{f\big(c(s_n)\big)}{f'\big(c(s_n)\big)\dot{c}(s_n)}.$$

Here we use the convention

$$f' = \left(\frac{\partial f}{\partial u_1}, \dots, \frac{\partial f}{\partial u_{N+1}} \right).$$

Equation (9.2.1) suggests that we can take the new steplength

(9.2.2)
$$h := -\frac{f\big(c(s_n)\big)}{f'\big(c(s_n)\big)\dot{c}(s_n)}$$

at $u := c(s_n)$ in order to obtain a predictor point $v = u + ht\big(H'(u)\big)$, which should lead to a better approximation of a zero point of f on c.

The following algorithm illustrates for a simple Euler-Newton method how a standard steplength adaptation can be switched to the above Newton-type steplength adaptation in order to approximate a zero point of f on c while traversing c.

(9.2.3) Newton Steplength Adaptation. *comment:*

input

  begin

  $u \in \mathbf{R}^{N+1}$ such that $H(u) = 0$;   *initial point*

  $h_{\min} > 0$;   *minimal steplength*

  $h > h_{\min}$;   *initial steplength*

  end;

 $\nu :=$ false;   *logical variable for switching to —*

repeat   *Newton-type steplength adaptation*

 $v := u + ht\big(H'(u)\big)$;   *predictor step*

 repeat

  $v := v - H'(v)^{+}H(v)$;   *corrector loop*

 until convergence;

 if $f(u)f(v) \le 0$ then $\nu :=$ true; *switching to Newton-type —*

 if $\nu =$ true then $h := -\dfrac{f(v)}{f'(v)t\big(H'(v)\big)}$ *steplength adaptation*

 else choose a new steplength $h > 0$; *see chapter 6*

 $u := v$;   *new point along $H^{-1}(0)$*

until $|h| < h_{\min}$.

A sufficient condition for a sequence of points u produced by the algorithm (9.2.3) to converge to a solution $\bar{u}$ of

$$H(u) = 0$$
$$f(u) = 0$$

is that the steplength h be sufficiently small and

$$\det \begin{pmatrix} H'(\bar{u}) \\ f'(\bar{u}) \end{pmatrix} \neq 0 .$$

Under these assumptions quadratic convergence can be shown.

 Algorithm (9.2.3) requires the quantity

$$\frac{d}{ds} f\big(c(s)\big) = f'\big(c(s)\big)\dot{c}(s),$$

and this may be inconvenient to obtain. As an example, we saw in chapter 8 that bifurcation points $c(\bar{s})$ are points where

$$f\big(c(s)\big) = \det \begin{pmatrix} H'\big(c(s)\big) \\ \dot{c}(s)^{*} \end{pmatrix} = 0$$

holds. In this case, furnishing $\frac{d}{ds}f(c(s))$ would be undesirable, since it would require at least formally, the calculation of H''. Thus it is reasonable to formulate the secant analogue of (9.2.1) which leads to the following Newton-type steplength adaptation:

(9.2.4)
$$h := -\frac{f(v)}{f(v) - f(u)}\, h\,,$$

which replaces the corresponding formula in (9.2.3). Of course, this reduces the above mentioned quadratic convergence to superlinear convergence, as is typical for the one-dimensional secant method.

For the case of calculating a simple bifurcation point, care should be taken since the augmented matrix

$$\begin{pmatrix} H'(u) \\ t(H'(u))^* \end{pmatrix}$$

is ill-conditioned near the bifurcation point, and hence the corrector iteration encounters instabilities. But the above mentioned superlinear convergence of the secant method generally overcomes this difficulty since the instability generally only manifests itself at a predictor point which can already be accepted as an adequate approximation of the bifurcation point.

Obviously, if one zero point $\bar{u}$ of the functional f on the curve c has been approximated, the predictor-corrector method can be restarted in order to seek additional zero points. In this case, the line of the algorithm (9.2.3) where the logical variable ν occurs should be activated only after the first accepted predictor-corrector step. This measure simply safeguards against returning to the already known zero point.

9.3 Calculating Extremal Points $\min_s f((c(s))$

The aim in this section is to give some specific details for calculating an extremal point on a curve $c(s) \in H^{-1}(0)$ for a smooth functional $f : \text{range}\, c \to \mathbf{R}$. Clearly, a necessary condition which must hold at a local extremum $c(\bar{s})$ of f is that the equation

(9.3.1)
$$f'(c(s))\dot{c}(s) = 0$$

holds. Following the same motivation as in section 9.2, we can formulate the analogous switchover to a Newton-type steplength adaptation:

$$h := -\frac{f'(c(s))\dot{c}(s)}{f'(c(s))\ddot{c}(s) + f''(c(s))[\dot{c}(s), \dot{c}(s)]}\,,$$

where $c(s)$ is the point currently approximated on c, see (9.2.3). Let us use the notation $v = c(s)$, $\dot{v} = \dot{c}(s) = t(H'(v))$ and $\ddot{v} = \ddot{c}(s)$. Then we have:

$$(9.3.2) \qquad h := -\frac{f'(v)\dot{v}}{f'(v)\ddot{v} + f''(v)[\dot{v}, \dot{v}]} \, .$$

The task which remains with this formula is that a numerical approximation of $\ddot{v}$ is needed. To obtain such an approximation, let us differentiate the equation $H(c(s)) \equiv 0$. We obtain $H'(v)\dot{v} = 0$ and

$$(9.3.3) \qquad H''(v)[\dot{v}, \dot{v}] + H'(v)\ddot{v} = 0.$$

Now $||\dot{c}(s)||^2 \equiv 1$ yields

$$\dot{v}^* \ddot{v} = 0.$$

This shows that $\ddot{v}$ is orthogonal to $\ker H'(v)$, and we obtain from (9.3.3) and (3.2.3)(1–2):

$$(9.3.4) \qquad \ddot{v} = -H'(v)^+ H''(v)[\dot{v}, \dot{v}].$$

To approximate $H''(v)[\dot{v}, \dot{v}]$ we can use the centered difference formula

$$(9.3.5) \qquad \frac{H(v + \varepsilon\dot{v}) - 2H(v) + H(v - \varepsilon\dot{v})}{\varepsilon^2} = H''(v)[\dot{v}, \dot{v}] + O(\varepsilon^2).$$

Now (9.3.4–5) provides an approximation of $\ddot{v}$ in the Newton-type steplength adaptation (9.3.2). If necessary, an extrapolation method may be used to obtain higher precision approximations of $\ddot{v}$.

The following example illustrates this approach for the case of calculating a turning point with respect to the last co-ordinate i.e. $f(x, \lambda) = \lambda$ in the case of a nonlinear eigenvalue problem $H(x, \lambda) = 0$. Then $f(v) = e_{N+1}^* v = v[N + 1]$, and the special form of (9.3.2) becomes

$$(9.3.6) \qquad h := -\frac{\dot{v}[N + 1]}{\ddot{v}[N + 1]} \, .$$

A second method for calculating a local extremal point of $f(c(s))$ is to use a secant steplength adaptation applied to equation (9.3.1). Analogously to the discussion in section 9.2 we obtain the Newton-type steplength adaptation

$$(9.3.7) \qquad h := -\frac{f'(v)\dot{v}}{f'(v)\dot{v} - f'(u)\dot{u}} \, h.$$

The advantage of using (9.3.7) in (9.2.3) is that the need to calculate $\ddot{v}$ is avoided. Under analogous assumptions to those following (9.2.3) superlinear convergence of the sequence u generated by the algorithm to a local extremum of $f(c(s))$ can be proven.

Chapter 10. Large Scale Problems

10.1 Introduction

As has been pointed out occasionally in the previous chapters, one of the primary applications of continuation methods involves the numerical solution of nonlinear eigenvalue problems. Such problems are likely to have arisen from a discretization of an operator equation in a Banach space context, and involving an additional *"eigenvalue"* parameter. Some examples were touched upon in Chapter 8. As a result of the discretization and the wish to maintain a reasonably low truncation error, the corresponding finite dimensional problem $H(u) = 0$ where $H : \mathbf{R}^{N+1} \to \mathbf{R}^{N}$, may require that N be quite large. This then leads to the task of solving large scale continuation problems.

The area in which perhaps the greatest amount of experience concerning large scale continuation methods exists is structural mechanics, see e.g. Rheinboldt (1986) and the further references cited therein. Recently too, there has been work on combining continuation methods with multigrid methods for solving large scale continuation problems arising from discretization of elliptic problems via finite differences, see e.g. Chan & Keller (1982), Bank & Chan (1983), Chan (1984), Mittelmann (1984), and some further literature cited therein. Another area where large scale continuation problems have been treated concerns finite element discretizations of elliptic problems, which are then combined with a conjugate gradient solver in the continuation algorithm, see Glowinski & Keller & Reinhart (1985). If the classical elimination theory of algebra is applied to the problem of finding the real zero points of systems of polynomials, large systems with special structure and sparsity arise. This is touched upon in section 11.6.

It seems clear that an endless variety of combinations can be made of continuation algorithms and sparse solvers. In view of this, we will discuss how in general any sparse solver process can be incorporated into the general scheme of continuation methods which we have been describing, and then indicate more specifically how to incorporate a conjugate gradient method.

10.2 General Large Scale Solvers

When dealing with large systems of equations in the context of continuation methods, it is very advisable to determine the general structure and sparseness properties of $H'(u)$, and to exploit them. It may thereby become possible to incorporate into the continuation method a special linear solver process possessing much better efficiency than any general linear solver (such as a QL decomposition) applied to the same problem. Such special solvers might be generically described as follows: Given $H'(u)$ and some vector $e \in \mathbf{R}^{N+1}$ which is not yet specified, we have an "efficient" method for obtaining the solution $x \in \mathbf{R}^{N+1}$ for the linear system

(10.2.1)
$$H'(u)x = y,$$
$$e^*x = 0,$$

whenever $y \in \mathbf{R}^N$ is given. Among such methods might be linear conjugate gradient methods, direct factorization methods exploiting bandedness or sparseness, multigrid, SOR, etc.

The choice of the vector e in (10.2.1) may be regarded as representing a local parametrization, which usually is changed in the process of numerically traversing a solution curve. Of primary importance in the choice of e is its influence upon the condition of the coefficient matrix in (10.2.1) viz. we should require that

(10.2.2)
$$\text{cond}\left(\begin{matrix} H'(u) \\ e^* \end{matrix} \right) \approx \sqrt{\text{cond}\big(H'(u)H'(u)^*\big)}$$

are approximately of the same order. Intuitively speaking, the vector e should be as parallel as possible to $\ker H'(u)$.

Very typical is the following

(10.2.3) Example. Let $e = e_i \in \mathbf{R}^{N+1}$ be the ith standard unit vector, $1 \le i \le N+1$. Then the linear system (10.2.1) reduces to

$$H_i'(u)x_i = y,$$
$$x[i] = 0$$

where $x[i]$ denotes the ith co-ordinate of x, $H_i'(u)$ is obtained from $H'(u)$ by deleting the ith column, and finally x_i is obtained from x by deleting the ith co-ordinate.

The above choice for e has frequently been used by several authors. The choice of index i may be governed by the following motivation. To conform with the requirement (10.2.2), choose i as the maximal index with respect to

(10.2.4)
$$\max_i \{\, e_i^* t\big(H'(u)\big) \mid i = 1, 2, \ldots, N+1 \}.$$

Of course, at any currently calculated point u_n, the tangent vector $t(H'(u_n))$ might not be available yet, so one might take instead i as the maximal index with respect to

(10.2.5) $$\max_i \{ e_i^* t(H'(u_{n-1})) \mid i = 1, 2, \ldots, N + 1 \}.$$

The use of (10.2.5) has been advocated by Rheinboldt (1980).

    We now want to show that given some efficient method for solving the system (10.2.1), then also the Euler step $t(H'(u))$ and the Newton step $u - H'(u)^+ H(u)$ can be cheaply computed. For convenience, let us denote by

$$x = By$$

the solution operator of (10.2.1). We emphasize that the $(N+1) \times N$-matrix B is not explicitly given, but instead we have some efficient means of calculating the **result** $x = By$.

    Let us first investigate the determination of the tangent vector $t(H'(u))$. By the definition of B, (10.2.1) implies $H'(u)x = H'(u)By = y$ and $e^*x = e^*By = 0$ for any $y \in \mathbf{R}^N$ and hence

(10.2.6)
$$H'(u)B = \text{Id};$$
$$e^*B = 0^*.$$

If we set

(10.2.7) $$\tau := e - BH'(u)e$$

then $H'(u)\tau = H'(u)e - H'(u)e = 0$ by (10.2.6). Furthermore $e^*\tau = e^*e > 0$ implies $\tau \neq 0$. Hence

(10.2.8) $$t(H'(u)) = \pm \frac{\tau}{||\tau||}$$

gives us the tangent vector. We note that the cost of calculating τ and hence $t(H'(u))$ requires essentially one calculation of $H'(u)e$ (which is free in the case of $e = i^{\text{th}}$ standard vector) and one solving of (10.1.1) i.e. $x := BH'(u)e$.

    In most applications, the choice of sign in (10.2.8) will be clear from the context e.g. we take the tangent which has a small angle with a previously obtained tangent along the curve. Occasionally, it may be desirable to explicitly calculate the sign of

$$\det \begin{pmatrix} H'(u) \\ \tau^* \end{pmatrix}$$

in order to obtain accurate information on the orientation of the curve, e.g. one may wish to check whether a simple bifurcation point has been encountered along the curve. To determine this sign, we note that

$$\begin{pmatrix} H'(u) \\ \tau^* \end{pmatrix} (B, \tau) = \begin{pmatrix} \text{Id} & 0 \\ \tau^* B & \tau^* \tau \end{pmatrix},$$

$$\begin{pmatrix} H'(u) \\ e^* \end{pmatrix} (B, \tau) = \begin{pmatrix} \text{Id} & 0 \\ 0 & e^* \tau \end{pmatrix}$$

and $\tau^* \tau \geq e^* \tau = e^* e > 0$ imply

(10.2.9) $$\text{sign det} \begin{pmatrix} H'(u) \\ \tau^* \end{pmatrix} = \text{sign det} \begin{pmatrix} H'(u) \\ e^* \end{pmatrix}.$$

In many cases, the right hand side is immediately available from the "*efficient*" linear equation solver we have chosen for (10.2.1).

Let us now consider how we can perform an operation involving the Moore-Penrose inverse. Using the tangent vector $t(H'(u))$ which we already obtained in the previous step, from (10.2.6) and (3.2.5)(3) it is readily seen that

$$H'(u)^+ = \left[\text{Id} - t(H'(u)) \, t(H'(u))^* \right] B.$$

Hence, once $t(H'(u))$ has been obtained, the cost of calculating $w := H'(u)^+ y$ amounts to one solving of (10.2.1) i.e. $x = By$, and then calculating $w = x - \left[t(H'(u))^* x \right] t(H'(u))$ which is essentially the cost of one scalar product.

Let us summarize the above discussion in the form of a pseudo code by sketching an example of a continuation method where the predictor step is given by Euler's method and the corrector consists of a simplified Newton method (Chord Method). It is assumed that a "fast linear equation solver" in the above sense has been selected.

(10.2.10) Euler-Newton Method With Fast Linear Solver. *comment:*

   input

      begin

      $u \in \mathbf{R}^{N+1}$ such that $H(u) = 0$;   *initial point*

      $h > 0$;   *initial steplength*

      $e \in \mathbf{R}^{N+1}$;   *vector for local parametrization*

      end;

   repeat

      solve $\left\{ \begin{array}{l} H'(u)\tau = H'(u)e \\ e^*\tau = 0 \end{array} \right\}$ for τ;   *apply fast solver*

      $\tau := e - \tau; \quad t := \dfrac{\tau}{\|\tau\|}$;   *tangent vector*

      fix orientation of t;

      $v := u + ht$;   *Euler predictor*

      repeat   *corrector loop*

         solve $\left\{ \begin{array}{l} H'(u)z = H(v) \\ e^*z = 0 \end{array} \right\}$ for z;   *apply fast solver*

         $z := z - (t^*z)t$;   *orthogonal projection*

         $v := v - z$;   *corrector point*

      until $\|z\|$ is sufficiently small;

      $u := v$;   *new point along $H^{-1}(0)$*

      choose a new steplength $h > 0$;   *steplength adaptation*

      choose a new direction $e \in \mathbf{R}^{N+1}$;   *the angle between e and t —*

   until traversing is stopped.   *should be small*

We have seen that any special linear solver can be cheaply and conveniently incorporated into the general Euler-Newton continuation method. In the next section we shall discuss as a particular example, some of the details concerning the integration of conjugate gradient methods into the numerical continuation procedure.

10.3 Nonlinear Conjugate Gradient Methods as Correctors

As we have already noted in the preceding sections, there are a number of candidates for combining special methods for solving large linear systems having special structure with numerical continuation methods. Among the ones which immediately come to mind are: multigrid, successive over relaxation and conjugate gradient methods. We choose to illustrate this point with conjugate gradient methods, because we regard them as being particularly versatile and important in this context. The only reference known to us to date using a combination of continuation and nonlinear conjugate gradient methods is the paper of Glowinski & Keller & Reinhart (1985), concerning the solution of certain nonlinear elliptic boundary value problems. Our discussion here will be somewhat more general.

We begin with a description of the nonlinear conjugate gradient method of Polak & Ribière (1969). This choice is based upon reports [cf. Powell (1977) or Bertsekas (1984)] that in numerical practice it has generally yielded the best results. To outline the method, let us assume that the problem to be solved is

$$\textbf{(10.3.1)} \qquad \min_{u} \{\varphi(u) \mid u \in \mathbf{R}^N \}$$

where $\varphi : \mathbf{R}^N \to \mathbf{R}$ is a smooth nonlinear functional, usually having an isolated local minimal point $\bar{u}$ which we desire to approximate.

The simplest example is a uniformly convex quadratic functional φ defined by

$$\textbf{(10.3.2)} \qquad \varphi(u) = \frac{1}{2}u^*Au - u^*b$$

where $b \in \mathbf{R}^N$ and A is a positive definite $N \times N$-matrix. Recalling that we have adopted the convention $\nabla\varphi = (\varphi')^*$ for the gradient of φ, it is consistent to denote the Hessian of φ by $\nabla\varphi'$. In the above example (10.3.2), we therefore have $\nabla\varphi(u) = Au - b$ and $\nabla\varphi'(u) = A$. The solution $\bar{u}$ to (10.3.1) for this functional φ is then clearly the solution $\bar{u} = A^{-1}b$ of the linear equation $Au = b$. The following is an outline of the conjugate gradient method due to Polak & Ribière (1969).

(10.3.3) Conjugate Gradient Algorithm.   *comment:*
  input $u_0 \in \mathbf{R}^N$;   *initial point*
  $g_0 := \nabla\varphi(u_0); \quad d_0 := g_0$;   *calculate initial gradients*
  repeat for $n = 0, 1, \ldots$
    $\rho_n := \arg\min_{\rho>0} \varphi(u_n - \rho d_n)$;   *line search*
    $u_{n+1} := u_n - \rho_n d_n$;
    $g_{n+1} := \nabla\varphi(u_{n+1})$;

$$\gamma_n := \frac{\left[(g_{n+1} - g_n)^* g_{n+1}\right]}{\|g_n\|^2};$$

    $d_{n+1} := g_{n+1} + \gamma_n d_n$;   *new conjugate gradient*
  until convergence.

Since our aim here is merely to make an application of a conjugate gradient method — and especially, in the context of an underdetermined nonlinear system of equations — we will not give a detailed account concerning conjugate gradient methods for nonlinear problems. However, we shall recall some of their properties. For more details we suggest the books of Fletcher (1980), Gill & Murray & Wright (1981), Hestenes (1980), McCormick (1983) or Polak (1971) and the survey paper of Stoer (1983).

The main theoretical justification of the conjugate gradient algorithm lies in its properties when $\varphi(u)$ is a uniformly convex quadratic functional as in (10.3.2). In this special case the algorithm becomes the familiar conjugate gradient method due to Hestenes & Stiefel (1952) for solving the linear system $Au = b$. For more discussion of this case, we suggest the books of Golub & Van Loan (1983) or Stoer & Bulirsch (1980).

In the special case (10.3.2) we obtain the following result for the choice of the steplength ρ_n in (10.3.3):

(10.3.4) Lemma. *Let φ be a uniformly convex quadratic form. Then the following statements are equivalent:*

(1) ρ_n *is a solution of the problem* $\min\limits_{\rho\in\mathbf{R}} \varphi(u_n - \rho d_n)$;

(2) $\varphi'(u_n - \rho_n d_n)d_n = 0$;

(3) $\left[\varphi'(u_n) - \rho_n d_n^* \nabla\varphi'(u_n)\right]d_n = 0$;

(4) $\rho_n = \dfrac{\varphi'(u_n)d_n}{d_n^* \nabla\varphi'(u_n)d_n}$.

The proof of the above lemma is immediate from the fact that $\nabla\varphi'$ is a constant positive definite matrix and (3) represents the Taylor expansion for $\varphi'(u_n - \rho_n d_n)d_n$ about u_n. The following theorem is the main result on conjugate gradient methods:

(10.3.5) Theorem. *Let φ be a uniformly convex quadratic form. Let $\lambda_1 > \lambda_2 > \cdots > \lambda_k$ be an enumeration of all distinct eigenvalues of the Hessian $\nabla\varphi'$. Then*

(1) *the conjugate gradient algorithm (10.3.3) stops with the solution $\bar{u}$ in k steps if the computations are exact;*

(2) $\|u_n - \bar{u}\| \leq 2 \left(\dfrac{\sqrt{\kappa} - 1}{\sqrt{\kappa} + 1}\right)^n \|u_0 - \bar{u}\|;$

where $\kappa = \|\nabla\varphi'\| \, \|{\nabla\varphi'}^{-1}\| = \lambda_1/\lambda_k$ is the condition number of the Hessian of φ.

The conclusion to be drawn from the above theorem is that initially, the convergence of the conjugate gradient method may be slow because of (2), but by the kth step a very substantial improvement in the approximation of the solution has been obtained. This appears to hold even in the general case where the functional φ is no longer quadratic. To be more specific, let us assume that $\bar{u}$ is a local minimal solution point of the problem (10.3.1) at which the Hessian $\nabla\varphi'(\bar{u})$ is positive definite. There are several results concerning the convergence of the conjugate gradient method which essentially state that local superlinear convergence towards $\bar{u}$ holds, see e.g. Cohen (1972) or McCormick & Ritter (1974). However it appears that as of this date, the convergence results are somewhat unsatisfactory. One of the difficulties is that there are various possibilities for obtaining the factors γ_n in (10.3.3), the one presented here is due to Polak & Ribière (1969). Another difficulty is that in practice, we do not want to perform a very precise one-dimensional minimization in (10.3.3) in order to obtain an acceptable ρ_n since this is costly. Most of the convergence rate proofs require cyclic reloading i.e. setting $\gamma_n = 0$ after every N steps. The general idea of such proofs involves the approximation of $\varphi(u)$ via Taylor's formula by

$$\varphi(u) \approx \varphi(\bar{u}) + \varphi'(\bar{u})(u - \bar{u}) + (u - \bar{u})^* \nabla\varphi'(\bar{u})(u - \bar{u}),$$

and then to use the convergence result (10.3.5) for the quadratic case. Actually, even in the quadratic case, because of the presence of rounding errors, we cannot expect that (1) will occur. Instead, we should regard the conjugate gradient method even in this case as an iterative method which makes a substantial improvement after k steps.

The ideal situation in (2) would occur when the condition number $\kappa = 1$ i.e. when all the eigenvalues of $\nabla\varphi'$ are equal. Intuitively, the next best situation would occur when the eigenvalues have as few "cluster points" as possible. We use this observation to motivate the idea of preconditioning for the conjugate gradient method. For more details (in case of a quadratic functional) the reader may refer to Golub & Van Loan (1983). Let us make the change of co-ordinates

(10.3.6) $$\tilde{u} = Bu$$

where B is an as yet unspecified nonsingular $N \times N$-matrix. We set

$$\tilde{\varphi}(\tilde{u}) := \varphi(B^{-1}\tilde{u}) = \varphi(u)$$

and consider the conjugate gradient method (10.3.3) for the new functional $\tilde{\varphi}$. It follows that

$$\nabla\tilde{\varphi}(\tilde{u}) = (B^*)^{-1}\nabla\varphi(\tilde{u});$$
$$\nabla\tilde{\varphi}'(\tilde{u}) = (B^*)^{-1}\nabla\varphi'(\tilde{u})B^{-1}.$$

In view of theorem (10.3.5) we would ideally like to choose B such that

$$(B^*)^{-1}\nabla\varphi'(\tilde{u}_n)B^{-1} = \text{Id}$$

at the current approximation point $\tilde{u}_n$ i.e. we would like to have the Cholesky factorization

$$\nabla\varphi'(\tilde{u}_n) = B^*B.$$

Using the above formulae for this choice of B, it is easily seen that the new gradient is obtained by transforming the Newton direction via the above tranformation (10.3.6):

$$\nabla\tilde{\varphi}(\tilde{u}_n) = B\nabla\varphi'(\tilde{u}_n)^{-1}\nabla\varphi(\tilde{u}_n).$$

Hence, this extreme case of preconditioning gives us a Newton-like step. If we recall however, that the Hessian $\nabla\varphi'(\tilde{u}_n)$ was to have been large and sparse, it is perhaps inefficient to perform the complete Cholesky factorization of $\nabla\varphi'(\tilde{u}_n)$ merely for the purpose of attaining the optimal conditioning. It is reasonable to compromise somewhat on the improvement of the conditioning in order to maintain a low computational cost when performing a step of the conjugate gradient method. For this purpose the strategy of incomplete Cholesky factorization may be adopted viz. one only calculates the entries of B for which the corresponding entries of the Hessian $\nabla\varphi'(\tilde{u}_n)$ are nonzero, and one regards all other entries of B as being equal to zero, see e.g. Gill & Murray & Wright (1981) and the papers on preconditioning cited therein. The incomplete Cholesky factorization is not always numerically stable. Manteuffel (1979) identifies classes of positive definite matrices for which incomplete Cholesky factorization is stable.

To continue our motivational discussion, let us now suppose that instead of using the quadratic functional (10.3.2) to solve the linear equations $Au = b$, we use

(10.3.7) $$\varphi(u) = \frac{1}{2}||Au - b||^2$$

in the minimization problem (10.3.1) where A is an $N \times (N+1)$-matrix with maximal rank, $b \in \mathbf{R}^N$ is given and u varies now in $\mathbf{R}^{N+1}$. We obtain

$$\varphi(u) = \frac{1}{2}||Au - b||^2;$$
$$\nabla\varphi(u) = A^*(Au - b);$$
$$\nabla\varphi'(u) = A^*A.$$

The first major difference from the previous discussion which we may note is that the solutions of (10.3.1) are not isolated but consist of a 1-dimensional linear space, and the Hessian $\nabla\varphi' = A^*A$ is not positive definite. However, by examining the conjugate gradient method (10.3.3), we observe immediately that all changes $u_{N+1} - u_N$ lie in the N-dimensional linear space range$A^* = (\ker A)^{\perp}$. Consequently, we may regard the entire algorithm (10.3.3) as if it is taking place in a hyperplane parallel to rangeA^*. However, on rangeA^*, the matrix A^*A is in fact positive definite. Thus, the earlier discussion for quadratic functionals applies again and we obtain from theorem (10.3.5) the

(10.3.8) Corollary. Let φ be the convex quadratic form (10.3.7). Let $\lambda_1 > \lambda_2 > \cdots > \lambda_k$ be an enumeration of all nonzero distinct eigenvalues of the Hessian $\nabla\varphi'(u) = A^*A$. Then

(1) the conjugate gradient algorithm (10.3.3) stops after k steps at the solution $\bar{u}$ such that $\bar{u} - u_0 \in$ rangeA^*, if the computations are exact;

(2) $||u_n - \bar{u}|| \leq 2 \left(\dfrac{\sqrt{\kappa} - 1}{\sqrt{\kappa} + 1} \right)^n ||u_0 - \bar{u}||;$

where $\kappa = ||AA^*|| \, ||(AA^*)^{-1}|| = \lambda_1/\lambda_k$ is the condition number of the nonsingular $N \times N$-matrix AA^*.

We note in passing that A^*A and AA^* have the same nonzero eigenvalues with the same multiplicities. This is a standard fact in linear algebra. A useful preconditioning is now given in a way slightly different from (10.3.6). The same minimal solution points are also obtained by the following transformation of φ:

(10.3.9) $$\tilde{\varphi}(u) = \frac{1}{2}||L^{-1}(Au - b)||^2$$

where L is again an as yet unspecified nonsingular $N \times N$-matrix. Then we have

$$\nabla\tilde{\varphi}(u) = A^*(L^*)^{-1}L^{-1}(Au - b);$$
$$\nabla\tilde{\varphi}'(u) = A^*(LL^*)^{-1}A.$$

Again, in view of corollary (10.3.8) we would ideally wish to choose L so that

$$LL^* = AA^*$$

for then

$$\nabla\tilde{\varphi}'(u) = A^*(AA^*)^{-1}A = A^+A$$

is the orthogonal projection onto rangeA^*, cf. (3.2.5)(1), and hence has one eigenvalue equal to 0 and N eigenvalues equal to 1. Thus if a QL factorization of A were available, so that $AQ = (L, 0)$, then

$$LL^* = AQQ^*A^* = AA^*$$

and hence this L would serve ideally as a transformation in (10.3.8). Of course, it is not intended to actually obtain the QL factorization of A as in our discussions in the preceding chapters, for then we would be abandoning the advantages offered by sparse solvers. Instead, analogously to the previous discussion concerning incomplete Cholesky factorization, one could obtain an L via a corresponding *"incomplete QL factorization"* of A, and then use this L as the transformation in (10.3.8). We shall give some further discussion of this below.

Let us now finally turn to the case which actually concerns us viz.

(10.3.10) $$\min_u \frac{1}{2}\|H(u)\|^2$$

where $H : \mathbf{R}^{N+1} \to \mathbf{R}^N$ is a smooth map characterizing a solution curve $H(u) = 0$. That is, in the context of (10.3.1), we are now considering the functional

(10.3.11) $$\varphi(u) := \frac{1}{2}\|H(u)\|^2.$$

The solutions of (10.3.10) form a 1-manifold if 0 is a regular value of H. We have

$$\nabla\varphi(u) = H'(u)^*H(u);$$
$$\nabla\varphi'(u) = H'(u)^*H'(u) + O(\|H(u)\|).$$

Hence the gradient $\nabla\varphi(u) = H'(u)^*H(u)$ is orthogonal to the tangent vector $t(H'(u))$. This motivates the idea for implementing the conjugate gradient method (10.3.3) as a corrector into a continuation method. Analogously to the case when the minimization problem has isolated solutions at which the Hessian is positive definite, we may expect local superlinear convergence of the conjugate gradient method (10.3.3) also for the functional (10.3.10). The solution will be a point $\bar{u} \in H^{-1}(0)$ which is essentially nearest to the starting point u_0. Our conjecture that superlinear convergence should occur is at this point only a conjecture. To our knowledge, no proof of this exists. We propose the above conjugate gradient method as a reasonable corrector procedure nevertheless, provided once again, that an effective preconditioning

is incorporated. In the present context, as a generalization of the preconditioning (10.3.9), this is now easy to describe. We consider the following
transformation of φ:

$$(10.3.12) \qquad \tilde{\varphi}(u) = \frac{1}{2}\|L^{-1}H(u)\|^2$$

where again L is an as yet unspecified nonsingular $N \times N$-matrix. Then we
have

$$(10.3.13) \qquad \nabla\tilde{\varphi}(u) = H'(u)^*(LL^*)^{-1}H(u);$$
$$(10.3.14) \qquad \nabla\tilde{\varphi}'(u) = H'(u)^*(LL^*)^{-1}H'(u) + O(\|H(u)\|).$$

If we assume that our continuation method furnishes predictor points which
are already near $H^{-1}(0)$, we may neglect the $O(\|H(u)\|)$ term in (10.3.13).
More precisely,

$$(10.3.15) \qquad H(u) = 0 \quad \Rightarrow \quad \nabla\tilde{\varphi}'(u) = H'(u)^*(LL^*)^{-1}H'(u).$$

Thus, from corollary (10.3.8) and the discussion after (10.3.9), an ideal choice
would be an L such that

$$(10.3.16) \qquad LL^* = H'(u)H'(u)^*$$

is the Cholesky decomposition for some current point u near the solution curve
$\mathcal{C} \subset H^{-1}(0)$. We then have

$$\begin{aligned}
\nabla\tilde{\varphi}(u) &= H'(u)^*(LL^*)^{-1}H(u) \\
&= H'(u)^*(H'(u)H'(u)^*)^{-1}H(u) \\
&= H'(u)^+H(u),
\end{aligned}$$

cf. (3.2.2). Hence in this case, the gradient $\nabla\tilde{\varphi}(u) = H'(u)^+H(u)$ coincides
with the usual Newton direction which has been used as a corrector in previous
chapters.

Of course, if we really want to use the Cholesky decomposition (10.3.16)
which can be obtained via a QL factorization of $H'(u)$, cf. the discussion
after (10.3.9), then we would relinquish whatever advantage sparseness may
have offered. Thus, we want to determine L also with a small computational
expense and in such a way that linear equations $LL^*x = y$ are cheaply solved
for x.

Let us describe the idea of the QL analogue of the incomplete Cholesky
factorization by means of an example. Let us suppose that $H'(u)$ has a band
structure with the exception of the last column e.g.

$$(10.3.17) \qquad H'(u) = \begin{pmatrix}
x & x & 0 & 0 & 0 & 0 & x \\
x & x & x & 0 & 0 & 0 & x \\
0 & x & x & x & 0 & 0 & x \\
0 & 0 & x & x & x & 0 & x \\
0 & 0 & 0 & x & x & x & x \\
0 & 0 & 0 & 0 & x & x & x
\end{pmatrix}.$$

Then $H'(u)^*$ is transformed to upper triangular form via e. g. Givens rotations so that

$$H'(u)^* = \begin{pmatrix} x & x & 0 & 0 & 0 & 0 \\ x & x & x & 0 & 0 & 0 \\ 0 & x & x & x & 0 & 0 \\ 0 & 0 & x & x & x & 0 \\ 0 & 0 & 0 & x & x & x \\ 0 & 0 & 0 & 0 & x & x \\ x & x & x & x & x & x \end{pmatrix} \longrightarrow \begin{pmatrix} x & x & x & z & z & z \\ 0 & x & x & x & z & z \\ 0 & 0 & x & x & x & z \\ 0 & 0 & 0 & x & x & x \\ 0 & 0 & 0 & 0 & x & x \\ 0 & 0 & 0 & 0 & 0 & x \\ 0 & 0 & 0 & 0 & 0 & 0 \end{pmatrix} = \begin{pmatrix} L^* \\ 0^* \end{pmatrix}.$$

The incomplete QL factorization would yield an upper triangular matrix L^* except that the elements designated by z are not calculated, but instead are set equal to zero. In general, if $H'(u)$ has k nonzero bands, then L should have k nonzero bands too.

We now outline an algorithm which incorporates a conjugate gradient corrector.

(10.3.18) Secant – Conjugate Gradient Algorithm.   *comment:*

> **input**
>> **begin**
>> $u \in \mathbf{R}^{N+1}$;   *approximate point on $H^{-1}(0)$*
>> $t \in \mathbf{R}^{N+1}$;   *approximation to $t(H'(u))$*
>> $h > 0$;   *steplength*
>> **end**;
> **repeat**
>> $v := u + ht$;   *predictor step*
>> calculate $LL^* \approx H'(v)H'(v)^*$ *preconditioner*
>>> such that L is lower triangular;
>> $g_v := H'(v)^*(LL^*)^{-1}H(v)$; $d := g_v$;   *gradients*
>> **repeat**   *corrector loop*
>>> $\bar\rho :\approx \arg\min_{\rho \geq 0} \frac{1}{2}\|L^{-1}H(v - \rho d)\|^2$;
>>> $w := v - \bar\rho d$;   *corrector step*
>>> $g_w := H'(w)^*(LL^*)^{-1}H(w)$;   *new gradient*
>>> $\gamma := \dfrac{(g_w - g_v)^* g_w}{\|g_v\|^2}$;
>>> $d := g_w + \gamma d$;   *new conjugate gradient*
>>> $v := w$; $g_v := g_w$;
>> **until** convergence;

adapt steplength $h > 0$;   *see chapter 6*

$$t := \frac{(w - u)}{||w - u||};$$ *approximation to $t\big(H'(w)\big)$*

$u := w;$ *new point approximately on $H^{-1}(0)$*

until traversing is stopped.

We conclude this chapter with a number of remarks concerning details and modifications of the above algorithm. First of all, if the evaluation of $H'(w)$ is very costly, one may prefer to hold it fixed in the corrector loop. Furthermore, let us mention several possibilities for solving the line search problem

(10.3.19) $$\min_{\rho \geq 0} ||L^{-1} H(v - \rho d)||^2.$$

Recalling (10.3.4), let us approximate the functional $\tilde{\varphi}(v - \rho d)$, which is to be minimized, by its truncated Taylor expansion:

$$\tilde{\varphi}(v) - \rho \tilde{\varphi}'(v)d + \frac{1}{2}\rho^2 d^* \nabla \tilde{\varphi}(v)'d.$$

This is minimized exactly when

(10.3.20) $$\rho = \frac{\tilde{\varphi}'(v)d}{d^* \nabla \tilde{\varphi}(v)'d}$$

provided $\nabla \tilde{\varphi}(v)'$ is positive definite. In particular, for

$$\tilde{\varphi}(v - \rho d) = \frac{1}{2}||L^{-1} H(v - \rho d)||^2$$

we have

(10.3.21) $$\tilde{\varphi}'(v)d = H(v)^*(LL^*)^{-1} H'(v)d = g_v^* d$$

and

$$d^* \nabla \tilde{\varphi}(v)'d = d^* H'(v)^*(LL^*)^{-1} H'(v)d + O(\,||H(v)||\,||d||\,)$$
(10.3.22) $$\approx ||L^{-1} H'(v)d||^2.$$

Furthermore, since the evaluation of $H'(v)d$ may be costly for large scale problems, an inexpensive approximation of $H'(v)d$ may be made by using the central difference formula

$$H'(v)d = (2\varepsilon)^{-1}\left(H(v + \varepsilon\frac{d}{||d||}) - H(v - \varepsilon\frac{d}{||d||})\right)||d||$$
(10.3.23) $$+ O(\varepsilon^2 ||d||)$$

for an appropriate discretization step $\varepsilon > 0$. Now items (10.3.21) – (10.3.23) can be used in (10.3.20) to approximate the solution $\bar{\rho}$ of (10.3.19). Let us finally note that this approach will require three evaluations of H viz. at v and $v \pm \varepsilon d/||d||$. We discuss below how an appropriate value for ε may be determined.

A second possibility for solving (10.3.19) is to merely use a standard line search algorithm which does not require the evaluation of $\nabla \tilde{\varphi}$ such as a quadratic fit or golden section algorithm. For more details on such methods any standard book on nonlinear optimization may be consulted e.g. Mc-Cormick (1983). The disadvantage of this approach is that it may require many evaluations of $L^{-1}H$.

Usually, the predictor-corrector steps of a continuation method are performed in such a way that all generated points are close to the solution curve in $H^{-1}(0)$. Hence, the quadratic approximation considered in (10.3.20) will give good results in the situation which we are presently considering. Thus we recommend the first approach using (10.3.20) – (10.3.23) for solving (10.3.19). We therefore only carry out an error analysis concerning the approximation (10.3.23) for determining the choice of ε. Hence let us consider the general approximation

$$(10.3.24) \qquad \psi'(0) \approx (2\varepsilon)^{-1}\big(\psi(\varepsilon) - \psi(-\varepsilon)\big)$$

where $\psi : \mathbf{R} \to \mathbf{R}$ is some smooth function. We must take two kinds of errors into consideration viz. the cancellation error and the truncation error. If we denote by δ the relative machine error, then the cancellation error for the difference in (10.3.24) can essentially be estimated by

$$(10.3.25) \qquad \frac{2C\delta}{2\varepsilon}.$$

Here we assume that the function ψ can be calculated within a precision $C\delta$ where C is a typical magnitude for ψ. Of course, this assumption may not hold for all types of functions. The leading term of the truncation error of the above approximation (10.3.24) is easily obtained by Taylor's expansion:

$$(10.3.26) \qquad \frac{|\psi^{(3)}(0)|}{6}\varepsilon^2.$$

Hence the optimal choice of ε is obtained by minimizing the sum of the estimates (10.3.25) and (10.3.26). This yields

$$\varepsilon^3 = \frac{C}{3|\psi^{(3)}(0)|}\delta.$$

If we neglect the factors which are likely to be $O(1)$, we finally obtain

$$(10.3.27) \qquad \varepsilon \approx \sqrt[3]{\delta}.$$

Recently, some classes of generalized conjugate direction methods have been developed to solve $N \times N$ systems of linear equations $Mx = b$ where the matrix M is not necessarily assumed to be positive definite or even symmetric, see Dennis & Turner (1987) for a unifying approach of convergence results. The generalized minimal residual algorithm of Saad & Schultz (1986), see also the more stable version of Walker (1988), seems to be of particular interest in our context, since it only uses multiplications by M. If we take

$$M = \begin{pmatrix} H'(u) \\ t^* \end{pmatrix},$$

where t is some suitable approximation of $t(H'(u))$ e.g. given by a secant, then it is easy to program a multiplication Mx. In fact, the multiplication $H'(u)x$ may be approximated by a forward or central difference formula for the directional derivative as in (10.3.23), so that one multiplication by M essentially involves one scalar product and one or two evaluations of the map H. The authors are currently investigating, how this linear solver should best be installed into the iterative Newton-type corrector process of (10.2.10). When this has been determined, it may turn out to be superior to using the nonlinear conjugate gradient method as a corrector as described in (10.3.18). The HOMPACK continuation package, see Watson & Billups & Morgan (1987), incorporates separate routines for dense and sparse Jacobians. Irani & Ribbens & Walker & Watson & Kamat (1989) implement and compare several preconditioned gradient variations in the context of HOMPACK.

Chapter 11. Numerically Implementable Existence Proofs

11.1 Preliminary Remarks

Existence theorems are among the most frequently invoked theorems of mathematics since they assure that a solution to some equation exists. Some of the celebrated examples are the fundamental theorem of algebra, the fixed point theorems of Banach, Brouwer, Leray & Schauder, and Kakutani. With the exception of the Banach fixed point theorem, the classical statements of the above theorems merely assert the existence of a fixed point or a zero point of a map, but their traditional proofs in general do not offer any means of actually obtaining the fixed point or zero point. Many of the classical proofs of fixed point theorems can be given via the concept of the Brouwer degree. We will not need this concept in our subsequent discussions. However, for readers wishing to read up on degree theory we can suggest the books of Amann (1974), Berger (1977), Cronin (1964), Deimling (1974) or Schwartz (1969).

Although many fixed point theorems are formulated in the context of maps on Banach spaces, our discussions will be primarily confined to finite dimensional spaces since we are mainly concerned with numerically calculating solution points. The following statement and proof of the well-known Leray & Schauder fixed point theorem gives a sample of the degree-theoretical proof.

(11.1.1) Theorem. *Let*

(1) $f : \mathbf{R}^N \to \mathbf{R}^N$ *be a continuous map; (for simplicity we assume f to be defined on all of $\mathbf{R}^N$)*

(2) $\Omega \subset \mathbf{R}^N$ *be an open, bounded, non-empty set;*

(3) $p \in \Omega$;

(4) $\lambda(f(x) - p) \neq (x - p)$ *for all $\lambda \in [0, 1)$, $x \in \partial\Omega$.*

*Then there exists a **fixed point** $x_0 \in \overline{\Omega}$ of f such that $f(x_0) = x_0$.*

Proof. We can assume that $f(x) \neq x$ for all $x \in \partial\Omega$ for otherwise we would already have a fixed point. Thus (4) also holds for $\lambda = 1$. Let us define a homotopy map

$$H : \mathbf{R}^N \times [0, 1] \to \mathbf{R}^N$$

by $H(x, \lambda) := (x - p) - \lambda(f(x) - p)$. Then $\deg(H(\cdot, \lambda), \Omega, 0)$ is well-defined and it remains constant with respect to λ. Since $\deg(H(\cdot, 0), \Omega, 0) = 1$, then $\deg(H(\cdot, 1), \Omega, 0) = 1$. Hence there is at least one $x_0 \in \Omega$ such that $H(x_0, 1) = 0$, i.e. $f(x_0) = x_0$. $\qquad\square$

The above proof is not "constructive" as such, since it doesn't tell us how to obtain x_0. However, it can be made constructive if we are able to numerically trace the implicitly defined curve $H^{-1}(0)$ from $\lambda = 0$ to $\lambda = 1$. Since we have described in the previous chapters numerical techniques for tracing such curves when f is smooth, the above proof can be made to be "constructive" or "implementable".

The idea that one could replace degree arguments by considering the inverse images of points of mappings was the theme of the book by Milnor (1965). The thought of actually numerically tracing inverse images of points came somewhat later although Haselgrove (1961) had already outlined the main ideas for doing it. The first numerically implementable proofs of the Brouwer fixed point theorem given by Scarf (1967) and Kuhn (1969) were not restricted to smooth maps f and were based more upon the Sperner's lemma approach to proving the Brouwer fixed point theorem, such as in the paper of Knaster & Kuratowski & Mazurkiewicz (1929). Eaves (1972) gave a "PL algorithm" which can be regarded as an implementation of the homotopy approach. The "restart method" of Merrill (1972) also may be considered to represent a homotopy approach.

A proof of the Brouwer fixed point theorem for smooth f involving the nonretraction proof of Hirsch (1963) and the tracing of the inverse image of a point for a mapping was given by Kellogg & Li & Yorke (1976). A proof using the numerical tracing of a homotopy curve was given by Chow & Mallet-Paret & Yorke (1978). In the latter approaches a general version of the Sard's theorem (see e.g. the books of Abraham & Robbins (1967) or Hirsch (1976)) played a crucial role for ruling out the presence of singular points on the homotopy paths. For the case that f is a smooth map the efficient predictor-corrector methods outlined in the earlier chapters are immediately available for numerically tracing a homotopy path.

For general discussions concerning the correspondence between degree arguments and numerical continuation algorithms we suggest the following articles: Alexander & Yorke (1978); Garcia & Zangwill (1979), (1981); Peitgen (1982). Further references are also cited in these articles. Finite-dimensional discretizations of continuation methods in Banach spaces have been studied by Brezzi & Rapaz & Raviart (1980), (1981), (1982).

Since the appearance of the constructive proofs of the Brouwer fixed point theorem many other constructive existence proofs have been described. To give a partial enumeration of just a few such examples we mention:

- antipodal theorems such as the Borsuk & Ulam theorem have been constructively proven in Alexander & Yorke (1978), Allgower & Georg (1980), Barany (1980), Meyerson & Wright (1979) and Todd & Wright (1980);
- fixed point theorems for multivalued maps such as the Kakutani theorem have been constructively proven in Eaves (1971), Eaves & Saigal (1972), Merrill (1972), Todd (1976) and Allgower & Georg (1980);
- economic equilibria existence has been constructively proven by Scarf & Hansen (1973);
- constructive existence proofs for nonlinear complementarity problems have been given by Eaves (1971), Kojima (1974), Kojima & Saigal (1979);
- implementable proofs of the fundamental theorem of algebra have been given by Chow & Mallet-Paret & Yorke (1978), Drexler (1977), Kojima & Nishino & Arima (1979), Kuhn (1974);
- continuation methods for finding all solutions to a system of complex polynomial equations have been published by Chow & Mallet-Paret & Yorke (1979), Garcia & Li (1980), Garcia & Zangwill (1979), Kojima & Mizuno (1983), Morgan (1983), (1987), Rosenberg (1983), Wright (1985);
- continuation methods for linking several solutions of a system of equations have been studied by Peitgen & Prüfer (1979), Jürgens & Peitgen & Saupe (1980), Allgower & Georg (1980), (1983) and Diener (1986), (1987).

11.2 An Example of an Implementable Existence Theorem

The specific example we give in this section is similar to a discussion given by Chow & Mallet-Paret & Yorke (1978). It deals with the case of smooth maps. We begin with the following

(11.2.1) Assumptions.

(1) $f : \mathbf{R}^N \to \mathbf{R}^N$ is a C^∞-map ;
(2) $\Omega \subset \mathbf{R}^N$ is bounded, open and non-empty;
(3) $p \in \Omega$;
(4) 0 is a regular value of $\mathrm{Id} - f$.

We have made the last assumption (4) in order to simplify the subsequent discussions. It is not difficult to discuss the general case along the same lines, but this would involve more technical details, see the remarks at the end of this section.

We will make repeated use of the following general version of Sard's theorem for maps with additional parameters, see e.g. Abraham & Robbin (1967) or Hirsch (1976). Yomdin (1990) has given a version of Sard's theorem which is adapted for numerical purposes.

(11.2.2) Sard's Theorem. *Let A, B, C be C^∞-manifolds of finite dimensions with $\dim A \geq \dim C$, and let $F : A \times B \to C$ be a C^∞-map. Assume that $c \in C$ is a regular value of F i.e. for $F(a,b) = c$ we have that the total derivative $F'(a,b) : T_a A \times T_b B \to T_c C$ has maximal rank. Here $T_a A$ denotes the tangent space of A at a etc. Then for almost all $b \in B$ (in the sense of some Lebesgue measure on B) the restricted map $F(\cdot, b) : A \to C$ has c as a regular value.*

Note that a value $c \in C$ which is not in the range of F is by definition a regular value. The following standard version of Sard's theorem can be obtained as a special case:

(11.2.3) Sard's Theorem. *Let A, C be C^∞-manifolds of finite dimensions such that $\dim A \geq \dim C$, and let $F : A \to C$ be a C^∞-map. Then almost all $c \in C$ are regular values of F.*

Let us define a homotopy map $H : \mathbf{R}^N \times \mathbf{R} \times \Omega \to \mathbf{R}^N$ by

(11.2.4) $$H(x, \lambda, p) := x - p - \lambda(f(x) - p).$$

For the "trivial level" $\lambda = 0$ we obtain the "trivial map" $H(x, 0, p) = x - p$ which has the unique zero point p, our "starting point". On the "target level" $\lambda = 1$ we obtain the "target map" $H(x, 1, p) = x - f(x)$ whose zero points are our points of interest i.e. the fixed points of f. The Jacobian of (11.2.4) with respect to all variables (x, λ, p) is given by

(11.2.5) $$H'(x, \lambda, p) = (\mathrm{Id} - \lambda f'(x), p - f(x), (\lambda - 1)\mathrm{Id}).$$

From (11.2.1) and (11.2.5) it follows that 0 is a regular value of H. In fact, the first N columns of (11.2.5) are linearly independent for $H(x, \lambda, p) = 0$ and $\lambda = 1$ due to (11.2.1)(4), and clearly the last N columns of (11.2.5) are linearly independent for $\lambda \neq 1$. Consequently, by Sard's theorem (11.2.2) we can conclude the following

(11.2.6) Proposition. *For almost all $p \in \Omega$ (in the sense of N-dimensional Lebesgue measure) 0 is a regular value of the restricted map $H(\cdot, \cdot, p)$.*

In view of this statement it is now reasonable to make the following

(11.2.7) Assumption. *Let us assume that the starting point $p \in \Omega$ is chosen in accordance with (11.2.6) i.e. in such a way that 0 is a regular value of the map $H(\cdot, \cdot, p)$.*

As a consequence of the preceding remarks let us give the following

(11.2.8) Summary. *For our choice of the starting point $p \in \Omega$, the connected component $\mathcal{C}_p$ of $H(\cdot, \cdot, p)^{-1}(0)$ which contains the point $(p, 0)$ represents a smooth curve $s \mapsto (x(s), \lambda(s))$ which can be regarded as being parametrized with respect to arclength s. That is*

(1) $C_p = \{(x(s), \lambda(s)) \mid s \in \mathbf{R}\}$;

(2) $(x(0), \lambda(0)) = (p, 0)$;

(3) furthermore, all points of C_p are regular points of the map $H(\cdot, \cdot, p)$.

We may now of course numerically trace the curve C_p by one of the predictor-corrector continuation methods described in some of our earlier chapters. A few items remain to be verified to see how a predictor-corrector method can be implemented. As far as the initial tangent vector is concerned, we can differentiate the equation $H(x(s), \lambda(s), p) = 0$ and evaluate the result at $s = 0$. Using (11.2.4) and (11.2.8)(2) we obtain

$$\dot{x}(0) + \dot{\lambda}(0)(p - f(p)) = 0.$$

Hence the initial tangent vector is given by

$$\begin{pmatrix} \dot{x}(0) \\ \dot{\lambda}(0) \end{pmatrix} = \alpha \begin{pmatrix} f(p) - p \\ 1 \end{pmatrix}$$

for some constant α. Since also $\|\dot{x}(0)\|^2 + |\dot{\lambda}(0)|^2 = 1$ must hold, we have

(11.2.9) $$\begin{pmatrix} \dot{x}(0) \\ \dot{\lambda}(0) \end{pmatrix} = \pm(1 + \|f(p) - p\|^2)^{-\frac{1}{2}} \begin{pmatrix} f(p) - p \\ 1 \end{pmatrix}.$$

Finally, since we want to numerically trace the curve C_p in the positive λ-direction, we must choose the positive sign in (11.2.9).

Now we see that at $s = 0$ i.e. at the starting point $(x(0), \lambda(0))$ the curve C_p is **transverse** to $\mathbf{R}^N \times \{0\}$ i.e. not tangential to the hyperplane $\mathbf{R}^N \times \{0\}$. From (11.2.4) it follows that $H(\cdot, \cdot, p)^{-1}(0)$ intersects $\mathbf{R}^N \times \{0\}$ only at $(p, 0)$, and since C_p is transverse to $\mathbf{R}^N \times \{0\}$, we have

(11.2.10) $$C_p \cong \mathbf{R}$$

i.e. C_p is homeomorphic to $\mathbf{R}$, because $H(\cdot, \cdot, p)^{-1}(0)$ consists only of components which are homeomorphic either to $\mathbf{R}$ or to the unit circle $S^1 \subset \mathbf{R}^2$. Using (11.2.8) it is now easy to see that

(11.2.11) $$C_p \cap \left[(\overline{\Omega} \times \{1\}) \cup (\partial\Omega \times (0, 1)) \right] \neq \emptyset$$

i.e. the curve C_p hits the boundary of $\Omega \times (0, 1)$ at some other point which is not on the trivial level $\mathbf{R}^N \times \{0\}$. The validity of (11.2.11) could be proven from elementary facts about ordinary differential equations concerning the maximal interval of existence of a solution ("a curve cannot just end"). However, we will argue as follows. If (11.2.11) did not hold, then

$$C_p = \left\{ (x(s), \lambda(s)) \mid s \in \mathbf{R} \right\}$$

would have an accumulation point $(x_0, \lambda_0) \in \overline{\Omega} \times [0,1]$ for $s \to \infty$, i.e. there would exists a sequence $\{s_n\}_{n \in \mathbb{N}} \subset \mathbb{R}$ with $\lim_{n \to \infty} s_n = \infty$ and $\lim_{n \to \infty} \big(x(s_n), \lambda(s_n)\big) = (x_0, \lambda_0)$. Then $H(x_0, \lambda_0, p) = 0$, because H is continuous. Since 0 is a regular value of $H(\cdot, \cdot, p)$, the 1-manifold $H(\cdot, \cdot, p)^{-1}(0)$ can be parametrized (say) with respect to arclength in a neighborhood of (x_0, λ_0). Since $\mathcal{C}_p$ is a connected component of $H(\cdot, \cdot, p)^{-1}(0)$, it must contain this parametrized curve near (x_0, λ_0). On the other hand, (x_0, λ_0) is an accumulation point for $s \to \infty$, and we conclude that $\mathcal{C}_p$ contains a closed loop, a contradiction to (11.2.10).

To assert the existence of a fixed point of f it is necessary to make an additional hypothesis concerning the behavior of f on the boundary $\partial\Omega$. We shall call this an LS boundary condition since it is similar to the condition of Leray & Schauder (1934) for infinite dimensional Banach spaces:

(11.2.12) $\lambda\big(f(x) - p\big) \neq (x - p)$ for $x \in \partial\Omega$ and $0 \leq \lambda < 1$.

This LS condition implies

$$\mathcal{C}_p \cap \big(\partial\Omega \times (0,1)\big) = \emptyset,$$

and from (11.2.11) we conclude

$$\mathcal{C}_p \cap \big(\overline{\Omega} \times \{1\}\big) \neq \emptyset.$$

Thus f has at least one fixed point. As a consequence of the preceding discussion we can conclude that a fixed point of f can be numerically obtained by tracing $\mathcal{C}_p$ until the target level $\lambda = 1$ is reached.

If all of the above hypotheses are fulfilled except the assumption that 0 is a regular value of $\mathrm{Id} - f$, cf. (11.2.1)(4), then the above argument does not work. The statement (11.2.6) can be modified in such a way that 0 is a regular value of the restricted map

$$(x, \lambda) \in \mathbb{R}^N \times (\mathbb{R} \setminus \{1\}) \longmapsto H(x, \lambda, p)$$

for almost all choices of the starting value $p \in \Omega$. Hence $\lambda = 1$ may possibly be an exceptional level. In this case, let

$$\mathcal{C}_p = \Big\{ \big(x(s), \lambda(s)\big) \mid 0 \leq s < s_0 \Big\}, \quad s_0 \in (0, \infty], \quad s = \text{arclength},$$

denote the connected component of

$$\{(x, \lambda) \mid H(x, \lambda, p) = 0, \ x \in \mathbb{R}^N, \ 0 \leq \lambda < 1\}$$

which contains $(p, 0)$. Only $\overline{\mathcal{C}_p} \cap (\overline{\Omega} \times \{1\}) \neq \emptyset$ can be shown, i.e. the curve $\mathcal{C}_p$ still reaches the target level $\lambda = 1$ at some point $(x_0, 1)$ such that $f(x_0) = x_0$, but only in the limit $s \to s_0$. More precisely, $(x(s), \lambda(s))$ has at least one accumulation point (x_0, λ_0) for $s \to s_0$, and it is not hard to see that we must have $\lambda_0 = 1$ and $f(x_0) = x_0 \in \overline{\Omega}$ for all such accumulation points. Let us indicate some possibilities.

(1) The curve C_p may end at the target level $\lambda = 1$ at some point

$$\big(x(s_0), \lambda(s_0)\big) = (x_0, 1)$$

such that $f(x_0) = x_0$. It may happen that x_0 is not a regular point of $\mathrm{Id} - f$ but however $(x_0, 1)$ is a regular point of the homotopy map H. In that case $\dot{\lambda}(s_0) = 0$ i.e. the tangent vector is in $\mathbf{R}^N \times \{0\}$.

(2) The curve $\big(x(s), \lambda(s)\big)$ may oscillate toward $\overline{\Omega} \times \{1\}$ as $s \to s_0 = \infty$. The set of accumulation points can be a singleton or a more complicated set of fixed points of f. The reader may construct some examples which demonstrate this.

In the next section we relate the above general discussion to the classical Brouwer fixed point theorem.

11.3 Several Implementations for Obtaining Brouwer Fixed Points

The implementation by Kellogg & Li & Yorke (1976) for obtaining a Brouwer fixed point was based upon Hirsch's theoretical proof using non-retractibility (1963). In this section we will describe the ideas of the Kellogg & Li & Yorke implementation and show that their approach is much nearer to the homotopy approach outlined in section 11.2 than it might appear at first glance. The main differences actually lie in certain technical considerations such as the numerical stability of the methods.

Let us begin this discussion with a simple hypothesis for the Brouwer fixed point theorem for smooth maps:

(11.3.1) Assumption.

(1) $f : \mathbf{R}^N \to \mathbf{R}^N$ is a C^∞-map;

(2) $\Omega = \{x \in \mathbf{R}^N \mid \|x\| < 1\}$ is the standard unit ball;

(3) $\overline{f(\mathbf{R}^N)} \subset \Omega$;

(4) 0 is a regular value of $\mathrm{Id} - f$.

Again, the somewhat stronger assumption (4) is made in order to simplify the subsequent discussion. It is clear that (11.3.1) implies the LS condition (11.2.12) for any starting point $p \in \Omega$ and hence it follows that f has a fixed point which can be numerically traced as described in section 11.2. Kellogg & Li & Yorke however originally used a different approach. They defined a map $\hat{H}$ in the following way:

(11.3.2) Definition. Let $C = \{x \in \mathbf{R}^N \mid f(x) = x\}$ be the fixed point set of f. Then $\hat{H} : \overline{\Omega} \setminus C \longrightarrow \partial\Omega$ is defined by $\hat{H}(x) = f(x) + \mu(x)(x - f(x))$ where $\mu(x) > 0$ is so chosen that $\|\hat{H}(x)\| = 1$.

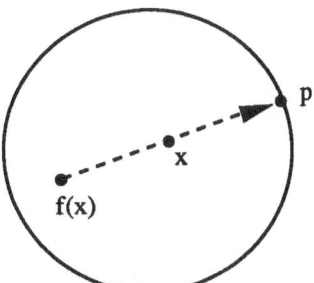

Figure 11.3.a The retraction
map $\hat{H}(x) = p$

Figure 11.3.a illustrates how $p = \hat{H}(x)$ is obtained. The idea now is to trace the component $\hat{C}_p \subset \hat{H}^{-1}(p)$ which contains the starting point $p \in \partial\Omega$ inward into Ω until a fixed point of f is reached. A straightforward application of Sard's theorem (11.2.3) shows that almost all points $p \in \partial\Omega$ are regular values of $\hat{H}$. It is therefore natural to make the following

(11.3.3) Assumption. *Let the starting point $p \in \partial\Omega$ be a regular value of* $\hat{H}$.

Let us investigate the tracing of the curve $\hat{C}_p$ more precisely. Clearly, for any $x \in \partial\Omega$ we have $\hat{H}(x) = x$ and $\mu(x) = 1$. Thus

(i) either $\hat{C}_p$ returns back to the boundary $\partial\Omega$ at a point $x = \hat{H}(x) = p$,

(ii) or $\hat{C}_p$ tends to the fixed point set C where the map $\hat{H}$ is not defined.

Let us show that case (i) is impossible. We assume, as we always do if convenient, that the curve $\hat{C}_p$ is parametrized according to arclength s. Using the general assumptions (11.3.1) on f, it is readily seen that case (i) implies that the curve $\hat{C}^p = \{x(s) \mid s \in \mathbf{R}\}$ is a closed loop which touches $\partial\Omega$ from the inside at the point p. We obtain the tangent by differentiating the equation $f(x(s)) + \mu(s)\big(x(s) - f(x(s))\big) = p$, cf. (11.3.2), with respect to s. For $p = x(s_0)$ and $\dot{p} := \dot{x}(s_0)$ we thus obtain $\mu(s_0) = 1$ and $\dot{\mu}(s_0)(p - f(p)) + \dot{p} = 0$. Since $\|\dot{p}\| = 1$, $\dot{p}$ is a nonzero multiple of $p - f(p)$. But $f(p) \in \Omega$ implies that $p - f(p)$ cannot be tangent to the boundary $\partial\Omega$ at p, and we have a contradiction. Hence, case (i) is impossible.

Thus, following the curve

$$\hat{C}_p = \{x(s) \mid 0 \leq s < s_0\}, \quad s_0 \in (0, \infty]$$

inward into Ω, any accumulation point of $x(s)$ as $s \to s_0$ is a fixed point of f. Also, since an accumulation point x_0 is an isolated fixed point of f by our assumption (11.3.1)(4) it is possible to show that $\lim_{s \to s_0} x(s) = x_0$. In

this sense, the Brouwer fixed point theorem is implementable by a numerical predictor-corrector continuation method if the above mild assumptions
(11.3.1) and (11.3.3) are verified.

From the numerical standpoint this approach has a serious disadvantage:
as can be seen from the definition of $\hat{H}$ (11.3.2) , see also Figure 11.3.a, for
x near the fixed point set C the evaluation of $\hat{H}(x)$ becomes numerically
unstable. Let show that this effect can be overcome. More exactly, we will
show that the same connected component $\hat{C}_p$ can be defined via a different
map $\tilde{H}$, which can be evaluated in a numerically stable way even for points
x which are near the fixed point set C.

Before doing this, let us examine the relationship of the Kellogg & Li &
Yorke method with the standard homotopy method outlined in section 11.2.
For the Kellogg & Li & Yorke method, following $\hat{C}_p \subset \hat{H}^{-1}(p)$ involves dealing
with the equation

$$\textbf{(11.3.4)} \qquad\qquad f(x) + \mu(x - f(x)) = p,$$

where $x \in \Omega$ and $\mu \geq 0$ are so taken that (11.3.4) is satisfied for the chosen
$p \in \partial\Omega$. By taking $\mu = (1 - \lambda)^{-1}$ we obtain a corresponding homotopy
equation $H(x, \lambda, p) = 0$ where $H : \overline{\Omega} \times \mathbf{R} \times \partial\Omega \to \mathbf{R}^N$ is defined by

$$\textbf{(11.3.5)} \qquad\qquad H(x, \lambda, p) = (x - p) - \lambda(f(x) - p).$$

We can now see that (11.3.5) corresponds exactly to the homotopy (11.2.4).
The only difference is that we take $p \in \partial\Omega$ instead of $p \in \Omega$. Let us assume
that 0 is a regular value of $H(\cdot, \cdot, p)$. In section 11.2 we saw that a smooth
curve of finite arclength in $H^{-1}(0)$ connects the point $(p, 0)$ to a point $(x_0, 1)$
such that x_0 is a fixed point of f if we make the additional hypothesis that the
fixed points of f are regular points of $\mathrm{Id} - f$. The reader may verify that the
arguments in section 11.2 can be modified in such a way that the case $p \in \partial\Omega$
is also covered, provided the LS condition (11.2.12) is modified to read

$$\textbf{(11.3.6)} \qquad
\begin{aligned}
&\textbf{(1)} \quad x - p \neq \lambda(f(x) - p) \quad \text{if} \quad \lambda \in (0, 1),\ x \in \partial\Omega,\ x \neq p; \\
&\textbf{(2)} \quad 0 \neq f(p) - p \quad \text{points into}\quad \Omega.
\end{aligned}$$

In order for the statement (11.3.6)(2) to have any meaning it is necessary
to assume that $(\Omega, \partial\Omega)$ forms a smooth manifold with boundary, but for
$\Omega = \{x \in \mathbf{R}^N \mid \|x\| < 1\}$ this is certainly satisfied (remember that we take
$\|\cdot\| = \|\cdot\|_2$ if not otherwise specified). Figure 11.3.b illustrates this modified
situation. Let us note that also the LS conditions (11.3.6) are implied by
our general hypothesis (11.3.1). Condition (11.3.6)(1) can be verified in the
following way: if $x - p = \lambda(f(x) - p)$ for some $\lambda \in (0, 1)$ and some $x \in \partial\Omega$,
then $\|f(x)\| > 1$.

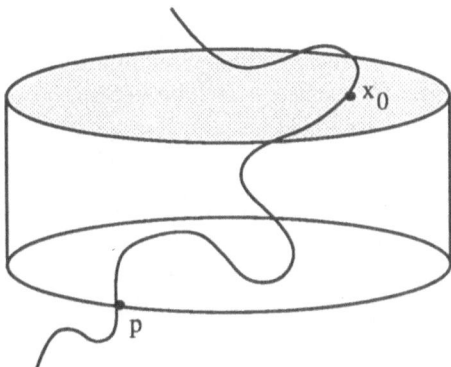

Figure 11.3.b Homotopy path from the boundary

Our argument in section 11.2 showed that due to the LS condition given in (11.2.12), the component $C_p \subset H(\cdot, \cdot, p)^{-1}(0)$ traced from $(p, 0)$ in the direction of increasing λ could exit the cylinder $\overline{\Omega} \times \mathbf{R}$ only after reaching the level $\lambda = 1$. Although we are now starting from the boundary $\partial\Omega \times \{0\}$, the same reasoning applies here. By (11.3.6)(2) we trace C_p from $(p, 0)$ into $\overline{\Omega} \times \mathbf{R}$, which is the direction of increasing λ.

Let us now summarize the relationship between the Kellogg & Li & Yorke method and the standard homotopy method of section 11.2 using an initial point p in the boundary $\partial\Omega$: The x-component of C_p (i.e. the projection of C_p onto any λ-hyperplane) coincides with $\hat{C}_p$. The particular parametrizations of these two curves may of course be different. We note in passing that it is technically more complicated to obtain a regularity result for (11.3.5) in the sense of (11.2.6) when considering starting points p which vary only over the boundary $\partial\Omega$.

One of the obvious differences between the standard homotopy method and the Kellogg & Li & Yorke method is that the latter works with one less variable. However, we note that the λ-variable in (11.3.5) can also be eliminated: It is clear that

$$p^*\big(f(x) - p\big) \neq 0$$

under our assumptions (11.3.1). Hence, if the homotopy equation

$$H(\lambda, x, p) = x - p - \lambda\big(f(x) - p\big) = 0$$

holds for $x \in \mathbf{R}^N$ and $\lambda \in \mathbf{R}$, then

(11.3.7)
$$\lambda(x, p) = \frac{p^*(x - p)}{p^*\big(f(x) - p\big)}.$$

The elimination step (11.3.7) would be particularly simple if p were chosen to be a unit co-ordinate vector. The preceding leads to a third implementation: Let the map $\tilde{H} : \mathbf{R}^N \times \partial\Omega \to \mathbf{R}^N$ be defined by

(11.3.8) $\tilde{H}(x,p) := x - p - \lambda(x,p)\big(f(x) - p\big).$

Actually, for fixed $p \in \partial\Omega$, the range of $\tilde{H}(\cdot,p)$ is contained in the tangent space $\{p\}^{\perp}$ of $\partial\Omega$ at p. Here $\{p\}^{\perp}$ denotes the linear space $\{x \in \mathbf{R}^N \mid p^*x = 0\}$ which is orthogonal to p. The Jacobian of $\tilde{H}$ i.e. the partial derivative of $\tilde{H}(x,p)$ with respect to x is obtained by a routine calculation:

(11.3.9) $\tilde{H}_x(x,p) = \left(\mathrm{Id} - \dfrac{(f(x) - p)p^*}{p^*(f(x) - p)}\right)\big(\mathrm{Id} - \lambda(x,p)f'(x)\big).$

Since

$$\inf_{x \in \mathbf{R}^N} \big| p^*(f(x) - p)\big| > 0$$

holds by our assumption (11.3.1)(3), we see from the above equations that the evaluations of $\tilde{H}(x,p)$ and $\tilde{H}_x(x,p)$, which would be required for a predictor-corrector tracing of $\tilde{H}(x,p) = 0$, can be performed in a numerically stable way. The following theorem, cf. Allgower & Georg (1988), assures that it is in general appropriate to assume that $\tilde{H}(\cdot,p)$ has 0 as a regular value:

(11.3.10) Theorem. *For almost all $p \in \partial\Omega$, the homotopy map $\tilde{H}(\cdot,p) :$ $\mathbf{R}^N \to \{p\}^{\perp}$ has 0 as a regular value.*

Now let us contrast the three methods we have just discussed.

(1) The standard homotopy method, cf. (11.2.4): Trace the component $\mathcal{C}_p \subset$ $H^{-1}(0)$ where $p \in \Omega$, $(p,0) \in \mathcal{C}_p$ and the homotopy map $H : \mathbf{R}^N \times \mathbf{R} \to$ $\mathbf{R}^N$ is given by $H(x,\lambda) = x - p - \lambda(f(x) - p)$.

(2) The Kellogg & Li & Yorke method, cf. (11.3.4): Trace the component $\hat{\mathcal{C}}_p \subset \hat{H}^{-1}(p)$. Here $p \in \partial\Omega$, $p \in \hat{\mathcal{C}}_p$ and the homotopy map $\hat{H} : \mathbf{R}^N \backslash C \to$ $\partial\Omega$ is given by $\hat{H}(x,\lambda) = f(x) + \mu(x)(x - f(x))$ where $\mu(x) > 0$ is chosen such that $\hat{H}(x,\lambda) \in \partial\Omega$ i.e. $\mu(x)$ is the positive solution of the quadratic equation

$$\mu^2\,\|x - f(x)\|^2 + 2\mu f(x)^*(x - f(x)) + \|f(x)\|^2 = 1.$$

(3) A numerically stable version of the Kellogg & Li & Yorke method, cf. (11.3.7): Trace the component $\tilde{\mathcal{C}}_p \subset \tilde{H}^{-1}(0)$. Here $p \in \partial\Omega$, $p \in \tilde{\mathcal{C}}_p$ and the homotopy map $\tilde{H} : \mathbf{R}^N \to \{p\}^{\perp}$ is given by $\tilde{H}(x,\lambda) = x - p - \lambda(x)(f(x) - p)$ where $\lambda(x)$ is chosen such that $\tilde{H}(x,\lambda) \in \{p\}^{\perp}$ i.e.

$$\lambda(x) = \frac{p^*(x - p)}{p^*(f(x) - p)}.$$

The advantage of the standard homotopy method (1) is that we are free to choose $p \in \Omega$ and thus possibly select a good starting guess for a fixed point. The advantage of the other two methods is that the co-ordinates μ or λ are implicit and thus we can perform our calculations with N instead of $N + 1$ variables. The particular choice $p \in \partial\Omega$ implies that the x-component of $\mathcal{C}_p$ in method (1) coincides with $\hat{\mathcal{C}}_p$ and $\tilde{\mathcal{C}}_p$ of methods (2) and (3) respectively, the parametrizations may of course be different. The components $\hat{\mathcal{C}}_p$ and $\tilde{\mathcal{C}}_p$ of methods (2) and (3) are the same. Under the Kellogg & Li & Yorke method (3) the curve $\hat{\mathcal{C}}_p$ has singularities at the fixed points of f which lie on $\hat{\mathcal{C}}_p$. However, as can be seen by our discussion of method (3), if the fixed points are regular points of $\mathrm{Id} - f$ e.g. under our assumption (11.3.1)(4), then these singularities are removable. A more serious drawback from the standpoint of numerical implementation is that the singularity of the fixed points for method (2) implies a numerically unstable calculation of the map $\hat{H}$ near the fixed points of f. For the application of a predictor-corrector method such as those discussed in the preceding chapters, a stable evaluation is needed. The initial values for the curve $\tilde{\mathcal{C}}_p$

$$x(0) = p; \quad \dot{x}(0) = \frac{f(p) - p}{\|f(p) - p\|}$$

can be obtained by differentiating the homotopy equation $x - p - \lambda\big(f(x) - p\big) = 0$ with respect to arclength and evaluating for $x = p$ and $\lambda = 0$.

11.4 Global Newton and Global Homotopy Methods

Newton's method is a favorite method for numerically calculating a zero point of a (nonlinear) C^∞-map $G : \mathbf{R}^N \to \mathbf{R}^N$. Recall that Newton's method is expressed as an iterative relation of the form

(11.4.1)
$$x_{n+1} = x_n - G'(x_n)^{-1}G(x_n), \quad n = 0, 1, \ldots;$$
$$x_0 = p \quad \{\text{starting point}\}.$$

As is well known, this method may diverge if the starting point p is not sufficiently near to a zero point $\bar{x}$ of G. Often one would like to determine whether a certain open bounded region $\Omega \subset \mathbf{R}^N$ contains a zero point $\bar{x}$ of G and furthermore, for which starting values p this solution $\bar{x}$ can be obtained by Newton's method. The so-called global Newton methods offer a possibility of answering such questions.

One may interpret (11.4.1) as the numerical integration of the differential equation

(11.4.2)
$$\dot{x} = -G'(x)^{-1}G(x)$$

using Euler's method with unit step size and initial point p. The idea of using the flow (11.4.2) to find zero points of G was exploited by Branin (1972). Smale (1976) gave boundary conditions on $\partial\Omega$ under which he could show, see theorem (11.4.8) below, that the flow (11.4.2) leads to a zero point $\bar{x}$ of G in Ω. Thus a numerically implementable existence theorem is obtained by integrating (11.4.2) using an appropriate starting point $p \in \partial\Omega$. Such numerical methods have been referred to as **global Newton methods**. In this section we will explore this and some related ideas.

First of all, let us note that in order to overcome simple singular points of G the equation for the flow defined by (11.4.2) can be modified by multiplying the right hand side by $\det G'(x)$. Nevertheless, near such singular points of G the evaluation of the right hand side remains numerically unstable. Keller (1978) observed that solutions of (11.4.2) can also be obtained from a homotopy equation which he consequently named the **global homotopy** method. Independently, Garcia and Gould (1978), (1980) obtained similar results. The global homotopy method involves tracing the curve defined by the equation

(11.4.3)
$$G(x) - \lambda G(p) = 0$$

starting from $(x, \lambda) = (p, 1) \in \partial\Omega \times \{1\}$ inward into $\Omega \times \mathbf{R}$. If the level $\Omega \times \{0\}$ is encountered, then a zero point of G has been found.

For an autonomous differential equation, changing the parameter of differentiation amounts to multiplying the right hand side by some smooth function of the new parameter. Say for example, s is replaced by $s(\xi)$ in the differential equation

$$\frac{d}{ds}x(s) = f\big(x(s)\big).$$

Then

$$\frac{d}{d\xi}x\big(s(\xi)\big) = \frac{d}{d\xi}s(\xi)\,f\Big(x\big(s(\xi)\big)\Big).$$

Keeping this in mind, let us now differentiate (11.4.3) with respect to say, arclength s. We obtain

$$G'(x)\dot{x} - \dot{\lambda}G(p) = 0.$$

Substituting $\lambda^{-1}G(x)$ for $G(p)$ yields

$$\dot{x} = (\dot{\lambda}/\lambda)G'(x)^{-1}G(x).$$

Hence we see that (11.4.2) and the x-component of (11.4.3) have the same solution curve — only the parametrizations are different. This holds so long as no singular point or zero point of G is encountered. However the global homotopy (11.4.3) handles such singularities in a more natural way. Hence we choose to present Keller's approach in our subsequent discussion. Let us now introduce the assumptions for this section.

(11.4.4) Assumptions.

(1) $G : \mathbf{R}^N \to \mathbf{R}^N$ is a C^∞-map;

(2) $\Omega \subset \mathbf{R}^N$ is open and bounded and $\partial\Omega$ is a connected C^∞-submanifold of $\mathbf{R}^N$;

(3) 0 is a regular value of G;

(4) $G(p) \neq 0$ for $p \in \partial\Omega$;

(5) the Jacobian $G'(p)$ is nonsingular for $p \in \partial\Omega$;

(6) the Newton direction $-G'(p)^{-1}G(p)$ is not tangent to $\partial\Omega$ at p.

The assumptions (11.4.4)(4)-(6) are Smale's boundary conditions. Let us define the global homotopy $H : \mathbf{R}^N \times \mathbf{R} \times \partial\Omega \longrightarrow \mathbf{R}^N$ by

$$(11.4.5) \qquad\qquad H(x, \lambda, p) := G(x) - \lambda G(p).$$

Since p varies over the $(N-1)$-dimensional surface $\partial\Omega$, it is somewhat difficult to apply Sard's theorem (11.2.2). This task was achieved by Percell (1980). We state his theorem under our somewhat stronger assumptions (11.4.4).

(11.4.6) Theorem. [Percell (1980)]
Let $\Omega \subset \mathbf{R}^N$ and $G : \mathbf{R}^N \to \mathbf{R}^N$ satisfy the assumptions (11.4.4). Then for almost all $p \in \partial\Omega$ the global homotopy $H(\cdot, \cdot, p) : \mathbf{R}^N \times \mathbf{R} \to \mathbf{R}^N$ defined by (11.4.5) has 0 as a regular value.

Hence, it is again reasonable to make the following

(11.4.7) Assumption. *Let the starting point $p \in \partial\Omega$ be chosen in accordance with (11.4.6) i.e. in such a way that 0 is a regular value of the map $H(\cdot, \cdot, p)$.*

As was already mentioned, the global homotopy method overcomes the numerical instabilities arising in the global Newton method near singular points of G. Keller (1978) has shown that this approach also leads to a simple geometrical proof of the following

(11.4.8) Theorem. [Smale (1976)]
Let $\Omega \subset \mathbf{R}^N$ and $G : \mathbf{R}^N \to \mathbf{R}^N$ satisfy the assumptions (11.4.4) and let $p \in \partial\Omega$ satisfy the assumption (11.4.7). Let $\mathcal{C}_p$ be the connected component of $\{(x, \lambda) \mid x \in \mathbf{R}^N, \lambda \in \mathbf{R}, H(x, \lambda, p) = 0\}$ which contains $(p, 1)$. Let $s \in \mathbf{R} \mapsto (x(s), \lambda(s))$ be a parametrization of $\mathcal{C}_p$ according to arclength s such that

(1) $x(0) = p$, $\lambda(0) = 1$;

(2) $\dot{x}(0)$ points into Ω.

Then there is a parameter $s_0 > 0$ such that

(3) $x(s) \in \Omega$ for $0 < s < s_0$;

(4) $x(s_0) \in \partial\Omega$;

(5) $\lambda(s_0) < 0$.

Consequently, the curve $\mathcal{C}_p$ hits the target level $\Omega \times \{0\}$ in an odd number of points $(\bar{x}, 0) \in \Omega \times \{0\}$ with $G(\bar{x}) = 0$.

Proof. Since $\partial\Omega$ is connected, c.f. (11.4.4)(2), we can assume without loss of generality that the Newton direction

(*) $-G'(p)^{-1}G(p)$ always points into Ω for $p \in \partial\Omega$.

The other case i.e. that the Newton direction $-G'(p)^{-1}G(p)$ always points out of Ω for $p \in \partial\Omega$, is treated in a similar way by switching a couple of signs. We differentiate the homotopy equation

(a) $G(x(s)) - \lambda(s)G(p),$

and by substituting $\lambda(s)^{-1}G(x(s))$ for $G(p)$ obtain

(b) $\lambda(s) \neq 0 \Rightarrow G'(x(s))\,\dot{x}(s) - \dfrac{\dot{\lambda}(s)}{\lambda(s)}\,G(x(s)) = 0.$

Since $\dot{x}(0)$ points into Ω and $\lambda(0) = 1$, the boundary condition (*) and (b) imply

(c) $\dot{\lambda}(0) < 0.$

Since $G(p) \neq 0$ and Ω is bounded, we see that the set

$$\{\lambda \mid G(x) = \lambda G(p), \ x \in \overline{\Omega}\}$$

is bounded. Hence the curve $\mathcal{C}_p$ must exit from $\Omega \times \mathbf{R}$ at some parameter $s_0 > 0$. All that remains to be shown is that

(d) $\lambda(s_0) < 0.$

Since $\dot{x}(s_0)$ points out of Ω, the boundary condition (*) and (b) imply that

(e) $\dfrac{\dot{\lambda}(s_0)}{\lambda(s_0)} > 0.$

Now consider the augmented Jacobian

$$A(s) := \begin{pmatrix} G'(x(s)) & -G(p) \\ \dot{x}(s)^* & \dot{\lambda}(s) \end{pmatrix}$$

of the homotopy (11.4.5). We obtain

$$A(s) \begin{pmatrix} \mathrm{Id} & \dot{x}(s) \\ 0^* & \dot{\lambda}(s) \end{pmatrix} = \begin{pmatrix} G'(x(s)) & 0 \\ \dot{x}(s)^* & 1 \end{pmatrix}$$

and consequently

(f) $\det A(s)\,\dot{\lambda}(s) = \det G'(x(s)).$

By (11.4.4)(5) and since $\partial\Omega$ is connected, the function $\det G'(x)$ does not change sign on $\partial\Omega$. On the other hand, the function $\det A(s)$ does not change sign along the path $\mathcal{C}_p$. Consequently (c) and (f) imply $\dot{\lambda}(s_0) < 0$, and from (e) we obtain the result (d). Hence the conclusion of the theorem follows. $\square$

Remarks.

- Given the assumptions (11.4.4)(1)-(3), the subsequent boundary conditions (11.4.4)(4)-(6) can be shown to hold for a sufficiently small ball Ω
  around a zero point of G. Thus, in a certain sense the above theorem
  generalizes the well-known Newton-Kantorovitch type theorems which
  discuss the local convergence of Newton's method, cf. Ortega & Rheinboldt (1970).
- If the global Newton method is implemented by numerically integrating
  (11.4.2), the evaluation of the right hand side can become numerically
  unstable. The global homotopy method of Keller (1978) overcomes this
  instability at the mild cost of increasing the dimension of the problem by
  one parameter.

We now indicate briefly how the numerical stability obtained by Keller can
also be maintained without increasing the dimension, cf. Allgower & Georg
(1988). We multiply (11.4.3) by $G(p)^*$ and obtain

$$(11.4.9) \qquad \lambda(x,p) = \frac{G(p)^* G(x)}{\|G(p)\|^2}.$$

This leads to a new homotopy $\tilde{H} : \mathbf{R}^N \times \partial\Omega \to \mathbf{R}^N$ defined by

$$(11.4.10) \qquad \tilde{H}(x,p) = G(x) - \lambda(x,p)G(p).$$

Note that the range of $\tilde{H}(\cdot,p)$ is contained in $\{G(p)\}^\perp$. We calculate

$$(11.4.11) \qquad \tilde{H}_x(x,p) = \left(\mathrm{Id} - \frac{G(p)G(p)^*}{\|G(p)\|^2}\right) G'(x)$$

which is the orthogonal projection of the Jacobian $G'(x)$ onto the tangent
space $\{G(p)\}^\perp$. Thus we see again that the evaluations of $\tilde{H}(x,p)$ and $\tilde{H}_x(x,p)$
which would be required for a predictor-corrector tracing of $\tilde{H}(x,p) = 0$ are
numerically stable. We conclude this section with an analogue of theorem
(11.3.10).

(11.4.12) Theorem. *Let $\Omega \subset \mathbf{R}^N$ and $G : \mathbf{R}^N \to \mathbf{R}^N$ satisfy the assumptions (11.4.4). Then for almost all $p \in \partial\Omega$, the homotopy map $\tilde{H}(\cdot,p) : \mathbf{R}^N \to
\{G(p)\}^\perp$ defined by (11.4.9) – (11.4.10) has 0 as a regular value.*

11.5 Multiple Solutions

In theorem (11.4.8) it was observed that the global homotopy method might actually yield more than one zero point of the map G in a bounded region Ω. This raises the question whether one might be able to compute more zero points of G in Ω beside those which lie on the global homotopy path defined by $H(x, \lambda, p) := G(x) - \lambda G(p) = 0$ for some fixed initial point $p \in \partial\Omega$.

To be more precise, let us suppose that $\Omega \subset \mathbf{R}^N$ is an open bounded region, and that $G : \mathbf{R}^N \to \mathbf{R}^N$ is a smooth map having a zero point $z_0 \in \Omega$. The task is now to find additional zero points of G in Ω provided they exist. One method which has often been used for handling this problem is deflation, see e.g. Brown & Gearhart (1971). In this method **a deflated map** $G_1 : \mathbf{R}^N \setminus \{z_0\} \to \mathbf{R}^N$ is defined by

(11.5.1) $G_1(x) := G(x)/\|x - z_0\|.$

One then applies an iterative method to try to find a zero of G_1. There are still a number of choices to be made viz.

- the choice of deflation functionals e.g. l_2, l_∞ norms or a "gradient deflation" studied by Brown & Gearhart;
- the choice of iterative solution e.g. Newton-like methods as described in Ortega & Rheinboldt (1970);
- the choice of starting value x_0 for the iterative method e.g. often the same x_0 is used which initially led to the zero point z_0.

The deflation device can be repeatedly applied by setting

$$G_k(x) := \frac{G(x)}{\prod_{j=0}^{k-1} \|x - z_j\|}$$

where $z_0, \ldots, z_{k-1}$ are zeros of G which have previously been found. Numerical experience with deflation has shown that it is often a matter of seeming chance whether one obtains an additional solution and if one is obtained, it is very often not the one which is nearest to z_0.

By utilizing homotopy-type methods we can give some conditions which will guarantee the existence of an additional solution. This additional solution will lie on a homotopy path, and we also obtain results on the topological index of zero points which are successively obtained on this path. We illustrate this approach with a discussion of the **"d-homotopy"**. Let us consider the homotopy map $H_d : \mathbf{R}^N \times \mathbf{R} \to \mathbf{R}$ defined by

(11.5.2) $H_d(x, \lambda) := G(x) - \lambda d$

where $d \in \mathbf{R}^N$ is some fixed vector with $d \neq 0$. Since we assume that a zero point z_0 is already given, we have $H_d(z_0, 0) = 0$. Let us further assume 0 is a regular value of G. Then it follows from Sard's theorem (11.2.2) that 0 is also a regular value of H_d for almost all $d \in \mathbf{R}^N$. Let us note here that the homotopy H_d is more general than the global homotopy (11.4.5), since d need not belong to the range of G. In order to assure that the curve $\mathcal{C} \subset H_d^{-1}(0)$ which contains $(z_0, 0)$ again reaches the level $\lambda = 0$, we need as usual to impose a boundary condition. The following proposition uses a boundary condition which is motivated by a simple degree consideration and has frequently been successfully used.

(11.5.3) Proposition. *Let the following hypotheses hold:*

(1) $G : \mathbf{R}^N \to \mathbf{R}^N$ *is a smooth map with regular value 0;*

(2) $d \in \mathbf{R}^N \setminus \{0\}$ *is a point such that the homotopy H_d also has regular value 0;*

(3) $\Omega \subset \mathbf{R}^N$ *is a bounded open set which contains a (known) initial zero point z_0 of G;*

(4) *the* **boundary condition** $H_d(x, \lambda) = G(x) - \lambda d \neq 0$ *holds for all $x \in \partial\Omega$, $\lambda \in \mathbf{R}$;*

Then the curve $\mathcal{C} \subset H_d^{-1}(0)$ which contains $(z_0, 0)$ intersects the level $\Omega \times \{0\}$ an even number of times in points $(z_i, 0)$, $i = 0, \ldots, n$, at which $G(z_i) = 0$.

Proof. The boundary condition (11.5.3)(4) implies that the curve $\mathcal{C}$ lies strictly inside the cylinder $\partial\Omega \times \mathbf{R}$. A solution $(x, \lambda) \in \mathcal{C}$ satisfies $G(x) = \lambda d$ and $x \in \Omega$, and hence $|\lambda| = \|G(x)\| / \|d\|$ remains bounded. Recalling that $\mathcal{C}$ is homeomorphic either to the line $\mathbf{R}$ or to the circle S^1, it follows that $\mathcal{C} \simeq S^1$. Since 0 is a regular value of G, it is easily seen that $\mathcal{C}$ intersects the level $\Omega \times \{0\}$ transversely, and the assertion of the above proposition follows immediately. $\square$

It is evident that the boundary condition can be relaxed.

(11.5.4) Corollary. *The conclusion of the above proposition (11.5.3) remains true if the boundary condition (4) is replaced by either of*

(4-1) $H_d(x, \lambda) = G(x) - \lambda d \neq 0$ *for all $x \in \partial\Omega$, $\lambda \geq 0$;*

(4-2) $H_d(x, \lambda) = G(x) - \lambda d \neq 0$ *for all $x \in \partial\Omega$, $\lambda \leq 0$.*

Proof. We consider only the first case (4-1). If $(x, \lambda) \in \mathcal{C}$ is a solution with $\lambda \geq 0$, the same argument as in the above proof shows that $\lambda = \|G(x)\| / \|d\|$ remains bounded. Hence, starting at a solution point and traversing the curve only in positive λ-direction gives the desired assertion. $\square$

To set the discussion forth we now need the following definition which describes the topological index of a zero point of a map in a very simple case.

(11.5.5) Definition. *Let z_0 be a zero point of the smooth map $G : \mathbf{R}^N \to$* $\mathbf{R}^N$ *such that $\det G'(z_0) \neq 0$. Then the index of z_0 is defined to be the sign of $\det G'(z_0)$.*

(11.5.6) Corollary. *Any two zero points of G which are consecutively obtained by traversing the curve $\mathcal{C}$ have opposite index.*

Proof. Let $s \in \mathbf{R} \mapsto \big(x(s), \lambda(s)\big) \in \mathbf{R}^{N+1}$ be a parametrization of $\mathcal{C}$ according to arclength s. Differentiating

$$H_d\big(x(s), \lambda(s)\big) = G\big(x(s)\big) - \lambda(s)d = 0$$

we obtain

$$G'\big(x(s)\big)\dot{x}(s) - \dot{\lambda}(s)d = 0$$

and hence

$$\begin{pmatrix} G'\big(x(s)\big) & -d \\ \dot{x}(s)^* & \dot{\lambda}(s) \end{pmatrix} \begin{pmatrix} \mathrm{Id} & \dot{x}(s) \\ 0^* & \dot{\lambda}(s) \end{pmatrix} = \begin{pmatrix} G'\big(x(s)\big) & 0 \\ \dot{x}(s)^* & 1 \end{pmatrix}.$$

We note that the left matrix in the latter equation is the augmented Jacobian of H_d. Its determinant is of constant sign on $\mathcal{C}$. Hence by the product rule of determinants we obtain that $\dot{\lambda}(s)$ and $\det G'\big(x(s)\big)$ change signs at exactly the same points $s \in \mathbf{R}$. Let $\big(x(s_1), 0\big)$ and $\big(x(s_2), 0\big)$ be two consecutive solution points on $\mathcal{C}$. It is clear that they are traversed in opposite λ-directions, i.e. $\dot{\lambda}(s_1)\dot{\lambda}(s_2) < 0$ and the assertion of the corollary follows. $\qquad\square$

(11.5.7) Example. To give an illustration of how corollary (11.5.5) can be applied, we consider a system of equations arising from a discretization of a nonlinear elliptic boundary value problem

$$\mathcal{L}u(\xi) = \mu f(u(\xi)), \quad \xi \in \mathcal{D};$$
$$u(\xi) = 0, \quad \xi \in \partial\mathcal{D}.$$

Here $\mathcal{D} \subset \mathbf{R}^m$ is a bounded domain, $\mathcal{L}$ is a linear elliptic differential operator, and f is a smooth nonlinear function which is bounded from below and satisfies

$$\lim_{u \to \infty} \frac{f(u)}{u} = \infty.$$

Problems of this general form are discussed in the survey paper of Amann (1976), the problem from which our particular example derives has been discussed by Ambrosetti & Hess (1980).

Discretizations of the above problem in general take the form

(11.5.8) $$G(x) := Ax - \mu F(x) = 0$$

where A is a positive definite $N \times N$-matrix such that A^{-1} has only positive entries, and $F : \mathbf{R}^N \to \mathbf{R}^N$ is a smooth map whose co-ordinates are bounded from below by a constant $-C < 0$ and satisfy

$$\lim_{x[i] \to \infty} \frac{F(x)[i]}{x[i]} = \infty.$$

From the contraction principle it follows for small $\mu > 0$ that the fixed point iteration

$$x_{k+1} = A^{-1}F(x_k), \quad x_0 = 0$$

converges to a zero point z_0 of G. We choose a directional vector $d \in \mathbf{R}^N$ whose co-ordinates are all positive, and a set

$$\Omega := \{x \in \mathbf{R}^N \mid ||x||_\infty < \beta\}$$

where $\beta > 0$ is chosen so large that the following conditions are satisfied:

$$\beta > \mu C||A^{-1}||_\infty;$$

$$F(x)[i] > \frac{\beta}{\mu a} \quad \text{for } x[i] = \beta \text{ and all } i.$$

In the latter inequality $a > 0$ denotes the smallest entry of A^{-1}. Both inequalities can be satisfied, because of the assumptions on F.

Let us show that for the above choices the boundary condition (11.5.4)(4-1) is satisfied. If $x \in \partial\Omega$, then either $x[i] = -\beta$ or $x[i] = \beta$ for some co-ordinate i. In the first case we estimate

$$A^{-1}\big(G(x) - \lambda d\big)[i] = -\beta - \mu A^{-1}F(x)[i] - \lambda A^{-1}d[i]$$
$$\leq -\beta + \mu C||A^{-1}||_\infty < 0.$$

In the second case we obtain

$$A^{-1}\big(G(x) - \lambda d\big)[i] = \beta - \mu A^{-1}F(x)[i] - \lambda A^{-1}d[i]$$
$$\leq \beta - \mu a F(x)[i] < 0.$$

Hence, in both cases we have $G(x) - \lambda d \neq 0$ for $\lambda \geq 0$. This application of the d-homotopy makes it possible to reach an additional solution z_1 via the homotopy equation

$$Ax - \mu F(x) - \lambda d = 0$$

for a fixed μ and varying λ. We emphasize that z_0, z_1 do not necessarily lie on the same solution branch of the equation

$$Ax - \mu F(x) = 0$$

for varying μ. Hence, the d-homotopy can permit moving between disjoint solution branches of the nonlinear eigenvalue problem (11.5.8). Such possibilities of moving between disjoint branches were discussed in a general context by Jürgens & Peitgen & Saupe (1980).

We conclude this section with a few remarks concerning the relationship between deflation and homotopy methods, see Allgower & Georg (1983) for a more detailed discussion. In the context of the discussion of deflation (11.5.1) let us consider the global homotopy $H : \left(\mathbf{R}^N \setminus \{z_0\}\right) \times \mathbf{R} \to \mathbf{R}^N$ defined by

$$(11.5.9) \qquad H(x, \lambda) := G_1(x) - \lambda G_1(x_0).$$

In view of our discussions following (11.4.2), performing Newton's method on G_1 starting at x_0 amounts to a particular integration of the global homotopy (11.5.9) and also starting at $(x_0, 1)$. Thus we see that in general successive deflation will at best produce the zeros of G which lie on $H^{-1}(0)$. However, because of the "crudeness" of the Newton steps, some iterate may get far enough away from $H^{-1}(0)$ that the Newton method might diverge or possibly accidently converge to a zero point not on $H^{-1}(0)$. Numerical experience, see Brown & Gearhart (1971), with deflation confirms the above analysis in the sense that in general zero points which are successively obtained via deflation have opposite index, and zero points which are in close proximity are not successively obtained if they have the same index. Recently Diener (1985), (1986) has given some conditions under which all of the zero points of certain maps can be linked on a single global homotopy path. However, the hypotheses concerning such maps appear to be severely restrictive.

11.6 Polynomial Systems

In the preceding section we considered the task of computing multiple zero points of general smooth maps. In the case of complex polynomial systems it is actually possible to compute (at least in principle) all of the zero points by means of homotopy methods. This subject has received considerable attention in recent years. The book of Morgan (1987) deals exclusively with this topic, using the smooth path tracing approach. It also contains most of the up to date references on this approach and a number of interesting applications to robotics and other fields.

We begin our discussion with the simplest general case viz. a complex polynomial $P : \mathbf{C} \to \mathbf{C}$. As is well known from the fundamental theorem of algebra, P has exactly $n = \deg P$ zero points if one accounts for multiplicities. Let us indicate how these zero points can be computed. Our discussion essentially follows an argument of Chow & Mallet-Paret & Yorke (1978).

Suppose $P(z) = z^n + \sum_{j=0}^{n-1} a_j z^j$ is a monic polynomial whose n zero points are to be calculated. Choosing $Q(z) = z^n + b_0$ where $b_0 \neq 0$ is arbitrary, we form the convex homotopy $H : \mathbf{C} \times [0,1] \to \mathbf{C}$ defined by

$$H(z,\lambda) = (1 - \lambda)Q(z) + \lambda P(z)$$

(11.6.1)
$$= z^n + \lambda \sum_{j=1}^{n-1} a_j z^j + (1 - \lambda)b_0 + \lambda a_0.$$

In the case of complex differentiation, we regard the complex field $\mathbf{C}$ as a corresponding 2-dimensional real space. Hence, if $f(z)$ is a holomorphic function, we consider its derivative as a 2×2-matrix

$$f'(z) = \begin{pmatrix} u_x & u_y \\ v_x & v_y \end{pmatrix} = \begin{pmatrix} u_x & u_y \\ -u_y & u_x \end{pmatrix}$$

where we write the real and imaginary parts as $f(z) = u(x,y) + iv(x,y)$ and $z = x + iy$. The latter equality is obtained from the Cauchy-Riemann equations.

The derivative of the right hand side of (11.6.1) with respect to (z, λ, b_0) is

$$\left((1 - \lambda)Q'(z) + \lambda P'(z) \quad , \quad -b_0 + \sum_{j=0}^{n-1} a_j z^j \quad , \quad (1 - \lambda)\mathrm{Id}_{\mathbf{C}} \right)$$

and, in the above context, it can be identified with a real 2×5-matrix. By Sard's theorem (11.2.2), zero is a regular value of the restricted map

$$\tilde{H} : \quad (z, \lambda) \in \mathbf{C} \times [0, 1) \quad \longmapsto \quad H(z, \lambda) \in \mathbf{C}$$

for almost all $b_0 \in \mathbf{C}$. Let us assume in the following that b_0 is so chosen. The level $\lambda = 1$ must be excluded in case the polynomial P has a multiple root.

At $\lambda = 0$, $\tilde{H}(z, 0) = z^n + b_0$ has n distinct roots $z_1, \ldots, z_n \in \mathbf{C}$. For each root z_j, $j = 1, \ldots, n$ the component $\mathcal{C}_j \subset \tilde{H}^{-1}(0)$ containing $(z_j, 0)$ is a smooth curve which may be parametrized according to arclength

$$s \in I_j \quad \longmapsto \quad \big(z_j(s), \lambda_j(s)\big) \in \mathbf{C} \times (-\infty, 1)$$

such that $z_j(0) = z_j$, $\lambda_j(0) = 0$ and $\dot{\lambda}_j(0) > 0$. As in the above, the Jacobian

$$\tilde{H}'(z, \lambda) = \left((1 - \lambda)Q'(z) + \lambda P'(z) \quad , \quad -b_0 + \sum_{j=0}^{n-1} a_j z^j \right)$$

can be considered as a real 2×3-matrix. By our choice of b_0, it has maximal rank two for $(z, \lambda) \in \tilde{H}^{-1}(0)$. Since by the Cauchy-Riemann equations, the leading 2×2-submatrix can only have rank 0 or 2, its determinant does not vanish. Differentiating the equation $\tilde{H}\big(z_j(s), \lambda_j(s)\big) = 0$ with respect to s

we obtain $\dot{\lambda}_j(s) > 0$ for all $s \in I_j$ by an argument analogous to the proof of (11.5.6). Thus the homotopy parameter $\lambda_j(s)$ is strictly increasing.

We now want to establish a boundary condition. Let us recall the following classical result concerning polynomials, see e.g. Marden (1949): If $p(z) = \sum_{j=0}^{n} c_j z^j = 0$, then

$$|z| < 1 + \max_{k \neq n} \left| \frac{c_k}{c_n} \right|.$$

In the present case for H as defined in (11.6.1), from $H\big(z_j(s), \lambda_j(s)\big) = 0$ we obtain

$$|z_j(s)| < M_1 + M_2 |\lambda_j(s)|$$

for some constants $M_1, M_2 > 0$.

It now is clear that the curves C_j are all disjoint and penetrate each level $\lambda < 1$ transversely. By continuity, if $\lambda_j(s) \to 1$ then $z_j(s) \to \tilde{z}_j$ for some root $\tilde{z}_j$ of P. For $\lambda \approx 1$ the homotopy $H(z, \lambda) = P(z) + (1 - \lambda)\big(P(z) + Q(z)\big)$ represents just a small perturbation of $P(z)$. Since $H(z, \lambda)$ has n distinct roots for $\lambda < 1$, a classical perturbation result of complex analysis implies: if $\tilde{z}_j$ is a root of P with multiplicity m_j then exactly m_j curves C_j converge to it in the above sense. However, only in the case of simple roots do the curves intersect the level $\lambda = 1$ transversely. The preceding discussion suggests how one could implement a predictor-corrector method for calculating all complex roots of a polynomial. Some authors, e.g. Kuhn (1974), (1977) and Kojima & Nishino & Arima (1979) have used piecewise-linear methods instead.

Let us stress that we do not claim that the above numerical methods are the most efficient methods to calculate roots of a polynomial. There are some popular methods in use, e.g. the method of Jenkins & Traub or Laguerre's method, which are usually combined with a deflation type process in order to obtain all roots. In fact, little effort has been made to compare the homotopy approach with these "direct methods". We refer the reader to standard books on numerical analysis, especially Henrici (1974), (1977) or to some of the library programs in current use such as IMSL. If the coefficients of the polynomial are real, one may be interested in only calculating the real roots. This can be done very efficiently using the idea of Sturm sequences, see e.g. Heindel (1971) or Collins & Loos (1976).

We gave the preceding discussion primarily as an introduction, since it suggests how we can proceed to find all of the zero points of a system of complex polynomials $P : \mathbf{C}^n \to \mathbf{C}^n$. The task is substantially more complicated than the case $n = 1$, because the equation $P(z_1, \ldots, z_n) = 0$ may have a continuum of solutions or no solution at all. To illustrate these situations we give some simple examples. The system

(11.6.2)
$$\begin{aligned} z_1 + z_2 &= 0 \\ z_1 + z_2 + 1 &= 0 \end{aligned}$$

has no solutions at all. The system

(11.6.3)
$$z_1^2 - 25 = 0$$
$$z_1 z_2 - z_1 - 5z_2 + 5 = 0$$

has $\{(-5,1)\} \cup \{(z_1, z_2) \in \mathbf{C}^2 \mid z_1 = 5\}$ as its complete solution set. Let us introduce some important notions arising in the theory of polynomial systems. If a term of the kth component P_k has the form

$$a z_1^{r_1} z_2^{r_2} \cdots z_n^{r_n} ,$$

then its degree is $r_1 + r_2 + \ldots + r_n$. The **degree** d_k of P_k is the maximum of the degrees of its terms. The **homogeneous part** $\hat{P}$ of P is obtained by deleting in each component P_k all terms having degree less than d_k. The **homogenization** $\tilde{P}$ of P is obtained by multiplying each term of each component P_k with an appropriate power z_0^r such that its degree is d_k. For example, the polynomial system

(11.6.4)(1)
$$z_1^3 - z_1 = 0$$
$$z_1^2 z_2 + 1 = 0$$

has the homogeneous part

(11.6.4)(2)
$$z_1^3 = 0$$
$$z_1^2 z_2 = 0$$

and the homogenization

(11.6.4)(3)
$$z_1^3 - z_0^2 z_1 = 0$$
$$z_1^2 z_2 + z_0^3 = 0 .$$

Note that the homogenization $\tilde{P} : \mathbf{C}^{n+1} \to \mathbf{C}^n$ involves one more variable. If

$$(w_0, \ldots, w_n) \neq 0$$

is a zero point of $\tilde{P}$, then the entire ray

$$[w_0 : \cdots : w_n] := \{(\xi w_0, \ldots, \xi w_n) \mid \xi \in \mathbf{C}\}$$

consists of zero points of $\tilde{P}$. Usually, $[w_0 : \cdots : w_n]$ is regarded as a point in the complex projective space $\mathbf{CP}^n$. There are two cases to consider:

(1) The solution $[w_0 : \cdots : w_n]$ intersects the hyperplane $z_0 = 0$ transversely i.e. without loss of generality $w_0 = 1$. This corresponds to a zero point $(w_1, \ldots, w_n)$ of P. Conversely, each zero point $(w_1, \ldots, w_n)$ of P corresponds to a solution $[1 : w_1 : \cdots : w_n]$ of $\tilde{P}$.

(2) The solution $[w_0 : \cdots : w_n]$ lies in the hyperplane $z_0 = 0$ i.e. $w_0 = 0$. This corresponds to a **nontrivial** solution $[w_1 : \cdots : w_n]$ of the homogeneous part $\hat{P}$, and such solutions are called **zero points of P at infinity**.

For example, the system (11.6.2) has only the solution $[0 : 1 : -1]$ at infinity. The system (11.6.4) has the two solutions $[1 : \pm 1 : -1]$ and in addition the solution $[0 : 0 : 1]$ at infinity.

As in the case of one variable, it is possible to define the multiplicity of a solution. However, this is a more complicated matter than it was in one dimension and requires some deeper ideas of algebra and analysis. We will give a brief sketch for doing this. If $[w_0 : \cdots : w_n]$ is an isolated solution of the homogenization $\tilde{P}(z_0, \ldots, z_n) = 0$ with respect to the topology of $\mathbf{CP}^n$, we can define a multiplicity of $[w_0 : \cdots : w_n]$ in two different ways. However, it is not a trivial exercise to show that these definitions are equivalent.

(1) Consider a co-ordinate w_k which is different from zero. Without loss of generality we can assume $w_k = 1$. If we fix the variable $z_k = 1$ in the the homogenization $\tilde{P}(z_0, \ldots, z_n) = 0$, then we have n complex equations in n complex variables or $2n$ real equations in $2n$ real variables with the complex solution $z_j = w_j$, $j = 0, \ldots, n$, $j \neq k$. The multiplicity is now defined by the local topological degree of this solution, see e.g. Milnor (1968). It can be shown that this definition is independent of the special choice of the non-vanishing co-ordinate w_k.

(2) As above, consider a co-ordinate $w_k = 1$. Again, we fix the variable $z_k = 1$ in the homogenization $\tilde{P}(z_0, \ldots, z_n) = 0$, and after a translation obtain new equations

$$F(z_0, \ldots, \widehat{z_k}, \ldots, z_n) := \tilde{P}(z_0 + w_0, \ldots, 1, \ldots, z_n + w_n) = 0$$

in the variables $z_0, \ldots, \widehat{z_k}, \ldots, z_n$ where $\widehat{}$ denotes omission of the term beneath it. These new equations have a zero point at the origin. Now, the multiplicity of the solution is defined as the dimension of the quotient

$$\frac{\mathbf{C}\big[[z_0, \ldots, \widehat{z_k}, \ldots, z_n]\big]}{(F_1, \ldots, \widehat{F_k}, \ldots, F_n)}$$

where $\mathbf{C}\big[[z_0, \ldots, \widehat{z_k}, \ldots, z_n]\big]$ is the usual power series ring and the symbol $(F_1, \ldots, \widehat{F_k}, \ldots, F_n)$ denotes the ideal generated by the corresponding polynomials $F_1, \ldots, \widehat{F_k}, \ldots, F_n$, see e.g. Fulton (1984) or van der Waerden (1953). It can be shown that also this definition is independent of the special choice of the non-vanishing co-ordinate w_k.

The higher dimensional analogue of the fundamental theorem of algebra is Bezout's theorem, which states that the number of zero points of P, counting their multiplicities, equals the product $d_1 d_2 \cdots d_n$, provided all solutions are isolated.

As an illustration, let us examine example (11.6.2) which had the three isolated solutions $[1 : \pm 1 : -1]$ and $[0 : 0 : 1]$. According to Bezout's theorem, this system has nine roots. It is routine to see that $\det P'(z) \neq 0$ at the zero points $(\pm 1, -1)$. Hence they are simple, and by Bezout's theorem the zero point at infinity has multiplicity seven. Let us show that this is true using the second definition of multiplicity. Setting $z_2 = 1$ in the homogenization (11.6.4)(3) we obtain

(11.6.4)(4)
$$z_1^3 - z_0^2 z_1 = 0$$
$$z_1^2 + z_0^3 = 0.$$

Since $z_1^3 - z_0^2 z_1 = z_1(z_1 + z_0)(z_1 - z_0)$, a factorization theorem yields

$$\dim \frac{\mathbf{C}[[z_0, z_1]]}{(z_1^3 - z_0^2 z_1,\, z_1^2 + z_0^3)} = \dim \frac{\mathbf{C}[[z_0, z_1]]}{(z_1,\, z_1^2 + z_0^3)} + \dim \frac{\mathbf{C}[[z_0, z_1]]}{(z_1 + z_0,\, z_1^2 + z_0^3)}$$
$$+ \dim \frac{\mathbf{C}[[z_0, z_1]]}{(z_1 - z_0,\, z_1^2 + z_0^3)}$$
$$= 3 + 2 + 2.$$

An attempt to treat the polynomial system for $n > 1$ by means of a homotopy H similar to (11.6.1) requires more care than in the case $n = 1$ since some solution paths may go to infinity. It is a trickier matter to construct perturbations which simultaneously yield zero as a regular value of H for all levels $\lambda < 1$ and also properly control the paths going to infinity.

A simple remedy for this problem was proposed by Garcia & Zangwill (1979). They use the homotopy

$$H(z_1, \ldots, z_n, \lambda) = (1 - \lambda)\left(z_k^{d_k + 1} - b_k\right) + \lambda P(z_1, \ldots, z_n).$$

It is routine to check that zero is a regular value of the restricted homotopy

$$z_1, \ldots, z_n \in \mathbf{C}, \lambda \in (-\infty, 1) \quad \longmapsto \quad H(z_1, \ldots, z_n, \lambda)$$

for almost all vectors $(b_1, \ldots, b_n) \in \mathbf{C}^n$. Since the homogeneous part of H has no nontrivial zero for $\lambda < 1$, the $(d_1 + 1) \cdots (d_n + 1)$, solution paths starting at the level $\lambda = 0$ can go off to infinity only for $\lambda \to 1$. Using the Cauchy-Riemann equations as in the 1-dimensional case, it can be shown that all solution paths are increasing with respect to the λ-co-ordinate. By a standard perturbation argument, it is also clear that all isolated zero points of P can be obtained in this way. However, this method has the disadvantage of introducing $(d_1 + 1) \cdots (d_n + 1) - d_1 \cdots d_n$ additional extraneous solution paths which diverge to infinity.

Chow & Mallet-Paret & Yorke (1979) introduced the following homotopy

$$H_k(z_1, \ldots, z_n, \lambda) = (1 - \lambda)\left(z_k^{d_k} - b_k\right) + \lambda P(z_1, \ldots, z_n) + \lambda(1 - \lambda) \sum_{j=1}^{n} a_{j,k} z_j^{d_k}$$

for $k = 1, \ldots, n$. They showed by using Sard type arguments that for almost all $b_k, a_{j,k} \in \mathbf{C}$ in the sense of $2n(n+1)$-dimensional Lebesgue measure, the restricted homotopy

$$z_1, \ldots, z_n \in \mathbf{C}, \lambda \in (-\infty, 1) \quad \longmapsto \quad H(z_1, \ldots, z_n, \lambda)$$

has zero as a regular value, and that $d_1 \cdots d_n$ solution paths starting at the level $\lambda = 0$ cannot go off to infinity unless $\lambda \to 1$. Also, by a perturbation argument, all isolated zero points of P can be obtained in this way.

A third method has been proposed by Wright (1985). He considers the following homotopy involving the homogenization $\tilde{P}$ of P

$$H_k(z_1, \ldots, z_n, \lambda) = (1-\lambda)(a_k z_k^{d_k} - b_k z_0^{d_k}) + \lambda \tilde{P}(z_0, \ldots, z_n).$$

He shows that for almost all coefficients $a_k, b_k \in \mathbf{C}$ the restricted homotopies

$$z_0, \ldots, \widehat{z_k}, \ldots, z_n \in \mathbf{C}, \lambda \in (-\infty, 1) \quad \longmapsto \quad H(z_0, \ldots, 1, \ldots, z_n, \lambda)$$

for $k = 0, \ldots, n$ have regular value zero. Hence, for fixed $\lambda < 1$, the homogeneous polynomial H has exactly $d_1 \cdots d_n$ simple zero points in the projective space $\mathbf{CP}^n$. Thus, in this approach solutions at infinity are treated no differently than finite solutions. The solution curves are traced in the projective space $\mathbf{CP}^n$, and from the numerical point of view we have the slight drawback that occasionally a chart in $\mathbf{CP}^n$ has to be switched.

Recently, attention has been given to the task of trying to formulate homotopies which eliminate the sometimes wasteful effort involved in tracing paths which go to solutions of $P(z_1, \ldots, z_n) = 0$ at infinity. Work in this direction has been done by Morgan & Sommese (1986) and by Li & Sauer & Yorke (1987). Morgan (1986) and Morgan & Sommese (1987) describe the easily implemented "projective transformation" which allows the user to avoid the drawback of changing co-ordinate charts on $\mathbf{CP}^n$. Morgan & Sommese (1988) shows how to exploit relations among the system coefficients, via "coefficient parameter continuation". Such relations occur commonly in engineering problems, as described in Wampler & Morgan & Sommese (1988). Li & Sauer (1987) investigate the application of homotopy methods for the solution of nonlinear matrix eigenvalue problems. The use of homotopy methods for solving analytic systems of equations has been considered by Carr & Mallet-Paret (1983), Allgower (1984) and Allgower & Georg (1983). Morgan & Sommese & Wampler (1989) combine a homotopy method with contour integrals to calculate singular solutions to nonlinear analytic systems.

We conclude this section with some remarks about "direct methods" for finding all roots of a system of polynomials. This idea seems to go back to Kronecker, see e.g. van der Waerden (1953), who observed that the resultant of polynomials can be used, at least in theory, as a device to eliminate variables in the polynomial system, somewhat similarly to the method of Gauss elimination for linear systems. Let us briefly indicate how this is done. Let

$$p(x) = \sum_{i=0}^{n} p_i x^i \text{ and}$$

(11.6.5)

$$q(x) = \sum_{i=0}^{m} q_i x^i$$

be two polynomials in the variable x with coefficients p_i, q_i respectively, and let us assume that $n \le m$. The resultant $r(p, q)$ is defined as the determinant of the so-called Sylvester matrix which is constructed from the coefficients p_i and q_i. We illustrate this construction for the special case $n = 3$ and $m = 5$:

$$(11.6.6) \qquad r(p, q) := \det \begin{pmatrix} p_3 & p_2 & p_1 & p_0 & 0 & 0 & 0 & 0 \\ 0 & p_3 & p_2 & p_1 & p_0 & 0 & 0 & 0 \\ 0 & 0 & p_3 & p_2 & p_1 & p_0 & 0 & 0 \\ 0 & 0 & 0 & p_3 & p_2 & p_1 & p_0 & 0 \\ 0 & 0 & 0 & 0 & p_3 & p_2 & p_1 & p_0 \\ q_5 & q_4 & q_3 & q_2 & q_1 & q_0 & 0 & 0 \\ 0 & q_5 & q_4 & q_3 & q_2 & q_1 & q_0 & 0 \\ 0 & 0 & q_5 & q_4 & q_3 & q_2 & q_1 & q_0 \end{pmatrix}.$$

If Gauss elimination is used to cancel the entry q_5 in row 6 via subtracting a q_5/p_3 multiple of row 1, then parallel operations can be performed for rows 7, 2 and rows 8, 3. This procedure can be repeated. As a result, we obtain a new matrix with the same determinant:

$$(11.6.7) \qquad r(p, q) := \det \begin{pmatrix} p_3 & p_2 & p_1 & p_0 & 0 & 0 & 0 & 0 \\ 0 & p_3 & p_2 & p_1 & p_0 & 0 & 0 & 0 \\ 0 & 0 & p_3 & p_2 & p_1 & p_0 & 0 & 0 \\ 0 & 0 & 0 & p_3 & p_2 & p_1 & p_0 & 0 \\ 0 & 0 & 0 & 0 & p_3 & p_2 & p_1 & p_0 \\ 0 & 0 & 0 & q_2' & q_1 & q_0 & 0 & 0 \\ 0 & 0 & 0 & 0 & q_2' & q_1 & q_0 & 0 \\ 0 & 0 & 0 & 0 & 0 & q_2' & q_1 & q_0 \end{pmatrix}.$$

The lower right hand 5×5-matrix can be rearranged:

$$(11.6.8) \qquad r(p, q) := \det \begin{pmatrix} p_3 & p_2 & p_1 & p_0 & 0 & 0 & 0 & 0 \\ 0 & p_3 & p_2 & p_1 & p_0 & 0 & 0 & 0 \\ 0 & 0 & p_3 & p_2 & p_1 & p_0 & 0 & 0 \\ 0 & 0 & 0 & q_2' & q_1 & q_0 & 0 & 0 \\ 0 & 0 & 0 & 0 & q_2' & q_1 & q_0 & 0 \\ 0 & 0 & 0 & 0 & 0 & q_2' & q_1 & q_0 \\ 0 & 0 & 0 & p_3 & p_2 & p_1 & p_0 & 0 \\ 0 & 0 & 0 & 0 & p_3 & p_2 & p_1 & p_0 \end{pmatrix}.$$

Now the same sort of procedure can be repeated on the lower right hand 5×5-matrix. This process can be repeated until the matrix has been reduced to upper triangular form.

The important feature of the resultant is that $r(p, q)$ vanishes at common zero points of p and q. The algorithm as sketched above for the calculation of the resultant is essentially equivalent to the Euclidean algorithm which calculates the greatest common divisor. Numerous modifications have been performed, see e.g. Barnett (1974). Of special importance is the case that the coefficients are in some integral domain, in particular that they are themselves polynomials over some other variables. This case is relevant for the present discussion. Efficient algorithms for the calculation of multivariate polynomial resultants have been given, e.g. Collins (1967).

Let us now sketch the elimination procedure via the resultant. Let

$$P_1(z_1, \ldots, z_n) = 0$$

(11.6.9)
$$\vdots$$

$$P_n(z_1, \ldots, z_n) = 0$$

be a polynomial system of equations. We can consider $P_1, \ldots, P_n$ as polynomials in the one variable z_n with coefficients in $\mathbf{C}[z_1, \ldots, z_{n-1}]$. Consequently, we can consider the resultants $r(P_1, P_2), \ldots, r(P_1, P_n)$ which are in $\mathbf{C}[z_1, \ldots, z_{n-1}]$, i.e. they can be considered as polynomials in the variables $z_1, \ldots, z_{n-1}$. Proceeding recursively we ultimately reach a single polynomial R_1 in z_1. It has the property that it vanishes at all points z_1 which are the first co-ordinates of the zeros of the system (11.6.9). Continuing to proceed in a fashion analogous to the Gaussian elimination method for linear systems, we may for example now perform a "backsolving" process. That is, inserting each of the roots of R_1 into one of the resultants involving only the variables z_1 and z_2, we can solve for the second co-ordinates of the zero points. Thus backsolving recursively, we obtain a set of points containing the zero points of the polynomial system. Now the actual zero points can be obtained by a final testing. A similar algorithm using (exact) integer arithmetic has been developed by Collins (1971), see also Ojika (1982).

It seems that to date the above direct methods and the homotopy methods have not been contrasted against each other. An obvious advantage of the direct methods is that it is possible to solve for only real zero points if the coefficients in (11.6.9) are real. Thus, if one is only interested in obtaining real zero points, it is possible to avoid tracing a large number of irrelevant paths which might arise in the homotopy method. This also pertains to solutions at infinity. On the other hand, the direct methods may incur serious numerical instability problems and extremely high degrees in the resultants. Recently, Lazard (1981) has given a new direct method based on deeper results in algebraic geometry which may make it possible to alleviate these drawbacks.

11.7 Nonlinear Complementarity

Let us next give a brief discussion of the nonlinear complementarity problem (NLCP), which is stated in the following form:

(11.7.1) The Nonlinear Complementarity Problem (NLCP). *Let* $g :$ $\mathbf{R}^N \to \mathbf{R}^N$ *be a continuous map. Find an* $x \in \mathbf{R}^N$ *such that*

$$(1) \quad x \in \mathbf{R}_+^N; \quad (2) \quad g(x) \in \mathbf{R}_+^N; \quad (3) \quad x^* g(x) = 0.$$

Here $\mathbf{R}_+$ *denotes the set of non-negative real numbers, and below we also denote the set of positive real numbers by* $\mathbf{R}_{++}$*. If* $g(0) \in \mathbf{R}_+^N$*, then* $x = 0$ *is a trivial solution to the problem. Hence this trivial case is always excluded and the additional assumption*

$$(4) \quad g(0) \notin \mathbf{R}_+^N$$

is made.

Nonlinear complementarity problems arise in nonlinear programming, suitable discretizations of variational inequalities and the determination of economic equilibria. Our reasons for including this topic here also stem from the historical fact that the papers of Lemke & Howson (1964) and Lemke (1965) gave an algorithm for solving the linear complementarity problem (where $g(x) = Ax + b$ is affine) by using complementary pivoting steps. These ideas were then utilized by Scarf (1967) in his constructive proof of the Brouwer fixed point theorem. These two papers in turn initiated a grat deal of research on complementary pivoting and piecewise linear fixed point algorithms. We cannot delve very deeply into the interrelationship between the NLCP and continuation methods. The following articles and books touch on this topic: Balinski & Cottle (1978), Cottle (1972), Doup & van der Elzen & Talman (1986), Eaves (1971), (1976), (1978), (1983) Eaves & Scarf (1976), Eaves & Gould & Peitgen & Todd (1983), Fisher & Gould (1974), Fisher & Tolle (1977), Garcia & Zangwill (1981), Gould & Tolle (1983), van der Heyden (1980), Karamardian (1977), Kojima (1975), (1978), van der Laan & Talman (1985), Lüthi (1976), Megiddo (1978), Megiddo & Kojima (1977), Saigal (1976), Saigal & Simon (1973), Scarf & Hansen (1973), Talman & Van der Heyden (1983), Todd (1976), Watson (1979).

We begin with a useful

(11.7.2) Definition. *For* $x \in \mathbf{R}^N$ *we define the positive part* $x_+ \in \mathbf{R}_+^N$ *by* $x_+[i] = \max\{x[i], 0\}$ *for* $i = 1, \ldots, N$ *and the negative part* $x_- \in \mathbf{R}_+^N$ *by* $x_- = (-x)_+$*. The following formulas are then obvious:*

$$(1) \quad x = x_+ - x_-; \quad (2) \quad (x_+)^*(x_-) = 0.$$

The next proposition which is not difficult to prove gives a simple and elegant equivalence between an NLCP and a zero point problem, cf. Megiddo & Kojima (1977):

(11.7.3) Proposition. *Let us define* $f : \mathbf{R}^N \to \mathbf{R}^N$ *by* $f(z) = g(z_+) - z_-$. *If* x *is a solution of the NLCP (11.7.1) then* $z := x - g(x)$ *is a zero point of* f. *Conversely, if* z *is a zero point of* f, *then* $x := z_+$ *solves the NLCP (11.7.1).*

As a consequence of the above proposition, we are naturally led to introduce the homotopy $H : \mathbf{R}^N \times [0,1] \to \mathbf{R}^N$ defined by

(11.7.4)
$$H(z,\lambda) = (1-\lambda)z + \lambda\big(g(z_+) - z_-\big)$$
$$= z + \lambda\big(g(z_+) - z_+\big)$$

in order to numerically solve the NLCP problem. In analogy to our discussion after (11.2.12) we need to establish a boundary condition

(11.7.5)
$$H(z,\lambda) \neq 0 \quad \text{for} \quad z \in \partial\Omega,\ \lambda \in [0,1]$$

for a suitable open bounded neighborhood $\Omega \subset \mathbf{R}^N$ of 0, in order to assure the success of this approach. We will derive (11.7.5) from a coercivity condition which is typically used to guarantee the existence of a solution of (11.7.1):

(11.7.6) Coercivity Condition. *Let* $V \subset \mathbf{R}^N$ *be a bounded open neighborhood of 0. Then a coercivity condition for* g *on* $\partial\Omega$ *can be stated in the following way:*
$$x^*g(x) > 0 \quad \text{for all}\ \ x \in \partial V\ \ \text{such that}\ \ x \in \mathbf{R}^N_+.$$

To see how (11.7.5) can be obtained from (11.7.6), let us choose a constant
$$\alpha > \beta := \max\{\|g(x)\|_\infty + \|x\|_\infty \mid x \in \overline{V},\ x \in \mathbf{R}^N_+\}$$

and define
$$\Omega := \{z \in \mathbf{R}^N \mid \|z_-\|_\infty < \alpha,\ z_+ \in V\}.$$

Then there are only two possibilities to consider for a point $z \in \partial\Omega$ and $\lambda \in [0,1]$:

$$z[i] = -\alpha \quad \text{for some}\ i \quad \Longrightarrow$$
(1)
$$H(z,\lambda)[i] = z[i] + \lambda\big(g(z_+)[i] - z_+[i]\big)$$
$$= -\alpha + \lambda g(z_+)[i] \leq -\alpha + \beta < 0.$$

$$z_+ \in \partial V \quad \Longrightarrow$$
(2)
$$(z_+)^*H(z,\lambda) = (z_+)^*\big[(1-\lambda)z + \lambda\big(g(z_+) - z_-\big)\big]$$
$$= (1-\lambda)\|z_+\|^2 + \lambda(z_+)^*g(z_+) > 0.$$

In both cases, the boundary condition (11.7.5) follows immediately. At this point we would normally try to give some argument that 0 is a regular value of H and that the curve $\mathcal{C} \subset H^{-1}(0)$ containing $(0,0)$ is a smooth curve which, because of (11.7.5), reaches a point $(\bar{z}, 1)$ such that $f(\bar{z}) = 0$. However, in the present case, since H is not smooth, we can only assert that $\mathcal{C}$ is a continuum, cf. Browder (1960) or Rabinowitz (1971) for general techniques to prove such connectedness assertions. The problem which remains is to discuss how to numerically implement a "tracing" of such continua. There are essentially three methods available:

(1) The piecewise linear or complementary pivoting methods can be viewed as a numerical tool to trace such continua. This observation is essentially due to Eaves (1971), (1972). In chapter 13 we show how the NLCP problem can be solved by some standard PL homotopy methods. More generally, PL methods may be viewed as numerical realizations of Leray-Schauder degree theory, as has been observed by Peitgen & Prüfer (1979), see also Peitgen (1982).

(2) As was shown by Alexander & Kellogg & Li & Yorke (1979), it is possible to construct a theory of piecewise smooth maps, including a version of the implicit function theorem and a Sard type theorem. Then $H^{-1}(0)$ can be traced by a predictor-corrector type algorithm where some minor updates have to be performed whenever a new piece of smoothness is encountered, see Kojima (1981) and Kojima & Hirabayashi (1984). To our knowledge, little work has been done concerning numerical implementations of this approach.

(3) It is possible to construct a more sophisticated homotopy $\tilde{H}(z, \lambda)$ than (11.7.4) such that the restriction to $\mathbf{R}^N \times (-\infty, 1)$ is smooth and such that $\tilde{H}(\cdot, 1) = f(z)$. Then standard numerical continuation methods can be used for a numerical tracing of $\tilde{H}^{-1}(0)$, cf. Watson (1979) for more details.

In (1979), Khachiyan started a new class of polynomial time algorithms for solving the linear programming problem. Karmarkar (1984) subsequently gave a much noted polynomial time algorithm based upon projective rescaling. Gill & Murray & Saunders & Tomlin & Wright (1986) noted that Karmarkar's algorithm is equivalent to a projected Newton barrier method which in turn is closely related to a recent class of polynomial time methods involving a continuation method, namely the tracing of the "path of centers". This last idea can be generalized to quadratic programming problems, and both linear and nonlinear complementarity problems. For details, we refer to e.g. Gonzaga (1987), Jarre & Sonnevend & Stoer (1987), Kojima & Mizuno & Noma (1987), Kojima & Mizuno & Yoshise (1987), Megiddo (1986), Renegar (1988), Sonnevend & Stoer (1988), Tanabe (1987). As an example, we outline the continuation approach of Kojima & Mizuno & Noma for tracing the path of centers for the general nonlinear complementarity problem (11.7.1). For the case of the linear complementarity problem, Mizuno & Yoshise & Kikuchi (1988) present several implementations and report computational experience which confirms the polynomial complexity.

The above mentioned continuation method actually turns out to be the global homotopy method, cf. section 11.4, for a map $G : \mathbf{R}^{2N} \to \mathbf{R}^{2N}$ defined by

$$(11.7.7) \qquad G(x, y) := \begin{pmatrix} \mathrm{diag}(x)y \\ g(x) - y \end{pmatrix} \quad \text{for} \quad x, y \in \mathbf{R}^N,$$

where $\text{diag}(x)$ denotes the $N \times N$ diagonal matrix whose diagonal entries are the components of x. Note that the NLCP (11.7.1) can equivalently be stated as the zero point problem

$$(11.7.8) \qquad G(x, y) = 0, \ x, y \in \mathbf{R}_+^N.$$

To solve this zero point problem, we introduce the global homotopy

$$(11.7.9) \qquad H(x, y, \lambda) := G(x, y) - \lambda G(x_1, y_1),$$

where

$$x_1, y_1 \in \mathbf{R}_{++}^N$$

are chosen starting points. Let us assume that g is a **uniform P-function** on $\mathbf{R}_+^N$, i.e. there exists an $\alpha > 0$ such that

$$(11.7.10) \qquad \max_{i=1,\dots,N} e_i^* \text{diag}\big(g(u) - g(v)\big)(u - v) \geq \alpha \|u - v\|^2$$

holds for all $u, v \in \mathbf{R}_+^N$. Then, using results of Moré (1974), it can be seen that the homotopy equation

$$H(x, y, \lambda) = 0$$

has exactly one solution $(x(\lambda), y(\lambda)) \in \mathbf{R}_+^{2N}$ for every $\lambda \geq 0$, and the map

$$\lambda \in [0, 1] \ \mapsto \ (x(\lambda), y(\lambda)) \in \mathbf{R}_+^{2N}$$

is a continuous curve which can be traced from the initial point $(x(1), y(1))$ downto the unique solution $(x(0), y(0))$ of the NLCP (11.7.1). In the special case of a linear complementarity problem, Kojima & Mizuno & Yoshise (1987) indicated a way to trace this curve to $\lambda = 0$ in polynomial time.

11.8 Critical Points and Continuation Methods

Among the applications of numerical continuation methods is the calculation of critical points of a smooth mapping $f : \mathbf{R}^N \to \mathbf{R}$. In general, one chooses a smooth mapping $g : \mathbf{R}^N \to \mathbf{R}$ with known regular **critical points** $a \in \mathbf{R}^N$ i.e. $\nabla g(a) = 0$ for the gradient and the Hessian $\nabla g'(a)$ has full rank. One then formulates a smooth homotopy map $H : \mathbf{R}^{N+1} \to \mathbf{R}^N$ such that

$$H(x,0) = \nabla g(x) \quad \text{and} \quad H(x,1) = \nabla f(x).$$

Typically, one uses the convex homotopy

(11.8.1) $$H(\lambda, x) := (1 - \lambda)\nabla g(x) + \lambda \nabla f(x).$$

The numerical aspect then consists of tracing a smooth curve

$$c(s) = (\lambda(s), x(s)) \in H^{-1}(0)$$

with starting point $c(0) = (0, a)$ for some given critical point a of g, and starting tangent $\dot{c}(0) = (\dot{\lambda}(0), \dot{x}(0))$ with $\dot{\lambda}(0) > 0$. The aim of course is to trace the curve c until the homotopy level $\lambda = 1$ is reached, at which a critical point of f is obtained. If all critical points of f are regular, then by Sard's theorem (11.2.2) it is generally possible to make a choice of g such that zero is a regular value of H. The following result of Allgower & Georg (1980) indicates that the continuation method has an appealing property which can permit targeting critical points having a specific Morse index.

(11.8.2) Theorem. *Let $f, g : \mathbf{R}^N \to \mathbf{R}$ be smooth functions and let H be the convex homotopy of (11.8.1) which has zero as a regular value. Let $c(s) = (\lambda(s), x(s)) \in H^{-1}(0)$ be the smooth curve obtained by the defining initial value problem*

$$\dot{c}(s) = \sigma t(H'(c(s))),$$
$$c(0) = (0, a),$$

where a is a regular critical point of g and $\sigma \in \{+1, -1\}$ is a fixed orientation. Suppose that $\lambda(s)$ is increasing for $s \in [0, \bar{s}]$, $\lambda(\bar{s}) = 1$, and that the critical point $b := x(\bar{s})$ of ∇f is regular. Then the critical points a, b of g and f respectively, have the same Morse index i.e. the Hessians $\nabla g'(a)$ and $\nabla f'(b)$ have the same number of negative eigenvalues.

Proof. From the defining initial value problem we obtain

$$\begin{pmatrix} \dot{\lambda}(s) & \dot{x}(s)^* \\ H_\lambda(c(s)) & H_x(c(s)) \end{pmatrix} \begin{pmatrix} \dot{\lambda}(s) \\ \dot{x}(s) \end{pmatrix} = \begin{pmatrix} 1 \\ 0 \end{pmatrix},$$

where subindices indicate partial derivatives. This implies that

$$(11.8.3) \qquad \begin{pmatrix} \dot{\lambda}(s) & \dot{x}(s)^* \\ H_\lambda(c(s)) & H_x(c(s)) \end{pmatrix} \begin{pmatrix} \dot{\lambda}(s) & 0^* \\ \dot{x}(s) & \mathrm{Id} \end{pmatrix} = \begin{pmatrix} 1 & \dot{x}(s)^* \\ 0 & H_x(c(s)) \end{pmatrix} .$$

Since the determinant of the augmented Jacobian never changes sign, cf. (2.1.5), it follows from the above equation that $\det H_x(c(s))$ changes sign exactly when $\dot{\lambda}(s)$ changes sign. The latter does not occur, and $H_x(c(s))$ can have at most one eigenvalue equal to zero, since all points $c(s)$ are regular points of H. Using a result of perturbation theory, cf. Dunford & Schwartz (1963), namely that an isolated eigenvalue of the symmetric matrix $H_x(c(s))$ depends smoothly on s, we conclude that $H_x(0,a) = \nabla g'(a)$ and $H_x(1,b) = \nabla f'(b)$ have the same number of negative eigenvalues. $\qquad \square$

The drawback to applying the above theorem is that in general it is difficult to choose the initial function g so that the curve c has a monotone λ co-ordinate and reaches the level $\lambda = 1$ in finite arclength. It has been observed by Allgower & Georg (1983), Allgower (1984) and Keller (1988) that turning points in the λ co-ordinate lead to bifurcation points in a setting of complex extension. Hence by considering the complexification of the homotopy map (and thereby doubling the number of variables), it is possible to extract a piecewise smooth curve which is always increasing in λ. It is nevertheless not assured that such a curve will reach $\lambda = 1$ unless some boundary condition such as

$$H(z,t) \neq 0 \quad \text{for} \quad z \in \partial\Omega,\ t \in [0,1]$$

is assumed for a suitable open bounded neighborhood $\Omega \subset \mathbf{R}^N$ of a. Furthermore, only real solutions correspond to critical points, and hence the method is only successful if a real solution is reached at level $\lambda = 1$. Even if the latter is achieved, the Morse index is not necessarily preserved if bifurcations have been traversed. Let us now briefly outline this approach.

We go into a more general setting and assume that $H : \mathbf{R} \times \mathbf{R}^N \to \mathbf{R}^N$ is a smooth homotopy. The important assumption which we need to make is that H is real analytic in the variables x. Hence it is meaningful to replace x in $H(\lambda,x)$ by $z \in \mathcal{C}^N$. In the following we use the notation $z = x+iy$ for $z \in \mathcal{C}^N$, where $x,y \in \mathbf{R}^N$ denote the real and imaginary parts of z respectively. Note that $\overline{H(\lambda,\overline{z})} = H(\lambda,z)$ since H is real analytic. Let us define the real and imaginary parts $H^r, H^i : \mathbf{R} \times \mathbf{R}^N \times \mathbf{R}^N \to \mathbf{R}^N$ by

$$(11.8.4) \qquad \begin{aligned} H^r(\lambda,x,y) &:= \frac{1}{2}\big(H(\lambda,z) + H(\lambda,\overline{z})\big), \\ H^i(\lambda,x,y) &:= \frac{-i}{2}\big(H(\lambda,z) - H(\lambda,\overline{z})\big), \end{aligned}$$

and the map $\hat{H} : \mathbf{R} \times \mathbf{R}^N \times \mathbf{R}^N \to \mathbf{R}^N \times \mathbf{R}^N$ by

(11.8.5)
$$\hat{H}(\lambda, x, y) := \begin{pmatrix} H^r(\lambda, x, y) \\ -H^i(\lambda, x, y) \end{pmatrix}.$$

Let $\hat{c} : s \mapsto (\lambda(s), x(s), y(s))$ be a smooth curve in $\hat{H}^{-1}(0)$, where for simplicity s is an arclength parameter. Differentiating $\hat{H}(\lambda(s), x(s), y(s)) = 0$ with respect to s yields

(11.8.6)
$$(\hat{H}_\lambda \quad \hat{H}_x \quad \hat{H}_y) \begin{pmatrix} \dot{\lambda} \\ \dot{x} \\ \dot{y} \end{pmatrix} = \begin{pmatrix} 0 \\ 0 \end{pmatrix}.$$

From (11.8.4) we obtain the Cauchy-Riemann equations

(11.8.7)
$$H^r_y = -H^i_x \quad \text{and} \quad H^i_y = H^r_x.$$

By using these and augmenting (11.8.6) in an obvious way, we obtain

(11.8.8)
$$\begin{pmatrix} \dot{\lambda} & \dot{x}^* & \dot{y}^* \\ \hat{H}_\lambda & \hat{H}_x & \hat{H}_y \end{pmatrix} \begin{pmatrix} \dot{\lambda} & 0^* & 0^* \\ \dot{x} & \text{Id} & 0 \\ \dot{y} & 0 & \text{Id} \end{pmatrix} = \begin{pmatrix} 1 & \dot{x}^* & \dot{y}^* \\ 0 & H^r_x & -H^i_x \\ 0 & -H^i_x & -H^r_x \end{pmatrix}$$

and

(11.8.9)
$$\dot{\lambda} \det \begin{pmatrix} \dot{\lambda} & \dot{x} & \dot{y} \\ \hat{H}_\lambda & \hat{H}_x & \hat{H}_y \end{pmatrix} = \det \begin{pmatrix} H^r_x & -H^i_x \\ -H^i_x & -H^r_x \end{pmatrix} = \det \hat{H}_{(x,y)}.$$

The Cauchy-Riemann equations (11.8.7) show that

$$\hat{H}_{(x,y)} = \begin{pmatrix} H^r_x & -H^i_x \\ -H^i_x & -H^r_x \end{pmatrix},$$

and hence $\hat{H}_{(x,y)}$ is symmetric. Furthermore, if μ is an eigenvalue of $\hat{H}_{(x,y)}$ having a corresponding eigenvector $\binom{u}{v}$, then so is $-\mu$ an eigenvalue having a corresponding eigenvector $\binom{v}{-u}$. Hence the eigenvalues of $\hat{H}_{(x,y)}$ occur in symmetric pairs about zero, and $\det \hat{H}_{(x,y)}$ never changes sign. Consequently, if U is a neighborhood of a parameter value $\bar{s}$ such that $\hat{c}(s)$ are regular points of $\hat{H}$ for $s \in U$, $s \neq \bar{s}$, then (11.8.9) shows that $\dot{\lambda}(\hat{c}(s))$ changes sign at $s = \bar{s}$ if and only if

$$\det \begin{pmatrix} \dot{\lambda} & \dot{x}^* & \dot{y}^* \\ \hat{H}_\lambda & \hat{H}_x & \hat{H}_y \end{pmatrix}\bigg|_{\hat{c}(s)}$$

does. Hence, a turning point of $\hat{c}$ with respect to the λ parameter is also a bifurcation point of the equation $\hat{H} = 0$. Let us now show that in the case of a real solution curve, the corresponding bifurcation point must be simple in the sense of (8.1.11).

(11.8.10) Proposition. *Under the above assumptions, let the curve $s \longmapsto \hat{c}(s) = (\lambda(s), x(s), 0)$ be a "real" solution curve of $\hat{H}^{-1}(0)$ such that the point $(\lambda(\bar{s}), x(\bar{s}))$ is a regular point of the real homotopy H. Suppose that $(\lambda(\bar{s}), x(\bar{s}))$ is a simple turning point of the equation $H = 0$ i.e. $\dot{\lambda}(\bar{s}) = 0$ and $\ddot{\lambda}(\bar{s}) \neq 0$. Then $\hat{c}(\bar{s})$ is a simple bifurcation point of the equation $\hat{H} = 0$.*

Proof. Since H is real analytic, it is easy to see that

(11.8.11) $H^i_x(\lambda, x, 0) = 0, \quad H^i_{xx}(\lambda, x, 0) = 0, \quad H^i_\lambda(\lambda, x, 0) = 0,$

etc. holds for $x \in \mathbf{R}^N$ and $\lambda \in \mathbf{R}$. Hence, using the Cauchy-Riemann equations (11.8.7), the augmented Jacobian takes the form

(11.8.12) $\begin{pmatrix} \dot{\lambda} & \dot{x}^* & \dot{y}^* \\ \hat{H}_\lambda & \hat{H}_x & \hat{H}_y \end{pmatrix}\Bigg|_{\hat{c}(\bar{s})} = \begin{pmatrix} 0 & \dot{x}^* & 0^* \\ H^r_\lambda & H^r_x & 0 \\ 0 & 0 & -H^r_x \end{pmatrix}\Bigg|_{\hat{c}(\bar{s})}.$

Since the rank of

$$\begin{pmatrix} 0 & \dot{x}^* \\ H_\lambda & H_x \end{pmatrix}\Bigg|_{(\lambda(\bar{s}), x(\bar{s}))}$$

is $N + 1$, we conclude that $\operatorname{rank} H^r_x(\hat{c}(\bar{s})) \geq N - 1$. On the other hand, the defining differential equation (11.8.6) implies that $H^r_x(\hat{c}(\bar{s}))\dot{x}(\bar{s}) = 0$, and hence $\operatorname{rank} H^r_x(\hat{c}(\bar{s})) = N - 1$. Therefore the matrix in (11.8.12) has rank $2N$, and the Jacobian $\hat{H}'(\hat{c}(\bar{s}))$ has a two-dimensional kernel spanned by the vectors

$$\begin{pmatrix} 0 \\ \dot{x}(\bar{s}) \\ 0 \end{pmatrix}, \begin{pmatrix} 0 \\ 0 \\ \dot{x}(\bar{s}) \end{pmatrix}.$$

It remains to show that the non-degeneracy conditions for the second derivatives hold, cf. (8.1.11)(3). Let e span the kernel of $H^r_x(\hat{c}(\bar{s}))^*$. Then $\binom{0}{e}$ spans the kernel of $\hat{H}'(\hat{c}(\bar{s}))^*$. Furthermore,

(11.8.13) $e^* H^r_\lambda(\hat{c}(\bar{s})) \neq 0$

since otherwise the above kernel would not have dimension one. We have to investigate whether the bilinear form

$$(\xi, \eta) \longmapsto (0, e^*)\hat{H}''(\hat{c}(\bar{s})) \left[\xi \begin{pmatrix} 0 \\ \dot{x}(\bar{s}) \\ 0 \end{pmatrix}, \eta \begin{pmatrix} 0 \\ 0 \\ \dot{x}(\bar{s}) \end{pmatrix} \right]$$

has one positive and one negative eigenvalue. Using (11.8.11) and the Cauchy-Riemann equations (11.8.7), a straightforward calculation shows that the above bilinear form reduces to

(11.8.14) $(\xi, \eta) \longmapsto 2\xi\eta \, e^* H^i_{xy}(\hat{c}(\bar{s}))[\dot{x}(\bar{s}), \dot{x}(\bar{s})].$

It is clear that the simple bilinear form (11.8.14) has one positive and one negative eigenvalue if and only if

(11.8.15) $$e^* H_{xy}^i(\hat{c}(\bar{s}))[\dot{x}(\bar{s}), \dot{x}(\bar{s})] \neq 0.$$

To show this, let us differentiate the equation $e^* H(\lambda(s), x(s)) = 0$ twice. Using the facts $e^* H_x(\lambda(\bar{s}), x(\bar{s})) = 0$, $H_x(\lambda(\bar{s}), x(\bar{s}))\dot{x}(\bar{s}) = 0$ and $\dot{\lambda}(\bar{s}) = 0$, we obtain

$$e^* H_\lambda(\lambda(\bar{s}), x(\bar{s}))\ddot{\lambda}(\bar{s}) + e^* H_{xx}(\lambda(\bar{s}), x(\bar{s}))[\dot{x}(\bar{s}), \dot{x}(\bar{s})] = 0.$$

Since $\ddot{\lambda}(\bar{s}) \neq 0$, we can conclude from (11.8.13) that $e^* H_{xx}(\lambda(\bar{s}), x(\bar{s})) \neq 0$. Now (11.8.15) follows from the Cauchy-Riemann equations (11.8.7). $\quad\square$

Let us finally show that at bifurcation points at which one of the two solution branches is real, the corresponding curvatures are of opposite sign, and hence at such points a choice of following branches is available so that the λ co-ordinate is increasing.

(11.8.16) Proposition. *Under the assumptions of (11.8.10), let us now denote the "real" solution curve in $\hat{H}^{-1}(0)$ by $c_1(s) =: (\lambda_1(s), x_1(s), 0)$ and the bifurcating solution curve in $\hat{H}^{-1}(0)$ by $c_2(s) =: (\tilde{\lambda}(s), x_2(s), y_2(s))$. The curves are defined for s near $\bar{s}$, and $\bar{u} := c_1(\bar{s}) = c_2(\bar{s})$ is the bifurcation point. Then $\ddot{\lambda}_1(\bar{s}) = -\ddot{\lambda}_2(\bar{s})$.*

Proof. Let us denote by $c(s) =: (\tilde{\lambda}(s), x(s), y(s))$ either of the two solution curves c_1 or c_2. Differentiating $\hat{H}(c(s)) = 0$ twice with respect to s and taking $\lambda(\bar{s}) = 0$ into account yields

(11.8.17)
$$\hat{H}_\lambda(\bar{u})\ddot{\lambda}(\bar{s}) + \hat{H}_{xx}(\bar{u})[\dot{x}(\bar{s}), \dot{x}(\bar{s})] + 2\hat{H}_{xy}(\bar{u})[\dot{x}(\bar{s}), \dot{y}(\bar{s})] +$$
$$\hat{H}_{yy}(\bar{u})[\dot{y}(\bar{s}), \dot{y}(\bar{s})] + \hat{H}_x(\bar{u})\ddot{x}(\bar{s}) + \hat{H}_y(\bar{u})\ddot{y}(\bar{s}) = 0.$$

Let e span the kernel of $H_x^r(\hat{c}(\bar{s}))^*$ as in the previous proof. Multiplying (11.8.17) from the left with $(e^*, 0)$ and taking the properties (11.8.7) and (11.8.11) into account, we obtain

(11.8.18) $e^* H_\lambda^r(\bar{u})\ddot{\lambda}(\bar{s}) + e^* H_{xx}^r(\bar{u})[\dot{x}(\bar{s}), \dot{x}(\bar{s})] + e^* H_{yy}^r(\bar{u})[\dot{y}(\bar{s}), \dot{y}(\bar{s})] = 0.$

Since

(11.8.19) $$\dot{c}_1(\bar{s}) = (0, \dot{x}_1(\bar{s}), 0)$$

holds, it can be seen from (11.8.14) and (8.1.13) that

(11.8.20) $$\dot{c}_2(\bar{s}) = (0, 0, \pm\dot{x}_1(\bar{s})).$$

Substituting (11.8.19–20) into (11.8.18) we obtain

(11.8.21)
$$e^* H_\lambda^r(\bar{u})\ddot{\lambda}_1(\bar{s}) + e^* H_{xx}^r(\bar{u})[\dot{x}_1(\bar{s}), \dot{x}_1(\bar{s})] = 0,$$
$$e^* H_\lambda^r(\bar{u})\ddot{\lambda}_2(\bar{s}) + e^* H_{yy}^r(\bar{u})[\dot{x}_1(\bar{s}), \dot{x}_1(\bar{s})] = 0,$$

respectively. Since $e^* H_\lambda^r(\bar{u}) \neq 0$, see (11.8.13), and since $H_{xx}^r(\bar{u}) = -H_{yy}^r(\bar{u})$ by (11.8.7), the assertion follows. $\quad\square$

More general studies of the behavior of critical points in parametric optimization are to be found in the books of Jongen & Jonker & Twilt. Also semi-infinite problems can in principle be regarded as parametric optimization problems, see e.g. Jongen & Zwier.

Chapter 12. PL Continuation Methods

12.1 Introduction

In previous chapters we assumed that the map $H : \mathbf{R}^{N+1} \to \mathbf{R}^N$ was smooth, that zero was a regular value, and that $H^{-1}(0)$ was a collection of disjoint smooth curves which could be numerically traced using PC-methods. Now we will discuss piecewise linear (**PL**) methods which can again be viewed as curve tracing methods, but the map H can now be arbitrary. The map H is approximated by a piecewise linear map $H_{\mathcal{T}}$ which affinely interpolates H at the nodes of a triangulation $\mathcal{T}$ of $\mathbf{R}^{N+1}$. The PL methods trace the piecewise linear 1-manifold $H_{\mathcal{T}}^{-1}(0)$. A connected component of the piecewise linear 1-manifold consists of a polygonal path which is obtained by successively stepping through certain "transverse" $(N+1)$-dimensional simplices of the triangulation. Although the PL method works for arbitrary maps, only under some smoothness assumptions on H can one obtain truncation error estimates in terms of the meshsize of the underlying triangulation. In order to be able to discuss these methods it is necessary to introduce a few combinatorial ideas. The first notions we need are those of a simplex and a triangulation.

(12.1.1) Definition. *A set of points $\{v_1, v_2, \ldots, v_{k+1}\}$ in $\mathbf{R}^m$ is said to be affinely independent (also called in general position) if the following matrix has full rank i.e. if its columns are linearly independent:*

$$(12.1.2) \qquad \begin{pmatrix} 1 & 1 & \ldots & 1 \\ v_1 & v_2 & \ldots & v_{k+1} \end{pmatrix}.$$

Equivalently, $v_1, v_2, \ldots, v_{k+1}$ are affinely independent if the differences $v_2 - v_1$, $v_3 - v_1, \ldots, v_{k+1} - v_1$ are linearly independent vectors in $\mathbf{R}^m$. Note that one point is always affinely independent.

The notion of a simplex is basic for the description of PL methods.

(12.1.3) Definition. *For any set of $k + 1$ affinely independent points*

$$\{v_1, v_2, \ldots, v_{k+1}\}$$

we define the k-dimensional **simplex** with **vertices** $\{v_1, v_2, \ldots, v_{k+1}\}$ to be the convex hull

$$(12.1.4) \qquad \left\{ v = \sum_{i=1}^{k+1} \alpha_i v_i \;\middle|\; \alpha_1 \geq 0,\, \alpha_2 \geq 0,\, \ldots,\, \alpha_{k+1} \geq 0,\, \sum_{i=1}^{k+1} \alpha_i = 1 \right\}.$$

Usually we will denote such simplices by Greek letters such as σ or τ, and in particular by $[v_1, v_2, \ldots, v_{k+1}]$ if we want to be specific about the vertices of (12.1.4). To abbreviate our notation, a k-dimensional simplex is often called a k-simplex. Note that a 0-simplex is a singleton. The above coefficients $\alpha_1, \ldots, \alpha_{k+1}$ are usually called the **barycentric co-ordinates** of v with respect to the affine basis $v_1, v_2, \ldots, v_{k+1}$.

(12.1.5) Definition. Let $\sigma = [v_1, v_2, \ldots, v_{k+1}]$ be a k-simplex and let $1 \leq l \leq k+1$. If $\{w_1, w_2, \ldots, w_{l+1}\} \subset \{v_1, v_2, \ldots, v_{k+1}\}$ is a subset of vertices of σ, we call the l-simplex $\tau = [w_1, w_2, \ldots, w_{l+1}]$ an l-dimensional face of σ, or simply an l-face. Of particular interest are:

(1) the 0-faces which are singletons containing one **vertex** of σ;
(2) the 1-faces which are also called the **edges** of σ;
(3) the $(k-1)$-faces which are called the **facets** of σ.

The latter play an important role in PL methods and are obtained by dropping just one vertex of σ, say v_i. Let us denote by the symbol $\hat{}$ the deletion of the element beneath it. Then the facet $\tau = [v_1, \ldots, \hat{v}_i, \ldots, v_{k+1}]$ of σ is called the facet of σ lying opposite the vertex v_i of σ. An important point of σ is its the **barycenter** $(k+1)^{-1} \sum_{i=1}^{k+1} v_i$.

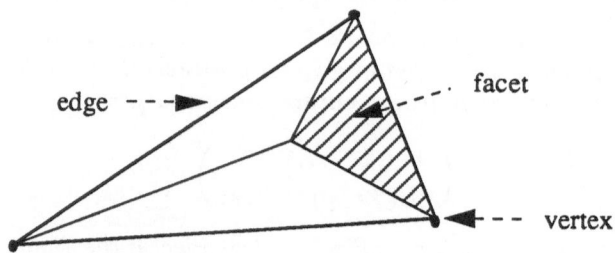

Figure 12.1.a A simplex of dimension 3

A simple example of a triangulation is obtained by triangulating the whole Euclidean space.

(12.1.6) Definition. Let $\mathcal{T}$ be a non-empty family of $(N+1)$-simplices in $\mathbf{R}^{N+1}$. We call $\mathcal{T}$ a triangulation of $\mathbf{R}^{N+1}$ if

(1)

$$\bigcup_{\sigma \in \mathcal{T}} \sigma = \mathbf{R}^{N+1};$$

(2) *the intersection $\sigma_1 \cap \sigma_2$ of two simplices σ_1, $\sigma_2 \in \mathcal{T}$ is empty or a **common** face of both simplices;*

(3) *the family $\mathcal{T}$ is locally finite i.e. any compact subset of $\mathbf{R}^{N+1}$ meets only finitely many simplices $\sigma \in \mathcal{T}$.*

Figure 12.1.b illustrates cases where condition (12.1.6)(2) is satisfied and not satisfied, respectively.

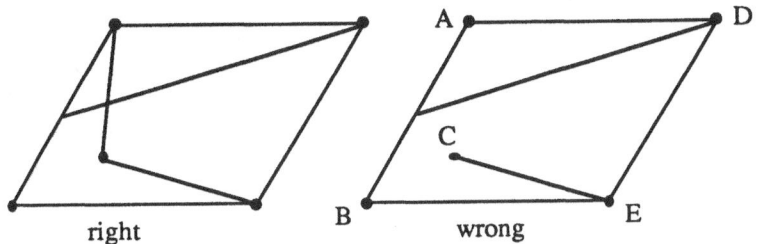

Figure 12.1.b The simplices [A,B,D] and [C,D,E] meet in [C,D] which is an edge of [C,D,E] but not of [A,B,D]

(12.1.7) Definition. *Let $\mathcal{T}$ be a triangulation of $\mathbf{R}^{N+1}$, and let $0 \leq k \leq N+1$. We denote by*

$$\mathcal{T}^k := \{\tau \mid \tau \text{ is a } k\text{-face of some simplex } \sigma \in \mathcal{T}\}$$

*the family of all k-faces in $\mathcal{T}$. Of special interest are the singletons $\mathcal{T}^0$ which contain the vertices or **nodes** of $\mathcal{T}$, and the facets $\mathcal{T}^N$ of $\mathcal{T}$. Naturally, we identify $\mathcal{T} = \mathcal{T}^{N+1}$. Two simplices σ_1, $\sigma_2 \in \mathcal{T}$ are called **adjacent** if they meet in a common facet.*

Triangulations are not stored in a computer by storing each individual simplex. Instead, only a current simplex is stored, together with information about how to obtain adjacent simplices as needed. The steps from one simplex to an adjacent one are called "**pivoting steps**". They are basic for the dynamics of PL methods. The following lemma prepares their introduction.

(12.1.8) Lemma. *Let $\mathcal{T}$ be a triangulation of $\mathbf{R}^{N+1}$. Consider a simplex $\sigma \in \mathcal{T}$ and let τ be a facet of σ. Then there is a unique simplex $\tilde{\sigma} \in \mathcal{T}$ such that*

(1) $\tilde{\sigma} \neq \sigma$;
(2) τ *is a facet of $\tilde{\sigma}$.*

Proof. Let $\mathcal{H} \subset \mathbf{R}^{N+1}$ denote the hyperplane which contains the facet τ of σ. Then σ must lie on one side of $\mathcal{H}$; let us call this side the "*left*" side of

$\mathcal{H}$. Consider a straight line $s \in \mathbf{R} \mapsto c(s) \in \mathbf{R}^{N+1}$ such that $c(0)$ is a point in the relative interior of τ and such that the tangent $\dot{c}(0)$ points into the "right" side of $\mathcal{H}$. By the properties (1) and (3) in the definition (12.1.6) of a triangulation, there must exist at least one simplex $\tilde{\sigma} \in \mathcal{T}$ which contains the interval $\{c(s) \mid 0 \leq s \leq \varepsilon\}$ for some small $\varepsilon > 0$. By property (2) the simplex $\tilde{\sigma}$ must meet σ in the common facet τ. Of course, $\tilde{\sigma}$ lies on the "right" side of $\mathcal{H}$. Finally, we observe that two simplices in $\mathcal{T}$ which have the same facet τ and lie on the same side of $\mathcal{H}$ must have a common interior point and hence coincide. This shows the uniqueness of the simplex $\tilde{\sigma}$.   □

(12.1.9) Definition. *Let* $\sigma = [v_1, v_2, \dots, v_{N+2}]$ *be an* $(N+1)$*-simplex of a triangulation* $\mathcal{T}$ *of* $\mathbf{R}^{N+1}$, *and let* $\tau = [v_1, \dots, \hat{v}_i, \dots, v_{N+2}]$ *be the facet of* σ *lying opposite the vertex* v_i. *By the preceding lemma, there must exist a unique node* $\tilde{v}_i$ *i.e.* $[\tilde{v}_i] \in \mathcal{T}^0$ *which is different from* v_i *and such that* $\tilde{\sigma} = [v_1, \dots, \tilde{v}_i, \dots, v_{N+2}] \in \mathcal{T}$. *The passage from* σ *to* $\tilde{\sigma}$ *is called a pivoting step. We say that the vertex* v_i *of* σ *is pivoted into* $\tilde{v}_i$, *and that the simplex* σ *is pivoted into the simplex* $\tilde{\sigma}$ *across the facet* τ.

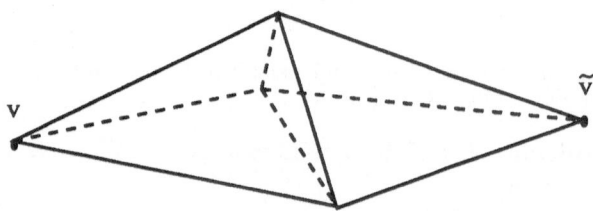

Figure 12.1.c The vertex v is pivoted into the vertex $\tilde{v}$

(12.1.10) Example. A following simple triangulation $\mathcal{T}$ of $\mathbf{R}^{N+1}$ which has been frequently used for computational purposes was already considered by Coxeter (1934) and Freudenthal (1942), see also Todd (1976). We will call this triangulation **Freudenthal's triangulation**. The nodes of $\mathcal{T}$ are those vectors $v \in \mathbf{R}^{N+1}$ which have integer co-ordinates $v[i] \in \mathbf{Z}$ for $i = 1, 2, \dots, N+1$. An $(N+1)$-simplex $\sigma \subset \mathbf{R}^{N+1}$ belongs to the triangulation $\mathcal{T}$ if the following rules are obeyed:

(1) the vertices of σ are nodes of $\mathcal{T}$ in the above sense;
(2) the vertices of σ can be ordered in such a way, say $\sigma = [v_1, v_2, \dots, v_{N+2}]$ that they are given by the following cyclic recursion formula

$$v_{i+1} = v_i + u_{\pi(i)}, \quad i = 1, \dots, N+1$$
$$v_1 = v_{N+2} + u_{\pi(N+2)},$$

where $u_1, \dots, u_{N+1}$ is the unit basis of $\mathbf{R}^{N+1}$, $u_{N+2} := -\sum_{i=1}^{N+1} u_i$ and

$$\pi : \{1, \dots, N+2\} \to \{1, \dots, N+2\}$$

is a permutation.

From the description of the pivoting rules below it is quite evident that these conditions define a triangulation $\mathcal{T}$ of $\mathbf{R}^{N+1}$. The formal proof is however somewhat technical, and we refer the reader to Todd (1976). The following diagram indicates how the vertices of a simplex σ are obtained:

$$\sigma: \qquad \cdots \quad \xrightarrow{u_{\pi(i-2)}} \quad v_{i-1} \quad \xrightarrow{u_{\pi(i-1)}} \quad v_i \quad \xrightarrow{u_{\pi(i)}} \quad v_{i+1} \quad \xrightarrow{u_{\pi(i+1)}} \quad \cdots$$

Then it is easy to see that the following diagram describes how the vertex v_i is pivoted into the vertex $\tilde{v}_i$:

$$\tilde{\sigma}: \qquad \cdots \quad \xrightarrow{u_{\pi(i-2)}} \quad v_{i-1} \quad \xrightarrow{u_{\pi(i)}} \quad \tilde{v}_i \quad \xrightarrow{u_{\pi(i-1)}} \quad v_{i+1} \quad \xrightarrow{u_{\pi(i+1)}} \quad \cdots$$

i.e. the directional vectors $u_{\pi(i-1)}$ and $u_{\pi(i)}$ are just switched.

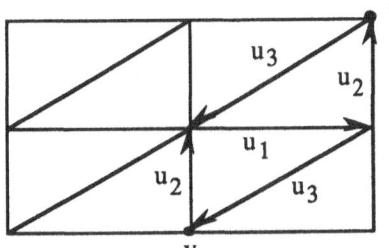

Figure 12.1.d The vertex v is pivoted into the vertex $\tilde{v}$ by reflection

This fact can also be expressed by the cyclic formulae

$$(12.1.11) \qquad \begin{aligned} \tilde{v}_i &= v_{i-1} - v_i + v_{i+1}, & i = 2, \ldots, N+1, \\ \tilde{v}_1 &= v_{N+2} - v_1 + v_2, \\ \tilde{v}_{N+2} &= v_{N+1} - v_{N+2} + v_1, \end{aligned}$$

which shows that $\tilde{v}_i$ is obtained by reflecting v_i across the center of the edge $[v_{i-1}, v_{i+1}]$ etc. This pivoting rule is to be understood in a **cyclic** way i.e. 0 has to be replaced by $N+2$ and $N+3$ has to be replaced by 1. It is easily programmed and has been called pivoting by reflection, cf. Allgower & Georg (1977). Figure 12.1.d illustrates a pivot by reflection. It has the advantage of being invariant under affine transformations, hence any triangulation which is obtained from Freudenthal's triangulation by some affine transformation obeys the above pivoting rule. Coxeter (1934) classified all triangulations $\mathcal{T}$ of $\mathbf{R}^{N+1}$ which are invariant under reflections across the hyperplanes which are generated by the N-faces of $\mathcal{T}$.

We summarize these facts in the following proposition, which was given in Allgower & Georg (1977).

(12.1.12) Proposition. *Let* $\sigma = [v_1, v_2, \ldots, v_{N+2}] \subset \mathbf{R}^{N+1}$ *be an* $(N+1)$-*simplex, and denote by* $\mathcal{T}$ *the family of all simplices which are obtained from* σ *by a repeated use of the pivoting rule (12.1.11). Then* $\mathcal{T}$ *is a triangulation of* $\mathbf{R}^{N+1}$, *in fact* $\mathcal{T}$ *is some affine image of Freudenthal's triangulation.*

12.2 PL Approximations

Let $H : \mathbf{R}^{N+1} \to \mathbf{R}^N$ be a map. We do not need any smoothness or continuity assumptions concerning H, unless we want to make precise statements about the truncation errors of the corresponding PL approximation. Given a triangulation $\mathcal{T}$ of $\mathbf{R}^{N+1}$, we intend to approximate the components of $H^{-1}(0)$ by using only the values of H on the nodes of $\mathcal{T}$. This leads us to the following

(12.2.1) Definition. *Let* $H : \mathbf{R}^{N+1} \to \mathbf{R}^N$ *be a map, let* $\mathcal{T}$ *be a triangulation of* $\mathbf{R}^{N+1}$, *and let* $\tau = [v_1, v_2, \ldots, v_{k+1}] \in \mathcal{T}^k$ *be a* k-*face of* $\mathcal{T}$ *where* $0 \leq k \leq N + 1$.

(1) *By* $H_\tau : \tau \to \mathbf{R}^N$ *we denote the uniquely defined affine map which coincides with* H *on the vertices* v_i *of* τ. *Using the notation of (12.1.4), we have*

$$H_\tau(v) := \sum_{i=1}^{k+1} \alpha_i H(v_i) \quad \text{for} \quad v = \sum_{i=1}^{k+1} \alpha_i v_i .$$

(2) *Let* $\operatorname{tng}(\tau)$ *("tangent space of* τ*") denote the linear space which contains all differences* $w_1 - w_2$ *with* $w_1, w_2 \in \tau$. *Then the Jacobian of* H_τ *is the linear map* $H'_\tau : \operatorname{tng}(\tau) \to \mathbf{R}^N$ *obtained by* $H'_\tau(w_1 - w_2) = H_\tau(w_1) - H_\tau(w_2)$ *for* $w_1, w_2 \in \tau$. *Since* H_τ *is affine, there is no ambiguity in this definition i.e.* $H_\tau(w_1) - H_\tau(w_2) = H_\tau(\tilde{w}_1) - H_\tau(\tilde{w}_2)$ *for* $w_1 - w_2 = \tilde{w}_1 - \tilde{w}_2$.

(3) *Finally, the* **PL approximation** *of* H *is obtained as the union*

$$H_\mathcal{T} = \bigcup_{\sigma \in \mathcal{T}} H_\sigma$$

i.e. $H_\mathcal{T}(v) = H_\sigma(v)$ *for* $v \in \sigma$ *and* $\sigma \in \mathcal{T}$. *There is no ambiguity in this definition, since it can be easily seen that* $H_{\sigma_1}(v) = H_{\sigma_2}(v)$ *for* $v \in \sigma_1 \cap \sigma_2$.

As in the case of smooth maps, it is also possible to avoid degenerate cases for PL maps by introducing corresponding concepts of regular points and regular values, cf. chapter 2.

(12.2.2) Definition. *A point* $x \in \mathbf{R}^{N+1}$ *is called a* **regular point** *of the PL map* $H_\mathcal{T}$ *if and only if*

(a) x *is not contained in any lower dimensional face* $\tau \in \mathcal{T}^k$ *for* $k < N$;

(b) H'_σ *has maximal rank for all* $\sigma \in T^N \cup T^{N+1}$ *such that* $x \in \sigma$.

A value $y \in \mathbf{R}^N$ *is a* **regular value** *of* H_T *if all points in* $H_T^{-1}(y)$ *are regular. By definition,* y *is vacuously a regular value if it is not contained in the range of* H_T. *If a point is not regular it is called* **singular**. *Analogously, if a value is not regular it is called singular.*

The perturbation arising in the following discussion can be viewed as a specific analogue for PL maps of the general perturbation c used in Sard's theorem (11.2.3), cf. Eaves (1976) and Peitgen & Siegberg (1981). For $\varepsilon > 0$, let us use the symbol

$$(12.2.3) \qquad \vec{\varepsilon} := \begin{pmatrix} \varepsilon^1 \\ \vdots \\ \varepsilon^N \end{pmatrix}$$

to denote the corresponding "ε-vector".

(12.2.4) Proposition. *For any compact subset* $C \subset \mathbf{R}^{N+1}$ *there are at most finitely many* $\varepsilon > 0$ *such that* $C \cap H_T^{-1}(\vec{\varepsilon})$ *contains a singular point of* H_T. *Consequently,* $\vec{\varepsilon}$ *is a regular value of* H_T *for almost all* $\varepsilon > 0$.

Proof. Let us assume that $\tau = [v_1, \ldots, v_{k+1}] \in T^k$ is a face for some $k \in \{0, 1, \ldots, N+1\}$ which contains solutions $z_j \in H_T^{-1}(\vec{\varepsilon}_j)$ for $j = 1, 2, \ldots, N+1$ such that

$$\varepsilon_1 > \varepsilon_2 > \cdots > \varepsilon_{N+1} > 0.$$

By using the notation

$$z_j = \sum_{i=1}^{k+1} \alpha_{j,i} v_i \in \tau,$$

definition (12.2.1) leads to the equations

$$(12.2.5) \qquad \begin{cases} \displaystyle\sum_{i=1}^{k+1} \alpha_{j,i} H(v_i) = \vec{\varepsilon}_j; \\ \displaystyle\sum_{i=1}^{k+1} \alpha_{j,i} = 1; \\ \alpha_{j,i} \geq 0 \quad \text{for } i = 1, 2, \ldots, k+1. \end{cases}$$

We introduce the matrices

$$\mathcal{L}(\tau) := \begin{pmatrix} 1 & \cdots & 1 \\ H(v_1) & \cdots & H(v_{k+1}) \end{pmatrix} ;$$

$$A_2 := \begin{pmatrix} \alpha_{1,1} & \cdots & \alpha_{N+1,1} \\ \vdots & \ddots & \vdots \\ \alpha_{1,k+1} & \cdots & \alpha_{N+1,k+1} \end{pmatrix} ;$$

$$A_3 := \begin{pmatrix} 1 & \cdots & 1 \\ \vec{\varepsilon}_1 & \cdots & \vec{\varepsilon}_{N+1} \end{pmatrix} .$$

$\mathcal{L}(\tau)$ is called the **labeling matrix** of H with respect to τ. The equations (12.2.5) can now be written as

(12.2.6) $\mathcal{L}(\sigma) A_2 = A_3 .$

Since A_3 is a nonsingular Van der Monde matrix, it follows that rank $\mathcal{L}(\sigma) = N + 1$. This implies $k \geq N$. Furthermore, it is an easy exercise to see that

$$\text{rank}\, \mathcal{L}(\tau) = \text{rank}\, H'_\tau + 1 .$$

Hence, rank $H'_\tau = N$. Now definition (12.2.2) implies that not all points $z_1, z_2, \ldots, z_{N+1}$ can be singular. The assertion now follows from the facts that a triangulation is locally finite, cf. (12.1.6)(3), and that $\mathbf{R}^{N+1}$ is a countable union of compact sets.   □

The ε-perturbation in (12.2.4) allows us to handle situations of degeneracy. This concept leads to the notion of lexicographically positive inverses of the labeling matrices, see section 12.4. Charnes (1952) seems to have been the first to use this idea in order to handle degeneracies in linear programming.

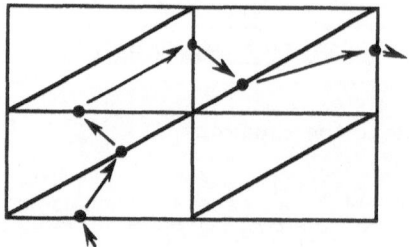

Figure 12.2.a The polygonal path $H_\mathcal{T}^{-1}(0)$

We make use of the PL approximation $H_\mathcal{T}$ of H by considering the approximation $H_\mathcal{T}^{-1}(0)$ of the solution manifold $H^{-1}(0)$. Except for degeneracies, we expect $H_\mathcal{T}^{-1}(0)$ to be a polygonal path with nodes on the N-faces $\tau \in \mathcal{T}^N$ of the triangulation $\mathcal{T}$, see figure 12.2.a for an illustration.

The PL continuation algorithm consists of successively generating these nodes of $H_T^{-1}(0)$ by steps which are similar to the pivoting steps in linear programming. Of course, this heuristic description seems to only make sense for smooth H. As we will see later however, the resulting combinatorial methods also have interesting applications in more general situations.

12.3 A PL Algorithm for Tracing H(u) = 0

Let us begin the description of the features of the PL continuation algorithm by pointing out the relationship between the ε-perturbation in (12.2.4) and the use of lexicographically positive matrices for dealing with degeneracies say, in the Simplex Method of linear programming, cf. Dantzig (1963). Again, we assume that a map $H : \mathbf{R}^{N+1} \to \mathbf{R}^N$ and a triangulation T of $\mathbf{R}^{N+1}$ are given.

(12.3.1) Definition. *We call an N-simplex $\tau \in T^N$* **completely labeled** *if and only if it contains solutions v_ε of the equation $H_\tau(v) = \vec{\varepsilon}$ for all sufficiently small $\varepsilon > 0$.*

In other words, we define an N-simplex τ to be completely labeled if it contains a zero point of the PL approximation H_τ and if this property of τ is stable under certain small perturbations in the above sense.

(12.3.2) Proposition. *Let $\tau = [v_1, \ldots, v_{N+1}] \in T^N$ be an N-simplex. Let*

(12.3.3)
$$\mathcal{L}(\tau) := \begin{pmatrix} 1 & \cdots & 1 \\ H(v_1) & \cdots & H(v_{N+1}) \end{pmatrix}$$

be the **labeling matrix** *on τ induced by H. Then τ is completely labeled if and only if the following two conditions hold:*

(a) *$\mathcal{L}(\tau)$ is nonsingular;*

(b) *$\mathcal{L}(\tau)^{-1}$ is* **lexicographically positive** *i.e. the first nonvanishing entry in any row of $\mathcal{L}(\tau)^{-1}$ is positive.*

Proof. We characterize a point $v_\varepsilon \in \tau$ by its barycentric coordinates

(12.3.4)
$$\begin{cases} v_\varepsilon = \displaystyle\sum_{i=1}^{N+1} \alpha_i(\varepsilon) v_i\,; \\ \displaystyle\sum_{i=1}^{N+1} \alpha_i(\varepsilon) = 1\,; \\ \alpha_i(\varepsilon) \geq 0 \ \text{ for } i = 1, 2, \ldots, N+1. \end{cases}$$

As in the proof of (12.2.4), we obtain the following characterization for v_ε to be a solution of the equation $H_\tau(v) = \vec{\varepsilon}$:

(12.3.5)
$$\mathcal{L}(\tau) \begin{pmatrix} \alpha_1(\varepsilon) \\ \vdots \\ \alpha_{N+1}(\varepsilon) \end{pmatrix} = \begin{pmatrix} 1 \\ \vec{\varepsilon} \end{pmatrix}.$$

Let τ be completely labeled. Then analogously to the proof of (12.2.4), we see that $\mathcal{L}(\tau)$ must be nonsingular. Multiplying (12.3.5) by $\mathcal{L}(\tau)^{-1}$, we see that the $\alpha_i(\varepsilon)$ must be polynomials in ε of degree N, hence

$$\alpha_i(\varepsilon) = \sum_{j=1}^{N+1} A[i,j]\, \varepsilon^j \quad \text{for } i = 1,\ldots,N+1,$$

where A is some $(N+1) \times (N+1)$-matrix. This leads to the linear equation

(12.3.6)
$$\begin{pmatrix} \alpha_1(\varepsilon) \\ \vdots \\ \alpha_{N+1}(\varepsilon) \end{pmatrix} = A \begin{pmatrix} 1 \\ \vec{\varepsilon} \end{pmatrix}.$$

From (12.3.5) we obtain
$$A = \mathcal{L}(\tau)^{-1}.$$

Now the condition $\alpha_i(\varepsilon) \geq 0$ for all sufficiently small $\varepsilon > 0$ and $i = 1,\ldots,N+1$ in equation (12.3.6) implies that A must be lexicographically positive.

Conversely, if $\mathcal{L}(\tau)^{-1}$ is lexicographically positive, then for $\varepsilon > 0$ being sufficiently small, the equation (12.3.6) provides us with the barycentric coordinates of a solution $v_\varepsilon \in \tau$ of the equation $H_\tau(v) = \vec{\varepsilon}$. ☐

As we will see below, keeping track of the labeling matrix $\mathcal{L}(\tau)$ is the basic means for numerically tracing $H_T^{-1}(0)$.

(12.3.7) Definition. *An $(N+1)$-simplex $\sigma \in T$ is called* **transverse** *(with respect to H) if it contains a completely labeled N-face.*

(12.3.8) Proposition (Door-In-Door-Out-Principle).
An $(N+1)$-simplex has either no or exactly two completely labeled N-faces.

Proof. We give a geometric proof. Let us assume that σ is transverse, and let us consider the equation $H_\sigma(v) = \vec{\varepsilon}$ for $v \in \sigma$. By proposition (12.2.4), for $\varepsilon > 0$ being sufficiently small, the solutions v form a line which does not intersect lower-dimensional faces of σ. Hence, the line intersects exactly two N-faces of σ. These two N-faces cannot change as $\varepsilon \to 0$, since otherwise a lower-dimensional face would be traversed and proposition (12.2.4) would be contradicted. In other words, exactly two N-faces of σ contain solutions of the equation $H_\sigma(v) = \vec{\varepsilon}$ for $\varepsilon > 0$ being sufficiently small. ☐

The PL continuation algorithm for tracing certain components of $H_\mathcal{T}^{-1}(0)$ can now be easily described via the above Door-In-Door-Out-Principle, cf. Eaves (1974). Heuristically, let us imagine that the $(N+1)$-simplices $\sigma \in \mathcal{T}$ are "rooms" in an "infinite" building $\mathcal{T}$, and the "walls" of a room σ are its N-faces τ. A wall has a "door" if it is completely labeled. Hence a room has either no or exactly two doors. The algorithm consists of passing from one room to the next, and the following rule must be obeyed: if a room is entered through one door, it must be exited through the other door, see figure 12.3.a for an illustration.

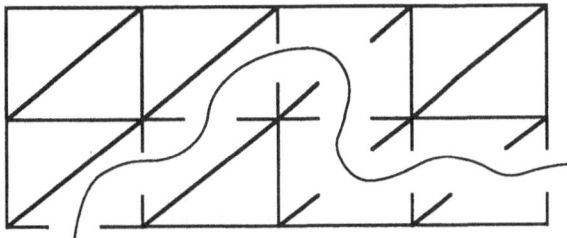

Figure 12.3.a Passing through the "doors"

This leads to the following

(12.3.9) Generic PL Continuation Algorithm.   *comment:*

    input

        begin

        $\sigma_0 \in \mathcal{T}$ transverse;   *starting simplex*

        τ_0 completely labeled N-face of σ_0;

        end;

    repeat for $n := 1, 2, \ldots$

        find $\sigma_n \in \mathcal{T}$, $\sigma_n \neq \sigma_{n-1}$ *pivoting step*

          such that $\tau_{n-1} = \sigma_n \cap \sigma_{n-1}$;

        find the completely labeled N-face τ_n of σ_n *door-in-door-out step*

          such that $\tau_n \neq \tau_{n-1}$;

    until traversing is stopped.

For the special case of homotopy methods which will be discussed in the next chapter, the choice of a starting simplex will be obvious. On the other hand, for general PL curve tracing methods, a suitable starting simplex has to be constructed. This can be done by the general methods described in section 15.4, see (15.4.3) and subsequent remarks. The numerical implementation of the door-in-door-out step usually amounts to solving linear equations in a manner analogous to the Simplex Method of linear programming. Therefore, we often refer to such steps as **linear programming (LP) steps**. This will be discussed in the next section.

The algorithm generates a sequence

(12.3.10)
$$\sigma_0 \supset \tau_0 \subset \sigma_1 \supset \tau_1 \subset \sigma_2 \supset \tau_2 \subset \sigma_3 \cdots$$

of transverse $(N+1)$-simplices σ_i containing the two completely labeled N-faces τ_{i-1} and τ_i. Since no $(N+1)$-simplex can have three completely labeled N-faces, the following classification is obvious.

(12.3.11) Proposition. *The sequence (12.3.10) has one of the following two properties:*

(a) **infinite:**   *Both infinite sequences $\sigma_0, \sigma_1, \ldots$ and $\tau_0, \tau_1, \ldots$ have pairwise different elements.*

(b) **cyclic:**   *There is a smallest integer $\tilde{n} > 0$ such that $\sigma_0 = \sigma_{\tilde{n}}$. Both finite sequences $\sigma_0, \sigma_1, \ldots, \sigma_{\tilde{n}-1}$ and $\tau_0, \tau_1, \ldots, \tau_{\tilde{n}-1}$ have pairwise different elements. The cyclic relations $\sigma_i = \sigma_{\tilde{n}+i}$ and $\tau_i = \tau_{\tilde{n}+i}$ hold for $i = 0, 1, \ldots$*

A formal proof of (12.3.11) can be given via graph theory. We consider a graph G whose nodes σ are the transverse $(N+1)$-simplices of $\mathcal{T}$. Two nodes σ_1, σ_2 of G are connected if they have a common completely labeled N-face. Proposition (12.3.8) states that all nodes of G have degree 2. Hence, all paths in G are either homeomorphic to $\mathbf{R}$ or S^1.

(12.3.12) The Polygonal Path. *Each completely labeled N-face τ_n generated by algorithm (12.3.9) possesses a unique zero point u_n of $H_{\mathcal{T}}$, which we call the H-center of τ_n. We will see in the next section that these centers are easy to calculate. Let $c_{\mathcal{T}} : [0, \infty) \to \mathbf{R}^{N+1}$ be the uniquely defined polygonal path, parametrized by arclength s, with nodes $c_{\mathcal{T}}(s_n) = u_n$ for $0 = s_0 < s_1 < s_2 < \ldots$.*

It is clear that $\text{range}(c_{\mathcal{T}}) \subset H_{\mathcal{T}}^{-1}(0)$. The PL algorithm can be considered as following this polygonal path. If H is smooth and has zero as regular value, this polygonal path approximates the smooth curve c in $H^{-1}(0)$, defined in (2.1.9), with initial value $c(0) = u_0$. As will be seen in sections 12.6 and 15.4, the approximation has truncation error $O(\delta^2)$, where δ is the meshsize of the triangulation.

12.4 Numerical Implementation of a PL Continuation Algorithm

Let us now discuss some details of numerically implementing the generic algorithm (12.3.9). To perform the linear programming steps, we have in essence to keep track of the inverse $\mathcal{L}(\tau)^{-1}$ of the labeling matrix of the current completely labeled N-face τ (= "door we are currently entering"), cf. (12.3.2). We investigate this in more detail by making use of the following identity which can be regarded as a special case of the formula of Sherman & Morrison (1949):

(12.4.1) Remark. Let A be an $(N+1) \times (N+2)$-matrix of maximal rank $N+1$, and B a $(N+2) \times (N+1)$-matrix such that $AB = \mathrm{Id}$ i.e. B is a right inverse of A. Since A has maximal rank, there is a vector $\gamma \in \mathbf{R}^{N+2}$, $\gamma \neq 0$, such that γ spans $\ker(A)$. For example, if $e \in \mathbf{R}^{N+2}$ does not belong to range(B) then $\gamma := e - BAe$ is a possible choice. For any unit basis vector e_i it is immediately seen that the rank-one modification

$$(12.4.2) \qquad\qquad B_i := B - \frac{\gamma e_i^* B}{\gamma[i]}$$

is the right inverse of A uniquely defined by the property $e_i^* B_i = 0$. In fact, the inverse of the matrix obtained by deleting the $i^{\,\mathrm{th}}$ column of A is the matrix obtained by deleting the $i^{\,\mathrm{th}}$ row of B_i. Of course, the equation (12.4.2) only makes sense under the assumption that $\gamma[i] \neq 0$, but it is clear that this condition is equivalent to the assumption that the matrix obtained by deleting the $i^{\,\mathrm{th}}$ column of A is invertible.

To simplify the following discussion, let us suppose, cf. (12.3.9), that we just performed a "pivoting step". We denote the current transverse $(N+1)$-simplex by $\sigma_n = [v_1, \ldots, v_{N+2}]$ and its current completely labeled N-face by $\tau_{n-1} = [v_1, \ldots, v_{N+1}]$. Hence, the vertex v_{N+2} was just pivoted in the preceding step. Our aim in the "linear programming step" is to find the completely labeled N-face τ_n of σ_n which is different from τ_{n-1} i.e. which is not opposite v_{N+2}. Assuming that a right inverse B of the labeling matrix $A = \mathcal{L}(\sigma_n)$ is known, from (12.4.1) it follows that we merely need to find an index $i \in \{1, \ldots, N+1\}$ such that the above B_i with the $i^{\,\mathrm{th}}$ row deleted is lexicographically positive. Hence we are seeking an index i such that the rows

$$e_j^* B - \frac{\gamma[j] e_i^* B}{\gamma[i]}$$

of B_i are lexicographically positive for all $j \neq i$. Dividing by $\gamma[j]$, we see that this is equivalent to the following three conditions:

$$(12.4.3)(1) \qquad \frac{e_j^* B}{\gamma[j]} \quad \text{lex. greater} \quad \frac{e_i^* B}{\gamma[i]} \quad \text{for} \quad j \neq i, \gamma[j] > 0;$$

(12.4.3)(2) $\dfrac{e_j^* B}{\gamma[j]}$ lex. smaller $\dfrac{e_i^* B}{\gamma[i]}$ for $j \neq i,\ \gamma[j] < 0$;

(12.4.3)(3) $e_j^* B$ lex. positive   for $j \neq i,\ \gamma[j] = 0$.

Since the N-face opposite v_{N+2} is completely labeled, we already know that (12.4.3) is satisfied for $i = N + 2$. In particular, this implies that condition (12.4.3)(3) holds, and combining the two conditions (12.4.3)(1)–(2) for $i = N + 2$, we also obtain

$$\frac{e_{j_1}^* B}{\gamma[j_1]} \quad \text{lex. smaller} \quad \frac{e_{j_2}^* B}{\gamma[j_2]} \quad \text{whenever} \quad \gamma[j_1] < 0 < \gamma[j_2].$$

Now lexicographically maximizing over the first terms and lexicographically minimizing over the second yields exactly the two indices corresponding to the two completely labeled N-faces. One of these two indices is of course the already known index, in our case $i = N + 2$, indicating the current completely labeled N-face τ_{n-1}, and the other is the new index we are seeking in the linear programming step. This leads to the numerical implementation of the door-in-door-out-principle. In our algorithms below, we always calculate the vector $\gamma \in \mathbf{R}^{N+2}$ representing $\ker(A)$ in such a way that the component of γ corresponding to the known index (assumed here to be $i = N + 2$) is negative. Then we have to always perform the lexicographical minimization described above in the linear programming step.

For numerical purposes, the lexicographical minimization in linear programming is usually performed only over the first co-ordinate of the rows. Theoretically, in cases of degeneracies, an algorithm based on this simplified test could cycle, but this is rarely observed in practice. For the simplex method of linear programming, sophisticated pivoting rules have been developed to avoid cycling, see e.g. Magnanti & Orlin (1988). Of course, if one wishes to test the complete lexicographic case, an implementation has to account for possible round-off errors. These also depend on the decomposition method which is updated together with the current labeling matrix, cf. chapter 16, and their magnitude is difficult to estimate.

Using this simplification, the numerical linear algebra of a step of (12.3.9) now consists of solving the equation

$$A\gamma = 0, \quad \gamma[j] = -1$$

for $\gamma \in \mathbf{R}^{N+2}$, where the index j is given and corresponds to the known completely labeled N-face, and of solving the equation

$$A\alpha = e_1$$

for $\alpha \in \mathbf{R}^{N+2}$. Note that the vector α is only determined up to one degree of freedom, which may be cut down by imposing the condition $\alpha[j] = 0$. Note also that α corresponds to the first column of the right inverse B used in the above discussion. Then a minimization

$$\min\left\{\frac{\alpha[i]}{\gamma[i]}\ \bigg|\ \gamma[i] > 0,\ i = 1,\ldots,N+2\right\}$$

is performed to find an index i, the vertex v_i is pivoted into the new vertex $\tilde{v}_i$, the corresponding label

$$y := \begin{pmatrix} 1 \\ H(\tilde{v}_i) \end{pmatrix}$$

is calculated, and the new labeling matrix $\tilde{A}$ is obtained by replacing the i th column of A by y:

$$\tilde{A} = A + (y - Ae_i)e_i^*.$$

Usually, at each step, a standard decomposition of $\tilde{A}$ is updated from a given decomposition of A in $O(N^2)$ flops, which enables us to solve the linear equations in each step. The cheapest such method directly updates the right inverse B of A such that $e_i^* B = 0$ for the current pivot index i, c.f. (12.4.1). A similar update is also used in the Revised Simplex Method of linear programming. Unfortunately, this method can produce numerical instabilities, due to the fact that the denominator in (12.4.1) may have a high relative error generated by cancellation of digits, see Bartels & Golub (1969). In chapter 16 we give a more thorough discussion and remedies. In the following algorithm we assume that the pivoting rules of a given triangulation T of $\mathbf{R}^{N+1}$ are easily performed, and that the linear equations described above are solved by updating the right inverse of the labeling matrix A, or by updating some standard decomposition of A in each step.

(12.4.4) General PL Continuation Algorithm.   *comment:*

    input

        begin

        $[v_1,\ldots,v_{N+2}] \in T$ transverse;   *starting simplex*

        $j \in \{1,\ldots,N+2\}$;   *the N-face opposite v_j —*

                                                      is known to be completely labeled

        end;

    $A := \begin{pmatrix} 1 & \cdots & 1 \\ H(v_1) & \cdots & H(v_{N+2}) \end{pmatrix}$;   *initialize labeling matrix*

repeat

  begin

    solve $A\alpha = e_1$, $\alpha[j] = 0$ for α;   *first linear equation*

    if $\alpha \not\geq 0$ **then** quit;   *failure, e.g. wrong starting simplex —*

                                                      or numerical instability

    solve $A\gamma = 0$, $\gamma[j] = -1$ for γ;   *second linear equation*

    find optimal index i with respect to

$$\min\left\{\frac{\alpha[i]}{\gamma[i]} \;\middle|\; \gamma[i] > 0, \; i = 1,\ldots,N+2\right\};$$ *door-in-door-out step*

    pivot v_i into $\tilde{v}_i$;   *pivoting step*

    $v_i := \tilde{v}_i$;

    $y := \begin{pmatrix} 1 \\ H(v_i) \end{pmatrix}$;   *new label on new vertex* v_i

    $A := A + (y - Ae_i)e_i^*$;   *update of labeling matrix*

    $j := i$;

  end.

Given the coefficients α in the "first linear equation", it is straightforward to obtain the H-center $u = H_\tau^{-1}(0)$ of the completely labeled N-face τ opposite the vertex v_j, namely:

(12.4.5)
$$u = \sum_{i=1}^{N+2} \alpha[i]\, v_i\,.$$

Recall that this H-center approximates a zero point of H in τ. Hence, the nodes of the polygonal path (12.3.12) generated by the PL algorithm are easily obtained.

    As an illustration of the above discussion, we conclude this section with a sketch of a customized version of (12.4.4) in which the right inverse B of the current labeling matrix A is always updated by using the formula (12.4.2). Let us again emphasize that this version may be numerically unstable, see the remarks preceding (12.4.4). Other update procedures are discussed in chapter 16.

(12.4.6) PL Algorithm Updating the Right Inverse. *comment:*

> **input**
>
>> **begin**
>>
>> $[v_1, \ldots, v_{N+2}] \in \mathcal{T}$;   *transverse starting simplex*
>>
>> $j \in \{1, \ldots, N+2\}$;   *the N-face opposite v_j —*
>>
>>   *is known to be completely labeled*
>>
>> $\varepsilon_{\text{tol}} > 0$;   *tolerance for avoiding —*
>>
>>   *division by zero in the update formula*
>>
>> **end**;   *initial labeling matrix —*
>>
> $A := \begin{pmatrix} 1 & \cdots & 1 \\ H(v_1) & \cdots & H(v_{N+2}) \end{pmatrix}$;   *in fact the jth column is not needed*
>
> find $(N+2) \times (N+1)$-matrix B such that
> $A\,B = \text{Id}$ and $e_j^* B = 0$;   *initial right inverse of A*
>
> **repeat**
>
>> **begin**
>>
>> **if** B is not lex. positive   *failure, e.g. wrong starting simplex —*
>>
>>> **then** quit;   *or numerical instability*
>>
>> $y := \begin{pmatrix} 1 \\ H(v_j) \end{pmatrix}$;   *new label*
>>
>> $\gamma := By - e_j$; $\alpha := Be_1$;   *solving the linear equations*
>>
>> find optimal index i with respect to
>>
>>> $\min\left\{ \dfrac{\alpha[i]}{\gamma[i]} \,\middle|\, \gamma[i] > \varepsilon_{\text{tol}},\ i = 1, \ldots, N+2 \right\}$;   *door-in-door-out step*
>>
>> pivot v_i into $\tilde{v}_i$;   *pivoting step*
>>
>> $v_i := \tilde{v}_i$;
>>
>> $B := B - \dfrac{\gamma e_i^* B}{\gamma[i]}$;   *update of B*
>>
>> $j := i$;
>>
>> **end**.

12.5 Integer Labeling

In the preceding sections we outlined a PL algorithm using a labeling matrix (12.3.3) based upon a map

$$\textbf{(12.5.1)} \qquad\qquad H : \mathbf{R}^{N+1} \to \mathbf{R}^N .$$

In this context, the map H is often referred to as **vector labeling**. Another class of labelings which have been used by several authors are **integer labelings**

$$\textbf{(12.5.2)} \qquad\qquad \ell : \mathbf{R}^{N+1} \mapsto \{1, 2, \ldots, N+1\} .$$

This was used in one of the first approaches to fixed point algorithms, see Kuhn (1968–69), and can be connected to Sperner's lemma (1920), see e.g. Todd (1976). Since integer labeling leads to a very coarse approximation of a given nonlinear problem, it is usually not recommended for smooth problems, but it may be interesting for problems of a more combinatorial nature. The great advantage of integer labeling is that numerical linear algebra is not required in order to drive the pivoting process.

Recall that an N-simplex $\tau = [v_1, \ldots, v_{N+1}] \subset \mathbf{R}^{N+1}$ is completely labeled with respect to H if the convex hull of $\{H(v_1), \ldots, H(v_{N+1})\}$ contains all ε-vectors $\vec{\varepsilon}$, see (12.2.3), for sufficiently small $\varepsilon > 0$. The analogous concept for integer labeling is as follows:

(12.5.3) Definition. *An N-simplex $\tau = [v_1, \ldots, v_{N+1}] \subset \mathbf{R}^{N+1}$ is said to be* **completely labeled** *with respect to a given integer labeling (12.5.2) if $\{\ell(v_1), \ldots, \ell(v_{N+1})\} = \{1, \ldots, N+1\}$.*

It is possible to unify these concepts by introducing a PL map $\ell_{\mathcal{T}}$ induced by an integer labeling (12.5.2).

(12.5.4) Definition. *Let ℓ be an integer labeling and $\mathcal{T}$ a triangulation of $\mathbf{R}^{N+1}$. We fix a standard simplex $\Sigma = [w_1, \ldots, w_{N+1}] \subset \mathbf{R}^N$ such that zero is an interior point of Σ. The particular choice of Σ is immaterial. For a node v of $\mathcal{T}$, we define*

$$\ell_{\mathcal{T}}(v) := w_{\ell(v)} .$$

Now $\ell_{\mathcal{T}}$ can be uniquely extended to a PL map $\ell_{\mathcal{T}} : \mathbf{R}^{N+1} \to \Sigma$ by the usual affine interpolation.

From the above definitions, it is immediately clear that an N-simplex $\tau \subset \mathbf{R}^{N+1}$ is completely labeled with respect to ℓ if and only if it is completely labeled with respect to $\ell_{\mathcal{T}}$. Hence, the dynamics of the PL algorithm for integer labelings can be regarded as a special case of the general discussion given in section 12.3. In particular, the door-in-door-out step of (12.3.9) is especially simple to program, since one needs to pivot only those vertices which have the same label ℓ. As an illustration, let us repeat algorithm (12.4.4) for the case of integer labeling:

(12.5.5) General PL Algorithm Using Integer Labeling.   *comment:*

input $[v_1, \ldots, v_{N+2}] \in \mathcal{T}$ and   *transverse starting simplex —*

                                the N-face opposite v_j —

$j \in \{1, \ldots, N+2\}$ such that   *is completely labeled*

$\{\ell(v_k) \mid k = 1, \ldots, N+1, \ k \neq j\} = \{1, 2, \ldots, N+1\};$

repeat

    find $i \in \{1, \ldots, N+2\}$, $i \neq j$ such that

    $\ell(v_i) = \ell(v_j);$ *door-in-door-out step*

    pivot v_i into $\tilde{v}_i;$ *pivoting step*

    $v_i := \tilde{v}_i; \ j := i;$

until traversing is stopped.

The following is an example of how an integer labeling ℓ may be defined in order to investigate a given zero point problem $H(u) = 0$ for a map $H : \mathbf{R}^{N+1} \to \mathbf{R}^N$. For $v \in \mathbf{R}^{N+1}$, we define $\ell(v) := 1 + m$ where m is the number of initial co-ordinates of $H(v)$ which are positive, i.e. $\ell(v)$ is defined by the following steps:

(12.5.6)
$$
\begin{aligned}
&i := 1; \\
&\text{while } i < N+1 \text{ and } e_i^* H(v) > 0 \text{ do } i := i+1; \\
&\ell(v) := i.
\end{aligned}
$$

To see the connection between the above integer labeling and the zero points of H, we note that the continuity of H implies

(12.5.7)
$$
\bigcap_{i=1}^{N+1} \overline{\ell^{-1}(i)} \subset H^{-1}(0).
$$

The reverse inclusion holds for regular zero points of H:

(12.5.8) Lemma. *Let H be differentiable at $u \in \mathbf{R}^{N+1}$ such that $H(u) = 0$ and $\mathrm{rank}\, H'(u) = N$. Then*

$$
u \in \bigcap_{i=1}^{N+1} \overline{\ell^{-1}(i)}.
$$

Proof. For $i = 1, 2, \ldots, N+1$ let q_i denote the i th column of the $N \times (N+1)$-matrix

$$
\begin{pmatrix}
-1 & 1 & 1 & \cdots & 1 \\
-1 & -1 & 1 & \cdots & 1 \\
\vdots & \vdots & \ddots & \ddots & \vdots \\
-1 & \cdots & \cdots & -1 & 1
\end{pmatrix}.
$$

Since $H'(u)$ has rank N, there exists a $C > 0$ such that the equation

$$H'(u)\,v_{n,i} = \frac{1}{n}\,q_i$$

has a solution $v_{n,i}$ for $i = 1, 2, \ldots, N+1$ and $n = 1, 2, \ldots$ such that $\|v_{n,i}\|_\infty \le C/n$. From the definition of differentiability it follows that $H(u + v) = H'(u)v + o(\|v\|)$ holds, and hence

$$\ell(u + v_{n,i}) = i \quad i = 1, 2, \ldots, N+1$$

holds for n sufficiently large. Since $\lim_{n=1}^{\infty} v_{n,i} = 0$, u is a limit point of $\ell^{-1}(i)$ for $i = 1, 2, \ldots, N+1$. □

The following lemma indicates how well a completely labeled simplex approximates a zero point of H:

(12.5.9) Lemma. Let $M \subset \mathbf{R}^{N+1}$ be a set such that $H(M)$ is bounded and $\ell(H(M)) = \{1, 2, \ldots, N+1\}$. Then $\|H(u)\|_\infty \le \mathrm{diam}_\infty H(M)$ holds for every $u \in M$.

Proof. For $i = 1, 2, \ldots, N$ we can find a $u_i \in M$ such that $\ell(u_i) = i$. This implies

$$e_i^* H(u) = e_i^*\big(H(u) - H(u_i)\big) + e_i^* H(u_i)$$

and $e_i^* H(u_i) \le 0$ yields

(12.5.10) $$e_i^* H(u) \le \mathrm{diam}_\infty H(M).$$

Furthermore, let $u_{N+1} \in M$ such that $\ell(u_{N+1}) = N+1$. Then we have for $j = 1, 2, \ldots, N$ that

$$e_j^* H(u) = e_j^*\big(H(u) - H(u_{N+1})\big) + e_j^* H(u_{N+1}),$$

and $e_j^* H(u_{N+1}) > 0$ yields

(12.5.11) $$e_j^* H(u) \ge -\mathrm{diam}_\infty H(M).$$

The inequalities (12.5.10–11) prove the assertion.   □

12.6 Truncation Errors

We conclude this chapter with some brief remarks on error estimates. A more comprehensive and general discussion will be given in section 15.5. Let us begin by defining the meshsize of a triangulation.

(12.6.1) Definition. *Let T be a triangulation of $\mathbf{R}^{N+1}$. The* **meshsize** *of T is defined by*

$$\delta := \sup_{\sigma \in T} \operatorname{diam}(\sigma).$$

For example, let us consider Freudenthal's triangulation T of $\mathbf{R}^{N+1}$, see (12.1.10). If we use the maximum norm $||\cdot||_\infty$, then the meshsize of T is $\delta = 1$, and if we use the Euclidean norm $||\cdot||_2$, then the meshsize is $\delta = \sqrt{N+1}$. Similarly, let us consider the affine image $T(\sigma)$ of Freudenthal's triangulation, see (12.1.12), obtained by the reflection rules (12.1.11), starting from the simplex $\sigma = [v_1, v_2, \ldots, v_{N+2}]$. Then the meshsize of $T(\sigma)$ is bounded by

$$\left\| \begin{pmatrix} v_2 - v_1 & \cdots & v_{N+2} - v_1 \end{pmatrix} \begin{pmatrix} 1 & 1 & \cdots & 1 \\ 0 & 1 & \cdots & 1 \\ \vdots & \ddots & \ddots & \vdots \\ 0 & \cdots & 0 & 1 \end{pmatrix}^{-1} \right\| \delta$$

$$= \left\| \begin{pmatrix} v_2 - v_1 & \cdots & v_{N+2} - v_1 \end{pmatrix} \begin{pmatrix} 1 & -1 & 0 & 0 & \cdots & 0 \\ 0 & 1 & -1 & 0 & \cdots & 0 \\ \vdots & \ddots & \ddots & \ddots & \ddots & \vdots \\ 0 & \cdots & 0 & 1 & -1 & 0 \\ 0 & \cdots & 0 & 0 & 1 & -1 \\ 0 & \cdots & 0 & 0 & 0 & 1 \end{pmatrix}^{-1} \right\| \delta$$

$$= \left\| \begin{pmatrix} v_2 - v_1, & v_3 - v_2, & v_{N+2} - v_{N+1} \end{pmatrix} \right\| \delta,$$

where δ is the meshsize of T according to whichever norm is used. It is not difficult to show that there exists a factor of proportionality $C > 0$, independent of the choice of σ, such that

$$\frac{1}{C} \operatorname{diam}(\sigma) \le \| \begin{pmatrix} v_2 - v_1 & \cdots & v_{N+2} - v_1 \end{pmatrix} \| \le C \operatorname{diam}(\sigma).$$

Hence, there exists a constant $K > 0$ such that the meshsize of $T(\sigma)$ can be bounded by $K \operatorname{diam}(\sigma)$. In particular, for the ∞-norm, Gnutzmann (1988) shows

$$K = \left[\frac{N+2}{2} \right].$$

The general aim of PL continuation methods is to obtain completely labeled N-faces τ in a triangulation T of meshsize δ. The idea is that such a τ approximates a zero point of the given map H. It is intuitively clear that the order of approximation depends on the labeling which is used (e.g. integer or vector labeling) and on the smoothness of H. For example, let us assume that H is Lipschitz continuous, i.e. there exists a constant $L > 0$ such that $\|H(u) - H(v)\| \leq L\,\|u - v\|$ holds for all $u, v \in \mathbf{R}^{N+1}$. Then it follows immediately from (12.5.9) that in the case of integer labeling (12.5.6), all points $u \in \tau$ approximate a zero point of H with order $O(\delta)$, more precisely:

$$\|H(u)\|_\infty \leq L\,\delta$$

holds for every $u \in \tau$. As will be seen in section 15.5 and in particular in proposition (15.5.2), vector labeling yields a second order approximation if H is sufficiently smooth:

$$\|H(u)\| \leq O(\delta^2)$$

for $u \in \tau$ such that $H_T(u) = 0$.

Chapter 13. PL Homotopy Algorithms

In the last chapter we discussed the general features of PL continuation methods. In this chapter we will apply them to find a zero point of a map $G : \mathbf{R}^N \to \mathbf{R}^N$. We will see that it is possible to greatly relax the smoothness hypotheses regarding the map G, which are usually assumed for numerically solving such problems. In fact, the map G may even be set-valued. Eaves (1971) showed that it is possible to calculate by PL algorithms the fixed points which are guaranteed to exist by a theorem of Kakutani (1941). Merrill (1972) gave a more general boundary condition for set-valued maps G, which is similar to the Leray-Schauder condition (11.2.12). In this chapter we present two PL algorithms due to Merrill (1972) and Eaves & Saigal (1972) which can be regarded as PL implementations of the homotopy method which was sketched generally in section 11.2. To insure success of the algorithms, we will follow a presentation of Georg (1982) which used a quite general boundary condition extending somewhat that used by Merrill.

13.1 Set-Valued Maps

Let us begin with a description of some properties of set-valued maps which we will use in the sequel. More details can be found in the books of Berge (1963), Rockafellar (1970), Stoer & Witzgall (1970) and Todd (1976). To motivate the ideas, we first consider a simple example. Let $G : \mathbf{R} \to \mathbf{R}$ be defined by

(13.1.1)
$$G(x) = \begin{cases} 1 & \text{if } x \geq 1, \\ -1 & \text{if } x < 1. \end{cases}$$

Although G changes sign at $x = 1$, due to its discontinuity it does not have a zero point. However, in a more general sense, it is useful to regard the point $x = 1$ as a "generalized zero point". In fact, $G(x)$ has two accumulation points as $x \to 1$, namely ± 1, and hence the convex hull of these accumulation points is $[-1, 1]$, which contains the point zero. This example motivates us to give the following definitions.

(13.1.2) Definition. *A map $G : \mathbf{R}^N \to \mathbf{R}^N$ is called* **locally bounded** *if each point $x \in \mathbf{R}^N$ has a neighborhood U_x such that $G(U_x)$ is a bounded set.*

For example, the map $G : \mathbf{R} \to \mathbf{R}$ defined by

$$G(x) = \begin{cases} x^{-1} & \text{if } x \neq 0, \\ 0 & \text{if } x = 0, \end{cases}$$

is not locally bounded at zero.

(13.1.3) Definition. *Let $G : \mathbf{R}^N \to \mathbf{R}^N$ be locally bounded. Denote by $\mathbf{R}^{N\#}$ the family of nonempty compact convex subsets of $\mathbf{R}^N$. We define the* **set-valued hull** $G^\# : \mathbf{R}^N \to \mathbf{R}^{N\#}$ *by*

$$G^\#(x) := \bigcap_{U \in \mathcal{U}_x} \overline{\mathrm{co}}\, G(U),$$

where $\mathcal{U}_x$ denotes the family of neighborhoods of $x \in \mathbf{R}^N$ and $\overline{\mathrm{co}}$ represents the operation of taking the closed convex hull.

For the function (13.1.1) we obtain

(13.1.4) $$G^\#(x) = \begin{cases} \{\ 1\} & \text{if } x > 1, \\ \{-1\} & \text{if } x < 1, \\ [-1,1] & \text{if } x = 1, \end{cases}$$

see figure 13.1.a.

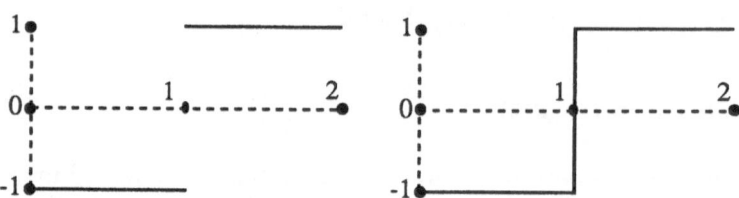

Figure 13.1.a The map G of (13.1.1) and its set-valued hull $G^\#$ with a "generalized zero point" at $x = 1$

(13.1.5) Lemma. *The set-valued hull $G^\# : \mathbf{R}^N \to \mathbf{R}^{N\#}$ of definition (13.1.3) is* **upper semi-continuous**, *i.e. for each point $x \in \mathbf{R}^N$ and each open subset $V \subset \mathbf{R}^N$ such that $G^\#(x) \subset V$ there exists a neighborhood $W \subset \mathbf{R}^N$ of x such that $G^\#(w) \subset V$ for all $w \in W$.*

Proof. Since G is locally bounded, we can find a closed ball $B \subset \mathbf{R}^N$ and a neighborhood W_o of x such that $G(W_o) \subset B$. Let us denote by $\mathcal{W}_x$ the family of all neighborhoods of x which are contained in W_o. Suppose the assertion is not true for a pair x and V as in the hypothesis. Then it follows that

(13.1.6) $$\overline{\mathrm{co}}\, G(W) \cap (B - V) \neq \emptyset$$

holds for all $W \in \mathcal{W}_x$. The system $\{\overline{\text{co}}\, G(W)\}_{W \in \mathcal{W}_x}$ is downward directed by inclusion, i.e. if $W_1, \ldots, W_k \in \mathcal{W}_x$, then $W_1 \cap \ldots \cap W_k \in \mathcal{W}_x$, and

$$\overline{\text{co}}\, G(W_1 \cap \ldots \cap W_k) \subset \overline{\text{co}}\, G(W_1) \cap \ldots \cap \overline{\text{co}}\, G(W_k).$$

It follows that the system of compact sets

$$(13.1.7) \qquad \left\{ \overline{\text{co}}\, G(W) \cap (B - V) \right\}_{W \in \mathcal{W}_x}$$

is also downward directed by inclusion. Now (13.1.6) implies that it has the finite intersection property. By a standard compactness argument we have

$$(13.1.8) \qquad \bigcap_{W \in \mathcal{W}_x} \overline{\text{co}}\, G(W) \cap (B - V) \neq \emptyset.$$

But by definition

$$G^{\#}(x) = \bigcap_{W \in \mathcal{W}_x} \overline{\text{co}}\, G(W),$$

and hence (13.1.8) contradicts the choice of V in the hypothesis.   □

It is now easy to see from the definition (13.1.3) and the above lemma that the map $G^{\#} : \mathbf{R}^N \to \mathbf{R}^{N\#}$ is the smallest upper semi-continuous map which contains G. We also obtain the following

(13.1.9) Corollary. *Let* $G : \mathbf{R}^N \to \mathbf{R}^N$ *be locally bounded and* $x \in \mathbf{R}^N$. *Then* G *is continuous at* x *if and only if* $G^{\#}(x) = \{ G(x) \}$ *is a singleton.*

Proof. Let us denote by $B_\rho(y)$ the closed ball with radius ρ and center y. If G is continuous at x, then for every $\varepsilon > 0$ there is a $\delta(\varepsilon) > 0$ such that $G(B_{\delta(\varepsilon)}(x)) \subset B_\varepsilon(G(x))$. Hence

$$G^{\#}(x) \subset \overline{\text{co}}\, G(B_{\delta(\varepsilon)}(x)) \subset B_\varepsilon(G(x)).$$

Intersecting both sides of the above relation over all $\varepsilon > 0$ and using the continuity of G at x yields

$$G^{\#}(x) \subset \bigcap_{\varepsilon > 0} B_\varepsilon(G(x)) = \{ G(x) \}$$

and hence $G^{\#}(x) = \{ G(x) \}$. Conversely, suppose that $G^{\#}(x) = \{ G(x) \}$ is a singleton. Then by lemma (13.1.5), for every $\varepsilon > 0$ there is a $\delta(\varepsilon) > 0$ such that $G^{\#}(w) \subset B_\varepsilon(G(x))$ for all $w \in B_{\delta(\varepsilon)}(x)$. Since $G(w) \in G^{\#}(w)$, it follows immediately that $G(B_{\delta(\varepsilon)}(x)) \subset B_\varepsilon(G(x))$ holds, which states the continuity of G at x.   □

In many applications G is defined in a piecewise fashion over sets which subdivide $\mathbf{R}^N$. For example, G may be defined by considering different cases. Let us describe such a general scenario.

(13.1.10) Definition. *A family of maps* $\left\{ G_\iota : M_\iota \to \mathbf{R}^N \right\}_{\iota \in I}$ *is called a* **locally bounded partition** *of* $G : \mathbf{R}^N \to \mathbf{R}^N$ *if the following conditions hold:*

(1) $\{M_\iota\}_{\iota \in I}$ form a disjoint partition of $\mathbf{R}^N$ i.e. $\bigcup_{\iota \in I} M_\iota = \mathbf{R}^N$ and $M_\alpha \cap M_\beta = \emptyset$ for $\alpha, \beta \in I$ with $\alpha \neq \beta$;

(2) $\{M_\iota\}_{\iota \in I}$ is locally finite i.e. for every $x \in \mathbf{R}^N$ there is a neighborhood $U \in \mathcal{U}_x$ such that only finitely many M_ι meet U, where $\mathcal{U}_x$ again denotes the family of neighborhoods of x;

(3) all $G_\iota : M_\iota \to \mathbf{R}^N$ are locally bounded maps;

(4) $G(x) = G_\iota(x)$ for $x \in M_\iota$ and $\iota \in I$.

Clearly, any map G satisfying the above definition is locally bounded, and hence has a set-valued hull. Analogously to (13.1.3), we can define the set-valued hull $G_\iota^\# : \overline{M_\iota} \to \mathbf{R}^{N\#}$ of a component G_ι by

$$G_\iota^\#(x) := \bigcap_{U \in \mathcal{U}_x} \overline{\mathrm{co}}\, G_\iota(U \cap M_\iota)$$

for any point x in the closure $\overline{M_\iota}$ of M_ι. We call an index $\iota \in I$ **active at** $x \in \mathbf{R}^N$ if x belongs to the closure $\overline{M_\iota}$. Let us denote by $I(x)$ the set of indices which are active at x, see figure 13.1.b. By the condition (13.1.10)(2) it is clear that $I(x)$ is finite. The following lemma is useful for obtaining the set-valued hull of a locally bounded partition.

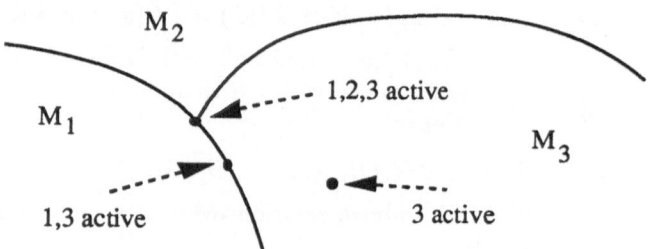

Figure 13.1.b   Active indices of a partitioned map

(13.1.11) Lemma. *Let* $\left\{ G_\iota : M_\iota \to \mathbf{R}^N \right\}_{\iota \in I}$ *be a locally bounded partition of the map* $G : \mathbf{R}^N \to \mathbf{R}^N$. *Then*

$$G^\#(x) = \mathrm{co}\left\{ G_\iota^\#(x) \mid \iota \in I(x) \right\}$$

holds for $x \in \mathbf{R}^N$.

Proof. By condition (13.1.10)(2) we can find a $U_o \in \mathcal{U}_x$ such that $G(U_o)$ is bounded and

$$U_o \cap M_\iota \neq \emptyset \iff \iota \in I(x).$$

For $U \in \mathcal{U}_x$ with $U \subset U_o$ we have

$$G(U) = \bigcup_{\iota \in I(x)} G_\iota(U \cap M_\iota)$$

and consequently

$$\overline{\mathrm{co}}\, G(U) = \mathrm{co}\Big(\bigcup_{\iota \in I(x)} \overline{\mathrm{co}}\, G_\iota(U \cap M_\iota)\Big).$$

Here we have used the well known fact that the convex hull of a finite union of closed convex sets is closed. Intersecting over all U on both sides of the above relation yields the assertion. $\square$

In many cases, the component maps $G_\iota : M_\iota \to \mathbf{R}^N$ are continuous and can be continuously extended to the closure $\overline{M_\iota}$, and then the set-valued hulls $G_\iota^\#(x)$ can be shown to reduce to the singletons $\{\, G_\iota(x)\,\}$ by an argument similar to that in corollary (13.1.9).

In the next sections we will present several PL homotopy algorithms. As we have seen in our discussions in section 11.2, in order for homotopy algorithms to be successful, a boundary condition is needed. The following definition gives rise to a boundary condition which is particularly well suited for PL homotopy algorithms.

(13.1.12) Definition. *A locally bounded map $G : \mathbf{R}^N \to \mathbf{R}^N$ is called* **asymptotically linear** *if there exists a nonsingular $N \times N$-matrix A such that*

$$\lim_{\|x\| \to \infty} \frac{\|G(x) - Ax\|}{\|x\|} = 0$$

holds. For obvious reasons we call $G'(\infty) := A$ the Jacobian of G at infinity.

It is possible to introduce a Leray-Schauder-type degree for set-valued maps, see e.g. Granas (1959), Górniewicz (1976) or Siegberg & Skordev (1982) for very general cases. The following theorem is a simple example of an existence theorem which can be proven by a degree argument.

(13.1.13) Theorem. *Let $G : \mathbf{R}^N \to \mathbf{R}^N$ be asymptotically linear. Then $G^\#$ has a zero point i.e. there exists a point $\bar{x} \in \mathbf{R}^N$ such that $0 \in G^\#(\bar{x})$.*

In fact, for a sufficiently large closed ball B with center zero, the degrees of $G^\#$ and $G'(\infty)$ coincide on B, and since the latter is different from zero, we have a zero point of $G^\#$ in B. A "constructive" proof of this theorem can be obtained by considering the homotopy algorithms of the next sections, i.e. these algorithms approximate a zero point $\bar{x}$ as above. However, our main point is that we do not regard (13.1.13) from a standpoint of an existence theorem, but rather as a general setting under which PL homotopy algorithms will be successful. This will be elaborated in the next sections. Hence we regard the above boundary condition viz. the asymptotic linearity of G as an appropriate boundary condition for these algorithms. Let us finally give some important examples where this boundary condition occurs in a natural way.

(13.1.14) Example. Let $C \subset \mathbf{R}^N$ be a compact convex set, and let $R : \mathbf{R}^N \to C$ be a retraction i.e. R is continuous and $R(x) = x$ for $x \in C$. Such a retraction always exists. For example, if C contains an interior point x_0 then $R(x)$ for $x \notin C$ may be defined as the intersection point of ∂C with the line connecting x_0 and x. Now let $F : C \to C$ be any map. We set $G := \mathrm{Id} - F \circ R$. It is immediately clear that G is locally bounded and asymptotically linear. In fact, $G'(\infty) = \mathrm{Id}$. Hence, there is an $\bar{x} \in \mathbf{R}^N$ such that $0 \in G^{\#}(\bar{x})$. It follows that $\bar{x} \in F^{\#}(R(\bar{x}))$, because R is continuous. Since C is compact and convex and $F(C) \subset C$, we obtain a fixed point $\bar{x} \in C$ such that $\bar{x} \in F^{\#}(\bar{x})$. This is a set-valued version of the fixed point theorem of Brouwer (1912), see Kakutani (1941) for a generalization to infinite dimensional spaces. Let us again remark: the discussions in the next sections will show that by using the above approach, such fixed points can be approximated via PL homotopy methods.

(13.1.15) Example. Let us now show how the Leray-Schauder condition (11.2.12) can be used to obtain an asymptotically linear map. Suppose that $\Omega \subset \mathbf{R}^N$ is a bounded open neighborhood of $p \in \mathbf{R}^N$, and let $F : \bar{\Omega} \to \mathbf{R}^N$ be a continuous map such that the following boundary condition holds:

(13.1.16) $\lambda(F(x) - p) \neq (x - p)$ for $x \in \partial\Omega$ and $0 \leq \lambda \leq 1$.

Let us define $G : \mathbf{R}^N \to \mathbf{R}^N$ by

$$G(x) = \begin{cases} x - p & \text{for } x \notin \Omega, \\ x - F(x) & \text{for } x \in \Omega. \end{cases}$$

Again G is obviously an asymptotically linear map with Jacobian $G'(\infty) = \mathrm{Id}$. Hence $G^{\#}$ possesses a zero point $\bar{x}$. Let us show that $\bar{x}$ is a fixed point of F in Ω by considering the three possible cases

1. $\bar{x} \in \Omega$: Then G is continuous at $\bar{x}$, and (13.1.9) implies that
   $$\{0\} = G^{\#}(\bar{x}) = \{G(\bar{x})\} = \{\bar{x} - F(\bar{x})\}.$$
   Hence $\bar{x}$ is a fixed point of F.
2. $\bar{x} \notin \bar{\Omega}$: Then we have $G^{\#}(\bar{x}) = \{\bar{x} - p\} = \{0\}$ by (13.1.9), and this contradicts $p \in \Omega$.
3. $\bar{x} \in \partial\Omega$: Then $0 \in G^{\#}(\bar{x})$, and (13.1.9), (13.1.11) imply that
   $$G^{\#}(\bar{x}) = \mathrm{co}\{\bar{x} - p, \bar{x} - F(\bar{x})\}.$$

Hence there is a convex combination: $\lambda_1, \lambda_2 \geq 0$, $\lambda_1 + \lambda_2 = 1$ such that $\lambda_1(\bar{x} - p) + \lambda_2(\bar{x} - F(\bar{x})) = 0$. Since $\bar{x} \neq p$, it follows that $\lambda_2 \neq 0$ and hence a simple manipulation of this last equation shows that

$$\left(\frac{\lambda_1}{\lambda_2} + 1\right)(\bar{x} - p) = F(\bar{x}) - \bar{x},$$

which can be seen to contradict the boundary condition (13.1.16).

Thus we have seen that a continuous map $F : \overline{\Omega} \to \mathbf{R}^N$ satisfying the boundary condition (13.1.16) has a fixed point in Ω. In section 11.2 we showed that such a fixed point could be approximated in the case of a smooth map F. In the following sections we will see that such fixed points can also be approximated for a continuous (not necessarily smooth) map F via PL homotopy algorithms.

(13.1.17) Example. The next example concerning constrained optimization essentially follows a discussion given by Merrill (1972), see also Todd (1976) and Georg (1980). A function $\theta : \mathbf{R}^N \to \mathbf{R}$ is called **convex** if

$$\lambda_1 \theta(x_1) + \lambda_2 \theta(x_2) \geq \theta(\lambda_1 x_1 + \lambda_2 x_2)$$

holds for all convex combinations: $\lambda_1, \lambda_2 \geq 0$, $\lambda_1 + \lambda_2 = 1$, $x_1, x_2 \in \mathbf{R}^N$. It is well known, see Rockafellar (1970), that a convex function is continuous and has an upper semi-continuous **subgradient** $\partial\theta : \mathbf{R}^N \to \mathbf{R}^{N\#}$ defined by

$$\partial\theta(x) := \left\{ y \in \mathbf{R}^N \mid \theta(z) - \theta(x) \geq y^*(z - x) \text{ for all } z \in \mathbf{R}^N \right\}.$$

A simple consequence of this fact is that a point $\bar{x} \in \mathbf{R}^N$ is a solution point of the minimization problem $\min_x \theta(x)$ if and only if $0 \in \partial\theta(\bar{x})$.

We now want to study the constrained minimization problem

(13.1.18)
$$\min_x \left\{ \theta(x) \mid \psi(x) \leq 0 \right\},$$

where $\theta, \psi : \mathbf{R}^N \to \mathbf{R}$ are convex. We assume the Slater condition

(13.1.19)
$$\left\{ x \mid \psi(x) < 0, \ ||x - x_0|| < r \right\} \neq \emptyset$$

and the boundary condition that the problem

(13.1.20)
$$\min_x \left\{ \theta(x) \mid \psi(x) \leq 0, \quad ||x - x_0|| \leq r \right\},$$

has no solution on the boundary $\left\{ x \mid ||x - x_0|| = r \right\}$ for some suitable $x_0 \in \mathbf{R}^N$ and $r > 0$. This boundary condition is satisfied for example if

$$\left\{ x \mid \psi(x) \leq 0 \right\} \subset \left\{ x \mid ||x - x_0|| < r \right\}$$

or more generally, if

$$\emptyset \neq \left\{ x \mid \psi(x) \leq 0 \right\} \cap \left\{ x \mid \theta(x) \leq C \right\} \subset \left\{ x \mid ||x - x_0|| < r \right\}.$$

Let us define the map $G : \mathbf{R}^N \to \mathbf{R}^N$ by

(13.1.21)
$$G(x) \in \begin{cases} \partial\theta(x) & \text{for } \psi(x) \leq 0 \text{ and } ||x - x_0|| < r, \\ \partial\psi(x) & \text{for } \psi(x) > 0 \text{ and } ||x - x_0|| < r, \\ \{x - x_0\} & \text{for } ||x - x_0|| \geq r. \end{cases}$$

Again it is obvious that G is asymptotically linear with Jacobian $G'(\infty) = \text{Id}$. Hence we obtain a zero point $\bar{x}$ of $G^{\#}$. We will show that $\bar{x}$ solves the minimization problem (13.1.18) by considering various possible cases. In doing so, we will repeatedly make use of the properties of convex functions as described above, and of (13.1.9), (13.1.11) without any further mention.

1. $||\bar{x} - x_0|| > r$: We obtain $\bar{x} - x_0 = 0$, which is a contradiction.
2. $\psi(\bar{x}) > 0$ and $||\bar{x} - x_0|| = r$: Then there exists a convex combination such that $0 \in \lambda_1 \partial \psi(\bar{x}) + \lambda_2 \{ x - x_0 \}$. Hence $\bar{x}$ solves $\min_x \lambda_1 \psi(x) + \frac{1}{2}\lambda_2||x - x_0||^2$. But by (13.1.19) there exists a v such that $\psi(v) < 0$ and $||v - x_0|| < r$, which yields a contradiction.
3. $\psi(\bar{x}) = 0$ and $||\bar{x} - x_0|| = r$: Then there exists a convex combination such that $0 \in \lambda_1 \partial \theta(\bar{x}) + \lambda_2 \partial \psi(\bar{x}) + \lambda_3 \{ x - x_0 \}$. Let us first observe that $\lambda_1 = 0$ yields the same contradiction as in the previous case. Hence $\lambda_1 \neq 0$, and we have that $\bar{x}$ solves $\min_x \lambda_1 \theta(x) + \lambda_2 \psi(x) + \frac{1}{2}\lambda_3||x - x_0||^2$. But by (13.1.20) there exists a v such that $\theta(v) < \theta(\bar{x})$, $\psi(v) \leq 0$ and $||v - x_0|| < r$, which yields a contradiction since $\lambda_1 \neq 0$.
4. $\psi(\bar{x}) < 0$ and $||\bar{x} - x_0|| = r$: Then there exists a convex combination such that $0 \in \lambda_1 \partial \theta(\bar{x}) + \lambda_2 \{ x - x_0 \}$. Hence $\bar{x}$ solves $\min_x \lambda_1 \theta(x) + \frac{1}{2}\lambda_2||x - x_0||^2$. But by (13.1.20) there exists a v such that $\theta(v) < \theta(\bar{x})$ and $||v - x_0|| < r$, which yields a contradiction.
5. $\psi(\bar{x}) > 0$ and $||\bar{x} - x_0|| < r$: Then $0 \in \partial \psi(\bar{x})$, and $\bar{x}$ solves $\min_x \psi(x)$ which contradicts (13.1.19).
6. $\psi(\bar{x}) = 0$ and $||\bar{x} - x_0|| < r$: Then there exists a convex combination such that $0 \in \lambda_1 \partial \theta(\bar{x}) + \lambda_2 \partial \psi(\bar{x})$. Let us first observe that $\lambda_1 = 0$ yields the same contradiction as in the previous case. Hence $\lambda_1 \neq 0$, and we have that $\bar{x}$ solves $\min_x \lambda_1 \theta(x) + \lambda_2 \psi(x)$. Let us assume that $\bar{x}$ is not a solution of the minimization problem (13.1.18). Then there exists a v such that $\theta(v) < \theta(\bar{x})$ and $\psi(v) \leq 0$, which yields a contradiction since $\lambda_1 \neq 0$.
7. $\psi(\bar{x}) < 0$ and $||\bar{x} - x_0|| < r$: Then $0 \in \partial \theta(\bar{x})$, and $\bar{x}$ solves $\min_x \theta(x)$.

So we have shown that the cases 1–5 are impossible and the cases 6–7 yield a solution of the minimization problem (13.1.18).

(13.1.22) Example. Let us return to the nonlinear complementarity problem (11.7.1): Find an $x \in \mathbf{R}^N$ such that

$$ x \in \mathbf{R}^N_+; \quad g(x) \in \mathbf{R}^N_+; \quad x^* g(x) = 0 \,, $$

where $g : \mathbf{R}^N \to \mathbf{R}^N$ is a continuous map. Because of (11.7.3), we seek a zero point $\bar{x}$ of the map $x \mapsto g(x^+) - x^-$, and then $\bar{x}^+$ solves the NLCP. We use the coercivity condition (11.7.6) and define $G : \mathbf{R}^N \to \mathbf{R}^N$ by

$$ G(x) = \begin{cases} x & \text{if } x \notin \Omega, \\ g(x^+) - x^- & \text{if } x \in \Omega. \end{cases} $$

Again, G is asymptotically linear and $G'(\infty) = \mathrm{Id}$. Hence we have a zero point $\bar{x}$ of $G^{\#}$. Since Ω is a bounded open neighborhood of zero, the case $\bar{x} \notin \overline{\Omega}$ is excluded, and the case $\bar{x} \in \partial \Omega$ is excluded by an argument very similar to the one following (11.7.6). The remaining case, namely $\bar{x} \in \Omega$, yields a zero point of the map $x \mapsto g(x^+) - x^-$.

The above examples are only a sample of the many possible applications of PL homotopy methods or more generally, of complementary pivoting methods. The reader can find many further cases in the bibliography.

13.2 Merrill's Restart Algorithm

In this section we describe a version of Merrill's algorithm which is a simple example of a "restart" method. For a historical account of early versions of PL methods we refer the reader to Todd (1976) and (1982). We begin with a useful

(13.2.1) Definition. Let $\mathcal{T}$ be a triangulation of $\mathbf{R}^{N+1}$ and let $C \subset \mathbf{R}^{N+1}$ be a closed convex set with nonempty interior $\text{int}(C)$. We call the triangulation $\mathcal{T}$ **compatible with** C if

$$\bigcup \{ \sigma \in \mathcal{T} \mid \sigma \cap \text{int}(C) \neq \emptyset \} = C.$$

It can be shown that $\mathcal{T}$ induces a triangulation of ∂C. For the sake of completeness, we sketch a proof of this in the following

(13.2.2) Lemma. Let $\mathcal{T}$ be a triangulation of $\mathbf{R}^{N+1}$ and let $C \subset \mathbf{R}^{N+1}$ be a closed convex set with nonempty interior $\text{int}(C)$. Furthermore, let $\mathcal{T}$ be compatible with C. Recalling the definition (12.1.7) of the system of N-faces $\mathcal{T}^N$, we define

$$\mathcal{T}_{\partial C} := \{ \tau \in \mathcal{T}^N \mid \tau \subset \partial C \neq \emptyset \}.$$

Then $\mathcal{T}_{\partial C}$ is a triangulation of ∂C, more precisely:

(1)
$$\bigcup_{\tau \in \mathcal{T}_{\partial C}} \tau = \partial C;$$

(2) the intersection $\tau_1 \cap \tau_2$ of two N-simplices $\tau_1, \tau_2 \in \mathcal{T}_{\partial C}$ is empty or a common face of both simplices;

(3) the family $\mathcal{T}_{\partial C}$ is locally finite;

(4) for any $\tau \in \mathcal{T}_{\partial C}$ and any $(N-1)$-face ξ of τ, there exists a unique $\tilde{\tau} \in \mathcal{T}_{\partial C}$ such that $\tilde{\tau} \neq \tau$ and $\xi \subset \tilde{\tau}$.

Sketch of Proof. Properties (1)–(3) are an immediate consequence of the fact that $\mathcal{T}$ is a triangulation of $\mathbf{R}^{N+1}$, cf. (12.1.6), and that $\mathcal{T}$ is compatible with C. Property (4) corresponds to lemma (12.1.8). In fact, it is again possible to formally show that properties (1)–(3) imply (4). To do this, let us first observe that there is at least one $\tilde{\tau} \in \mathcal{T}_{\partial C}$ such that $\tilde{\tau} \neq \tau$ and $\xi \subset \tilde{\tau}$ hold. We consider the hyperplane $\mathcal{H} \subset \mathbf{R}^{N+1}$ containing τ, and consider a straight line $s \in \mathbf{R} \mapsto c(s) \in \mathbf{R}^{N+1}$ such that $c(0)$ is the barycenter of ξ and such that $c(-1)$ is the barycenter of τ. If we take an interior point b of C, then the line between b and $c(\varepsilon)$ for small $\varepsilon > 0$ must intersect ∂C in a unique point a_ε which is not in τ, because C is convex and $\mathcal{H}$ supports C at ξ. Hence it must be in some $\tilde{\tau} \in \mathcal{T}_{\partial C}$. Clearly, $\tilde{\tau}$ has the above properties. We now proceed to show that $\tilde{\tau}$ is unique. Suppose that we have a third $\hat{\tau} \in \mathcal{T}_{\partial C}$ such that $\hat{\tau} \neq \tau$ and $\xi \subset \hat{\tau}$ hold. If a_ε is in $\hat{\tau}$ for small $\varepsilon > 0$ or small $-\varepsilon > 0$ then $\hat{\tau}$ must coincide with τ and $\tilde{\tau}$ respectively, since it agrees an ξ and some additional point. Otherwise, the line from b to $c(0)$ must be in the hyperplane $\hat{\mathcal{H}} \subset \mathbf{R}^{N+1}$ containing $\hat{\tau}$. But since b is an interior point of C, this contradicts the fact that $\hat{\mathcal{H}}$ supports C at ξ.   □

(13.2.3) Example. Let $\delta > 0$. We call the image $\mathcal{T}_\delta$ of the Freudenthal triangulation $\mathcal{T}$, cf. (12.1.10), under the dilation map $u \in \mathbf{R}^{N+1} \mapsto \delta u$ the **Freudenthal triangulation of meshsize** δ. We note that in this context, for simplicity we use the norm $\| \cdot \|_\infty$. Then $\mathcal{T}_\delta$ is compatible with the δ-slab $\mathbf{R}^N \times [0, \delta] \subset \mathbf{R}^{N+1}$. Hence, by lemma (13.2.2), $\mathcal{T}_\delta$ also induces triangulations of the levels $\mathbf{R}^N \times \{0\}$ and $\mathbf{R}^N \times \{\delta\}$. In this particular case, it is in fact easy to see that the induced triangulations are again Freudenthal triangulations with meshsize δ.

Now more generally, let $\mathcal{T}$ be any triangulation of $\mathbf{R}^{N+1}$ which is compatible with the δ-slab $\mathbf{R}^N \times [0, \delta] \subset \mathbf{R}^{N+1}$, and let $G : \mathbf{R}^N \to \mathbf{R}^N$ be an asymptotically linear map. For a starting point $x_0 \in \mathbf{R}^N$, one cycle of Merrill's algorithm can be viewed as following a polygonal path in $\tilde{H}_\mathcal{T}^{-1}(0)$ from the level $\lambda = 0$ to the level $\lambda = \delta$, where $\tilde{H}_\mathcal{T}$ denotes the PL approximation of the convex homotopy $\tilde{H} : \mathbf{R}^N \times \mathbf{R} \to \mathbf{R}^N$ defined by $\tilde{H}(x, \lambda) = (1 - \delta^{-1}\lambda)G'(\infty)(x - x_0) + \delta^{-1}\lambda G(x)$, see figure 13.2.a.

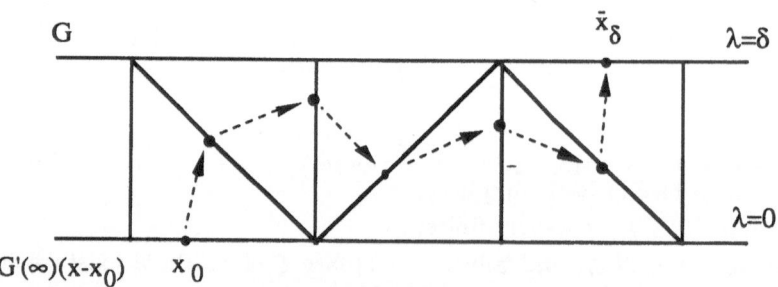

Figure 13.2.a One cycle of Merrill's algorithm. In the next cycle, the starting point is the approximate solution obtained from the previous cycle, i.e. $x_0 := \bar{x}_\delta$

Since a PL algorithm only makes use of the values of the map $\tilde{H}$ at the nodes of the triangulation, it suffices to consider the labeling

$$(13.2.4) \qquad H(x, \lambda) = \begin{cases} G'(\infty)(x - x_0) & \text{for } \lambda \leq 0, \\ G(x) & \text{for } \lambda > 0. \end{cases}$$

Given an asymptotically linear map $G : \mathbf{R}^N \to \mathbf{R}^N$, we now sketch Merrill's algorithm using the labeling (13.2.4). The reader should bear in mind that this labeling changes in each cycle on the trivial level $\lambda = 0$, since the starting point x_0 changes at each cycle to the approximate solution found in the previous cycle. The algorithm may be stopped if the meshsize δ gets smaller than a given minimal meshsize.

(13.2.5) Merrill's Algorithm. *comment:*

    input $x_0 \in \mathbf{R}^N$, $\delta > 0$;   *starting point and initial meshsize*

    repeat

        find a triangulation   *e.g. a shifted* —

          $\mathcal{T}_\delta$ of $\mathbf{R}^{N+1}$ which is compatible   *Freudenthal triangulation*

            with the δ-slab $\mathbf{R}^N \times [0,\delta] \subset \mathbf{R}^{N+1}$ and has meshsize δ;

        find a completely labeled N-face $\tau \in \mathcal{T}_\delta^N$ *τ is unique* —

          such that $\tau \subset \mathbf{R}^N \times \{0\}$;   *by lemma (13.2.6) below*

        determine $\sigma \in \mathcal{T}_\delta$ such that

          $\tau \subset \sigma$ and $\sigma \subset \mathbf{R}^N \times [0,\delta] \subset \mathbf{R}^{N+1}$;

        while $\tau \not\subset \mathbf{R}^N \times \{\delta\}$ **do**

          begin

          find the completely labeled N-face   *door-in-door-out step*

            $\tilde{\tau}$ of σ such that $\tilde{\tau} \neq \tau$;

          find $\tilde{\sigma} \in \mathcal{T}_\delta$ such that $\tilde{\tau} \subset \tilde{\sigma}$ and $\tilde{\sigma} \neq \sigma$;   *pivoting step*

          $\sigma := \tilde{\sigma}$; $\tau := \tilde{\tau}$;

          end;

        calculate the H-center (x,δ) of τ;   *see the remark (12.4.5)*

        $x_0 := x$; output x_0;   *cycle successful*

        $\delta := \dfrac{\delta}{2}$;   *reduce meshsize*

    until traversing is stopped.

The usual way to obtain a completely labeled N-face τ at the start of the cycle is to shift the triangulation $\mathcal{T}_\delta$ in such a way that the starting point x_0 coincides with the barycenter of a standard N-face $\tau \in \mathcal{T}_\delta^N$. Then τ is completely labeled. More generally, we have the following result:

(13.2.6) Lemma. *Let $A : \mathbf{R}^N \to \mathbf{R}^N$ be an affine map such that the Jacobian A' is nonsingular, and let $\mathcal{T}_0$ be a triangulation of $\mathbf{R}^N$. Then there is exactly one simplex $\tau \in \mathcal{T}_0$ which is completely labeled with respect to the labeling A.*

Proof. If $A^{-1}(0)$ is in the interior of some simplex τ, then the assertion is trivial. Of course, this is the typical case in applications. The following proof also incorporates the degenerate case. Using the notation (12.2.3), let $x_\varepsilon := A^{-1}(\bar\varepsilon)$. Since A is affine, it coincides with its PL approximation, and by an argument similar to the proof of (12.2.4), we see that there is an $\tilde\varepsilon > 0$ such that $\{\, x_\varepsilon \mid 0 < \varepsilon < \tilde\varepsilon \,\}$ does not intersect a lower dimensional face of $\mathcal{T}_o$. Hence, $\{\, x_\varepsilon \mid 0 < \varepsilon < \tilde\varepsilon \,\}$ must be contained in the interior of a unique simplex $\tau \in \mathcal{T}_o$, which clearly is completely labeled.   □

Let us now show that each cycle of the algorithm (13.2.5) succeeds in finding a completely labeled N-face in the level $\mathbf{R}^N \times \{\delta\}$ in finitely many steps. Thus, if G is smooth, then the discussion in section 12.6 shows that the new point x_0 approximates a zero point of G with truncation error $O(\delta^2)$. We first give a technical lemma which will also be useful in subsequent discussions.

(13.2.7) Lemma. Let $G : \mathbf{R}^N \to \mathbf{R}^N$ be an asymptotically linear map, and define the labeling H as in (13.2.4). If $\mathcal{T}$ is a triangulation of $\mathbf{R}^{N+1}$ with finite meshsize, then the PL approximation $H_{\mathcal{T}}$ is also asymptotically linear, more precisely:

$$\lim_{\|x\| \to \infty} \frac{H_{\mathcal{T}}(x, \lambda) - G'(\infty)x}{\|x\|} = 0 \quad \text{uniformly in } \lambda.$$

Proof. Let $\pi : \mathbf{R}^N \times \mathbf{R} \to \mathbf{R}^N$ denote the canonical projection, i.e. $\pi(x, \lambda) = x$. For $u \in \mathbf{R}^N \times \mathbf{R}$, we find a convex combination

$$
\textbf{(13.2.8)} \qquad\qquad u = \sum_{i=1}^{N+2} \alpha_i(u)\, v_i(u),
$$

where $[\, v_1(u), \ldots, v_{N+2}(u)\,] \in \mathcal{T}$ is a suitable simplex depending on u. Since $\mathcal{T}$ has a finite meshsize δ, we have

$$\|\pi(v_i(u))\| - \delta \le \|\pi(u)\| \le \|\pi(v_i(u))\| + \delta,$$

and hence the following asymptotic relationships hold for $\|\pi(u)\| \to \infty$ or equivalently, $\|\pi(v_i(u))\| \to \infty$:

(13.2.9) $o(\|\pi(u)\|) = o(\|\pi(v_i(u))\|)$ and $\pi(u - v_i(u)) = o(\|\pi(u)\|)$.

Since $G(x)$ and $G'(\infty)(x - x_0)$ are asymptotically linear, it follows from the definition of H that

$$\textbf{(13.2.10)} \qquad\qquad H(u) - G'(\infty)\pi(u) = o(\|\pi(u)\|).$$

This and (13.2.9) imply

$$
\begin{aligned}
\textbf{(13.2.11)} \qquad H(u) - H(v_i(u)) &= G'(\infty)\pi(u) + o(\|\pi(u)\|) \\
&\quad - G'(\infty)\pi(v_i(u)) - o(\|\pi(v_i(u))\|) \\
&= o(\|\pi(u)\|).
\end{aligned}
$$

Using the convex combination

$$H_T(u) = \sum_{i=1}^{N+2} \alpha_i(u)\, H\big(v_i(u)\big)$$

over (13.2.11), we obtain

$$H(u) - H_T(u) = o\big(\|\pi(u)\|\big),$$

and by (13.2.10) the assertion follows.   □

(13.2.12) Lemma. *Each cycle of algorithm (13.2.5) generates a sequence, cf. (12.3.10), of simplices in the δ-slab $\mathbf{R}^N \times [0,\delta]$ and stops with a completely labeled N-face $\tau \in \mathbf{R}^N \times \{\delta\}$ after finitely many steps.*

Proof. Let us denote by $c_{T_\delta}(s) = (x(s), \lambda(s))$ the polygonal path generated by one cycle of algorithm (13.2.5), cf. (12.3.12). Then we have $c_{T_\delta}(0) = (x_0, 0)$, and by lemma (13.2.6) this path cannot cross the level $\mathbf{R}^N \times \{0\}$ any more. Since $H_{T_\delta}(x(s), \lambda(s)) = 0$, the preceding lemma yields

(13.2.13) $G'(\infty)\, x(s) = o\big(\|x(s)\|\big).$

From the nonsingularity of $G'(\infty)$ we obtain a $\gamma > 0$ such that $\|G'(\infty)x\| \geq \gamma\|x\|$ for all $x \in \mathbf{R}^N$, and therefore (13.2.13) implies that $x(s)$ stays bounded. Hence $c_{T_\delta}(s)$ hits the level $\mathbf{R}^N \times \{\delta\}$ for some $s = \bar{s}$ i.e. $\lambda(\bar{s}) = \delta$, and thus $x(\bar{s})$ is the H-center of a completely labeled N-face in $\mathbf{R}^N \times \{\delta\}$, which is generated by one cycle of algorithm (13.2.5) after a finite number of steps.   □

By the above lemma, it is now clear that algorithm (13.2.5) generates a sequence x_n of approximate zero points of $G^\#$: each cycle terminates with an H-center, say $(x_n, \delta/2^n)$, of a completely labeled N-face $\tau_n = \xi_n \times \{\delta/2^n\}$ in the level $\mathbf{R}^N \times \{\delta/2^n\}$ for $n = 0, 1, 2, \ldots$ where $\delta > 0$ is the initial meshsize. Projecting the sequences $(x_n, \delta/2^n)$ and τ_n onto $\mathbf{R}^N$, we obtain G-centers x_n of G-completely labeled N-simplices $\xi_n \subset \mathbf{R}^N$ with $\text{diam}\,\xi_n = \delta/2^n$. In the next two lemmata we show in what sense the sequence x_n approximates a zero point $\bar{x}$ of $G^\#$.

(13.2.14) Lemma. *The sequence x_n is bounded and hence has at least one accumulation point.*

Proof. We use asymptotic arguments analogous to those in the proof of lemma (13.2.7). Let the sequence x_n be given by the convex combinations $x_n = \sum_{i=1}^{N+1} \alpha_i(n)\, v_i(n)$, where $\xi_n = [v_1(n), \ldots, v_{N+1}(n)]$. Then

$$0 = \sum_{i=1}^{N+1} \alpha_i(n)\, G\big(v_i(n)\big)$$

$$= \sum_{i=1}^{N+1} \alpha_i(n)\Big(G'(\infty) v_i(n) + o(\|v_i(n)\|)\Big)$$

$$= G'(\infty)\, x_n + o\big(\|x_n\|\big).$$

Arguing as in the discussion following (13.2.13), we see that the sequence x_n remains bounded.   □

(13.2.15) Lemma. *Each accumulation point $\bar{x}$ of the sequence x_n is a zero point of $G^\#$.*

Proof. Since x_n is a G-center of ξ_n, we have $0 \in \mathrm{co} G(\xi_n)$. Let $\mathcal{U}_{\bar{x}}$ be the family of neighborhoods of $\bar{x}$, and let $U \in \mathcal{U}_{\bar{x}}$. Since $\lim_{n \to \infty} \mathrm{diam} \xi_n = 0$ and $\bar{x}$ is an accumulation point of x_n, we find an n such that $\xi_n \subset U$. This implies $0 \in \overline{\mathrm{co}}\, G(U)$. Intersecting the last relation over all $U \in \mathcal{U}_{\bar{x}}$ yields $0 \in G^\#(\bar{x})$.
                                                                    □

It is now clear that algorithm (13.2.5) can be considered as providing a "constructive proof" of theorem (13.1.13): if G is asymptotically linear, then $G^\#$ has at least one zero point $\bar{x}$, which is approximated by the sequence generated by algorithm (13.2.5) in the sense of the above two lemmata.

Generally, we cannot prove that the sequence x_n converges. On the other hand, if we assume that $G^\#$ has only one zero point, then the convergence follows trivially. However, especially in this context of general applicability, the latter assumption is very restrictive. We will see in section 13.4 that the Eaves-Saigal algorithm and related continuous deformation algorithms allow a proof of convergence under the reasonable assumption that all zero points of $G^\#$ are isolated.

13.3 Some Triangulations and their Implementations

Up to now we have given essentially only one example of a triangulation, namely Freudenthal's triangulation, cf. (12.1.10), and affine images of it, cf. (12.1.12). In this section we present two more important examples of triangulations: the triangulation J_1 of Todd (1976) and the refining triangulation J_3, which is Todd's modification of a refining triangulation introduced by Eaves (1972) and Eaves & Saigal (1972). We will describe these triangulations and formulate their pivoting rules in pseudo codes. For an extensive treatment of triangulations which are particularly useful in the context of PL homotopy methods, we refer the reader to the monograph of Eaves (1984).

In order to familiarize the reader with the approach which will be adopted in this section, we will first review Freudenthal's triangulation, or more precisely, an affine image of it, by describing its pivoting rules in the form of a pseudo code. The code is based on a starting simplex $[v_1, v_2, \ldots, v_{N+2}] \subset \mathbf{R}^{N+1}$ which must be furnished by the user. This starting simplex defines a triangulation $\mathcal{T}$ as the image of Freudenthal's triangulation under the affine map $A : \mathbf{R}^{N+1} \to \mathbf{R}^{N+1}$ which maps the respective columns of the matrix

$$\begin{pmatrix} 0 & 1 & 1 & \cdots & 1 \\ 0 & 0 & 1 & \cdots & 1 \\ \vdots & \vdots & \ddots & \ddots & \vdots \\ 0 & \cdots & \cdots & 0 & 1 \end{pmatrix}$$

onto the columns of the matrix $(v_1, v_2, \ldots, v_{N+2})$, see also (12.1.12).

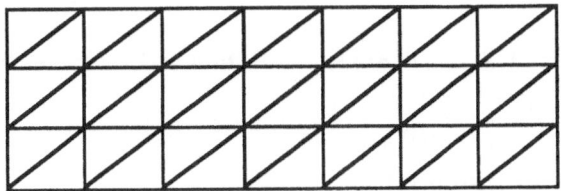

Figure 13.3.a Freudenthal's triangulation in $\mathbf{R}^2$

The codes assume at each step that a decision has been made for determining which vertex is to be pivoted next e.g. the door-in-door-out steps of chapter 12 may furnish such a decision. Our first code is based on pivoting by reflection, see (12.1.11):

**(13.3.1) Pivoting by Reflection
in Freudenthal's Triangulation.**   *comment:*

   input $[v_1, v_2, \ldots, v_{N+2}] \subset \mathbf{R}^{N+1}$;   *starting simplex*

$$\rho(j) := \begin{cases} j+1 & \text{for } j = 1, \ldots, N+1, \\ 1 & \text{for } j = N+2; \end{cases}$$ *cyclic right shift*

   repeat

      enter $i \in \{1, 2, \ldots, N+2\}$;   *index of vertex to be pivoted next*

      $v_i := v_{\rho^{-1}(i)} - v_i + v_{\rho(i)}$ *reflection rule*

   until pivoting is stopped.

Equivalently, these pivoting rules can also be obtained by interchange permutations, see the discussion preceding (12.1.11):

**(13.3.2) Pivoting by Interchange Permutations
in Freudenthal's Triangulation.**   *comment:*

   input $[v_1, v_2, \ldots, v_{N+2}] \subset \mathbf{R}^{N+1}$;   *starting simplex*

$$\rho(j) := \begin{cases} j+1 & \text{for } j = 1, \ldots, N+1, \\ 1 & \text{for } j = N+2; \end{cases} \qquad \textit{cyclic right shift}$$

$$u_j := \begin{cases} v_{j+1} - v_j & \text{for } j = 1, \ldots, N+1, \\ v_1 - v_{N+2} & \text{for } j = N+2; \end{cases} \qquad \textit{standard axes}$$

   for $j = 1, \ldots, N+2$ **do** $\pi(j) := j$;   *initial permutation*

   repeat

      enter $i \in \{1, 2, \ldots, N+2\}$;   *index of vertex to be pivoted next*

      $v_i := v_{\rho^{-1}(i)} + u_{\pi(i)}$;   *pivoting rule*

      interchange $\pi\big(\rho^{-1}(i)\big)$ and $\pi(i)$;

   until pivoting is stopped.

The above codes have been given in order to acquaint the reader with our method of presentation in this section and to emphasize again that at any given stage of a PL algorithm, only one simplex has to be stored. As our next example, we give similar descriptions of the triangulation J_1. One of the advantageous features of J_1 over Freudenthal's triangulation is that it carries less directional bias. The nodes of J_1 are again given by the points $v \in \mathbf{R}^{N+1}$ which have integer co-ordinates. An $(N+1)$-simplex $\sigma \subset \mathbf{R}^{N+1}$ belongs to the triangulation J_1 if the following rules are obeyed:

(1) the vertices of σ are nodes of J_1 in the above sense;
(2) the vertices of σ can be ordered in such a way, say $\sigma = [v_1, v_2, \ldots, v_{N+2}]$
   that they are given by the following recursion formula

$$v_{j+1} = v_j + s(j)\, e_{\pi(j)}, \quad j = 1, \ldots, N+1$$

where $e_1, \ldots, e_{N+1}$ is the standard unit basis of $\mathbf{R}^{N+1}$,

$$\pi : \{1, 2, \ldots, N+1\} \to \{1, 2, \ldots, N+1\}$$

is a permutation and

$$s : \{1, 2, \ldots, N+1\} \to \{+1, -1\}$$

is a sign function;
(3) the **central vertex** v_1 has odd integer co-ordinates.

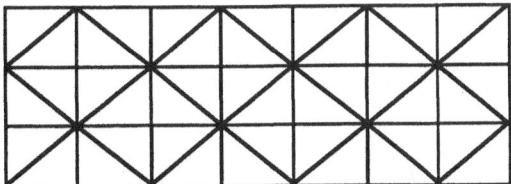

Figure 13.3.b The triangulation J_1 of $\mathbf{R}^2$

From the description of the pivoting rules below it is quite evident that these conditions define a triangulation T of $\mathbf{R}^{N+1}$. The formal proof is however somewhat technical, and we refer the reader to Todd (1976). Analogously to (13.3.1) and (13.3.2), let us now describe the pivoting rules in an affine image of J_1:

(13.3.3) Pivoting by Reflection in J_1.   *comment:*
> **input** $[v_1, v_2, \ldots, v_{N+2}] \subset \mathbf{R}^{N+1}$;   *starting simplex*
> **repeat**
>> **enter** $i \in \{1, 2, \ldots, N+2\}$;   *index of vertex to be pivoted next*
>>
>> $$v_i := \begin{cases} 2v_2 - v_1 & \text{for } i = 1, \\ 2v_{N+1} - v_{N+2} & \text{for } i = N+2, \\ v_{i-1} - v_i + v_{i+1} & \text{else}; \end{cases}$$
>
> **until** pivoting is stopped.

Similarly to the discussion for pivoting in Freudenthal's triangulation, the pivoting rules for J_1 can also be obtained by interchange permutations:

(13.3.4) Pivoting by Interchange Permutations in J_1.   *comment:*
> **input** $[v_1, v_2, \ldots, v_{N+2}] \subset \mathbf{R}^{N+1}$;   *starting simplex*
> **for** $j = 1$ **to** $N+1$ **do**
>> **begin**
>>> $u_j := v_{j+1} - v_j$;   *standard axes*
>>> $\pi(j) := j$;   *initial permutation*
>>> $s(j) := 1$;   *initial sign function*
>>
>> **end**;
>
> **repeat**
>> **enter** $i \in \{1, 2, \ldots, N+2\}$;   *index of vertex to be pivoted next*
>> **case** $i = 1$:   *consider different cases*
>>> $v_1 := v_2 + s(1)\, u_{\pi(1)}$;
>>> $s(1) := -s(1)$;

case $i = N + 2$:

$$v_{N+2} := v_{N+1} - s(N+1)\, u_{\pi(N+1)};$$
$$s(N+1) := -s(N+1);$$

case else:

$$v_i := v_{i-1} + s(i)\, u_{\pi(i)};$$
interchange $s(i-1)$ and $s(i)$;
interchange $\pi(i-1)$ and $\pi(i)$

end cases ;

until pivoting is stopped.

We call the next triangulation J_3 a **refining triangulation** of $\mathbf{R}^N \times \mathbf{R}$ since it induces triangulations $\mathcal{T}_i$ on each level $\mathbf{R}^N \times \{i\}$ such that the meshsize $\delta(\mathcal{T}_i) \to 0$ as $i \to \infty$. We will see in the next section that such refining triangulations are very useful in the context of PL homotopy methods. The nodes of J_3 are given by the points $(x, \lambda) \in \mathbf{R}^N \times \mathbf{R}$ such that $\lambda = k$ for some integer k and such that all co-ordinates of x are integer multiples of 2^{-k}. An $(N+1)$-simplex $\sigma \subset \mathbf{R}^{N+1}$ belongs to the triangulation J_3 if the following rules are obeyed:

(13.3.5) J_3-Rules.

(1) the vertices of σ are nodes of J_3 in the above sense;
(2) there exists an ordering

$$\sigma = \big[(x_1, \lambda_1), (x_2, \lambda_2), \ldots, (x_{N+2}, \lambda_{N+2})\big]$$

of the vertices of σ, a permutation

$$\pi : \{1, 2, \ldots, N+1\} \to \{1, 2, \ldots, N+1\}$$

and a sign function

$$s : \{1, 2, \ldots, N+1\} \to \{+1, -1\}$$

such that the following conditions hold for $q := \pi^{-1}(N+1)$ ("last index on the fine level"):

(a) $\qquad \lambda_{j+1} = \lambda_j \quad \text{for} \quad j = 1, \ldots, N+1, \ j \neq q;$

(b) $\qquad \lambda_{q+1} = \lambda_q - 1;$

(c) $\qquad x_{j+1} = x_j + s(j)\, 2^{-\lambda_j}\, e_{\pi(j)}$

$$\text{for} \quad j = 1, \ldots, q-1, q+1, \ldots, N+1;$$

(d) $\qquad x_{q+1} = x_q - \displaystyle\sum_{r=q+1}^{N+1} s(r)\, 2^{-\lambda_1}\, e_{\pi(r)} \, ;$

(3) the central point x_1 has odd integer multiples of the finer meshsize $2^{-\lambda_1}$ as co-ordinates;

(4) the first point x_{q+1} on the coarser grid has a maximal number of odd integer multiples of the coarser meshsize $2^{-\lambda_q+1}$ as co-ordinates, i.e.

$$2^{\lambda_q+1} x_{q+1}\big[\pi(r)\big]$$

is odd for $r = q+1, \ldots, N+1$.

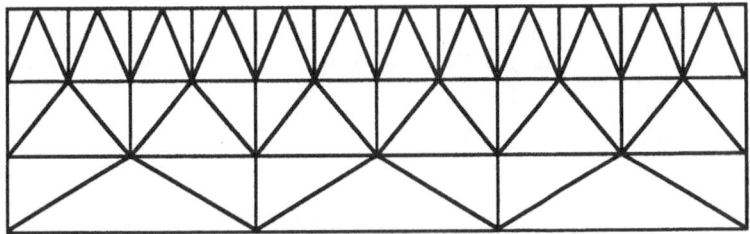

Figure 13.3.c The triangulation J_3 of $\mathbf{R} \times \mathbf{R}$

Here $e_1, \ldots, e_N$ denotes the standard unit basis of $\mathbf{R}^N$. From the description of the pivoting rules below it is quite evident that these conditions define a triangulation $\mathcal{T}$ of $\mathbf{R}^{N+1}$. The formal proof is however very technical, and we refer the reader to Todd (1976). An affine map $A : \mathbf{R}^N \to \mathbf{R}^N$ extends in a natural way to an affine map $\tilde{A} : \mathbf{R}^N \times \mathbf{R} \to \mathbf{R}^N \times \mathbf{R}$ by setting $\tilde{A}(x, \lambda) = (Ax, \lambda)$. We now describe the pivoting rules in such an affine image of J_3. At each step, the following data are updated:

(13.3.6) J_3-Data.

(i)   the vertices (z_j, λ_j), $j = 1, 2, \ldots, N+2$ of the current simplex $\tilde{A}\sigma$;

(ii) the central point x_1 of the current simplex $\sigma \in J_3$;

(iii) the permutation $\pi : \{1, 2, \ldots, N+2\} \to \{1, 2, \ldots, N+2\}$;

(iv) the sign function $s : \{1, 2, \ldots, N+2\} \to \{+1, -1\}$;

(v) a cyclic permutation $\rho : \{1, 2, \ldots, N+2\} \to \{1, 2, \ldots, N+2\}$ describing the ordering of the vertices as referred to in (13.3.5)(2) (note that this order may change after a pivoting step).

For the permutation π and the ordering ρ we use the convention that $Ax_1 = z_{\rho(j)}$ corresponds to the central point x_1 of the current simplex $\sigma \in J_3$ if $\pi(j) = N+2$. Instead of updating λ_j and $s(j)$ separately, we update the useful "steplength" $d(j) := s(j) 2^{-\lambda_j}$ for $j = 1, 2, \ldots, N+2$, see (13.3.5)(2c).

(13.3.7) Pivoting in J_3.
$\quad$ *comment:*
$\quad\quad$ **input** $[z_2, z_3, \ldots, z_{N+2}] \subset \mathbf{R}^N;$
$\quad\quad$ *corresponds —*

$\quad\quad\quad\quad\quad$ *to a starting N-face in $\mathbf{R}^N \times \{0\}$*

$z_1 := \frac{1}{2}(z_2 + z_{N+2});$

$$\rho(j) := \begin{cases} j+1 & \text{for } j = 1, \ldots, N+1, \\ 1 & \text{for } j = N+2; \end{cases} \quad\quad \textit{starting cyclic ordering}$$

$$d(j) := \begin{cases} \frac{1}{2} & \text{for } j = 1, \\ -1 & \text{for } j = 2, \ldots, N+1; \end{cases} \quad\quad \textit{starting steplengths}$$

$$\pi(j) := \begin{cases} N+1 & \text{for } j = 1, \\ j-1 & \text{for } j = 2, \ldots, N+1, \\ N+2 & \text{for } j = N+2; \end{cases} \quad \textit{starting permutation}$$

$x_1(j) := 0.5$ for $j = 1, \ldots N;$ $\quad\quad\quad\quad\quad$ *central point*

repeat

$\quad$ **enter** $i \in \{1, 2, \ldots, N+2\};$ $\quad\quad$ *index of vertex to be pivoted next*

$\quad$ **1:** $\quad\quad\quad\quad\quad$ *reference point for automatic pivots, see section 13.6*

$\quad$ $i_- := \rho^{-1}(i); \quad i_+ := \rho(i);$ $\quad\quad$ *cyclic left and right neighbors*

$\quad$ **case** $|d(i_-)| = |d(i)| = |d(i_+)|:$ $\quad\quad$ *neighbors on same level*

$\quad\quad$ $z_i := z_{i_-} - z_i + z_{i_+};$ $\quad\quad\quad\quad\quad$ *reflection pivot*

$\quad\quad$ interchange $d(i_-)$ and $d(i);$

$\quad\quad$ interchange $\pi(i_-)$ and $\pi(i);$

$\quad$ **case** $|d(i_-)| > |d(i)| = |d(i_+)|:$ $\quad\quad$ *z_i is first point on fine level*

$\quad\quad$ $z_i := 2z_{i_+} - z_i;$ $\quad\quad\quad\quad\quad$ *reflection pivot*

$\quad\quad$ $x_1[\pi(i)] := x_1[\pi(i)] + 2d(i);$ $\quad\quad$ *new central point*

$\quad\quad$ $d(i) := -d(i);$

$\quad$ **case** $|d(i_-)| < |d(i)| = |d(i_+)|:$ $\quad\quad$ *z_i is first point on coarse level*

$\quad\quad$ $z_i := z_{i_-} - \frac{1}{2}z_i + \frac{1}{2}z_{i_+};$ $\quad\quad\quad\quad$ *skew pivot*

$\quad\quad$ $d(i_-) := \frac{1}{2}d(i); \quad d(i) := d(i_-);$

$\quad\quad$ interchange $\pi(i_-)$ and $\pi(i);$

$\quad$ **case** $|d(i_-)| = |d(i)| > |d(i_+)|:$ $\quad\quad$ *z_i is last point on coarse level*

$\quad\quad$ find q with $\pi(q) = N+1; \quad q_+ := \rho(q);$ $\quad\quad$ *index of last point —*

$\quad\quad\quad\quad\quad\quad\quad\quad\quad\quad\quad\quad\quad\quad$ *on fine level*

$\quad\quad$ $z_i := z_q - \frac{1}{2}z_i + \frac{1}{2}z_{i_-};$ $\quad\quad\quad\quad$ *skew pivot*

$\quad\quad$ $d(q) := -\frac{1}{2}d(i_-); \quad d(i) := d(q);$

$\quad\quad$ $\pi(q) := \pi(i_-); \quad \pi(i_-) := N+2; \quad \pi(i) := N+1;$

$\quad\quad$ $\rho(i_-) := i_+; \quad \rho(i) := q_+; \quad \rho(q) := i;$ $\quad\quad$ *new cyclic ordering*

case $|d(i_-)| > |d(i)| < |d(i_+)|$: $\quad z_i$ is the only point on fine level

$z_i := z_{i_-}$;

interchange $\pi(i_-)$ and $\pi(i)$;

$d(i) := 4d(i)$;

for $j = 1$ to $N + 2$ such that $\pi(j) \neq N + 1, N + 2$ **do**

$\quad x_1[\pi(j)] := x_1[\pi(j)] - \frac{1}{2}d(j)$; $\qquad$ new central point

case $|d(i_-)| < |d(i)| > |d(i_+)|$: $\quad z_i$ is the only point on coarse level

$z_i := \frac{1}{2}(z_{i_-} + z_{i_+})$;

interchange $\pi(i_-)$ and $\pi(i)$;

$d(i) := \frac{1}{4}d(i)$;

for $j = 1$ to $N + 2$ such that $\pi(j) \neq N + 1, N + 2$ **do**

$\quad x_1[\pi(j)] := x_1[\pi(j)] + \frac{1}{2}d(j)$; $\qquad$ new central point

case $|d(i_-)| = |d(i)| < |d(i_+)|$ **and** $\dfrac{x_1(\pi(i_-)) + d(i_-)}{|d(i_+)|} \equiv 0 \pmod 2$:

$\qquad\qquad\qquad\qquad\qquad\qquad z_i$ is last point on fine level —

$\qquad\qquad\qquad$ no change of cyclic ordering, see (13.3.5)(4)

$z_i := 2z_{i_-} - 2z_i + z_{i_+}$; $\qquad\qquad\qquad\qquad$ skew pivot

interchange $\pi(i_-)$ and $\pi(i)$;

$d(i) := 2d(i_-)$;

case else: $\qquad\qquad\qquad\qquad z_i$ is last point on fine level —

$\qquad\qquad\qquad$ cyclic ordering will be changed, see (13.3.5)(4)

find q with $\pi(q) = N + 2$; $q_+ := \rho(q)$; $\qquad$ index of last point —

$\qquad\qquad\qquad\qquad\qquad\qquad\qquad\qquad\qquad$ on coarse level

$z_i := 2z_{i_-} - 2z_i + z_q$; $\qquad\qquad\qquad\qquad$ skew pivot

$d(q) := -2d(i_-)$; $d(i) := d(q)$;

$\pi(q) := \pi(i_-)$; $\pi(i) := N + 2$; $\pi(i_-) := N + 1$;

$\rho(i_-) := i_+$; $\rho(q) := i$; $\rho(i) := q_+$; $\qquad$ new cyclic ordering

end cases

until pivoting is stopped.

Code (13.3.7) is in the spirit of the codes (13.3.1) and (13.3.3) using a more geometric approach of the reflection rule. It is easy to convert (13.3.7) to the spirit of codes (13.3.2) and (13.3.4). Code (13.3.7) has been implemented in the program of the Eaves-Saigal algorithm in the appendix. The above code has been presented more for readability than efficiency. The triangulation J_3 presented here is the simplest case of an already efficient refining triangulation: it reduces the meshsize on each successive level by a factor $\frac{1}{2}$. It is possible to construct refining triangulations with arbitrary refining factors, see Bárány (1979), Shamir (1979), Engles (1980), van der Laan & Talman (1980), Kojima & Yamamoto (1982), Broadie & Eaves (1987). However, Todd(1978) and Todd & Acar (1980) show that attempts to decrease the meshsize to fast may result in a larger number of PL steps.

It is possible to compare different triangulations via various ways of measuring their efficiency. Such results can be found in Todd (1976), Saigal (1977), van der Laan & Talman (1980), Alexander & Slud (1983), Eaves & Yorke (1984), Eaves (1984).

13.4 The Homotopy Algorithm of Eaves & Saigal

In this section we describe a version of the Eaves & Saigal (1972) algorithm which is a simple example of a PL homotopy deformation algorithm. The earliest such algorithm is due to Eaves (1972). We return to the problem of seeking a zero point of the set-valued hull $G^\#$ of an asymptotically linear map $G : \mathbf{R}^N \to \mathbf{R}^N$, see (13.1.3–12). The idea of the Eaves-Saigal algorithm is to apply a PL algorithm for the labeling H as given in (13.2.4):

$$H(x, \lambda) = \begin{cases} G'(\infty)(x - x_0) & \text{for } \lambda \leq 0, \\ G(x) & \text{for } \lambda > 0, \end{cases}$$

using some refining triangulation for $\mathbf{R}^N \times \mathbf{R}$. We will illustrate the algorithm for an affine image $\mathcal{T}$ of the refining triangulation J_3. But let us emphasize here that other refining triangulations such as those cited above may be used as well. A starting face $\tau \in \mathbf{R}^N \times \{0\}$ is chosen so that the starting point $(x_0, 0)$ is the barycenter of τ. By lemma (13.2.6), τ is the only completely labeled N-face of $\mathcal{T}$ lying in $\mathbf{R}^N \times \{0\}$. The algorithm starts from τ and enters $\mathbf{R}^N \times \mathbf{R}_+$. It may be stopped when a specified level $\mathbf{R}^N \times \{\lambda_{\max}\}$ is traversed. As an affine image of J_3, the refining triangulation $\mathcal{T}$ is completely determined by the starting N-face τ and the pivoting rules in (13.3.7).

(13.4.1) Eaves-Saigal Algorithm.   *comment:*

> **input** $\tau_0 \subset \mathbf{R}^N$, an N-simplex with barycenter x_0 ;   *starting point*
>
> output $(x_0, 0)$;   *first node of polygonal path, cf. (12.3.12)*
>
> $\tau := \tau_0 \times \{0\}$;   *starting N-face*
>
> get $\sigma \in \mathcal{T}$ with $\tau \subset \sigma$ and $\sigma \subset \mathbf{R}^N \times \mathbf{R}_+$;

repeat

 find the completely labeled N-face $\tilde{\tau}$ of σ

  such that $\tilde{\tau} \neq \tau$;   *door-in-door-out step*

 find $\tilde{\sigma} \in \mathcal{T}$ such that $\tilde{\tau} \subset \tilde{\sigma}$ and $\tilde{\sigma} \neq \sigma$;   *pivoting step, see (13.3.7)*

 $\sigma := \tilde{\sigma}$; $\tau := \tilde{\tau}$;

 calculate the H-center (x, λ) of τ;   *see the remark (12.4.5)*

 output (x, λ);   *next node of polygonal path, cf. (12.3.12)*

until traversing is stopped.

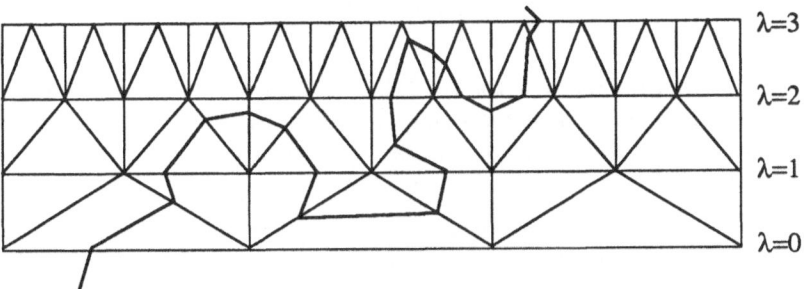

Figure 13.4.a The polygonal path of the Eaves & Saigal algorithm

The algorithm (13.4.1) generates a polygonal path $c_{\mathcal{T}}(s) = \big(x(s), \lambda(s)\big) \in H_{\mathcal{T}}^{-1}(0)$, cf. (12.3.12), whose nodes are given by the output. We have $x(0) = x_0$, $\lambda(0) = 0$ and $\lambda(s) > 0$ for $s > 0$. For simplicity, we assume that s represents the arclength of $c_{\mathcal{T}}$. Recall from (13.2.7) that the PL approximation $H_{\mathcal{T}}$ is also asymptotically linear. This was the basis for proving (13.2.12), (13.2.14), (13.2.15). Using similar arguments, we obtain the following

(13.4.2) Lemma. *Let* $G : \mathbf{R}^N \to \mathbf{R}^N$ *be asymptotically linear. Then the polygonal path* $c_{\mathcal{T}}(s) = \big(x(s), \lambda(s)\big) \in H_{\mathcal{T}}^{-1}(0)$ *generated by the Eaves-Saigal algorithm (13.4.1) without stopping (i.e. for* $\lambda_{\max} = \infty$*) has the following properties:*

(1) *since no stopping is allowed,* $c_{\mathcal{T}}(s)$ *is generated for all* $s \geq 0$;

(2) $\lambda(s) \to \infty$ *as* $s \to \infty$;

(3) $x(s)$ *is bounded for* $s \to \infty$ *and hence has at least one accumulation point;*

(4) *each accumulation point* $\bar{x}$ *of* $x(s)$ *as* $s \to \infty$ *is a zero point of* $G^{\#}$.

As in the case of Merrill's algorithm, it is clear again in the sense of the above lemma that algorithm (13.4.1) can be considered as providing a "constructive proof" of theorem (13.1.13): if G is asymptotically linear, then $G^{\#}$ has at least one zero point $\bar{x}$, which is approximated by an accumulation point of the path $x(s)$ generated by algorithm (13.4.1).

Generally, we cannot prove that the path $x(s)$ converges as $s \to \infty$, see also the remarks at the end of section 13.2. However, we can now obtain a convergence result under a reasonable additional condition:

(13.4.3) Corollary. *If $G^{\#}$ has only isolated zero points in $\mathbf{R}^N$, then the path $x(s)$ converges to a zero point of $G^{\#}$ as $s \to \infty$.*

Proof. A straightforward argument of point set topology shows that the set of accumulation points of $x(s)$ as $s \to \infty$ can be expressed as

(13.4.4)
$$\bigcap_{s>0} \overline{x\big((s,\infty)\big)}.$$

Since $x(s)$ is continuous in s, the set (13.4.4) is connected. Since it contains only isolated points, it must be a singleton $\{\bar{x}\}$. Since $x(s)$ is bounded and has exactly one accumulation point $\bar{x}$ as $s \to \infty$, we obtain the convergence $\lim_{s \to \infty} x(s) = \bar{x}$. □

More generally, the above proof also shows: if $\bar{x}$ is an isolated zero point of $G^{\#}$ and $\bar{x}$ is an accumulation point of $x(s)$, then $x(s)$ converges to $\bar{x}$. The assumption in (13.4.3) is reasonable, since it essentially means that the problem "find x such that $0 \in G^{\#}(x)$" is well formulated.

13.5 Mixing PL and Newton Steps

As we have seen above, the Eaves-Saigal algorithm (13.4.7) is more convenient for discussing the question of convergence. If no stopping is allowed, it generates a sequence of nodes (x_n, λ_n) for $n = 0, 1, 2, \ldots$. We have seen that x_n converges to a zero point $\bar{x}$ of $G^{\#}$ under reasonable and very weak assumptions. Without additional assumptions on G however, nothing can be said about the rate of convergence of x_n. Brooks (1980) has shown that infinite retrogression can occur. To ensure linear convergence, assumptions in the spirit of the Newton-Kantorovitch theorems, see Ortega & Rheinboldt (1970), are necessary. Such convergence discussions have been given by Saigal (1977) and Cromme (1980). Saigal was the first to see the close interrelationship between PL steps and Newton's method. Several papers discuss techniques of mixing PL and Newton steps in order to accelerate a PL homotopy algorithm, see e.g. Saigal (1977), Saigal & Todd (1978) and Todd, M. J. (1978), (1980). In the context of PL continuation methods, i.e. when a whole curve $c(s)$ is to be approximated by a polygonal path $c_T(s)$, Saupe (1982) has considered a mixing of PL and predictor-corrector steps.

We describe here an elementary way of mixing PL and Newton steps given by Georg (1982). It is based on the simple observation that a modified

Newton's method expressed in barycentric co-ordinates leads to a system of linear equations which is closely related to the linear equations obtained in the "door-in-door-out step" of complementary pivoting as given in (12.4.4) and (12.4.5).

(13.5.1) Lemma. *Let* $G : \mathbf{R}^N \to \mathbf{R}^N$ *be a map and* $\tau = [z_1, z_2, \ldots, z_{N+1}] \subset \mathbf{R}^N$ *an N-simplex. Let* $B := G'_\tau$ *denote the Jacobian of the PL approximation* G_τ *of G, see (12.2.1)(1), i.e. B is the finite difference approximation of G' using the values of G on $z_1, z_2, \ldots, z_{N+1}$. We assume that B is nonsingular and define a modified Newton step* $\mathbf{N} : \mathbf{R}^N \to \mathbf{R}^N$ *by*

$$(1) \qquad\qquad \mathbf{N}(x) := x - B^{-1}G(x).$$

Then

$$(2) \qquad\qquad \mathbf{N}(z_i) = b \quad \text{for} \quad i = 1, 2, \ldots, N+1,$$

where b denotes the G-center of τ, which is characterized by $G_\tau(b) = 0$, cf. (12.3.12). Furthermore, for any $z_{N+2} \in \mathbf{R}^N$, let

$$(3) \qquad \mathcal{L}(\tau, z_{N+2}) := \begin{pmatrix} 1 & \cdots & 1 & 1 \\ G(z_1) & \cdots & G(z_{N+1}) & G(z_{N+2}) \end{pmatrix}$$

be the labeling matrix, defined analogously to (12.2.5–6), and consider the barycentric co-ordinates

$$\alpha = \big(\alpha[1], \alpha[2], \ldots, \alpha[N+2] \big)^*, \qquad \gamma = \big(\gamma[1], \gamma[2], \ldots, \gamma[N+2] \big)^*$$

defined by the equations

$$(4) \qquad \begin{array}{ll} \mathcal{L}(\tau, z_{N+2})\,\alpha = e_1, & \alpha[N+2] = 0, \\ \mathcal{L}(\tau, z_{N+2})\,\gamma = 0, & \gamma[N+2] = -1. \end{array}$$

Then $\alpha - \gamma$ represents the barycentric co-ordinates of $\mathbf{N}(z_{N+2})$, i.e.

$$(5) \qquad\qquad \mathbf{N}(z_{N+2}) = \sum_{j=1}^{N+2} (\alpha[j] - \gamma[j]) z_j.$$

Proof. Note first that the nonsingularity of B implies that G_τ is bijective i.e. the linear equations (4) can be uniquely solved. Since G_τ is affine, we have $B = G_\tau + C$ for some constant $C \in \mathbf{R}^N$, and consequently $\sum_j \xi_j B z_j = \sum_j \xi_j G_\tau(z_j)$ for coefficients ξ_j such that $\sum_j \xi_j = 0$. This will be used in the sequel.

From equation (1) we have $G(x) = Bx - BN(x) = G_\tau(x) - G_\tau(N(x))$. If x is a vertex of τ, then $G(x) = G_\tau(x)$ and hence $G_\tau(N(x)) = 0$. Since G_τ is bijective and $b = G_\tau^{-1}(0)$ by definition, assertion (2) follows.

Combining the two equations in (4) we easily see that

$$\alpha[N+2] - \gamma[N+2] = 1\,,$$

$$\sum_{j=1}^{N+1} \alpha[j] - \gamma[j] = 0\,,$$

$$\sum_{j=1}^{N+1} (\alpha[j] - \gamma[j])G(z_j) = -G(z_{N+2})\,.$$

This implies

$$-G(z_{N+2}) = \sum_{j=1}^{N+1} (\alpha[j] - \gamma[j])G_\tau(z_j) = \sum_{j=1}^{N+1} (\alpha[j] - \gamma[j])Bz_j$$

and consequently

$$-B^{-1}G(z_{N+2}) = \sum_{j=1}^{N+1} (\alpha[j] - \gamma[j])z_j$$

or

$$N(z_{N+2}) = x - B^{-1}G(z_{N+2}) = \sum_{j=1}^{N+2} (\alpha[j] - \gamma[j])z_j\,.$$

This proves (5).   □

The above lemma suggests, how the PL steps of the Eaves-Saigal algorithm (13.4.1) can be combined with Newton steps. Let us assume that a level $\mathbf{R}^N \times \{k\}$ for $k > 0$ is encountered for the first time. In this case we have a completely labeled N-face

$$\tau = [(z_1, k), (z_2, k), \ldots, (z_{N+1}, k)]$$

and an $(N+1)$-simplex

$$\sigma = [(z_1, k), (z_2, k), \ldots, (z_{N+1}, k), (z_{N+2}, k-1)]$$

such that the vertex $(z_{N+2}, k-1)$ has to be pivoted next. Since the labeling on the level $\mathbf{R}^N \times \{k\}$ is given by G, see (13.2.4), apart from the unimportant level co-ordinate $\lambda = k$, we are exactly in the situation of the preceding lemma. In particular, the first Newton point $N(z_1) = b$ is given by the

G-center (b, k) of τ, and subsequent Newton points $\mathbf{N}(b)$, $\mathbf{N}^2(b)$, ... can be obtained by replacing the vertex $(z_{N+2}, k-1)$ by $(b, k), (\mathbf{N}(b), k), \ldots$ and solving linear equations which are very similar to the ones for the PL steps. In fact, we continue to update a labeling matrix (13.5.1)(4) by replacing the column corresponding to the vertex $(z_{N+2}, k-1)$ in each Newton step with the labels of the Newton points. Hence, we continue to use updating techniques which will be described in chapter 16 and which result in $O(N^2)$ flops step, regardless of whether it is a PL or a Newton step. If the Newton steps are not successful, we simply perform the pivot of the vertex $(z_{N+2}, k-1)$ and resume the updating of the PL method. The following pseudo code sketches this technique. As a simple example, we use updating of the right inverse as in (12.4.6).

(13.5.2) Eaves-Saigal Algorithm With Newton Steps.   *comment:*

   input

      begin

         $0 < \kappa < 1$;   *maximal admissible contraction rate for Newton steps*

         $\varepsilon_G > 0$;   *stopping tolerance for $\|G\|$*

         $\lambda_{\max} > 0$;   *stopping level*

         $\varepsilon_{\text{tol}} > 0$; *tolerance for avoiding division by zero in the update formula*

         $\tau_0 = [z_2, \ldots, z_{N+2}] \subset \mathbf{R}^N$;   *starting N-face*

      end;

      $z_1 := \frac{1}{2}(z_2 + z_{N+2})$; *loading starting simplex $[(z_1, \lambda_1), \ldots, (z_{N+2}, \lambda_{N+2})]$*

      $\lambda_k := \begin{cases} 1 & \text{for } k = 1, \\ 0 & \text{for } k = 2, \ldots, N+1; \end{cases}$

      $x_0 := \dfrac{1}{N+1} \displaystyle\sum_{k=2}^{N+2} z_k$;   *starting point makes $\tau_0 \times \{0\}$ completely labeled*

      $j = 1$;   *N-face opposite (z_1, λ_1) is completely labeled*

      $A := \begin{pmatrix} 1 & \cdots & 1 \\ H(z_1, \lambda_1) & \cdots & H(z_{N+2}, \lambda_{N+2}) \end{pmatrix}$;   *labeling matrix*

      find $(N+2) \times (N+1)$-matrix B such that

         $AB = \text{Id}$ and $e_j^* B = 0$;   *initial right inverse of A*

      $\bar{\lambda} := 1$;   *highest level currently encountered*

   repeat

      if B is not lex. positive **then** quit;   *failure —*

         e.g. wrong starting simplex or numerical instability

      $y := \begin{pmatrix} 1 \\ H(z_j, \lambda_j) \end{pmatrix}$;   *new label w.r.t. (13.2.4)*

$$\gamma := By - e_j ; \quad \alpha := Be_1 ; \qquad\qquad \textit{solving the linear equations}$$

$$i := \operatorname*{arg\,min}_{k=1,\dots,N+2} \left\{ \frac{\alpha[k]}{\gamma[k]} \,\middle|\, \gamma[k] > \varepsilon_{\text{tol}} \right\} ; \qquad \textit{door-in-door-out step}$$

$$\text{pivot } (z_i , \lambda_i); \qquad\qquad \textit{pivoting step w.r.t. (13.3.7)}$$

$$B := B - \frac{\gamma e_i^* B}{\gamma[i]} ; \qquad\qquad\qquad\qquad \textit{update of } B$$

if $\lambda_i > \bar{\lambda}$ **then** $\qquad\qquad$ *new level traversed: tentative Newton steps*
$\quad$ **begin**
$\quad$ $\bar{\lambda} := \lambda_i ;$ $\qquad\qquad\qquad\qquad\qquad$ *new highest level*

$$w_1 := \sum_{k=1}^{N+2} \alpha[k]\, z_k ; \qquad\qquad\qquad \textit{first Newton point}$$

$\quad$ **if** $\|G(w_1)\| < \varepsilon_G$ **then** stop;
$\quad$ **repeat** $\qquad\qquad\qquad\qquad\qquad\qquad$ *Newton iteration*

$$y := \begin{pmatrix} 1 \\ G(w_1) \end{pmatrix} ; \qquad\qquad\qquad\qquad \textit{new label}$$

$$\gamma := By - e_i ;$$

$$w_2 := \sum_{k=1}^{N+2} (\alpha[k] - \gamma[k])\, z_k ; \quad \textit{second Newton point, see (13.5.1)(5)}$$

$\quad$ **if** $\|G(w_2)\| < \varepsilon_G$ **then** stop;

$\quad$ **if** $\dfrac{\|G(w_2)\|}{\|G(w_1)\|} > \kappa$ **then go to 1;** $\qquad$ *Newton steps failed —*

$\qquad\qquad\qquad\qquad\qquad\qquad\qquad\qquad$ *back to PL steps*

$\qquad$ $w_2 := w_1 ;$
$\quad$ **until** stopped;
end{if};
1: $\qquad\qquad\qquad\qquad$ *reference point for continuing PL steps*
$\quad$ $j := i;$
until $\lambda_j > \lambda_{\max} .$

As each new level $\mathbf{R}^N \times \{k\}$ is reached in the above algorithm, some tentative Newton steps are performed. If the contraction rate of the Newton steps is sufficiently strong, the algorithm continues to perform modified Newton steps until it stops at an approximate solution. This means of course that some conditions typical for Newton's method must be satisfied. For example, if $G^{\#}$ has only isolated zero points, and if G is in fact smooth at these points with a nonsingular Jacobian G', then it is not difficult to see by combining the convergence result (13.4.3) and the classical Newton-Kantorovitch theory, that the above algorithm will eventually converge to a solution via Newton steps. Hence, we can look at the above algorithm as a type of globalization of Newton's method, where in addition the algorithm reveals whether the assumptions for the Newton-Kantorovitch theory are verified.

It is evident that a treatment similar to that in algorithm (13.5.2) can be formulated for restart algorithms such as that of Merrill. We have illustrated the mixing of Newton steps for the case of the update method (12.4.6). Other update methods which will be discussed in more detail in chapter 16, can be treated in an analogous way.

13.6 Automatic Pivots for the Eaves-Saigal Algorithm

By noticing the special form of the labeling (13.2.4) which is used in the Eaves-Saigal algorithm, it becomes clear that the label on two vertices coincides i.e. $H(x_i, \lambda_i) = H(x_j, \lambda_j)$ if $x_i = x_j$ and $\lambda_i, \lambda_j > 0$. This fact can be used to occasionally perform **automatic pivots** and thereby reduce the computational effort. More precisely, if the vertex (x_i, λ_i) has been pivoted in at the last step, then the N-face opposite (x_i, λ_i) is completely labeled. If $H(x_i, \lambda_i) = H(x_j, \lambda_j)$ holds, then it is immediately clear that also the N-face opposite (x_j, λ_j) is completely labeled. Hence, the next vertex to be pivoted must be (x_j, λ_j). Thus the linear algebra for the door-in-door-out step need not be performed. We only have to keep track of the interchange $i \leftrightarrow j$.

Let us now illustrate this effect by describing in more detail the automatic pivots for the case that the Eaves-Saigal algorithm is used in conjunction with an affine image of the refining triangulation J_3 for $\mathbf{R}^N \times \mathbf{R}$. Since the pivoting code (13.3.7) for J_3 involves occasional reorderings of the vertices via the cyclic permutation ρ anyway, it is convenient to perform the interchange by changing ρ instead of performing the interchange in the update of the linear equations.

The automatic pivots in the Eaves-Saigal algorithm can occur in the following three cases of code (13.3.7):

$$\textbf{case } |d(i_-)| < |d(i)| = |d(i_+)|$$
$$\textbf{case } |d(i_-)| = |d(i)| > |d(i_+)|$$
$$\textbf{case } |d(i_-)| > |d(i)| < |d(i_+)|$$

As an example, let us illustrate the steps of the automatic pivot after the pivot in the first case has been performed, i.e. between the two lines

$$\text{interchange } \pi(i_-) \text{ and } \pi(i);$$
$$\textbf{case } |d(i_-)| = |d(i)| > |d(i_+)|$$

in (13.3.7), we simply add the following steps.

if $|d(\rho(i))| > 0$ **and** $\pi(\rho(i)) = N + 2$ **then**
begin
$i_- := \rho^{-1}(i);\ \ i_+ := \rho(i);\ \ i_{++} := \rho(i_+);$
interchange $d(i)$ and $d(i_+);$
interchange $\pi(i)$ and $\pi(i_+);$
$\rho(i_-) := i_+;\ \ \rho(i_+) := i;\ \ \rho(i) := i_{++};$
go to 1
end

The other two cases are handled in a similar way. They are incorporated in the implementation of the Eaves-Saigal algorithm in the appendix.

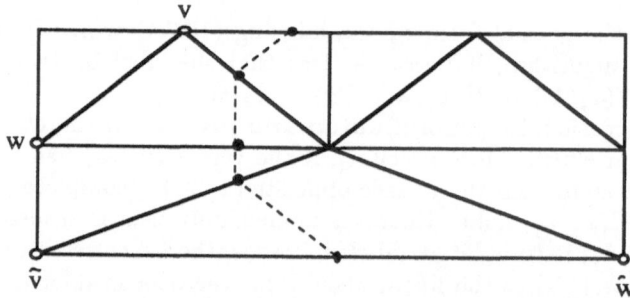

Figure 13.6.a The pivoting step $v \to \tilde{v}$ automatically implies the pivoting step $w \to \tilde{w}$. Conversely, the pivoting step $\tilde{w} \to w$ automatically implies the pivoting step $\tilde{v} \to v$

Chapter 14. General PL Algorithms on PL Manifolds

In the last 20 years, a vast variety of algorithms have been developed which are based on the concept of complementary pivoting. Many of these are listed in our bibliography. The PL continuation and homotopy algorithms described in the last two chapters are important examples. In order to give a better idea of the flexibility which is possible and to describe the construction of such algorithms for special purposes, we are now going to cast the notion of PL algorithms into the more general setting of PL manifolds. Eaves (1976) has given a very elegant geometric approach to general PL methods which has strongly influenced the writing of this chapter, see also Eaves & Scarf (1976). In the first two sections we give a general formulation of PL algorithms in the context of PL manifolds which will then allow us to describe and study a variety of sophisticated PL algorithms in a unified framework.

14.1 PL Manifolds

The notion of complementary pivoting can be roughly described by looking at the following system of linear equations and inequalities

$$(14.1.1) \qquad\qquad Ax = b, \quad Lx \geq c,$$

where $x \in \mathbf{R}^{N+1}$, $b \in \mathbf{R}^N$, A is an $N \times (N + 1)$-matrix, $c \in \mathbf{R}^K$ and L is an $K \times (N + 1)$-matrix for some integer $K > 0$. Let us assume a simple but important case, namely that A has maximal rank N and that the line $\{x \in \mathbf{R}^{N+1} \mid Ax = b\}$ intersects the convex polyhedral set of feasible points $\{x \in \mathbf{R}^{N+1} \mid Lx \geq c\}$ in exactly two relative boundary points x_1 and x_2. Then the step $x_1 \to x_2$ is considered a step of **complementary pivoting**. This notion is closely related to the door-in-door-out steps described in chapters 12 and 13. In fact, by taking $L = \mathrm{Id}$ and $c = 0$ and using barycentric coordinates, we see that the latter is a special case, cf. (12.2.5). It is important to realize that also the more general steps considered here can be numerically implemented in a fashion similar to the implementation of the door-in-door-out steps in section 12.4. In fact, if γ denotes a nontrivial solution of $A\gamma = 0$, then the above assumptions imply the existence of an index $i \in \{1, \ldots, N+1\}$ such that $e_i^*(Lx_1 - c) = 0$ and $e_i^* L\gamma \neq 0$. Without loss of generality we may assume that $e_i^* L\gamma > 0$. From the ansatz $x_2 = x_1 - \varepsilon\gamma$ we obtain

$$(14.1.2) \qquad \varepsilon := \min_{j=1,\ldots,N+1} \left\{ \frac{e_j^*(Lx_1 - c)}{e_j^* L\gamma} \ \middle|\ e_j^* L\gamma > 0 \right\}.$$

In view of this numerical step, it is thus interesting to consider a class of algorithms which trace a solution curve in $H^{-1}(0)$ where $H : \mathcal{M} \to \mathbf{R}^N$ is a PL map on an $(N+1)$-dimensional PL manifold $\mathcal{M}$, and the pieces of linearity (cells) $\{\sigma\}_{\sigma \in \mathcal{M}}$ are given by linear inequalities such as $Lx \geq c$. This leads us to the notion of general PL manifolds $\mathcal{M}$ and PL maps which will be discussed in this chapter. First we have to introduce some preliminary terminology. Throughout the rest of this chapter, E denotes an ambient finite dimensional Euclidean space which contains all points which arise in the sequel.

(14.1.3) Notation. *If $\sigma \subset E$ is any subset, then*

(1) $\mathrm{co}(\sigma)$ *denotes the convex hull of σ;*

(2) $\overline{\mathrm{co}}(\sigma)$ *denotes the closed convex hull of σ;*

(3) $\mathrm{aff}(\sigma)$ *denotes the affine hull of σ;*

(4) $\mathrm{tng}(\sigma) := \mathrm{aff}(\sigma) - \sigma$ *denotes the tangent space of σ which is the linear space obtained by translating $\mathrm{aff}(\sigma)$ to the origin;*

(5) $\mathrm{int}(\sigma)$ *denotes the relative interior of σ with respect to the space $\mathrm{aff}(\sigma)$;*

(6) $\partial(\sigma)$ *denotes the relative boundary of σ with respect to the space $\mathrm{aff}(\sigma)$.*

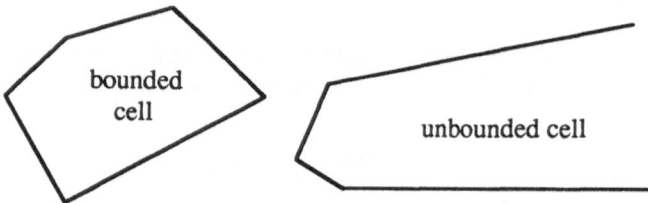

Figure 14.1.a Bounded and unbounded cell

(14.1.4) Cells. If $u \in E$, $u \neq 0$ and $\gamma \in \mathbf{R}$, then the set $\{x \in E \mid u^*x \geq \gamma\}$ is called a **half-space** and its boundary $\{x \in E \mid u^*x = \gamma\}$ a **hyperplane**. A finite intersection of half-spaces is called a convex polyhedral set or **cell**. Hence cells are closed convex sets, they may be bounded or unbounded, see figure 14.1.a, and trivially any affine space of finite dimension is a cell. The **dimension** of a cell σ is identified with the dimension of its tangent space i.e. $\dim(\sigma) := \dim\big(\mathrm{tng}(\sigma)\big)$, and we call an m-dimensional cell also simply an m-cell. A **face** τ of a cell σ is a convex subset $\tau \subset \sigma$ such that for all x, y, λ

$$(14.1.5)(a) \qquad x, y \in \sigma, \quad 0 < \lambda < 1, \quad (\lambda - 1)x + \lambda y \in \tau \quad \Rightarrow \quad x, y \in \tau$$

holds. Trivially, the cell σ is a face of itself. All other faces of σ are called **proper** faces. In the theory of convex sets, the above definition of a face coincides with that of an **extremal set**. By using separation theorems for convex sets, it can be shown that a subset $\tau \subset \sigma$, $\tau \neq \sigma$ is a face of σ if and only if there is a half-space ξ such that $\sigma \subset \xi$ and

(14.1.5)(b) $\tau = \sigma \cap \partial \xi,$

see any book including an introduction into the theory of convex sets, e.g. Dunford & Schwartz (1963), Rockafellar (1970), Stoer & Witzgall (1970). To include the trivial case $\sigma = \tau$, we have to assume here that $\dim \sigma < \dim E$ which can always be arranged. In the language of the theory of convex sets, $\partial \xi$ is called a hyperplane **supporting** σ at τ. Figure 14.1.b illustrates this characterization. From this it follows immediately that faces are cells, and that any cell has only finitely many faces. Furthermore, any finite intersection of faces is again a face. A proper face of maximal dimension i.e. $\dim(\tau) = \dim(\sigma) - 1$ is called a **facet** of σ. The 0-faces are singletons containing one **vertex** of σ, 1-faces are also called **edges** of σ. Simplices are particularly simple cells, and for this case the definitions given here are compatible with those in (12.1.5).

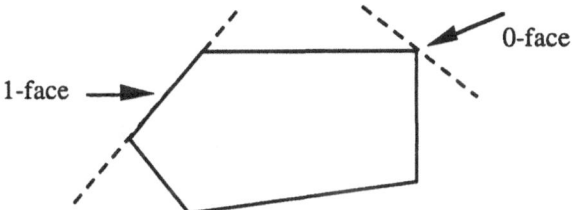

Figure 14.1.b Supporting hyperplanes

(14.1.6) PL manifolds. Let $\mathcal{M}$ be a non-empty family of $(N+1)$-cells. For $0 \leq k \leq N+1$, the following notation is introduced as in (12.1.7):

$$\mathcal{M}^k := \{\tau \mid \tau \text{ is a } k\text{-face of some cell } \sigma \in \mathcal{M}\}.$$

Furthermore, we set

$$|\mathcal{M}| := \bigcup_{\sigma \in \mathcal{M}} \sigma.$$

We call $\mathcal{M}$ a **PL manifold** of dimension $N+1$ if and only if the following conditions hold:

(1) the intersection $\sigma_1 \cap \sigma_2$ of two cells σ_1, $\sigma_2 \in \mathcal{M}$ is empty or a **common** face of both cells;

(2) a facet $\tau \in \mathcal{M}^N$ is common to at most two cells of $\mathcal{M}$;

(3) the family $\mathcal{M}$ is locally finite i.e. any relatively compact subset of $|\mathcal{M}|$ meets only finitely many cells $\sigma \in \mathcal{M}$.

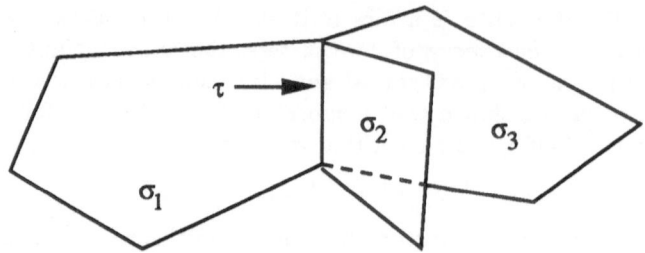

Figure 14.1.c The facet τ belongs to the three simplices σ_1, σ_2 and σ_3

Figure 14.1.c illustrates an example where condition (2) is not met. We denote by the **boundary** $\partial \mathcal{M}$ of $\mathcal{M}$ the system of facets $\tau \in \mathcal{M}^N$ which are common to exactly one cell of $\mathcal{M}$. Figure 14.1.d illustrates two PL manifolds of dimension 2: one possesses a boundary and one does not. A PL manifold $\mathcal{M}$ is called a **pseudo manifold** if and only if all cells of $\mathcal{M}$ are actually simplices, see again figure 14.1.d for an illustration. The triangulations of $\mathbf{R}^{N+1}$ which we introduced in section 12.1 and in particular in section 13.3 are all pseudo manifolds without boundary.

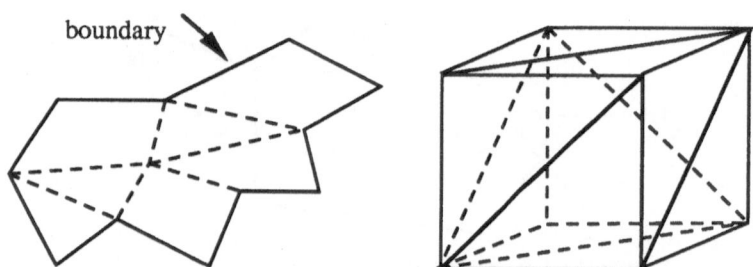

Figure 14.1.d A 2-dimensional PL manifold with boundary and a 2-dimensional pseudo manifold without boundary

We have to distinguish between a PL manifold $\mathcal{M}$ and the set $|\mathcal{M}|$ which is **subdivided** by the PL manifold $\mathcal{M}$. In the case that $\mathcal{M}$ is a pseudo manifold, we also say that $\mathcal{M}$ **triangulates** the set $|\mathcal{M}|$.

(14.1.7) PL maps. Let $\mathcal{M}$ be a PL manifold of dimension $N + 1$. We call $H : \mathcal{M} \to \mathbf{R}^N$ a PL map if and only if

(1) $H : |\mathcal{M}| \to \mathbf{R}^N$ is a continuous map;

(2) the restriction $H_\sigma : \sigma \to \mathbf{R}^N$ of H to σ is an affine map for all $\sigma \in \mathcal{M}$.

Analogously to (12.2.1) we can again speak of the **Jacobian** $H'_\tau : \text{tng}(\tau) \to \mathbf{R}^N$ of the affine map $H_\tau : \tau \to \mathbf{R}^N$ where $\tau \in \mathcal{M}^k$ is a face for some $0 \le k \le N + 1$. Again as in (12.2.2), a point $x \in |\mathcal{M}|$ is called a **regular point** of the PL map H if and only if

(a) x is not contained in any lower dimensional face $\tau \in \mathcal{M}^k$ for $k < N$;

(b) H'_σ has maximal rank for all $\sigma \in \mathcal{M}^N \cup \mathcal{M}^{N+1}$ such that $x \in \sigma$.

A value $y \in \mathbf{R}^N$ is a **regular value** of H if all points in $H^{-1}(y)$ are regular. By definition, y is vacuously a regular value if it is not contained in the range of H. If a point is not regular it is called **singular**. Analogously, if a value is not regular it is called singular.

We can once more use the perturbation

$$(14.1.8) \qquad\qquad \vec{\varepsilon} := \begin{pmatrix} \varepsilon^1 \\ \vdots \\ \varepsilon^N \end{pmatrix}$$

in order to prove a Sard type theorem.

(14.1.9) Proposition. *Let $H : \mathcal{M} \to \mathbf{R}^N$ be a PL map where $\mathcal{M}$ is a PL manifold of dimension $N+1$. Then for any relatively compact subset $C \subset |\mathcal{M}|$ there are at most finitely many $\varepsilon > 0$ such that $C \cap H^{-1}(\vec{\varepsilon})$ contains a singular point of H. Consequently, $\vec{\varepsilon}$ is a regular value of H for almost all $\varepsilon > 0$.*

Proof. The above proposition is a generalization of (12.2.4). Unfortunately, the technique of the proof has to be modified to include this more general situation. Since the proof of (12.2.4) was quite explicit, we will now give a concise proof by contradiction. Hence let us assume there is a strictly decreasing sequence $\{\varepsilon_i\}_{i \in \mathbf{N}}$ of positive numbers, converging to zero, for which a bounded sequence $\{x_i\}_{i \in \mathbf{N}} \subset |\mathcal{M}|$ of singular points can be found such that the equations

$$(14.1.10) \qquad\qquad H(x_i) = \vec{\varepsilon}_i$$

for $i \in \mathbf{N}$ are satisfied. For any subset $I \subset \mathbf{N}$ of cardinality $N+1$ we see that the $\{\vec{\varepsilon}_i\}_{i \in I}$ are affinely independent, and by the above equations (14.1.10) and the piecewise linearity (14.1.7)(2) of H the $\{x_i\}_{i \in I}$ cannot all be contained in the same lower dimensional face $\tau \in \mathcal{M}^k$ for $k < N$. Since this holds for all index sets I, we use this argument repeatedly, and the local finiteness (14.1.6)(3) of $\mathcal{M}$ permits us to find a strictly increasing function $\nu : \mathbf{N} \to \mathbf{N}$ (to generate a subsequence), and to find a face $\sigma \in \mathcal{M}^{N+1} \cup \mathcal{M}^N$ such that the subsequence $\{x_{\nu(i)}\}_{i \in \mathbf{N}}$ is contained in σ, but no point of $\{x_{\nu(i)}\}_{i \in \mathbf{N}}$ is contained in any lower dimensional face $\tau \in \mathcal{M}^k$ for $k < N$. But now we can again use the above argument: for an index set $I \subset \nu(\mathbf{N})$ of cardinality $N+1$ the $\{\vec{\varepsilon}_i\}_{i \in I}$ are affinely independent, and we conclude that H'_σ has maximal rank N. However, this means that all points $\{x_{\nu(i)}\}_{i \in \mathbf{N}}$ are regular, a contradiction to the choice of $\{x_i\}_{i \in \mathbf{N}}$. The last assertion of the proposition follows since $|\mathcal{M}|$ can be written as a countable union of relatively compact subsets. $\square$

We shall first discuss the PL algorithm for regular values. Then we will use the above proposition to show that, similarly to section 12.3, a PL algorithm can as well be discussed for singular values. The following lemma shows that for regular values the solution path is always transverse to the facets $\tau \in \mathcal{M}^N$ of $\mathcal{M}$. For the rest of this section, we assume that $H : \mathcal{M} \to \mathbf{R}^N$ is a PL map where $\mathcal{M}$ is a PL manifold of dimension $N + 1$.

(14.1.11) Lemma. *Let zero be a regular value of H, and let $\tau \in \mathcal{M}^N$ be a facet such that $H^{-1}(0) \cap \tau \neq \emptyset$. Then $H^{-1}(0) \cap \tau$ contains exactly one point p, and $p \in \mathrm{int}(\tau)$.*

Proof. Any $p \in H^{-1}(0) \cap \tau$ must be interior to τ since otherwise p would lie in a lower dimensional face and could not be a regular zero point, contrary to the assumption. Let us now assume that $H^{-1}(0) \cap \tau$ contains more than one point. Then there is a $p_0 \in \tau$ and a $t \in E$, $t \neq 0$ such that $x(s) := p_0 + st$ is contained in $H^{-1}(0) \cap \tau$ for small $s \geq 0$. Differentiating $H_\tau(x(s)) = 0$ with respect to s at $s = 0$ yields $H'_\tau t = 0$, hence the Jacobian H'_τ does not have maximal rank, and p_0 is not a regular zero point of H, contrary to our assumption. $\square$

(14.1.12) Corollary. *If zero is a regular value of H, then*

$$\mathcal{N}_H := \{H^{-1}(0) \cap \sigma \mid \sigma \in \mathcal{M},\ H^{-1}(0) \cap \sigma \neq \emptyset\}$$

is a one-dimensional PL manifold which subdivides $H^{-1}(0)$.

It is clear that each connected component is a polygonal path which is either isomorphic to the circle $\{z \in \mathcal{C} \mid ||z|| = 1\}$, the line $\mathbf{R}$, the ray $[0, \infty)$ or the segment $[0, 1]$. Contrary to the case of pseudo manifolds which has been essentially discussed in chapter 12, now a 1-cell $\xi \in \mathcal{N}_H$ may not only be a **segment** i.e. $\xi = \{x + st \mid 0 \leq s \leq 1\}$ where $x, t \in E$ with $t \neq 0$ are suitably chosen, but can also be a **ray** i.e. $\xi = \{x + st \mid s \geq 0\}$ or a **line** i.e. $\xi = \{x + st \mid s \in \mathbf{R}\}$. However, this latter case is not interesting and will be omitted from the subsequent discussion.

(14.1.13) Definition of a PL Algorithm. For the case that zero is a regular value of H, a PL algorithm consists of traversing a connected component (path) of the above one-dimensional solution manifold $\mathcal{N}_H$. Excluding the uninteresting case that the whole path consists of just one line, the algorithm may be started in three different ways:

(1) **Ray start.** We start in a ray $\xi \in \mathcal{N}_H$ such that its only vertex is traversed next.
(2) **Boundary start.** We start in a segment $\xi \in \mathcal{N}_H$ such that one vertex of ξ is on the boundary $\partial \mathcal{M}$ and the other vertex is traversed next.
(3) **Interior start.** We start in a segment $\xi \in \mathcal{N}_H$ which has no vertex in the boundary $\partial \mathcal{M}$. One of the two vertices of ξ is traversed next, hence in this case an orientation of the path has to be chosen initially.

There are also different ways in which the algorithm may terminate:

(a) Boundary termination. The vertex of the current 1-cell $\xi \in \mathcal{N}_H$ which has to be traversed next lies in the boundary $\partial \mathcal{M}$.

(b) Ray termination. The current 1-cell $\xi \in \mathcal{N}_H$ which has been entered through one vertex is actually a ray.

(c) Loop termination. The algorithm enters a 1-cell $\xi \in \mathcal{N}_H$ which was already traversed. In this case an interior start was performed, and the whole path is isomorphic to a circle.

(d) No termination. The algorithm never stops and never repeats a cell. In this case the whole path is isomorphic either to a ray or to a line.

Figure 14.1.e illustrates two cases.

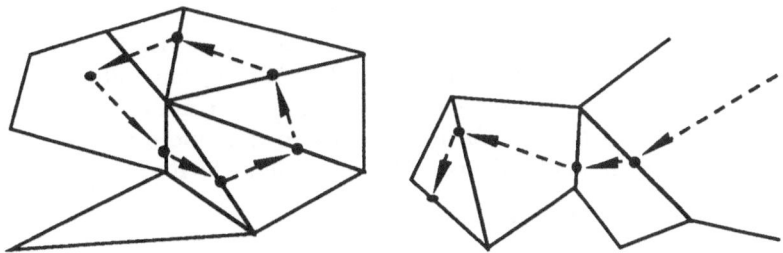

Figure 14.1.e A loop, and a ray start with boundary termination or a boundary start with ray termination

Let us now consider the more general case that zero is possibly a singular value of H. As in chapter 12, we can avoid this difficulty by looking at the perturbation $H^{-1}(\vec{\varepsilon})$ where $\vec{\varepsilon}$ is a regular value for almost all $\varepsilon > 0$, see (14.1.9), and then considering the limiting procedure $\varepsilon \to 0$. We are hence tracing a path in

$$(14.1.14) \qquad \bigcap_{\vec{\varepsilon} > 0} \overline{\bigcup_{0 < \varepsilon < \vec{\varepsilon}} H^{-1}(\vec{\varepsilon})} \quad \subset \quad H^{-1}(0)$$

which is unfortunately not necessarily subdivided by a one-dimensional PL manifold, e.g. it may contain bifurcation points. Nevertheless, the above idea carries through here too, as we will now show.

(14.1.15) Definition. *We call a cell $\sigma \in \mathcal{M}$ **transverse** (with respect to H) if and only if $\sigma \cap H^{-1}(\vec{\varepsilon}) \neq \emptyset$ for all sufficiently small $\varepsilon > 0$. We call a facet $\tau \in \mathcal{M}^N$ **completely labeled** (with respect to H) if and only if $\tau \cap H^{-1}(\vec{\varepsilon}) \neq \emptyset$ for all sufficiently small $\varepsilon > 0$. It is clear that these definitions are consistent with the corresponding definitions given in chapter 12.*

From the discussion in (14.1.9–13) we conclude:

(14.1.16) **Lemma.** *Let $\sigma \in \mathcal{M}$ be transverse. Then one of the following three cases holds:*

(1) *The transverse cell σ does not contain any completely labeled facet. Then σ possesses a line of solutions. This case is not interesting from an algorithmic point of view, since this cell indicates simultaneously the start and the termination of the algorithm.*

(2) *The transverse cell σ contains exactly one completely labeled facet. Then σ possesses a ray of solutions. This case can only occur at the start or at the termination of the algorithm.*

(3) *The transverse cell σ contains exactly two completely labeled facets. Then σ possesses a segment of solutions. This case corresponds to the door-in-door-out principle (12.3.8), and it is the only case which can occur for pseudo manifolds $\mathcal{M}$.*

Of course, if τ is a completely labeled facet of a cell σ, then σ is transverse by definition (14.1.15). Hence, as in chapter 12, we can follow a solution path by going from one transverse cell to the next transverse cell through a completely labeled facet. We merely need to take more possibilities into account. Thus we are led to a general PL algorithm which we will now sketch.

(14.1.17) **General PL Algorithm.**   *comment:*

> **input**
>> **begin**
>> $\sigma_1 \in \mathcal{M}$ transverse;   *starting cell*
>> $\tau_1 \in \mathcal{M}^N$ completely labeled facet of σ_1;   *τ_i is the facet —*
>> **end**;   *through which σ_i is exited*
> **for** $i = 1, 2, \ldots$ **do**
>> **begin**
>> **if** $\tau_i \in \partial\mathcal{M}$ **then** stop;   *termination on the boundary of $\mathcal{M}$*
>> find the unique $\sigma_{i+1} \in \mathcal{M}$
>>> such that $\sigma_{i+1} \neq \sigma_i$ and $\tau_i \subset \sigma_{i+1}$;   *pivoting step*
>> **if** τ_i is the only completely labeled
>>> facet of σ_{i+1} **then** stop;   *termination on a ray in σ_{i+1}*
>> find the unique completely labeled facet
>>> τ_{i+1} of σ_{i+1} such that $\tau_{i+1} \neq \tau_i$;   *general door-in-door-out step*
>> **end.**

The algorithm is not stopped in the cases (14.1.13)(c–d). Additional features have to be incorporated if those events are suspected.

(14.1.18) Note On Numerical Implementations. As the reader may already have imagined, the numerical implementation of the above general algorithm is more complex than in the case of triangulations which was discussed in chapters 12–13. The underlying principle is the discussion of (14.1.1–2). A cell is represented by a system such as (14.1.1), and the *"general door-in-door-out step"* is performed by a minimization as in (14.1.2). The *"pivoting step"* is performed by generating a new system (14.1.1), and hopefully we can make use of the numerical effort such as decomposing the old system to cheaply decompose the new system by update procedures. The special features of these techniques vary with the type of PL manifold under consideration. In some important cases such as standard pseudo manifolds triangulating $\mathbf{R}^{N+1}$ (see chapters 12–13) or other simple subdivisions by PL manifolds, the numerical details are efficiently implemented. We refer to our extensive bibliography for numerical details in such cases and will only give some hints in the following sections.

14.2 Orientation and Index

Keeping track of orientations in PL methods furnishes some useful information, e.g. it is possible to define and study an index for zero points on the boundary of the PL manifold $\mathcal{M}$. This is the object of the current section. Nearly all manifolds which are of importance for practical implementations, are orientable. Index and orientability have been studied in the context of PL methods by Allgower & Georg (1980), Eaves (1976), Eaves & Scarf (1976), Garcia & Zangwill (1979), van der Laan & Talman (1981), Lemke & Grotzinger (1976), Peitgen (1982). Peitgen & Prüfer (1979), Peitgen & Siegberg (1981), Prüfer (1978), Prüfer & Siegberg (1979), (1981), Shapley (1974), Todd (1976), Yamamoto (1988). We begin with some basic definitions.

(14.2.1) Orientation. Let F be a linear space of dimension k. An **orientation** of F is a function

$$\text{or} : F^k \to \{-1, 0, 1\}$$

such that the following conditions hold:

(1) $\text{or}(b_1, \ldots, b_k) \neq 0$ if and only if $b_1, \ldots, b_k$ are linearly independent,

(2) $\text{or}(b_1, \ldots, b_k) = \text{or}(c_1, \ldots, c_k) \neq 0$ if and only if the transformation matrix $(\alpha_{i,j})_{i,j=1,\ldots,k}$ defined by $b_i = \sum_{j=1}^{k} \alpha_{i,j} c_j$ has positive determinant.

It is clear from the basic facts of linear algebra that any finite dimensional linear space permits exactly two orientations. A k-cell σ is oriented by orienting its tangent space $\text{tng}(\sigma)$. Such an orientation or_σ of σ induces an orientation $\text{or}_{\tau,\sigma}$ on a facet τ of σ by the following convention:

(3) $\mathrm{or}_{\tau,\sigma}(b_1,\ldots,b_{k-1}) := \mathrm{or}_\sigma(b_1,\ldots,b_k)$ whenever b_k points from τ into the cell σ.

It is routine to check that the above definition of $\mathrm{or}_{\tau,\sigma}$ verifies the conditions (1)–(2).

(14.2.2) Oriented PL Manifolds. Let $\mathcal{M}$ be a PL manifold of dimension $N+1$. Then an orientation of $\mathcal{M}$ is a choice of orientations $\{\mathrm{or}_\sigma\}_{\sigma\in\mathcal{M}}$ such that

(*)
$$\mathrm{or}_{\tau,\sigma_1} = -\mathrm{or}_{\tau,\sigma_2}$$

for each facet $\tau \in \mathcal{M}^N$ which is adjacent to two different cells $\sigma_1, \sigma_2 \in \mathcal{M}$. By making use of the standard orientation

$$\mathrm{or}(b_1,\ldots,b_{N+1}) := \mathrm{sign\,det}\,(b_1,\ldots,b_{N+1})$$

of $\mathbf{R}^{N+1}$, it is clear that any PL manifold of dimension $N+1$ which subdivides a subset of $\mathbf{R}^{N+1}$ is oriented in a natural way. But many less trivial oriented PL manifolds are known.

(14.2.3) Orientation of $\mathcal{N}_H$. If $H : \mathcal{M} \to \mathbf{R}^N$ is a PL map on an oriented PL manifold of dimension $N+1$ such that zero is a regular value of H, then the orientation of $\mathcal{M}$ and the natural orientation of $\mathbf{R}^N$ induces a natural orientation of the 1-dimensional solution manifold $\mathcal{N}_H$. Namely, for $\xi \in \mathcal{N}_H$, $b_{N+1} \in \mathrm{tng}(\xi)$ and $\sigma \in \mathcal{M}$ such that $\xi \subset \sigma$, the definition

(14.2.4) $\mathrm{or}_\xi(b_{N+1}) := \mathrm{or}_\sigma(b_1,\ldots,b_{N+1})\,\mathrm{sign\,det}\,(H'_\sigma b_1,\ \ldots\ ,H'_\sigma b_N)$

is independent of the special choice of $b_1,\ldots,b_N \in \mathrm{tng}(\sigma)$, provided the $b_1,\ldots,b_N$ are linearly independent. Clearly, an orientation of the 1-dimensional solution manifold $\mathcal{N}_H$ is just a rule which indicates a direction for traversing each connected component of $\mathcal{N}_H$. Keeping this in mind, we now briefly indicate why the above definition indeed yields an orientation for $\mathcal{N}_H$. Let $\tau \in \mathcal{M}^N$ be a facet which meets $H^{-1}(0)$ and does not belong to the boundary $\partial\mathcal{M}$, let $\sigma_1, \sigma_2 \in \mathcal{M}$ be the two cells adjacent to τ, and let $\xi_j := H^{-1}(0) \cap \sigma_j \in \mathcal{N}_H$ for $j = 1,2$. If $b_1,\ldots,b_N$ is a basis of $\mathrm{tng}(\tau)$, and if $b_{j,N+1} \in \mathrm{tng}(\xi_j)$ points into σ_j, then from condition (14.2.2)(*) it follows that

$$\mathrm{or}_{\sigma_1}(b_1,\ldots,b_N,b_{1,N+1}) = -\mathrm{or}_{\sigma_2}(b_1,\ldots,b_N,b_{2,N+1}),$$

and hence (14.2.4) implies that

$$\mathrm{or}_{\xi_1}(b_{1,N+1}) = -\mathrm{or}_{\xi_2}(b_{2,N+1}),$$

which is exactly the right condition in the sense of (14.2.2)(*) to ensure that the manifold $\mathcal{N}_H$ is oriented. The definitions given here are related in a natural way to the orientation of smooth curves defined in the remarks following (2.1.5).

(14.2.5) Index of a Boundary Solution. Let us again assume that $H :$ $\mathcal{M} \to \mathbf{R}^N$ is a PL map on an oriented PL manifold of dimension $N + 1$ such that zero is a regular value of H. If x is a boundary solution of $H(x) = 0$ i.e. $x \in |\partial \mathcal{M}| \cap H^{-1}(0)$, then there exists exactly one connected component $\mathcal{C}$ of $\mathcal{N}_H$ containing x. We let $\mathcal{C}$ inherit the natural ordering of $\mathcal{N}_H$ and distinguish between two cases: $\mathcal{C}$ starts from x into $\mathcal{M}$, or $\mathcal{C}$ terminates in x. Accordingly, we define

(14.2.6)
$$\mathrm{index}_H(x) := \begin{cases} 1 & \text{for } \mathcal{C} \text{ starting in } x, \\ -1 & \text{for } \mathcal{C} \text{ terminating at } x. \end{cases}$$

The following result is, then an obvious but a surprisingly powerful topological tool for investigating nonlinear equations.

(14.2.7) Theorem. *Let $H : \mathcal{M} \to \mathbf{R}^N$ be a PL map on an oriented PL manifold of dimension $N + 1$ such that zero is a regular value of H. If $\mathcal{N}_H$ is compact, then*

(14.2.8)
$$\sum_{x \in |\partial \mathcal{M}| \cap H^{-1}(0)} \mathrm{index}_H(x) = 0.$$

Proof. Let $\mathcal{C}$ be a connected component of $\mathcal{N}_H$. Since $\mathcal{C}$ is bounded by our assumption, either $\mathcal{C}$ is isomorphic to a circle or a segment, see (14.1.13). In the first case $\mathcal{C}$ does not hit the boundary $\partial \mathcal{M}$ at all, and in the second case $\mathcal{C}$ begins and terminates at two points in $|\partial \mathcal{M}|$ which must have opposite indices. Hence formula (14.2.8) follows immediately. $\qquad\Box$

The two most common contexts in which formula (14.2.8) is applied, are the case in which the PL manifold $\mathcal{M}$ itself is compact, or the case that some boundary conditions for H hold which result in a priori estimates for the solutions $\mathcal{N}_H$ and thus imply its compactness. By making use of the ε-perturbations following (14.1.8), it is again possible to also include the case that zero is a singular value of H, but the statements have to be slightly revised.

By making PL approximations of continuous maps (cf. section 12.2) or upper semi-continuous set valued maps (cf. section 13.1), it is not difficult to introduce the Brouwer degree via the above index and to establish its properties by making use of (14.2.8), see Peitgen & Siegberg (1981) for an elegant presentation of this. The relations between the above index and the Brouwer degree of a map have been investigated also by Prüfer (1978), Garcia & Zangwill (1979), Peitgen & Prüfer (1979), Prüfer & Siegberg (1979), (1981), Peitgen (1982).

14.3 Lemke's Algorithm for the Linear Complementarity Problem

The first and most prominent example of a PL algorithm was designed by Lemke to calculate a solution of the linear complementarity problem, see Lemke & Howson (1964), Lemke (1965) and the survey of Lemke (1980). This algorithm played a crucial role in the development of subsequent PL algorithms. The linear complementarity problem is a special case of the non-linear complementarity problem discussed in section 11.7. Initially, some of our terminology and results will be carried over. For the rest of this section we consider the following

(14.3.1) Linear Complementarity Problem (LCP). *Let* $g : \mathbf{R}^N \to \mathbf{R}^N$ *be an affine map. Find an* $x \in \mathbf{R}^N$ *such that*

$$\textbf{(1)} \quad x \in \mathbf{R}^N_+; \quad \textbf{(2)} \quad g(x) \in \mathbf{R}^N_+; \quad \textbf{(3)} \quad x^* g(x) = 0.$$

Here $\mathbf{R}_+$ *denotes the set of non-negative real numbers, and in the sequel we also denote the set of positive real numbers by* $\mathbf{R}_{++}$. *If* $g(0) \in \mathbf{R}^N_+$, *then* $x = 0$ *is a trivial solution to the problem. Hence this trivial case is always excluded and the additional assumption*

$$\textbf{(4)} \quad g(0) \notin \mathbf{R}^N_+$$

is made.

LCPs arise in quadratic programming, bimatrix games, variational in-equalities and economic equilibria problems, and numerical methods for their solution have been of considerable interest. We cite the following books and articles for further references: Asmuth & Eaves & Peterson (1979), Balinski & Cottle (1978), Doup & van der Elzen & Talman (1986), Cottle (1972), (1974), Cottle & Dantzig (1968), (1970), Cottle & Gianessi & Lions (1980), Cottle & Golub & Sacher (1978), Cottle & Stone (1983), Doup & van der Elzen & Talman (1986), Eaves (1971), (1976), (1978), (1983) Eaves & Scarf (1976), Eaves & Gould & Peitgen & Todd (1983), Garcia & Gould & Turnbull (1984), Garcia & Zangwill (1981), van der Heyden (1980), Karamardian (1977), Kojima & Mizuno & Noma (1987), Kojima & Mizuno & Yoshise (1987), Kojima & Nishino & Sekine (1976), van der Laan & Talman (1985), Lemke (1965), (1968), (1980), Lemke & Howson (1964), Lüthi (1976), Megiddo (1986), Mizuno & Yoshise & Kikuchi (1988), van der Panne (1974), Saigal (1971), (1972), (1976) Saigal & Simon (1973), Scarf & Hansen (1973), Talman & Van der Heyden (1983), Todd (1976), (1978), (1980), (1984), (1986).

As in (11.7.2) we introduce the

(14.3.2) Definition. *For* $x \in \mathbf{R}^N$ *and* $i = 1, \ldots, N$ *we define* $x_+ \in \mathbf{R}^N_+$ *by* $x_+[i] := max\{x[i], 0\}$, *and* $x_- \in \mathbf{R}^N_+$ *by* $x_- := (-x)_+$. *The following formulas are then obvious:*

$$(1) \quad x = x_+ - x_-; \quad (2) \quad (x_+)^*(x_-) = 0.$$

Again, the next proposition is not difficult to prove and reduces the LCP to a zero point problem in a simple way:

(14.3.3) Proposition. *Let us define* $f : \mathbf{R}^N \to \mathbf{R}^N$ *by* $f(z) = g(z_+) - z_-$. *If* x *is a solution of the LCP (14.3.1), then* $z := x - g(x)$ *is a zero point of* f. *Conversely, if* z *is a zero point of* f, *then* $x := z_+$ *solves the LCP (14.3.1).*

The advantage which f provides is that it is obviously a PL map if we subdivide $\mathbf{R}^N$ into orthants. This is the basis for our description of Lemke's algorithm. For a fixed $d \in \mathbf{R}^N_{++}$ we define the homotopy $H : \mathbf{R}^N \times [0, \infty) \to \mathbf{R}^N$ by

$$(14.3.4) \qquad\qquad H(x, \lambda) := f(x) + \lambda d.$$

For a given subset $I \subset \{1, 2, \dots, N\}$ we introduce the complement $I' := \{1, 2, \dots, N\} \setminus I$, and furthermore we introduce the power set

$$(14.3.5) \qquad\qquad \mathcal{P}_N := \Big\{ I \mid I \subset \{1, 2, \dots, N\} \Big\}.$$

Then an orthant in $\mathbf{R}^N \times [0, \infty)$ can be written in the form

$$(14.3.6) \quad \sigma_I := \{ (x, \lambda) \mid \lambda \geq 0, \ x[i] \geq 0 \text{ for } i \in I, \ x[i] \leq 0 \text{ for } i \in I' \},$$

and the family

$$(14.3.7) \qquad\qquad \mathcal{M} := \big\{ \sigma_I \big\}_{I \in \mathcal{P}_N}$$

is a PL manifold (of dimension $N+1$) which subdivides $\mathbf{R}^N \times [0, \infty)$. Furthermore it is clear from (14.3.2–4) that $H : \mathcal{M} \to \mathbf{R}^N$ is a PL map since $x \mapsto x_+$ switches its linearity character only at the hyperplanes $\{ x \in \mathbf{R}^N \mid x[i] = 0 \}_{i=1,2,\dots N}$. Let us assume for simplicity that zero is a regular value of H. We note however, that the case of a singular value is treated in the same way by using the techniques described in the discussion beginning at (14.1.14).

Lemke's algorithm is started on a ray: if $\lambda > 0$ is sufficiently large, then

$$\big(-g(0) - \lambda d \big)_+ = 0 \quad \text{and} \quad \big(-g(0) - \lambda d \big)_- = g(0) + \lambda d \in \mathbf{R}^N_{++},$$

and consequently

$$H\big(-g(0) - \lambda d, \lambda \big) = f\big(-g(0) - \lambda d \big) + \lambda d = g(0) - \big(g(0) + \lambda d \big) + \lambda d = 0.$$

Hence, the ray $\xi \in \mathcal{N}_H$ defined by

$$\lambda \in [\lambda_0, \infty) \longmapsto -g(0) - \lambda d \in \sigma_\emptyset$$

$$(14.3.8) \qquad\qquad \text{for} \quad \lambda_0 := \max_{i=1,\dots,N} \frac{-g(0)[i]}{d[i]}$$

is used (for decreasing λ-values) for the ray start. This ray is usually called the **primary ray**, and all other rays in $\mathcal{N}_H$ are called **secondary rays**. Note that $\lambda_0 > 0$ by assumption (14.3.1)(4). Since the PL manifold $\mathcal{M}$ consists of the orthants of $\mathbf{R}^N \times [0, \infty)$, it is finite i.e. $\#\mathcal{M} = 2^N$, there are only two possibilities left:

(1) The algorithm terminates on the boundary $|\partial\mathcal{M}| = \mathbf{R}^N \times \{0\}$ at a point $(z,0)$. Then z is a zero point of f, and (14.3.3) implies that z_+ solves the LCP (14.3.1).

(2) The algorithm terminates on a secondary ray. Then the LCP (14.3.1) has no solution if the Jacobian g' belongs to a certain class of matrices, see the literature cited above.

Let us illustrate the use of index and orientation by showing that the algorithm generates a solution in the sense of (1) under the assumption that all principle minors of the Jacobian g' are positive. Note that the Jacobian g' is a constant matrix since g is affine. For $\sigma_I \in \mathcal{M}$, see (14.3.5–6), we immediately calculate the Jacobian

$$H'_{\sigma_I} = (f'_{\sigma_I}, d),$$

(14.3.9)

$$\text{where} \quad f'_{\sigma_I} e_i = \begin{cases} g'e_i & \text{for } i \in I, \\ e_i & \text{for } i \in I'. \end{cases}$$

If $\xi \in \mathcal{N}_H$ is a solution path in σ_I, then formula (14.2.4) immediately yields

$$\text{or}_\xi(v) = \text{sign det } f'_{\sigma_I} \text{ or}_{\sigma_I}(e_1, \dots, e_N, v),$$

and since $\text{or}_{\sigma_I}(e_1, \dots, e_N, v) = \text{sign} v^* e_{N+1}$ by the standard orientation in $\mathbf{R}^{N+1}$, det f'_{σ_I} is positive or negative if and only if the λ-direction is increasing or decreasing, respectively, while ξ is traversed according to its orientation. It is immediately seen from (14.3.9) that det f'_{σ_I} is obtained as a **principle minor of** g' i.e. by deleting all columns and rows of g' with index $i \in I'$ and taking the determinant of the resulting matrix (where the determinant of the "empty matrix" is assumed to be 1). Since we start in the negative orthant $\sigma_\emptyset$ where the principle minor is 1, we see that the algorithm traverses the primary ray against its orientation, because the λ-values are initially decreased. Hence, the algorithm continues to traverse $\mathcal{N}_H$ against its orientation. For the important case that all principle minors of g' are positive, the algorithm must continue to decrease the λ-values and thus stops in the boundary $|\partial\mathcal{M}| = \mathbf{R}^N \times \{0\}$. Hence, in this case the algorithm finds a solution. Furthermore, it is clear that this solution is unique, since $\mathcal{N}_H$ can contain no other ray than the primary ray. The next lemma shows that positive definite matrices represent a special case.

(14.3.10) Definition. *Let A be an $N \times N$-matrix which is not necessarily symmetric. Then A is called **positive definite** if and only if $x^* A x > 0$ for all $x \in \mathbf{R}^N$ with $x \neq 0$.*

(14.3.11) Lemma. *Let A be positive definite. Then all principle minors are positive.*

Proof. It follows immediately that all real eigenvalues of all principal submatrices of A are positive. Since the complex eigenvalues $\lambda \in \mathbf{C}\backslash\mathbf{R}$ occur in pairs, and since the determinant of a submatrix is the product of its eigenvalues, the assertion follows.   □

In view of the condition (11.7.6) and the subsequent discussion of the nonlinear complementarity problem (NLCP), it would be tempting to use again the homotopy (11.7.4). But let us point out that this homotopy is not a PL map. In (14.4.15) we will see how we can overcome this difficulty by using a formal cone construction.

Let us discuss how Lemke's algorithm may be implemented numerically. For $I \in \mathcal{P}_N$ we follow the path in the cell σ_I by considering the following system

$$\sum_{i \in I} g'e_i x[i] + \sum_{i \in I'} e_i x[i] + \lambda d + g(0) = 0,$$

(14.3.12)

$$x[i] \geq 0 \text{ for } i \in I, \quad x[i] \leq 0 \text{ for } i \in I', \quad \lambda \geq 0.$$

The path goes through a facet if $x[j]$ changes sign for some $j \in \{1, 2, \ldots, N\}$. Then the pivoting into the adjacent cell σ_I is performed by adding or subtracting j from I, and hence the new system differs from the old system in that one column of the system matrix changes. The algorithm stops successfully if the level $\lambda = 0$ is reached. More formally, we define $y \in \mathbf{R}^{N+1}$ by setting

(14.3.13)
$$y[i] := \begin{cases} x[i] & \text{for } i \in I, \\ -x[i] & \text{for } i \in I', \\ \lambda & \text{for } i = N+1, \end{cases}$$

and defining an $N \times (N+1)$-matrix A_I such that

(14.3.14)
$$A_I e_i := \begin{cases} g'e_i & \text{for } i \in I, \\ -e_i & \text{for } i \in I', \\ d & \text{for } i = N+1. \end{cases}$$

Then the system (14.3.12) can be written as

(14.3.15)
$$A_I y + g(0) = 0, \quad y \in \mathbf{R}_+^{N+1},$$

and a pivoting from one facet to the next facet can be described by considering a solution y of (14.1.15) and a solution $w \in \mathbf{R}^{N+1}$ of $A_I w = 0$ such that $w[j] = -1$. In fact, if we set

$$\varepsilon_0 := \max_{\varepsilon} \{\varepsilon \mid \varepsilon > 0, \ y - \varepsilon w \in \mathbf{R}_+^{N+1}\},$$

then $y - \varepsilon_0 w$ is on a facet of σ_I. The algorithm is now sketched by the following pseudo-code:

(14.3.16) Lemke's Algorithm.   *comment:*

$$j := \operatorname*{arg\,min}_{i=1,2,\ldots,N} \frac{-g(0)[i]}{d[i]};$$

$I := \{j\};$ *the starting ray hits the facet of σ_0 corresponding to j*

repeat

    begin

    solve $A_I y + g(0) = 0$, $y[j] = 0$ for y;   *linear system*

    solve $A_I w = 0$, $w[j] = -1$ for w;   *linear system*

    if $-w \in \mathbf{R}_+^{N+1}$ **then** stop;   *ray termination, no solution found*

$$j := \operatorname*{arg\,min}_{i=1,2,\ldots,N} \left\{ \frac{y[i]}{w[i]} \ \middle| \ w[i] > 0 \right\};$$ *next facet is found*

    if $j = N + 1$ **then**   *solution of LCP found*

        begin

        solve $A_I y + g(0) = 0$, $y[j] = 0$ for y;

        define $x \in \mathbf{R}^N$ by $x[i] := \begin{cases} y[i] & \text{for } i \in I; \\ 0 & \text{for } i \in I'; \end{cases}$ *solution vector*

        print x; stop;

        end;

    if $j \in I$ **then** $I := I \setminus \{j\}$ **else** $I := I \cup \{j\}$;

    end.

As in section 12.4, lexicographic ordering could have been incorporated to also handle the case that zero is a singular value of H. The linear systems are typically solved by using some update method, since each system matrix differs from the preceding one in just one column, see section 12.4 and chapter 16.

14.4 Variable Dimension Algorithms

In recent years, a new class of PL algorithms has attracted considerable attention. They are called **variable dimension algorithms** since they all start from a single point, a zero dimensional simplex, and successively generate simplices of varying dimension, until a completely labeled simplex is found. Numerical results indicate that these algorithms improve the computational efficiency of PL homotopy methods, see van der Laan & Talman (1979), (1981), Kojima & Yamamoto (1984).

The first variable dimension algorithm is due to Kuhn (1969) and is best illustrated for Sperner's lemma, see the discussion below. However, this algorithm had the disadvantage that it could only be started from a vertex of a large triangulated standard simplex S, and therefore PL homotopy algorithms were preferred. By increasing the sophistication of Kuhn's algorithm considerably, van der Laan & Talman (1978) developed an algorithm which could start from any point inside S. It soon became clear, see Todd (1978), that this algorithm could be interpreted as a homotopy algorithm. Numerous other variable dimension algorithms were developed by Doup (1988), Doup & van der Elzen & Talman (1987), Doup & van der Laan & Talman (1987) Doup & Talman (1986), Van der Elzen & van der Laan & Talman (1985), Freund (1984–86), Kojima (1980–81), Kojima & Oishi & Sumi & Horiuchi (1985), Kojima & Yamamoto (1982–84), van der Laan (1980–84), van der Laan & Seelen (1984), van der Laan & Talman (1978–87), van der Laan & Talman & van der Heyden (1987), Reiser (1981), Talman (1980), Talman & Van der Heyden (1983), Talman & Yamamoto (1986), Todd (1978), (1980), Todd & Wright (1980), Wright (1981), Yamamoto (1981–86), Yamamoto & Kaneko (1986), Yamamura & Horiuchi (1988).

In this section, we will give a brief introduction into the very complex field of variable dimension algorithms. Two unifying approaches have been given, one due to Kojima & Yamamoto (1982), the other due to Freund (1984). We present here a modified version of the first approach. The modification consists of introducing a cone construction for dealing with the homotopy parameter. In a special case, this construction was also used by Kojima & Yamamoto, see their lemma 5.13.

Let us first illustrate the original versions of the variable dimension algorithms of Kuhn and van der Laan & Talman for the celebrated lemma of Sperner (1928):

(14.4.1) Sperner's Lemma. *Let* $S = [p_1, p_2, \ldots, p_{N+1}]$ *be a fixed (standard) N-simplex and $\mathcal{T}$ a pseudo manifold triangulating S. Let $\ell : S \rightarrow \{1, 2, \ldots, N+1\}$ be an **integer labeling** such that for all $x \in S$ and all $i \in \{1, 2, \ldots, N+1\}$*

$$(14.4.2) \qquad \beta_i(x) = 0 \quad \Rightarrow \quad \ell(x) \neq i$$

*holds, where $\beta_i(x)$ denotes the **barycentric co-ordinates** of x with respect to S i.e.*

$$(14.4.3) \qquad x = \sum_{i=1}^{N+1} \beta_i(x) p_i, \quad \sum_{i=1}^{N+1} \beta_i(x) = 1.$$

*Then there is a **completely labeled simplex** $\sigma \in \mathcal{T}$, i.e. a simplex $\sigma = [v_1, v_2, \ldots, v_{N+1}]$ such that $\ell\{v_1, v_2, \ldots, v_{N+1}\} = \{1, 2, \ldots, N+1\}$.*

Figure 14.4.a illustrates Kuhn's algorithm for obtaining a completely labeled simplex for the case $N = 2$. The dots correspond to the barycenters of the faces (i.e. vertices, edges, 2-simplices) which are generated.

Before we can give a description of these algorithms, we introduce the notion of a primal-dual pair of PL manifolds due to Kojima & Yamamoto (1982). In fact, we only need a special case.

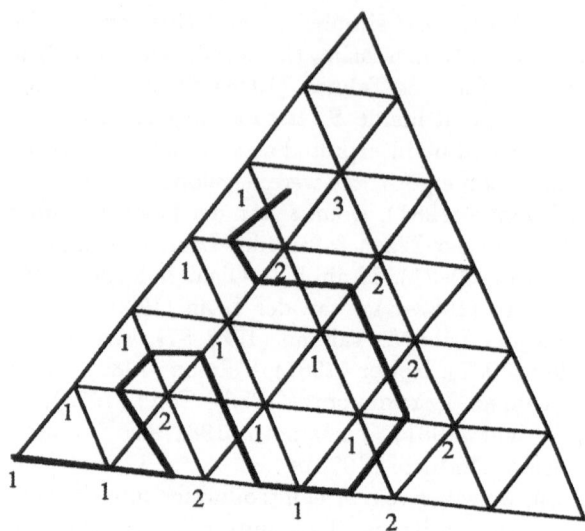

Figure 14.4.a The algorithm of Kuhn

(14.4.4) Definition. *Let $\mathcal{P}$ and $\mathcal{D}$ be two PL manifolds of dimension N. We call $(\mathcal{P}, \mathcal{D})$ a* **primal-dual pair** *if there is a bijective map*

$$\tau \in \mathcal{P}^k \longmapsto \tau^d \in \mathcal{D}^{N-k}, \quad k = 0, 1, \ldots, N,$$

such that

(14.4.5) $\tau_1 \subset \tau_2 \quad \Leftrightarrow \quad \tau_2^d \subset \tau_1^d$

holds for all $\tau_1 \in \mathcal{P}^{k_1}$ and $\tau_2 \in \mathcal{P}^{k_2}$.

   We will deal with a homotopy parameter via the following cone construction. Throughout the rest of this section, ω denotes a point which is affinely independent from all cells under consideration. The introduction of ω is only formal and may be obtained by e.g. increasing the dimension of the ambient finite dimensional Euclidean space E introduced in the remarks preceding (14.1.3).

(14.4.6) Cone Construction. If σ is a cell, then $\omega \bullet \sigma := \{(1-\lambda)\omega + \lambda x \mid x \in \sigma, \lambda \geq 0\}$ denotes the cone containing σ with vertex ω. Clearly, $\omega \bullet \sigma$ is again a cell and $\dim \omega \bullet \sigma = \dim \sigma + 1$. If $H : \sigma \to \mathbf{R}^k$ is an affine map, then the affine extension $\omega \bullet H : \omega \bullet \sigma \to \mathbf{R}^k$ is defined by $\omega \bullet H((1-\lambda)\omega + \lambda x) := \lambda H(x)$ for $x \in \sigma$ and $\lambda \geq 0$. If $\mathcal{M}$ is a PL manifold of dimension N, then $\omega \bullet \mathcal{M} := \{\omega \bullet \sigma\}_{\sigma \in \mathcal{M}}$ is a PL manifold of dimension $N+1$, and a PL map $H : \mathcal{M} \to \mathbf{R}^k$ is extended to a PL map $\omega \bullet H : \omega \bullet \mathcal{M} \to \mathbf{R}^k$.

We will be interested below in rays traversing a cone, and we therefore collect some properties in the following remark.

(14.4.7) Rays Traversing a Cone. Let σ be a cell. We consider a ray

$$\{(1-\varepsilon)z_1 + \varepsilon z_2 \mid \varepsilon \geq 0\} \subset \omega \bullet \sigma,$$

where $\quad z_1 = (1-\lambda_1)\omega + \lambda_1 x_1 \quad$ and $\quad z_2 = (1-\lambda_2)\omega + \lambda_2 x_2$

for some suitable $\lambda_1, \lambda_2 \geq 0$ and $x_1, x_2 \in \sigma$. A simple calculation using the affine independence of ω yields

$$(1-\varepsilon)z_1 + \varepsilon z_2 = (1-\lambda_\varepsilon)\omega + \lambda_\varepsilon x_\varepsilon,$$

where $\quad \lambda_\varepsilon = (1-\varepsilon)\lambda_1 + \varepsilon\lambda_2$

and $\quad x_\varepsilon = \dfrac{(1-\varepsilon)\lambda_1 x_1 + \varepsilon\lambda_2 x_2}{\lambda_\varepsilon}.$

Since $\lambda_\varepsilon \geq 0$ for all $\varepsilon \geq 0$, it follows that $\lambda_2 \geq \lambda_1$. This leaves two cases to consider:

$$\lambda_2 > \lambda_1 \geq 0 \quad \Rightarrow \quad \lim_{\varepsilon \to \infty} x_\varepsilon = \frac{\lambda_2 x_2 - \lambda_1 x_1}{\lambda_2 - \lambda_1} \in \sigma,$$

$$\lambda_2 = \lambda_1 > 0 \quad \Rightarrow \quad x_1 \neq x_2,$$

$$x_\varepsilon = (1-\varepsilon)x_1 + \varepsilon x_2 \in \sigma \quad \text{for} \quad \varepsilon \geq 0.$$

The second case is only possible if the cell σ is unbounded.

The last notion we need is that of a refining PL manifold.

(14.4.8) Definition. Let $\mathcal{T}$ and $\mathcal{M}$ be manifolds of dimension N. We call $\mathcal{T}$ a **refinement** of $\mathcal{M}$ if for all $\sigma \in \mathcal{M}$ the restricted PL manifold $\mathcal{T}_\sigma :=$ $\{\xi \mid \xi \in \mathcal{T},\ \xi \subset \sigma\}$ subdivides σ.

We are now in a position to introduce

(14.4.9) Primal-Dual Manifolds. Let $(\mathcal{P}, \mathcal{D})$ be a primal-dual pair of N-dimensional PL manifolds, and let $\mathcal{T}$ be a refinement $\mathcal{P}$. Then

(14.4.10) $\mathcal{T} \otimes \mathcal{D} := \{\xi \times \tau^d \mid k \in \{0, 1, \ldots, N\},\ \xi \in \mathcal{T}^k,\ \tau \in \mathcal{M}^k,\ \xi \subset \tau\}$

is an N-dimensional PL manifold with empty boundary. A proof of this and related results was given by Kojima & Yamamoto (1982). We call $\mathcal{T} \otimes \mathcal{D}$ the **primal-dual manifold** generated by $\mathcal{T}$ and $\mathcal{D}$. An essential part of the proof consists of discussing the possible pivoting steps. Let $\xi \times \tau^d \in \mathcal{T} \otimes \mathcal{D}$ with $k = \dim \xi$ as above, and let κ be a facet of $\xi \times \tau^d$. We now describe the pivoting of $\xi \times \tau^d$ across the facet κ, see (14.1.17), i.e. we have to find a cell $\eta \in \mathcal{T} \otimes \mathcal{D}$ such that $\eta \neq \xi \times \tau^d$ and $\kappa \subset \eta$. There are three possible cases:

(14.4.11) Increasing The Dimension. Let $\kappa = \xi \times \sigma^d$ such that $\sigma \in \mathcal{M}^{k+1}$ contains τ. Then there is exactly one $\rho \in \mathcal{T}^{k+1}$ such that $\xi \subset \rho$ and $\rho \subset \sigma$. This is a consequence of definition (14.4.8) and is not difficult to prove. Then $\eta := \rho \times \sigma$ is the desired second N-cell. In this case the dimension k of the primal cell ξ is increased.

(14.4.12) Decreasing The Dimension. Let $\kappa = \delta \times \tau^d$ such that $\delta \in \mathcal{T}^{k-1}$ is a facet of ξ. If $\delta \subset \partial\tau$, then there exists exactly one facet $\nu \in \mathcal{M}^{k-1}$ of τ such that $\delta \subset \nu$, and $\eta := \delta \times \nu^d$ is the desired second cell. In this case the dimension k of the primal cell ξ is decreased.

(14.4.13) Keeping The Dimension. Let $\kappa = \delta \times \tau^d$ such that $\delta \in \mathcal{T}^{k-1}$ is a facet of ξ. If $\delta \not\subset \partial\tau$, then there exists exactly one cell $\xi' \in \mathcal{T}^k$ such that $\xi' \neq \xi$, $\xi' \subset \tau$ and $\delta \subset \xi'$. This is again a consequence of definition (14.4.8) and is not difficult to prove. Now $\eta := \xi' \times \tau$ is the desired second cell. In this case the dimension k of the primal cell ξ is left invariant.

The main point for practical purposes is that the above three different kinds of pivoting steps must be easy to implement on a computer. This is of course mainly a question of choosing a simple primal-dual pair $(\mathcal{P}, \mathcal{D})$ and either $\mathcal{T} = \mathcal{P}$ or some standard refinement $\mathcal{T}$ of $\mathcal{P}$ which can be handled well. We do not wish to go into these details which vary for different choices, and instead refer the reader to the above mentioned special literature.

(14.4.14) Primal-Dual Manifolds With Cone. We now slightly modify the construction (14.4.9) of primal-dual manifolds to include cones for the refinement $\mathcal{T}$ of the primal manifold:

(14.4.15) $\quad \omega \bullet \mathcal{T} \otimes \mathcal{D} := \{\omega \bullet \xi \times \tau^d \mid k = 0, 1, \ldots, N, \; \xi \in \mathcal{T}^k, \; \tau \in \mathcal{M}^k, \; \xi \subset \tau\}.$

If $\dim \xi = k > 0$, then the facets of $\omega \bullet \xi$ are simply the $\omega \bullet \rho$ where $\rho \in \mathcal{T}^{k-1}$ is a facet of ξ, and it is readily seen that the pivoting steps (14.4.11–13) apply. The only exception is the case $\dim \xi = k = 0$. In this case it follows that $\xi = \tau$, and ξ is a vertex of the primal manifold $\mathcal{P}$, but $\omega \bullet \xi$ is a ray which has one vertex, namely $\{\omega\}$. Hence, we now have a boundary

(14.4.16) $\qquad \partial(\omega \bullet \mathcal{T} \otimes \mathcal{D}) = \{\{\omega\} \times \{v\}^d \mid \{v\} \in \mathcal{P}^0\}.$

Clearly, such a boundary facet $\{\omega\} \times \{v\}^d$ belongs to the $(N+1)$-cell $\omega \bullet \{v\} \times \{v\}^d \in \omega \bullet \mathcal{T} \otimes \mathcal{D}$. We will later see that such boundary facets are used for starting a PL algorithm. This corresponds to starting a homotopy method on the trivial level $\lambda = 0$ at the point v.

We will now apply the above concept of primal-dual manifolds in order to describe some PL algorithms. We begin with

(14.4.17) Lemke's Algorithm Revisited. We consider again the LCP (14.3.1) and introduce a primal-dual pair $(\mathcal{P}, \mathcal{D})$ by defining

$$\text{for } I \subset \{1, 2, \ldots, N\} \text{ and } I' := \{1, 2, \ldots, N\} \setminus I$$

the primal and dual faces

(14.4.18)
$$\alpha_I := \{x \in \mathbf{R}^N \mid x[i] \geq 0 \text{ for } i \in I, \ x[i] = 0 \text{ for } i \in I'\},$$
$$\alpha_I^d := \alpha_{I'}.$$

The primal and dual manifolds consist of just one cell: $\mathcal{P} = \mathcal{D} = \{\mathbf{R}_+^N\}$. We now define a PL map $H : \mathcal{P} \otimes \mathcal{D} \times [0, \infty) \longrightarrow \mathbf{R}^N$ by $H(x, y, \lambda) := y - g(x) - \lambda d$ where $d \in \mathbf{R}_{++}^N$ is fixed. Note that the variables x and y are placed into complementarity with each other by the construction of $\mathcal{P} \otimes \mathcal{D}$, and hence a more complex definition of H as in (14.3.4) is not necessary. For sufficiently large $\lambda > 0$ the solutions of $H(x, y, \lambda) = 0$ are given by the primary ray $(x, y, \lambda) = (0, \ g(0) + \lambda d, \ \lambda)$. Here the PL algorithm following $H^{-1}(0)$ is started in negative λ-direction. If the level $\lambda = 0$ is reached, a solution $H(x, y, 0) = 0$ solves the LCP since the complementarity $x \in \mathbf{R}_+^N$, $y = g(x) \in \mathbf{R}_+^N$, $x^* y = 0$ holds by the construction of $\mathcal{P} \otimes \mathcal{D}$.

(14.4.19) A Different LCP Method. Let us now show that an analogue of the homotopy (11.7.4) can also be used for the LCP (14.3.1). We consider the same primal-dual pair $(\mathcal{P}, \mathcal{D})$ as in (14.4.17). Let $y_0 \in \mathbf{R}_{++}^N$ be a fixed starting point. We define the homotopy $H : \mathcal{P} \otimes \mathcal{D} \times [0, \infty) \longrightarrow \mathbf{R}^N$ by setting

(14.4.20)(a)
$$H(x, y, \lambda) := (y - y_0) + \lambda g(x).$$

Unfortunately, H is not PL. Hence, we use the cone construction to identify H with a PL map $H_\omega : \omega \bullet \mathcal{P} \otimes \mathcal{D} \longrightarrow \mathbf{R}^N$ by collecting the variables in a different way:

(14.4.20)(b)
$$H_\omega(z, y) := (y - y_0) + \omega \bullet g(z).$$

For $z = \omega$, which corresponds to $\lambda = 0$, there is exactly one solution of $H_\omega(z, y) = 0$, namely $(z, y) = (\omega, y_0)$. Hence $H_\omega^{-1}(0)$ intersects the boundary $\partial(\omega \bullet \mathcal{P} \otimes \mathcal{D})$ in just one point. This is the starting point for our PL algorithm which traces $H_\omega^{-1}(0)$. Initially, the solution path coincides with the segment

$$\left\{ \left((1 - \varepsilon)\omega + \varepsilon 0, \ y_0 - \varepsilon g(0)\right) \mid 0 \leq \varepsilon \leq \varepsilon_0 \right\} \subset \omega \bullet \mathcal{P} \otimes \mathcal{D}$$

for sufficiently small $\varepsilon_0 > 0$. Since there are only finitely many cells in $\omega \bullet \mathcal{P} \otimes \mathcal{D}$ and the solution on the boundary is unique, the algorithm can terminate in only one way: on a ray

(14.4.21) $\left\{ \left((1 - \varepsilon)z_1 + \varepsilon z_2, \ (1 - \varepsilon)y_1 + \varepsilon y_2\right) \mid \varepsilon \geq 0 \right\} \subset \omega \bullet \alpha_I \times \alpha_{I'} \in \omega \bullet \mathcal{P} \otimes \mathcal{D}.$

Here I denotes some subset of $\{1, 2, \ldots, N\}$ and $I' = \{1, 2, \ldots, N\} \setminus I$, see (14.4.18). Using the notation and remarks of (14.4.7), it follows that

(14.4.22) $\quad ((1 - \varepsilon)y_1 + \varepsilon y_2 - y_0) + \lambda_\varepsilon g(x_\varepsilon) = 0 \quad \text{for} \quad \varepsilon \geq 0.$

We have to consider two possible cases:

1: $\lambda_2 > \lambda_1 \geq 0$: Dividing equation (14.4.22) by $\varepsilon > 0$ and letting $\varepsilon \to \infty$ yields

$$y + \lambda g(x) = 0\,,$$

(14.4.23)   where $y := y_2 - y_1 \in \alpha_{I'}\,, \quad \lambda := \lambda_2 - \lambda_1 > 0$

$$\text{and} \quad x := \frac{\lambda_2 x_2 - \lambda_1 x_1}{\lambda_2 - \lambda_1} \in \alpha_I\,.$$

From the complementarity $x \in \mathbf{R}_+^N$, $y = \lambda g(x) \geq 0 \in \mathbf{R}_+^N$, $x^* y = 0$ it follows that x solves the LCP (14.3.1).

2: $\lambda_2 = \lambda_1 > 0$: From (14.4.22) and the fact that g is PL we obtain the equation

$$((1-\varepsilon)y_1 + \varepsilon y_2 - y_0) + \lambda_1(1-\varepsilon)g(x_1) + \lambda_1 \varepsilon g(x_2) = 0 \quad \text{for} \quad \varepsilon \geq 0\,.$$

Dividing by $\varepsilon > 0$ yields

$$y + \lambda_1 g' x = 0\,,$$

(14.4.24)   where $y := y_2 - y_1 \in \alpha_{I'}$

$$\text{and} \quad x := x_2 - x_1 \in \alpha_I \setminus \{0\}\,.$$

For some classes of matrices g', this last conclusion leads to a contradiction, and then only case 1 is possible i.e. the algorithm finds a solution of the LCP. Let us mention two such classes of matrices. Multiplying (14.4.24) from the left with x^* yields

(14.4.25) $x^* g(x) = 0\,,$

since $x^* y = 0$. Hence, if g' is positive definite, see (14.3.10), or if all entries of g' are positive, we obtain such a contradiction from (14.4.25).

(14.4.26) The Van Der Laan & Talman Algorithm. We now sketch the original algorithm of van der Laan & Talman (1978–79). Let

$$S = [p_1, p_2, \ldots, p_{N+1}]$$

be some fixed (standard) N-simplex in $\mathbf{R}^N$. We again introduce the barycentric co-ordinates

$$\beta_i : \mathbf{R}^N \to \mathbf{R} \quad \text{for} \quad i = 1, 2, \ldots, N+1$$

with respect to the vertices of S via the equations

(14.4.27) $x = \sum_{i=1}^{N+1} \beta_i(x) p_i\,, \quad \sum_{i=1}^{N+1} \beta_i(x) = 1 \quad \text{for} \quad x \in \mathbf{R}^N\,.$

We define the directional vectors

$$(14.4.28) \qquad d_i := \begin{cases} p_{i+1} - p_i & \text{for } i = 1, 2, \ldots, N, \\ p_1 - p_{N+1} & \text{for } i = N + 1. \end{cases}$$

For a given meshsize $\delta > 0$ such that δ^{-1} is a natural number, we consider the simplex $\sigma_0 = [v_1, v_2, \ldots, v_{N+1}]$ such that $v_1 := p_1$ and $v_{i+1} := v_i + \delta d_i$ for $i = 1, 2, \ldots, N$. Let $\mathcal{T}$ be the triangulation of $\mathbf{R}^N$ obtained from σ_0 via reflections, see (12.1.11–12). Then the restriction $\mathcal{T}_S := \{\sigma \in \mathcal{T} \mid \sigma \subset S\}$ is a pseudo manifold which triangulates S. Let $\{x_0\}$ be a fixed vertex (starting point) in $\mathcal{T}$ such that $x_0 \in S$. Then we define a primal-dual pair $(\mathcal{P}, \mathcal{D})$ of N-dimensional manifolds by introducing the following duality: For $I \subset \{1, 2, \ldots, N+1\}$ and $I' := \{1, 2, \ldots, N+1\} \setminus I$ such that $\#I \leq N$ we consider

$$(14.4.29) \qquad \begin{aligned} \alpha_I &:= \left\{ x_0 + \sum_{i \in I} \lambda_i d_i \; \middle| \; \lambda_i \geq 0 \text{ for } i \in I \right\}, \\ \alpha_I^d &:= \left\{ \sum_{i \in I'} \lambda_i p_i \; \middle| \; \lambda_i \geq 0 \text{ for } i \in I', \; \sum_{i \in I'} \lambda_i = 1 \right\}, \end{aligned}$$

Hence, the primal manifold $\mathcal{P}$ subdivides $\mathbf{R}^N$ into $N + 1$ cones with vertex x_0, and the triangulation $\mathcal{T}$ is a refinement of $\mathcal{P}$. The dual manifold is simply $\mathcal{D} = \{S\}$.

Let us consider the integer labeling ℓ of Sperner's lemma (14.4.1) for an illustration. First we choose some extension $\ell : \mathbf{R}^N \to \{1, 2, \ldots, N+1\}$ such that $\beta_i(x) \leq 0 \Rightarrow \ell(x) \neq i$ for $i = 1, 2, \ldots, N + 1$. This extension is only formal and will not be needed in the actual computations by the algorithm. For each vertex $\{v\} \in \mathcal{T}^0$ such that $v \in S$, let us define $g(v) := p_{\ell(v)}$. Then we extend $g : \mathcal{T} \to \mathbf{R}^N$ in a unique way such that it is PL. From the properties of ℓ we conclude that $g(\mathbf{R}^N) = S$, and moreover $\beta_i(x) \leq 0 \Rightarrow \beta_i(g(x)) = 0$ for $i = 1, 2, \ldots, N + 1$ and $x \in \mathbf{R}^N$. Let b denote the barycenter

$$b := (N+1)^{-1} \sum_{k=1}^{N+1} p_k$$

of S. From the construction of g it is clear that $g(x) = b$ if and only if x is the barycenter of a simplex $\sigma \in \mathcal{T}$, $\sigma \subset S$ which is completely labeled in the sense of Sperner's lemma (14.4.1). Hence we define the homotopy $H : \mathcal{T} \otimes \mathcal{D} \times [0, \infty) \to \mathbf{R}^N$ by setting

$$(14.4.30)(a) \qquad H(x, y, \lambda) := (y - b) + \lambda(g(x) - b).$$

Again, to obtain a PL map, we actually have to identify H with $H_\omega : \omega \bullet \mathcal{T} \otimes \mathcal{D} \to \mathbf{R}^N$ by setting

$$(14.4.30)(b) \qquad H_\omega(z, y) := (y - b) + \omega \bullet (g(z) - b).$$

For $z = \omega$, which corresponds to $\lambda = 0$, there is exactly one solution of $H_\omega(z, y) = 0$, namely $(z, y) = (\omega, b)$. Hence $H_\omega^{-1}(0)$ intersects the boundary $\partial(\omega \bullet \mathcal{T} \otimes \mathcal{D})$ at just one point. This is the starting point for our PL algorithm which traces $H_\omega^{-1}(0)$. Initially, the solution path coincides with the segment

$$\left\{ \left((1 - \varepsilon)\omega + \varepsilon x_0 \,,\, b - \varepsilon\big(g(x_0) - b\big) \right) \;\Big|\; 0 \leq \varepsilon \leq \varepsilon_0 \right\} \subset \omega \bullet \mathcal{T} \otimes \mathcal{D}$$

for sufficiently small $\varepsilon_0 > 0$. Let us now show that the algorithm cannot generate a point $(z, y) = \big((1 - \lambda)\omega + \lambda x \,,\, y \big) \in H_\omega^{-1}(0)$ such that $x \notin S$.

Indeed, otherwise we could find an index j such that $\beta_j(x) < 0$. Let $x \in \alpha_I$, see (14.4.29). From

$$\beta_j(x) = \beta_j\left(x_0 + \sum_{i \in I} \lambda_i d_i \right) = \beta_j(x_0) + \sum_{i \in I} \lambda_i \beta_j(d_i) < 0$$

we see that $j \notin I$, and hence $\beta_j(y) = 0$ by the duality in $\mathcal{P} \otimes \mathcal{D}$. Furthermore $\beta_j(g(x)) = 0$ by the construction of g. Now $y + \lambda[g(x) - b] = b$ implies that $\beta_j(y) + \lambda[\beta_j(g(x)) - \beta_j(b)] = \beta_j(b)$, and since $\beta_j(b) = (N + 1)^{-1}$, we obtain a contradiction by looking at the sign of the various barycentric co-ordinates.

Now, since the algorithm can only traverse finitely many cells, and since the solution on the boundary $\partial(\omega \bullet \mathcal{T} \otimes \mathcal{D})$ is unique, it can only terminate in a ray

$$(14.4.31) \quad \{((1 - \varepsilon)z_1 + \varepsilon z_2 \,,\, (1 - \varepsilon)y_1 + \varepsilon y_2) \mid \varepsilon \geq 0\} \subset \omega \bullet \tau \times \alpha_I^d \in \omega \bullet \mathcal{T} \otimes \mathcal{D},$$

where $\tau \in \mathcal{T}^k$ such that $\tau \subset \alpha_I$ and $k = \#I$. We again use the notation and remarks of (14.4.7). It follows that

$$(14.4.32) \qquad (1 - \varepsilon)y_1 + \varepsilon y_2 - y_0 + \lambda_\varepsilon\big(g(x_\varepsilon) - b\big) = b \quad \text{for} \quad \varepsilon \geq 0.$$

Since the k-cell τ is bounded, we only have to consider the case $\lambda_2 > \lambda_1 \geq 0$. Dividing equation (14.4.32) by $\varepsilon > 0$ and letting $\varepsilon \to \infty$ yields

$$\lambda\big(g(x) - b\big) = 0,$$

$$(14.4.33) \qquad \text{where} \quad \lambda := \lambda_2 - \lambda_1 > 0 \quad \text{and} \quad x := \frac{\lambda_2 x_2 - \lambda_1 x_1}{\lambda_2 - \lambda_1} \in \tau,$$

and the completely labeled simplex τ in the sense of Sperner's lemma (14.4.1) is found. Note that $\dim \tau = N$ follows from this.

Figure 14.4.b illustrates the algorithm of van der Laan & Talman for obtaining a completely labeled simplex for the case $N = 2$. The dots correspond to the barycenters of the faces (i.e. vertices, edges, 2-simplices) which are generated.

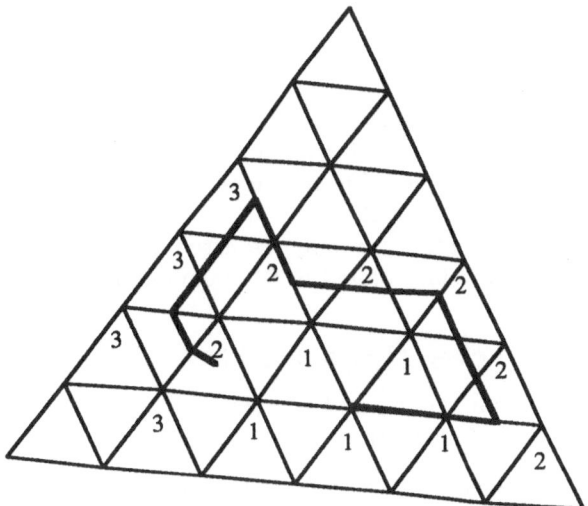

Figure 14.4.b The algorithm of van der Laan & Talman

(14.4.34) The $(N+1)$-algorithm. More generally, the above setting can be used to calculate an approximate zero point of a map $G : \mathbf{R}^N \to \mathbf{R}^N$ which is asymptotically linear, see (13.1.12). Since the primal manifold $\mathcal{P}$ of (14.4.29) contains $(N+1)$ ray edges, this method is often called the $(N+1)$-algorithm. We consider the PL approximation $G_T : T \to \mathbf{R}^N$ of G, see section 12.2. It is clear, see (13.2.7), that G_T is also asymptotically linear and $G'_T(\infty) = G'(\infty)$. A homotopy $H : T \otimes \mathcal{D} \times [0, \infty) \to \mathbf{R}^N$ is introduced by setting

(14.4.35a) $$H(x, y, \lambda) := G'(\infty)(y - b) + \lambda G_T(x).$$

Again, a PL map $H_\omega : \omega \bullet T \otimes \mathcal{D} \to \mathbf{R}^N$ is obtained by setting

(14.4.35b) $$H_\omega(z, y) := G'(\infty)(y - b) + \omega \bullet G_T(z).$$

The algorithm is started as before, see the remarks following (14.4.30).

Let us first show that there is a constant $C > 0$ such that $H(x, y, \lambda) = 0$ implies $\|x\| < C$. Indeed, otherwise we could find a sequence

$$\left\{ (x_n, y_n, \lambda_n) \right\}_{n=1,2,\ldots} \subset H^{-1}(0)$$

such that $\lim_{n \to \infty} \|x_n\| = \infty$. It follows from $H(x_n, y_n, \lambda_n) = 0$ that

(14.4.36) $$\lambda_n^{-1}(y_n - b) + G'(\infty)^{-1} G_T(x_n) = 0,$$

and consequently the barycentric co-ordinates verify the equation

(14.4.37) $$\lambda_n^{-1}\big(\beta_j(y)_n - \beta_j(b)\big) + \beta_j\big(G'(\infty)^{-1} G_T(x_n)\big) = 0.$$

However, since $\lim_{n\to\infty} \|x_n\| = \infty$, by possibly choosing a subsequence, we may without loss of generality assume that there is a $j \in \{1, 2, \ldots, N+1\}$ such that $\beta_j(x_n) < 0$ for $n = 1, 2, \ldots$. By the asymptotic linearity of G_T we hence have $\beta_j\big(G'(\infty)^{-1}G_T(x_n)\big) < 0$ for large enough n. Furthermore $\beta_j(b) = (N+1)^{-1}$ and $\beta_j(y) = 0$ by the duality of $\mathcal{P} \otimes \mathcal{D}$, see the remarks preceding (14.3.31). Now checking the signs of the barycentric co-ordinates in (14.4.37) leads to a contradiction.

Now, since the algorithm can only traverse finitely many cells, and since the solution on the boundary $\partial(\omega \bullet T \otimes \mathcal{D})$ is unique, it can only terminate in a ray

$$(14.4.38) \quad \big\{\big((1-\varepsilon)z_1 + \varepsilon z_2,\, (1-\varepsilon)y_1 + \varepsilon y_2\big) \mid \varepsilon \geq 0\big\} \subset \omega \bullet \tau \times \alpha_I^d \in \omega \bullet T \otimes \mathcal{D},$$

where $\tau \in T^k$ such that $\tau \subset \alpha_I$ and $k = \#I$. We again use the notation and remarks of (14.4.7). It follows that

$$(14.4.39) \qquad (1 - \varepsilon)y_1 + \varepsilon y_2 + \lambda_\varepsilon G'(\infty)^{-1}G_T(x_\varepsilon) = b \quad \text{for} \quad \varepsilon \geq 0.$$

Since the k-cell τ is bounded, we only have to consider the case $\lambda_2 > \lambda_1 \geq 0$. Dividing equation (14.4.39) by $\varepsilon > 0$ and letting $\varepsilon \to \infty$ yields

$$G_T(x) = 0, \quad \text{where} \quad x := \frac{\lambda_2 x_2 - \lambda_1 x_1}{\lambda_2 - \lambda_1} \in \tau$$

is the desired approximate zero point of G.

(14.4.40) The octahedral algorithm. There are many examples of variable dimension algorithms in the literature which can be efficiently described by our concept of primal-dual manifolds. We refer to Kojima & Yamamoto (1982–84) for more details. Here, we will only give one last example, namely the octahedral method of Wright (1981), since numerical experiments indicate that it performs favorably, see e.g. Kojima & Yamamoto (1984).

We denote by $\Sigma := \{+1, 0, -1\}^N \setminus \{0\}$ the set of all nonzero sign vectors. For two vectors $s, t \in \Sigma$ we introduce the relation

$$(14.4.41) \qquad s \prec p \quad :\Longleftrightarrow \quad \forall_{i=1,\ldots,N}\Big(s[i] \neq 0 \Rightarrow s[i] = p[i]\Big).$$

Then we define a primal-dual pair $(\mathcal{P}, \mathcal{D})$ of N-dimensional manifolds by introducing the following duality:

$$\alpha_0 := \{0\}, \quad \alpha_0^d := \Big\{y \in \mathbf{R}^N \,\Big|\, \|y\|_1 \leq 1\Big\},$$

and for $s \in \Sigma$ we consider

$$(14.4.42) \qquad \alpha_s := \left\{\sum_{\substack{p \in \Sigma \\ s \prec p}} \lambda_p\, p \;\middle|\; \lambda_p \geq 0\right\},$$

$$\alpha_s^d := \Big\{y \in \mathbf{R}^N \,\Big|\, \|y\|_1 \leq 1,\ s^* y = 1\Big\}.$$

Hence, the primal manifold $\mathcal{P}$ subdivides $\mathbf{R}^N$ into $2N$ cones centered around the unit base vectors $\pm e_i$ for $i = 1, 2, \ldots, N$, and the dual manifold $\mathcal{D}$ just consists of the unit ball with respect to the $\|.\|_1$-norm. We easily check that

$$y \in \alpha_s^d, \ s \prec p \quad \Rightarrow \quad y^* p \geq 0$$

and hence

(14.4.43) $(x, y) \in \mathcal{P} \otimes \mathcal{D} \quad \Rightarrow \quad x^* y \geq 0 \,.$

We now consider a triangulation $\mathcal{T}$ which is a refinement of $\mathcal{P}$, for example it is easy to see that the triangulation J_1 introduced in the remarks preceding (13.3.3) has this property. Again our aim is to find an approximate zero point of an asymptotically linear map G by using the homotopy (14.4.35) where $b = 0$. The discussion following (14.4.35) applies also here, we only have to show again that (14.4.36) leads to a contradiction. Indeed the asymptotic linearity and $\lim_{n\to\infty} \|x_n\| = \infty$ implies that $x_n^* G'(\infty)^{-1} G_{\mathcal{T}}(x_n) > 0$ for all sufficiently large n. Hence, if we multiply (14.4.36) from the left with x_n^* and take $b = 0$ and (14.4.43) into account, the contradiction results.

(14.4.44) Concluding Remarks. We have seen that the concept of primal-dual manifolds $\mathcal{P} \otimes \mathcal{D}$ enables us to describe many variable dimension algorithms in a unified way. One class of such algorithms are the homotopy methods of chapter 13. An important feature of primal-dual manifolds is that a complementarity property of the variables (x, y) may be incorporated into the construction of $\mathcal{P} \otimes \mathcal{D}$ so that $(x, y) \in \mathcal{P} \otimes \mathcal{D}$ (automatically) implies this property. This is a very convenient trick for dealing with complementarity problems or related questions, and was illustrated here for the case of the linear complementarity problem, see (14.4.17) and (14.4.19), but many more applications have been considered in the literature.

14.5 Exploiting Special Structure

It is a general opinion that PL homotopy methods tend to be slower than more classical methods or even predictor-corrector methods in solving nonlinear problems such as $G(x) = 0$ for $G : \mathbf{R}^N \to \mathbf{R}^N$. One reason for this is the large number of simplices which have to be traversed. Indeed, if we consider triangulations such as Freudenthal's triangulation or Todd's triangulation J_1 of $\mathbf{R}^N$, then each unit cube contains $N!$ simplices, and though usually only a small portion of these simplices is actually traversed, this may still amount to a fair number of PL steps. In particular for large dimensions N, this may lead to an unacceptable amount of computational expense.

Now the main reason that classical methods perform so much better in higher dimensions (if they work at all!) is that they make an implicit and usually also an explicit use of the special structure of the problem, and it is a fact that all higher dimensional problems of practical importance carry some special structure which can be used to increase the computational efficiency considerably. For example, if G is affine, then Newton's method only makes one step, but a PL method still has to traverse many simplices. Of course, nobody would solve a linear problem by either way, but this is just an example to point out that something drastic has to be done in order to enable PL methods to be competitive.

The crucial point we want to make here is that for a general map $G : \mathbf{R}^N \to \mathbf{R}^N$ we cannot avoid triangulations by pseudo manifolds, since for an arbitrary PL manifold $\mathcal{M}$ subdividing $\mathbf{R}^N$ there usually exists no PL approximation $G_{\mathcal{M}} : \mathbf{R}^N \to \mathbf{R}^N$ i.e. a PL map such that G and $G_{\mathcal{M}}$ coincide on all vertices of $\mathcal{M}$. This is generally only true for pseudo manifolds $\mathcal{T}$. However, the picture changes if G has some structure. Then the "pieces of linearity" of $G_{\mathcal{T}}$ may be larger than simplices. In fact, we have already seen such examples:

1) The cells (actually cones) traversed by Lemke's algorithm, see section 14.3, can be taken so large since the complementarity problem under consideration is linear. The situation in (14.4.17) and (14.4.19) is similar.

2) We consider a pseudo manifold $\mathcal{T}$ such as Freudenthal's triangulation or Todd's J_1 triangulating a slab $\mathbf{R}^N \times [0, \delta]$ and a standard homotopy $H : \mathbf{R}^N \times [0, \delta] \to \mathbf{R}^N$ such as (13.2.4) for calculating a zero point of an asymptotically linear map $G : \mathbf{R}^N \to \mathbf{R}^N$. Then the restriction of H to the trivial level $\mathbf{R}^N \times \{0\}$ is affine. Hence, if σ_1 and σ_2 are two $(N + 1)$-simplices in $\mathbf{R}^N \times [0, \delta]$ which have a common N-face such that the vertices which are not common lie in the trivial level $\mathbf{R}^N \times \{0\}$, then the PL approximation $H_{\mathcal{T}}$ is affine on the piece $\sigma_1 \cup \sigma_2$. This fact is incorporated into the variable dimension algorithms such as (14.4.34) and (14.4.40), since the primal-dual manifolds $\omega \bullet \mathcal{T} \otimes \mathcal{D}$ are a convenient tool for producing such larger pieces of linearity.

(3) The automatic pivots of section 13.6 use the fact that the homotopy map H of (13.2.4) has the same values on vertices of the triangulation J_3 which differ only in the λ-co-ordinate. However, this does not correspond to a larger piece of linearity.

In this section we want to make it clear that the efficiency of PL homotopy algorithms can be greatly enhanced if they are modified to exploit special structure of maps such as sparsity and separability. The usual approach is to show that the PL approximations of such maps are in fact affine on larger pieces which fit together to form a PL manifold. Such pieces can be traversed in one pivoting step where the numerical linear algebra is only slightly more complex than in the standard case for simplices as described in chapter 12. This observation is essentially due to Kojima (1978) and has been further studied by Todd (1980),(1983), see also Allgower & Georg (1980), Awoniyi

& Todd (1983), Saigal (1983), Sagara & Fukushima (1984), Yamamura & Horiuchi (1988).

We illustrate the essential idea for a simple but important example, namely for a separable map.

(14.5.1) Definition. *A map* $G : \mathbf{R}^N \to \mathbf{R}^N$ *is called* **separable** *if there are functions* $G_i : \mathbf{R} \to \mathbf{R}^N$ *such that* $G(x) = \sum_{i=1}^{N} G_i(x)$ *holds for* $x \in \mathbf{R}^N$.

The following lemma is trivial but provides the essential argument for dealing with this special structure.

(14.5.2) Lemma. *Let* $G : \mathbf{R}^N \to \mathbf{R}^N$ *be separable, and let* $\sigma := \prod_{i=1}^{N} [a_i, b_i]$ *with* $a_i < b_i$ *be a cube. Then there is a unique affine map* $G_\sigma : \sigma \to \mathbf{R}^N$ *such that* $G(v) = G_\sigma(v)$ *for all vertices* $\{v\}$ *of* σ.

Let us now assume that $G : \mathbf{R}^N \to \mathbf{R}^N$ is a separable and asymptotically linear map. The following primal-dual pair has been used by Kojima & Yamamoto (1982) to provide an efficient implementation of Merrill's method (13.2.5), and a related subdivision of $\mathbf{R}^N \times [0,1]$ has been used by Todd (1980) to give an efficient implementation of Merrill's method for the case of separable mappings. For a fixed meshsize $\delta > 0$ we consider a vector $q \in \mathbf{R}^N$ such that $q[i]$ is an integer multiple of δ for $i = 1, 2, \ldots, N$. Then we define a primal-dual pair $(\mathcal{P}, \mathcal{D})$ of N-dimensional manifolds by introducing the following

(14.5.3) Duality.

$$\alpha_q := \left\{ x \in \mathbf{R}^N \ \middle| \ x[i] = q[i] \ \text{for} \ \frac{x[i]}{\delta} \ \text{odd}, \right.$$

$$\left. x[i] \in [q[i] - \delta, \, q[i] + \delta] \ \text{for} \ \frac{x[i]}{\delta} \ \text{even} \right\},$$

$$\alpha_q^d := \left\{ x \in \mathbf{R}^N \ \middle| \ x[i] = q[i] \ \text{for} \ \frac{x[i]}{\delta} \ \text{even}, \right.$$

$$\left. x[i] \in [q[i] - \delta, \, q[i] + \delta] \ \text{for} \ \frac{x[i]}{\delta} \ \text{odd} \right\}.$$

Hence, the primal manifold $\mathcal{P}$ represents a subdivision of $\mathbf{R}^N$ into cubes of length 2δ with centers q such that all co-ordinates of q are odd multiples of δ, and the dual manifold $\mathcal{D}$ is similar, only that now the centers q are even multiples of δ. The following lemma follows immediately from the construction of the duality (14.5.3):

(14.5.4) Lemma. *Let* $\{ (x_n, y_n) \}_{n \in \mathbf{N}} \subset \mathcal{P} \otimes \mathcal{D}$ *be a sequence. Then*

$$\lim_{n \to \infty} \|x_n\| = \infty \quad \Longleftrightarrow \quad \lim_{n \to \infty} \|y_n\| = \infty,$$

$$\lim_{n \to \infty} \|x_n\| = \infty \quad \Longrightarrow \quad \lim_{n \to \infty} \frac{x_n^* y_n}{\|x_n\|^2} = 1.$$

If $G : \mathbf{R}^N \to \mathbf{R}^N$ is any asymptotically linear map and $\mathcal{T}$ a pseudo manifold which refines $\mathcal{P}$ such as Todd's triangulation J_1, by using (14.5.4) we can proceed as in the octahedral method (14.4.40), in particular as in the remarks following (14.4.43), to show that the corresponding variable dimension algorithm finds an approximate zero point of G. However, if G is also separable, then we do not need the refinement $\mathcal{T}$. In fact, there is a unique PL approximation $G_{\mathcal{P}} : \mathcal{P} \to \mathbf{R}^N$ such that G and $G_{\mathcal{P}}$ coincide on all vertices of $\mathcal{P}$. Analogously to (14.4.35), we define a PL map $H_\omega : \omega \bullet \mathcal{P} \otimes \mathcal{D} \to \mathbf{R}^N$ by setting

$$(14.5.5) \qquad H_\omega(z, y) := G'(\infty)y + \omega \bullet G_{\mathcal{P}}(z).$$

The corresponding variable dimension algorithm which follows a solution curve in $H_\omega^{-1}(0)$, traverses much larger pieces than the general algorithm described above.

Similar, but more complex subdivision techniques can be used if G has a Jacobian which respects a certain sparsity pattern. Furthermore, such sparsity can also be used to decrease the computational effort involved in the numerical linear algebra of each PL step (door-in-door-out step). We refer the reader to Todd (1980) for more details.

Chapter 15. Approximating Implicitly Defined Manifolds

15.1 Introduction

Up to now we have been discussing ways of numerically tracing a curve $C \subset H^{-1}(0)$ where $H : \mathbf{R}^{N+1} \to \mathbf{R}^N$. We have outlined both the predictor-corrector methods which rest rather strongly upon smoothness properties of H and the piecewise-linear methods which require less smoothness, but then on the other hand are generally less flexible as far as allowing general step-lengths is concerned. The latter methods are also no longer viable for large N. The behavior of the two methods was also seen to be considerably different in the presence of singular points on the curve.

The reader will recall some version of the classical result known as the implicit function theorem, which enables one under certain hypotheses to conclude the existence of a manifold $M \subset H^{-1}(0)$ where $H : \mathbf{R}^{N+K} \to \mathbf{R}^N$. The question therefore arises whether we cannot analogously approximate M via numerical methods as we have done for curves in our previous chapters. It turns out that both the PC and the PL methods can be modified to apply to this case. In the actual implementations, certain technical complications can arise. In this chapter we will describe some of the details for approximating M and related questions. In reading this chapter the reader may occasionally need to refer to the notation and terminology of chapter 12.

Before proceeding, let us for the sake of completeness recall a theorem upon which much of our discussion in this chapter will rest.

(15.1.1) Implicit Function Theorem. *Let $H : \mathbf{R}^{N+K} \to \mathbf{R}^N$ be a smooth map such that $0 \in \mathrm{range}(H)$. Then*

$$M = \{x \in \mathbf{R}^{N+K} \mid H(x) = 0, \ x \text{ is a regular point of } H\}$$

is a smooth K-dimensional manifold.

The case $K = 1$ is what our earlier chapters concerning implicitly defined curves have rested upon. For the case $K = 2$ the manifold M is a surface. Let us consider a few simple examples.

(15.1.2) Example. Let $H : \mathbf{R}^3 \to \mathbf{R}^1$ be defined by

$$H(x_1, x_2, x_3) = x_1^2 + x_2^2 + x_3^2 - 1.$$

In this case, 0 is a regular value of H, and the manifold M is the unit sphere in $\mathbf{R}^3$.

(15.1.3) Example. Let $H : \mathbf{R}^3 \to \mathbf{R}^1$ be defined by

$$H(x_1, x_2, x_3) = x_1^2 + x_2^2 - x_3^2.$$

Since $H'(x_1, x_2, x_3) = (2x_1, \ 2x_2, \ -2x_3)$, we see that $0 \in \mathbf{R}^3$ is the only point where H is not regular. Now M consists of the two halves of a cone without the vertex (which is the only singular point of H).

(15.1.4) Example. Let $H : \mathbf{R}^3 \to \mathbf{R}^1$ be defined by

$$H(x_1, x_2, x_3) = x_1 \cdot x_2.$$

Since $H'(x_1, x_2, x_3) = (x_2, \ x_1, \ 0)$, we see that H is not regular when both $x_1 = 0$ and $x_2 = 0$. Now M consists of the two co-ordinate planes $\{x_1 = 0\}$ and $\{x_2 = 0\}$ without the x_3-axis (which is the set of singular points of H).

It is easy to see that given a point on M, then further points on M can be obtained by reducing the number of free variables (e.g. via holding $K - 1$ variables fixed) and then applying one of the earlier continuation methods. However, one usually wants to approximate the manifold in a way which gives some insight into its structure. For example, we might want to obtain an approximate local parametrization, or an idea of its qualitative global shape. In the case $K = 1$ approximate local parametrizations of the curve can easily be obtained from the output of the methods we described previously. If the curve is compact, the PC and PL continuation methods can be easily modified so that overlapping approximations are avoided.

For $K > 1$ it becomes apparent that the naive approach mentioned above is unsatisfactory if one wishes to obtain approximations to M which reveal structure. In most applications it is desirable or often satisfactory to obtain a pseudo manifold which approximates M locally or even globally. The latter is important for example, if one wishes to use a finite element method for some problem defined on the manifold. In this case it is crucial that no overlappings of the approximating pieces occur. So when one attempts to approximate M with such objectives in mind the task becomes harder and requires a more sophisticated approach.

To date two different approaches for obtaining piecewise linear approximations of M have been given. One of them is the "moving frame algorithm" given by Rheinboldt (1987) which we will discuss first. The other approach which we will discuss generalizes the PL approximation of curves. These generalizations have been given by Allgower & Schmidt (1985) for integer labeling and Allgower & Gnutzmann (1987) for vector labeling.

A different approach using polygonizations of implicit surfaces in $\mathbf{R}^3$ has recently been given by Bloomenthal (1988).

The moving frame algorithm offers the advantage of allowing relatively large values for N. The algorithm yields a piecewise linear approximation of a covering of M and hence reveals its structure. The objective of having non-overlapping is at this time not completely met in general. The generalized PL algorithm meets both objectives but is not viable for large N. It can also be used for computer graphics and finite element methods over compact manifolds.

15.2 Newton's Method and Orthogonal Decompositions Revisited

Before we discuss the methods for obtaining PL approximations of M, let us briefly indicate that Newton's method can be used as a corrector in the sense of chapter 3 also in this more general setting. For simplicity, we assume throughout this chapter that zero is a regular value of the smooth map $H : \mathbf{R}^{N+K} \to \mathbf{R}^N$. Hence $M = H^{-1}(0)$ is a smooth K-dimensional manifold.

If B is an $N \times (N + K)$-matrix with maximal rank, then in analogy to section 3.1, the Moore-Penrose inverse B^+ of B is given by e.g. $B^+ = B^*(BB^*)^{-1}$. The product BB^+ is the identity on R^N, and $\mathrm{Id} - B^+B$ is the orthogonal projection onto $\ker(B)$.

In analogy to theorem (3.3.1), there exists an open neighborhood U of M such that Newton's method

$$(15.2.1) \qquad v_{i+1} = v_i - H'(v_i)^+ H(v_i), \ i = 0, 1, \dots$$

converges quadratically to a point $v_\infty \in M$ whenever the starting point v_0 is in U. Since the evaluation and decomposition of the Jacobian matrix $H'(v_i)$ may be costly, one often modifies (15.2.1) to the so-called chord method

$$(15.2.2) \qquad v_{i+1} = v_i - B^+ H(v_i), \ i = 0, 1, \dots$$

where B is some fixed approximation of $H'(v_0)$. It is well known that the above mentioned quadratic convergence reduces to linear convergence in the latter case, see e.g. Ben-Israel (1966).

Orthogonal decompositions are particularly useful in this context. If Q is an orthogonal $(N + K) \times (N + K)$-matrix such that $BQ = (L, 0)$ for some lower triangular $N \times N$-matrix L, and if we split the orthogonal matrix $Q = (Q_N, Q_K)$ into the first N and the last K columns, then it is straightforward to see that $B^+ = Q_N L^{-1}$, and the columns of Q_K provide an orthonormal basis for $\ker(B)$.

Unfortunately, for all known decomposition methods, this basis matrix Q_K does not depend continuously on the choice of the matrix B, and this is a fact which complicates matters in constructing the moving frame algorithm, see Rheinboldt (1986). The remedy is to introduce a reference $(N+K) \times K$-matrix T_K whose columns form an orthonormal system (i.e. $T_K^* T_K = \mathrm{Id}$) and to use the singular value decomposition $V_1^* T_K^* Q_K V_2 = \Sigma$, see e.g. Golub & Van Loan (1983). Rheinboldt shows that the map $B \mapsto W_K := Q_K V_1 V_2^*$ is smooth if B varies over the open set of $(N+K) \times K$-matrices which have full rank and a kernel such that $T_K^* Q_K$ is non-singular. We simplify our discussion by slightly abusing the notation of Rheinboldt and calling the new matrix W_K the **moving frame** of the kernel of B with respect to the **reference matrix** T_K.

15.3 The Moving Frame Algorithm

To motivate the idea of the moving frame algorithm, we first give a very heuristic description. At some starting point $p \in M$, we triangulate the tangent space of M at p using some standard triangulation (e.g. Freudenthal's). We now imagine that the manifold is "rolled" over the tangent space, thereby "imprinting" a triangulation on M. This is used to provide an approximation of some part of M by a pseudo manifold. For $K > 1$, the moving frame idea prevents twisting in the "rolling" process which would mess up the fitting of the imprinted triangulation. The following pseudo algorithm sketches the essential ideas of Rheinboldt's method. Given a triangulation $\mathcal{T}$ of R^K, the algorithm constructs an "imprint" $\varphi : \mathcal{X} \to M$ where $\mathcal{X} \subset \mathcal{T}^0$ is a subset of "marked" nodes of $\mathcal{T}$ which is successively enlarged.

(15.3.1) Moving Frame Algorithm.   *comment:*

    input

        begin

        $s \in \mathcal{T}^0$;   *initial marked node*

        $\varphi(s) \in M$;   *starting point on M, imprint of s*

        T_K;   *reference matrix*

        $h > 0$;   *steplength for moving frame —*
            should be much bigger than the meshsize of the triangulation

        end;

    $\mathcal{X} := \{s\}$;   *initial set of marked nodes*

    repeat

        get $x \in \mathcal{X}$ such that   *begin building a new frame*

            $\{y \in \mathcal{T}^0 \mid y \notin \mathcal{X}, \; \|y - x\| < h\} \neq \emptyset$;

$B := H'(\varphi(x));$ *new Jacobian —*

which will generally be decomposed at this point

calculate W_K;   *moving frame of* $\ker(B)$ *with respect to* T_K

while $\{y \in T^0 \mid y \notin \mathcal{X},\ \|y - x\| < h\} \neq \emptyset$ **do**

   begin

   get $y \in \{y \in T^0 \mid y \notin \mathcal{X},\ \|y - x\| < h\};$ *new marked node*

   $v := W_K(y - x) + \varphi(x);$ *predictor for imprint of* y

   repeat $v := v - B^+ H(v)$ *chord corrector method*

   until convergence;

   $\varphi(y) := v;$ *imprint of* y

   $\mathcal{X} := \mathcal{X} \cup \{y\};$ *set of marked nodes is augmented*

   end{while};

until a stopping criterion is satisfied.

By examining the construction of the moving frame in section 15.2 it becomes evident that we have to make the following technical restriction for nodes $x \in \mathcal{X}$ where we begin a new frame: let M_0 be the set of points where the reference matrix T_K induces a local co-ordinate system on M i.e.

$$M_0 := \left\{ z \in M \ \middle| \ \det \left(\begin{array}{c} H'(z) \\ T_K^* \end{array} \right) \neq 0 \right\}.$$

Then a point x is only permitted if its imprint $\varphi(x)$ is in the connected component of M_0 which contains the starting point $\varphi(s)$. It is possible to relax this restriction, but this is usually done at the cost of having an overlapping approximation of M by a pseudo manifold.

In typical applications of the above method, the dimension N will be significantly larger than K, and hence the computational cost of the singular value decomposition is comparatively small.

The above algorithm can be regarded as a higher dimensional analogue of the PC continuation methods. The predictor step is more complicated than for $K = 1$, since a triangulation of $\mathbf{R}^K$ is mapped onto the tangent space of M at x via the moving frame device. For the case $K = 1$, the moving frame idea coincides with the concept of orientation as described e.g. in chapter 2. For $K > 1$ however, the moving frame device induces more structure than just orientation. The corrector process is quite analogous to the case $K = 1$. Some topics which remain to be investigated further are:

- Globalization. If M is a compact manifold, it would be desirable to adapt the construction of the marked nodes $\mathcal{X}$ and the imprint $\varphi(\mathcal{X})$ in such a way that $\varphi(\mathcal{X})$ can be regarded as a compact pseudo manifold (by adding appropriate edges).

- Steplength adaptation. As in the case $K = 1$, it is possible to vary the steplength h in the above algorithm according to the performance of the Newton corrector and possibly other factors.
- Handling singular points. It would be desirable to incorporate techniques for detecting, classifying and handling singularities on the manifold (e.g. bifurcation points). This is a much more complex problem than even for the case $K = 1$.

15.4 Approximating Manifolds by PL Methods

In the previous section it was seen that PC methods can be adapted to the task of approximating an implicitly defined manifold M of higher dimension. In this section we will describe an algorithm which yields an approximation of M by a piecewise linear K-manifold. The algorithm is based on the same kinds of ideas as the PL continuation method as described in chapter 12, and as before the smoothness assumptions on H can be considerably relaxed. We essentially present some of the ideas given in Allgower & Schmidt (1985), Allgower & Gnutzmann (1987) and Gnutzmann (1988). The latter reference contains a rather sophisticated PASCAL program which is too lengthy to be reproduced here.

We begin with a description of the underlying ideas. Let us suppose that the space $\mathbf{R}^{N+K}$ is triangulated by a triangulation $\mathcal{T}$. In our earlier PL algorithms there was not much reason to store any simplices. In the present situation however, we will need for certain reasons to store some of the simplices. An important advantage of the usual standard triangulations is that any simplex can be very compactly stored and cheaply recovered by means of an $(N + K)$-tuple of integers corresponding to its barycenter. Let us illustrate this for the example of Freudenthal's triangulation $\mathcal{T}$. We use the notation of (12.1.10) and the subsequent discussion.

Let the diagram

$$\sigma: \qquad v_1 \xrightarrow{u_{\pi(1)}} v_2 \xrightarrow{u_{\pi(2)}} \cdots \xrightarrow{u_{\pi(N+K)}} v_{N+K+1}$$

characterize a simplex $\sigma = [v_1, v_2, \ldots, v_{N+K+1}] \in \mathcal{T}$ where

$$\pi : \{1, 2, \ldots, N + K + 1\} \longrightarrow \{1, 2, \ldots N + K + 1\}$$

is a permutation. We make the additional assumption that $\pi(N + K + 1) = N + K + 1$. Summing the vertices of σ yields the integer vector m with components $m_1, \ldots, m_{N+K}$. Each component can be decomposed in the form $m_q = \kappa_q (N + K + 1) + \lambda_q$ where the remainder terms $0 < \lambda_q < (N + K + 1)$ are all distinct integers for $q = 1, \ldots, N + K$. On the other hand, given an integer vector m with these properties, the leading vertex of the corresponding simplex σ is given by $v_1 = (\kappa_1, \ldots, \kappa_{N+K})$, and the corresponding permutation π is obtained by ordering the $N + K$ distinct remainder terms

$$\lambda_{\pi(1)} > \lambda_{\pi(2)} > \cdots > \lambda_{\pi(N+K)}.$$

It is also possible to perform the pivoting steps directly on the integer vector m and thereby to save some arithmetic operations. The following rules are immediately recovered by translating the pivoting rules (12.1.11) for m:

1. Pivoting the leading vertex v_1 of σ generates a simplex $\tilde{\sigma}$ whose integer vector $\tilde{m}$ is obtained by adding 1 to all components of m and an additional 1 to the component $m_{\pi(1)}$, which otherwise would have a remainder 0.
2. Conversely, if the last vertex v_{N+K+1} of σ is pivoted, a simplex $\tilde{\sigma}$ is generated whose integer vector $\tilde{m}$ is obtained by subtracting 1 from all components of m and an additional 1 from the component $m_{\pi(N+K)}$, which otherwise would have a remainder 0.
3. Pivoting one of the other vertices v_q, $1 < q < N + K + 1$, of σ generates a simplex $\tilde{\sigma}$ whose integer vector $\tilde{m}$ is obtained by adding 1 to the component $m_{\pi(q)}$ and subtracting 1 from the component $m_{\pi(q-1)}$.

To take advantage of the above described compact storing, the reader may assume that we are considering Freudenthal's triangulation in the sequel. As in (12.2.1) we let $H_{\mathcal{T}}$ denote the PL approximation of H with respect to $\mathcal{T}$. The definition (12.2.2) of regular points and regular values extend to this context. We again obtain a Sard type theorem i.e. the proof of proposition (12.2.4) involving ε-perturbations, generalizes verbatim if 1 is replaced by K. Hence, if zero is a regular value of $H_{\mathcal{T}}$, the zero set $H_{\mathcal{T}}^{-1}(0)$ carries the structure of a K-dimensional PL manifold. We formulate this last remark more precisely:

(15.4.1) Theorem. *Let zero be a regular value of $H_{\mathcal{T}}$. If $\sigma \in \mathcal{T}$ has a nonempty intersection with $H_{\mathcal{T}}^{-1}(0)$, then $M_\sigma := \sigma \cap H_{\mathcal{T}}^{-1}(0)$ is a K-dimensional polytope, and the family*

$$M_{\mathcal{T}} := \{ M_\sigma \mid \sigma \in \mathcal{T}, \ \sigma \cap H_{\mathcal{T}}^{-1}(0) \neq \emptyset \}$$

is a K-dimensional PL manifold.

The following algorithm describes the fundamental steps of a PL algorithm for obtaining the PL manifold $M_{\mathcal{T}}$ approximating M. We again make the assumptions that $H : \mathbf{R}^{N+K} \to \mathbf{R}^N$ is a smooth map, $\mathcal{T}$ is a triangulation of $\mathbf{R}^{N+K}$, and zero is a regular value of both H and its PL approximation $H_{\mathcal{T}}$. Analogously to (12.3.7) we call a simplex $\sigma \in \mathcal{T}$ **transverse** if it contains an N-face which is completely labeled with respect to H. In the algorithm, the dynamically varying set $V(\sigma)$ keeps track of all vertices of the transverse simplex σ which remain to be checked in order to find all possible new transverse simplices by pivoting, cf. "update" in the algorithm below.

(15.4.2) Generic PL Approximation of a Manifold.   *comment:*

input

    begin

      $\sigma \in \mathcal{T}$ transverse;   *starting simplex*

      $D \subset \mathbf{R}^{N+K}$ compact;   *bounds the region where M is approximated*

    end;

    $\Sigma := \{\sigma\}$;   *current list of transverse simplices*

    $V(\sigma) :=$ set of vertices of σ;

    while $V(\sigma) \neq \emptyset$ for some $\sigma \in \Sigma$ **do**   *since D is compact —*

    begin   *the algorithm will eventually stop via this line*

    get $\sigma \in \Sigma$ such that $V(\sigma) \neq \emptyset$;

    get $v \in V(\sigma)$;

    obtain σ' from σ by pivoting the vertex v into v';

    if σ' is not transverse **or** $\sigma' \cap D = \emptyset$ *σ' is not of interest in this case*

      then drop v from $V(\sigma)$ *update*

    else   *in this case σ' is transverse*

      if $\sigma' \in \Sigma$ **then**   *check whether σ' is new*

        drop v from $V(\sigma)$ and v' from $V(\sigma')$ *update*

      else   *σ' is added to the list Σ in this case*

        begin

        $\Sigma := \Sigma \cup \{\sigma'\}$;

        $V(\sigma') :=$ set of vertices of σ';

        drop v from $V(\sigma)$ and v' from $V(\sigma')$;   *update*

      end{else};

    end{while}.

For purposes of exposition we have formulated the above generic algorithm in a very general way. One may regard the algorithm as a draft for the "outer loop" of the method. A number of items remain to be clarified and elaborated. We will show below how a starting simplex can be obtained in the neighborhood of a point $x \in M$. The list Σ can be used to generate a K-dimensional connected PL manifold

$$\mathcal{M}_0 := \{M_\sigma\}_{\sigma \in \Sigma},$$

cf. (15.4.1). This PL manifold approximates M quadratically, as will be seen from the error estimates in the next section. If M is compact, the restriction imposed by the bounding region D can be dropped, and the generated PL manifold will be compact with no boundary, provided the mesh of the triangulation is sufficiently small. It is not really necessary to perform the pivot $\sigma \to \sigma'$ if σ' is not transverse, since it will already be known from the current data whether the facet $\sigma \cap \sigma'$ is transverse. In the above comparing process called "check whether σ' is new", it is crucial that compact exact storing is possible by standard triangulations such as that of Freudenthal. The list searching can be performed via efficient binary tree searching. An implementation using such ideas has been given by Gnutzmann (1988).

The above PL manifold $\mathcal{M}_0$ furnishes a first coarse PL approximation of M. Several improvements are possible. The first is quite obvious in view of section 15.2: some version of Newton's method can be used to project the nodes of $\mathcal{M}_0$ onto M. Thus a new PL manifold $\mathcal{M}_1$ is generated which inherits the adjacency structure of the nodes from $\mathcal{M}_0$ and has nodes on M. A next step which would be important for certain applications (e.g. finite element methods) might be to subdivide the cells of the PL manifolds $\mathcal{M}_0$ or $\mathcal{M}_1$ into simplices in such a way that the resulting manifold can be given the structure of a pseudo manifold $\mathcal{M}_2$. This is a technical problem which for $K = 2$ is easy to implement, but is more complex for $K > 2$, and although it is in principle solvable, it has not yet been satisfactorily implemented. When all of this has been done, we may be left with a pseudo manifold $\mathcal{M}_2$ which contains some "flat" simplices. These can be eliminated by "identifying" certain nodes. Here too, there has not yet been given an implementation which is in general satisfactory. Once an approximating pseudo manifold $\mathcal{M}_2$ has been generated, it is easy to refine it by e.g. the well-known construction of halving all edges of each simplex $\tau \in \mathcal{M}_2$, triangulating it into 2^K subsimplices and projecting the new nodes back onto M. The above subtriangulation can be performed by using combinatorial ideas similar to those for generating Freudenthal's triangulation.

We have assumed that zero is a regular value of H_T. In fact, similarly to chapter 12, $\bar{\varepsilon}$-perturbations and the corresponding use of the lexicographically positive inverse of the labeling matrix automatically resolves singularities even if zero is not a regular value of H_T. The situation is similar to the case $K = 1$ which has been explained by Peitgen (1982) and Peitgen & Schmitt (1983), see also Gnutzmann (1988) where the general case is treated.

Let us now address the question of obtaining a transverse starting simplex. If we assume that a point x on M is given, then it can be shown that any $(N + K)$-simplex with barycenter x and sufficiently small diameter is transverse, see (15.5.6). Error estimates implying such facts will be given in the next section. Since we may not know a priori how small to choose the diameter, let us indicate how to check whether any simplex $\sigma = [v_1, v_2, \ldots, v_{N+K+1}] \subset \mathbf{R}^{N+K}$ is transverse and if so, how to obtain a completely labeled N-face of σ. We formulate the following auxiliary linear programming problem which is motivated by the well-known "First Phase" of the Simplex Method:

$$\textbf{(15.4.3)} \qquad \min_{y,\lambda} \left\{ \sum_{i=1}^{N+1} y[i] \;\middle|\; y + \mathcal{L}(\sigma)\lambda = e_1, \;\; y, \lambda \geq 0 \right\},$$

where

$$\textbf{(15.4.4)} \qquad \mathcal{L}(\sigma) := \begin{pmatrix} 1 & \cdots & 1 \\ H(v_1) & \cdots & H(v_{N+K+1}) \end{pmatrix}$$

is the usual labeling matrix. If no degeneracies are present, then an optimal solution (y_0, λ_0) leads to a completely labeled N-face τ of σ if and only if $y_0 = 0$ i.e. the optimal value is zero.

On the other hand, there is a direct way to construct a completely labeled N-face and a transverse simplex containing it. Let us assume that x is a point in M. The normal space of M at x is given by the orthogonal complement $\ker H'(x)^\perp$ of $\ker H'(x)$. From the Inverse Function Theorem, it is clear that the restriction $H : x + \ker H'(x)^\perp \to \mathbf{R}^N$ has x as a regular isolated zero point. Hence, if $\tau \subset x + \ker H'(x)^\perp$ is an N-simplex with barycenter x and sufficiently small diameter, then it is completely labeled. Error estimates implying this will be given in the next section. Hence, we only have to construct an affine map T sending an N-face $\tilde{\tau} \in \tilde{\mathcal{T}}^N$ of some standard triangulation $\tilde{\mathcal{T}}$ of $\mathbf{R}^{N+K}$ onto such a τ. This can be achieved in the following way. For simplicity, let us assume that $\tilde{\mathcal{T}}$ is Freudenthal's triangulation, cf. (12.1.10). Let the simplex $\tilde{\sigma} = [v_1, v_2, \ldots, v_{N+K+1}] \in \mathcal{T}$ be defined by $v_1 = 0$, $v_{i+1} - v_i = e_i$ for $i = 1, 2, \ldots, N + K$. We consider the N-face $\tilde{\tau} = [v_1, v_2, \ldots, v_{N+1}]$ which has the barycenter $\tilde{b}$. Let $H'(x)Q = (L, 0)$ be a factorization of the Jacobian $H'(x)$ such that Q is an orthogonal $(N + K) \times (N + K)$-matrix and L is a lower triangular $N \times N$-matrix. We will use the obvious fact that the first N columns of Q span the normal space of M at x. Corresponding to the meshsize $\delta > 0$, the affine map $T(u) := \delta Q(u - \tilde{b}) + x$ sends the triangulation $\tilde{\mathcal{T}}$ onto a triangulation $\mathcal{T}$, and in particular, T maps the N-face $\tilde{\tau}$ of the simplex $\tilde{\sigma}$ onto an N-face τ of the simplex σ. It is clear that τ has the barycenter x, and the meshsize of the new triangulation $\mathcal{T}$ is δ since Q is an orthogonal matrix.

As we have previously remarked, the algorithm (15.4.2) merely generates the list Σ of transverse simplices. For particular purposes like finite element methods, computer graphics etc., a user will wish to have more information concerning the structure of the PL manifold $M_\mathcal{T}$ e.g. all nodes of the PL manifold $M_\mathcal{T}$ together with their adjacency structure. Hence, to meet such requirements, it is necessary to "customize" the above algorithm by e.g. incorporating inner loops which serve to yield such information. As examples of what we have in mind, we present two algorithms. The first shows how one may obtain all completely labeled N-faces of a transverse $N + K$-simplex if a completely labeled N-face is already given (see the above constructions). The second shows how algorithm (15.4.2) can be customized to efficiently obtain all completely labeled N-faces of all transverse simplices of the list Σ. These algorithms can in turn be easily adapted for special purposes.

Before presenting these modifications, let us introduce the following notation and remark. Following terminology from linear programming, we call a set

$$\beta = \{b_1, b_2, \ldots, b_{N+1}\} \subset \mathbf{R}^{N+K}$$

an **LP basis** if

$$\begin{pmatrix} 1 & \cdots & 1 \\ H(b_1) & \cdots & H(b_{N+1}) \end{pmatrix}^{-1}$$

exists and is lexicographically positive. If β consists of the vertices of an N-simplex τ, then from (12.3.1)–(12.3.2) it is clear that β is an LP basis if and only if τ is completely labeled. Furthermore, the Door-In-Door-Out-Principle (12.3.8) generalizes in the following way:

(15.4.5) LP Step. *If $\beta \subset \mathbf{R}^{N+K}$ is an LP basis with respect to a map $H : \mathbf{R}^{N+K} \to \mathbf{R}^N$ and $v \in \mathbf{R}^{N+K} - \beta$, then there exists exactly one $v' \in \beta$ such that $\beta \cup \{v\} - \{v'\}$ is again an LP basis.*

This modification of the Door-In-Door-Out-Principle (12.3.8) is necessary since a general transverse $(N+K)$-simplex may have many completely labeled N-faces. Gnutzmann (1988) gave the following sharp upper bound for this number:

(15.4.6) $$\binom{N + K - \left[\frac{K+1}{2}\right]}{N+1} + \binom{N + K - \left[\frac{K+2}{2}\right]}{N+1}.$$

An obvious sharp lower bound is given by $K + 1$.

The following algorithm describes how one may obtain all completely labeled N-faces of a transverse $N + K$-simplex if a completely labeled N-face is already given.

(15.4.7) Completely Labeled Faces of One Simplex.   *comment:*

    input

        begin

        $\sigma \in \mathcal{T}$ a transverse $N + K$-simplex;

        τ a completely labeled N-face of σ;

    end;

    $\beta :=$ all vertices of τ;   *starting LP basis*

    $\alpha :=$ all vertices of σ;

    $W(\beta) := \alpha - \beta$;   *dynamic trial set for LP steps*

    $B := \{\beta\}$;   *dynamic list of LP bases*

    while $W(\beta) \neq \emptyset$ for some $\beta \in B$ **do**
        begin
        get $\beta \in B$ such that $W(\beta) \neq \emptyset$;
        get $v \in W(\beta)$;
        find $v' \in \beta$ such that $\beta' := \beta \cup \{v\} - \{v'\}$ is an LP basis;   *LP step*
        if $\beta' \in B$ **then**   *LP basis is already listed*
           drop v from $W(\beta)$ and v' from $W(\beta')$ *updates*
        else   *new LP basis*
           begin
           $B := B \cup \{\beta'\}$;
           $W(\beta') := \alpha - \beta'$;
           drop v from $W(\beta)$ and v' from $W(\beta')$;   *updates*
           end{else};
        end{while};
    output all $\beta \in B$.   *list of all LP bases in σ*

The next algorithm shows how to efficiently obtain all completely labeled N-faces of all transverse simplices of a connected component.

(15.4.8) All Completely Labeled Faces.   *comment:*
    input
        begin
        $\sigma \in T$ transverse;   *starting simplex*
        $D \subset \mathbf{R}^{N+K}$ compact;   *D bounds the region —*
        end;   *where M is approximated*
    $B(\sigma) :=$ all LP bases of σ;   *c.f. (15.4.7)*
    $\Sigma := \{\sigma\}$;   *current list of transverse simplices*
    $V(\sigma) :=$ set of vertices of σ;
    while $V(\sigma) \neq \emptyset$ for some $\sigma \in \Sigma$ **do**   *since D is compact —*
        begin   *the algorithm will eventually stop via this line*
        get $\sigma \in \Sigma$ such that $V(\sigma) \neq \emptyset$;
        get $v \in V(\sigma)$;
        if $v \in \beta$ for all $\beta \in B(\sigma)$ *pivoting v would not generate —*
           then drop v from $V(\sigma)$ *a transverse simplex*

   else
     begin   *v is pivoted in this case*
     obtain σ' from σ by pivoting the vertex v into v';
     if $\sigma' \cap D = \emptyset$ σ' *is not of interest in this case*
       then drop v from $V(\sigma)$ *update*
     else   *in this case* σ' *is transverse*
      if $\sigma' \in \Sigma$ *check whether* σ' *is new*
       then drop v from $V(\sigma)$ and v' from $V(\sigma')$ *update*
      else
       begin σ' *is added to the list* Σ *in this case*
       $\Sigma := \Sigma \cup \{\sigma'\}$;
       $V(\sigma') :=$ set of vertices of σ';
       drop v from $V(\sigma)$ and v' from $V(\sigma')$;   *update*
       $B(\sigma') := \{\beta \in B(\sigma) \mid v \notin \beta\}$;   *LP bases common to* σ *and* σ'
       for all $\beta \in B(\sigma)$ **such that** $v \notin \beta$ **do**
        begin   *generate all LP bases of* σ'
        find $w \in \beta$ *LP step*
         such that $\tilde{\beta} := \beta \cup \{v'\} \setminus \{w\}$ is an LP basis;
        if $\tilde{\beta} \notin B(\sigma')$ **then** $B(\sigma') := B(\sigma') \cup \{\tilde{\beta}\}$.
       end{for};
      end{else};
    end{else};
  end{while}.

15.5 Approximation Estimates

We conclude this chapter with some error estimates concerning the quality of the preceding PL approximations. These estimates also pertain to the approximations described in chapters 12 and 13. Although some of the PL algorithms are useful under much weaker assumptions on the map H, in order to obtain error estimates, it is necessary to make some smoothness assumptions regarding the first and second derivatives of H. The results in this section are analogous to results given in Gnutzmann (1988) and Allgower & Georg (1989). For reasons of simplicity, in this section we make the following

(15.5.1) Assumptions. Let $H : \mathbf{R}^{N+K} \to \mathbf{R}^N$ be a smooth map with zero a regular value. We assume that the following bounds hold:

(1) $\|H'(x)^+\| \leq \kappa$ for all $x \in M := H^{-1}(0)$;

(2) $\|H''(x)\| \leq \alpha$ for all $x \in \mathbf{R}^{N+K}$.

In fact, these bounds need only to hold in a convex region containing all of the points considered in the sequel. We remark also that it would be sufficient to assume that the Jacobian $H'(x)$ is Lipschitz continuous with constant α. The above assumptions only serve to make our proofs less technical, however the results are essentially the same.

Let T be a triangulation of $\mathbf{R}^{N+K}$ having mesh size $\delta > 0$, see definition (12.6.1). As in the preceding section we let H_T denote the PL approximation of H with respect to T. Our first result concerns the accuracy with which H_T approximates H.

(15.5.2) Proposition. $\|H(x) - H_T(x)\| \leq \frac{1}{2}\alpha\delta^2$ for $x \in \mathbf{R}^{N+K}$.

Proof. Let $\sigma = [v_1, v_2, \ldots, v_{N+K+1}] \in T$ be an $(N+K)$-simplex such that

$$x = \sum_{i=1}^{N+K+1} \gamma_i v_i \in \sigma.$$

From Taylor's formula we have

$$H(v_i) = H(x) + H'(x)(v_i - x) + \frac{1}{2}A_i[v_i - x, v_i - x]$$

for $i = 1, 2, \ldots, N + K + 1$ where we use the mean values $A_i := \int_0^1 H''(x + t(v_i - x))2(1-t)dt$ of H''. Multiplying these equations with the corresponding barycentric co-ordinates γ_i, summing and taking norms yields

$$\left\|H(x) - \sum_{i=1}^{N+K+1} \gamma_i H(v_i)\right\| \leq \frac{1}{2}\alpha\delta^2$$

as a consequence of (15.5.1)(2). The result now follows since $H_T(x) = \sum_{i=1}^{N+K+1} \gamma_i H(v_i)$. ☐

In the next estimate the thickness of a simplex has a meaningful role. One possible measure of thickness is the following

(15.5.3) Definition. *Let σ be a simplex with diameter δ and barycenter x. Let ρ be the radius of the largest ball having center x and being contained in σ. Then the **measure of thickness** of σ is defined by*

$$\theta(\sigma) := \frac{\rho}{\delta}.$$

The measure of thickness of a triangulation T is defined by

$$\theta(T) := \inf\{\theta(\sigma) \mid \sigma \in T\}.$$

For standard triangulations such as affine images of Freudenthal's triangulation, such measures are well-known and > 0, see e.g. Kojima (1978) or Saigal (1978). For example, the standard Freudenthal triangulation of R^q has thickness $\theta = 1/((q+1)\sqrt{2})$.

(15.5.4) Proposition. *Let* $\sigma \subset \mathbf{R}^{N+K}$ *be an* $(N + K)$-*simplex having diameter* δ *and thickness* θ. *If* $x \in \sigma$, *then* $\|H'(x) - H'_\sigma(x)\| \leq \delta\alpha/\theta$.

Proof. Let $\sigma = [v_1, v_2, \ldots, v_{N+K+1}]$. From Taylor's formula we have

$$H'(x)(v_i - v_j) = H'(x)(v_i - x) - H'(x)(v_j - x)$$
$$= H(v_i) - H(v_j) - \frac{1}{2}A_i[v_i - x, v_i - x] + \frac{1}{2}A_j[v_j - x, v_j - x]$$

for $i, j = 1, 2, \ldots, N + K + 1$, where the mean values A_i of H'' are defined as in the previous proof. From the definition of the PL approximation we immediately obtain

$$H'_\sigma(x)(v_i - v_j) = H(v_i) - H(v_j).$$

Subtracting corresponding sides of the above equations and taking norms and using (15.5.2) yields

$$\|(H'(x) - H'_\sigma(x))(v_i - v_j)\| \leq \alpha\delta^2.$$

By making convex combinations with this last estimate, we obtain

$$\|(H'(x) - H'_\sigma(x))(u - v)\| \leq \alpha\delta^2$$

for all $u, v \in \sigma$. From the definition (15.5.3) it follows that the set $\{u - v \mid u, v \in \sigma\}$ contains the ball with radius $\theta\delta$ and center zero. Thus the above estimate extends to the corresponding matrix norms

$$\theta\delta\|(H'(x) - H'_\sigma(x))\| \leq \alpha\delta^2,$$

and the assertion follows.   □

The next proposition is a useful characterization of transverse simplices. We employ the notation of (12.3.1).

(15.5.5) Proposition. *A simplex* $\sigma \in \mathcal{T}$ *is transverse if and only if it contains solutions* v_ε *of* $H_\mathcal{T}(v) = \vec{\varepsilon}$ *for sufficiently small* $\varepsilon > 0$.

Proof. The proof is obtained by modifying the arguments in (12.2.4) and (12.3.8). If σ does not contain the asserted solutions v_ε for sufficiently small $\varepsilon > 0$, then by definition (12.3.1) it cannot be transverse. On the other hand, if σ contains solutions v_ε for sufficiently small $\varepsilon > 0$, then by an obvious generalization of (12.2.4), the solution set consists of regular points of $H_\mathcal{T}$ for sufficiently small $\varepsilon > 0$. Hence, if ε varies, no faces of σ of dimension $< N$ can be intersected, and hence always the same N-faces of σ have to be intersected by this solution set. Clearly, those are the completely labeled N-faces of σ.
                                                                               □

The following proposition guarantees that all regular zero points of H can be approximated by transverse simplices. In particular, such estimates may be used for obtaining the starting simplices for the PL algorithms of sections 15.4, 12.3 and 12.4.

(15.5.6) Proposition. *Let* $H(x) = 0$, *and let* $\sigma \subset \mathbf{R}^{N+K}$ *be an* $(N + K)$-*simplex having barycenter* x, *diameter* δ *and thickness* θ. *If*

$$\frac{\kappa\alpha\delta}{\theta} < \frac{1}{2},$$

then σ *is transverse.*

Proof. In view of (15.5.5), it suffices to show that the affine approximation H_σ has a solution point $x_\varepsilon \in \sigma$ such that

(15.5.7)
$$H_\sigma(x_\varepsilon) = \vec{\varepsilon}$$

for sufficiently small $\varepsilon > 0$. Since H_σ is affine, any point given by a generalized Newton step

$$x_\varepsilon := x - B\big(H_\sigma(x) - \vec{\varepsilon}\big)$$

satisfies the equation (15.5.7), provided that B is a right inverse of H'_σ. If we show that the essential part of the Newton term satisfies the estimate

(15.5.8)
$$\|BH_\sigma(x)\| < \theta\delta$$

for a particular B, then we conclude from definition (15.5.3) that $x_\varepsilon \in \sigma$ for sufficiently small $\varepsilon > 0$, and the assertion follows. From proposition (15.5.4) we have

$$\|H'(x) - H'_\sigma(x)\| \leq \frac{\delta\alpha}{\theta}$$

and hence by (15.5.1)(1) and the hypothesis,

$$\|H'(x)^+\big(H'(x) - H'_\sigma(x)\big)\| \leq \frac{\kappa\delta\alpha}{\theta} < \frac{1}{2}.$$

We can now define B via the Neumann series

$$B := \sum_{i=0}^{\infty} \Big(H'(x)^+\big(H'(x) - H'_\sigma(x)\big)\Big)^i H'(x)^+.$$

Multiplying the identity

$$H'_\sigma(x) = H'(x)\Big(\mathrm{Id} - H'(x)^+\big(H'(x) - H'_\sigma(x)\big)\Big)$$

from the right by B verifies that B is indeed a right inverse of H'_σ. From the Neumann series we can also see that the estimate

$$\|B\| \leq \frac{\kappa}{1 - \frac{\kappa\alpha\delta}{\theta}} < 2\kappa$$

holds. On the other hand, proposition (15.5.2) implies

$$\|H_\sigma(x)\| = \|H_\sigma(x) - H(x)\| \leq \frac{1}{2}\alpha\delta^2.$$

Combining the last two estimates yields the estimate (15.5.8) and hence the assertion follows. □

The next proposition shows that the PL manifold $M_T = H_T^{-1}(0)$ approximates the given manifold $M = H^{-1}(0)$ quadratically in the meshsize.

(15.5.9) Proposition. *Let* $x \in \mathbf{R}^{N+K}$ *be such that* $\mathrm{dist}(x, M) < (\kappa\alpha)^{-1}$. *Let* $w \in M$ *be a nearest point to* x *i.e.* $||x - w|| = \mathrm{dist}(x, M)$. *If* $H_T(x) = 0$ *then* $||x - w|| \leq \kappa\alpha\delta^2$.

Proof. Since w satisfies the optimization problem

$$\min_w \{ \, ||x - w|| \mid H(w) = 0 \, \},$$

the Lagrange equations yield $x - w \in \mathrm{range}(H'(w)^*)$ or equivalently, $(x-w) \perp \ker(H'(w))$. From Taylor's formula we have

$$H(x) - H(w) = H'(w)(x - w) + \frac{1}{2}A[x - w, x - w],$$

where

$$A = \int_0^1 H''(w + t(x - w))2(1 - t)\,dt$$

again denotes a mean value of H''. Since $(x - w) \perp \ker(H'(w))$, and since the Moore-Penrose inverse performs the inversion orthogonally to $\ker(H'(w))$, we have

$$H'(w)^+ H(x) = x - w + \frac{1}{2}H'(w)^+ A[x - w, x - w].$$

From (15.5.2) we have

$$||H(x)|| = ||H(x) - H_T(x)|| \leq \frac{1}{2}\alpha\delta^2.$$

From these last two statements and the assumptions (15.5.1) we obtain

$$||x - w|| \leq \frac{1}{2}\kappa\alpha\delta^2 + \frac{1}{2}\kappa\alpha||x - w||^2$$

$$\leq \frac{1}{2}\kappa\alpha\delta^2 + \frac{1}{2}||x - w||,$$

and the assertion follows.   □

Up to now our approximation estimates have been of a local nature. In order to obtain global approximation results we need to apply more sophisticated tools and technical arguments. One such tool is the Brouwer degree, which for $K = 1$ may be used in a manner similar to that of Rabinowitz (1971) to obtain the existence of global continua. Peitgen & Prüfer (1979) and also Peitgen (1982) have given extensive discussions of the constructive role the PL methods play in connection with such arguments. For our purpose the continuous Newton method seems to be a suitable tool. We consider the autonomous ODE

(15.5.10) $\dot{x} = -H'(x)^+ H(x).$

If an initial point x_0 for (15.5.10) is sufficiently near the manifold $M = H^{-1}(0)$, then the flow initiating at x_0 has an exponentially asymptotic limit $x_\infty \in M$, and the map $x_0 \mapsto x_\infty$ is smooth, see e.g. Tanabe (1979). Analogously, if zero is a regular value of H_T and the meshsize of T is sufficiently small, then we may consider the flow defined by

(15.5.11) $\dot{x} = -H'_T(x)^+ H_T(x).$

Note that the right hand of (15.5.11) is piecewise affine but not continuous, and that a solution path consists of a polygonal path having nodes on lower dimensional faces $\tau \in T^{N+K-1}$. Nevertheless, it is possible by use of some technical arguments to show that the analogous results hold here too i.e. if an initial point x_0 for (15.5.11) is sufficiently near the manifold $M_T = H_T^{-1}(0)$, then the flow initiating at x_0 has an exponentially asymptotic limit $x_\infty \in M_T$, and the map $x_0 \mapsto x_\infty$ is absolutely continuous. The detailed arguments concerning (15.5.11) are beyond our present scope and will be presented elsewhere. We merely sketch how this technique may be used to obtain the following two propositions.

(15.5.12) Proposition. *If $x_0 \in M$ and the meshsize divided by the measure of thickness δ/θ of T is sufficiently small, then there exists a transverse $\sigma \in T$ such that $\mathrm{dist}(x_0, \sigma) \le \kappa \alpha \delta^2$.*

Sketch of Proof. We consider the initial value problem (15.5.11) with initial value x_0 and asymptotic limit $x_\infty \in M_T$. A full initial Newton step is given by $-H'_T(x_0)^+ H_T(x_0)$. From (15.5.2) we obtain the estimate $\|H_T(x_0)\| \le \frac{1}{2}\alpha\delta^2$. From (15.5.4) and (15.5.1) we obtain $\|H'_T(x_0)^+\| \approx \|H'(x_0)^+\| \le \kappa$. Thus a rough bound for the full initial Newton step is given by $\frac{1}{2}\kappa\alpha\delta^2$. Hence to obtain the assertion we estimate $\|x_0 - x_\infty\|$ by twice this steplength.   □

The algorithms in section 15.4 generate connected components of the PL manifold M_T. The following proposition assures that such a connected component approximates the entire manifold M if it is compact and connected.

(15.5.13) Proposition. *Let zero also be a regular value of H_T. Let $C \subset M$ be a compact connected subset (which could be all of M). Then for any triangulation T for which the meshsize divided by the measure of thickness δ/θ is sufficiently small, there is a connected compact PL submanifold $C_T \subset M_T$ such that for every $x_0 \in C$ there is an $x_\infty \in C_T$ for which $\|x_0 - x_\infty\| < \kappa\alpha\delta^2$ holds.*

Sketch of Proof. Consider the Newton map $x_0 \in C \mapsto x_\infty \in M_T$ introduced above. Since this map is continuous, and since the continuous image of a compact and connected set is compact and connected, the PL submanifold

$$C_{\mathcal{T}} := \{M_\sigma \mid \sigma \in \mathcal{T} \text{ and } x_\infty \in \sigma \text{ for some } x_0 \in C\}$$

is compact and connected. Now the assertion follows from estimates in (15.5.12). □

It is now clear from the preceding discussion that if M is compact and connected, then a connected component of $M_{\mathcal{T}}$ approximates M globally and quadratically for sufficiently small meshsize, provided the measure of thickness of $\mathcal{T}$ stays bounded away from zero.

It is also possible to formulate measures of efficiency for piecewise linear approximations of k-manifolds. Analogously to corresponding results for $k = 1$ as cited at the end of section 13.3, Alexander (1987) has studied the average intersection density for several triangulations in the context of PL approximations of k-manifolds.

If zero is a regular value of H and $H_{\mathcal{T}}$, then the smooth manifold $\mathcal{M}$ and the approximating manifold $\mathcal{M}_{\mathcal{T}}$ inherit a natural orientation which in the former case is a basic concept of differential geometry and in the latter case is analogous to the orientation described in (14.2.3). It can be shown that these orientations are consistent with each other for sufficiently fine mesh size, see Gnutzmann (1988).

Chapter 16. Update Methods and their Numerical Stability

16.1 Introduction

In numerical continuation methods, we are usually confronted with the problem of solving linear equations such as

$$(16.1.1) \qquad\qquad Ax = y$$

at each step. Update methods can be applied when the matrix A is only slightly modified at each subsequent step. This is in particular the case for the update algorithms of chapter 7 and for the PL algorithms of chapters 12–15. As we have noted in those chapters, the modification of A is of the form

$$(16.1.2) \qquad\qquad \tilde{A} := A + (a - Ae)e^*,$$

where e is some vector of unit length. For example, see (12.4.4), if e denotes the ith unit basis vector, then the above formula indicates that the ith column of A is replaced by the column a. Similar formulae arise via Broyden's update in chapter 7, see (7.2.3). In order to solve linear equations such as (16.1.1), it is usually necessary to decompose A. In the present chapter we show that by making use of (16.1.2), such a decomposition can be cheaply updated in order to obtain a decomposition of $\tilde{A}$. A simple example is provided by (12.4.6) where a certain right inverse of A was updated at each step. However, as was pointed out there, this update is not always stable, see Bartels & Golub (1968–69). Thus, the question arises whether cheap numerically stable updates of a decomposition are possible.

We outline some of the most commonly used procedures for updating certain decompositions of A, see the survey of Gill & Golub & Murray & Saunders (1974), and address the question of numerical stability for these procedures. Such update methods and their numerical stability have been extensively studied in the context of the simplex method of linear programming and in the context of quasi-Newton methods, see e.g. Gill & Murray & Wright

(1981). We will only give a short account, specifically for the purposes of the algorithms presented in this book. However, we will not assume that A is an $N \times (N + 1)$-matrix, i.e. that the system (16.1.1) is overdetermined as was the case in some examples of this book see e.g. chapters 4 and 7. Instead, we leave such slight modifications to the reader in order to keep the discussion as simple as possible. Our presentation is strongly influenced by the investigation of Georg & Hettich (1987). To make the description clearer, we will always refer to the following

(16.1.3) Standard Update Algorithm.   *comment:*

> **input** A;   *an initial $N \times N$-matrix*
>
> **repeat**
>
>> **enter**   *new data to be changed in each cycle of the method*
>> $a, e, y \in \mathbf{R}^N$ such that $||e|| = 1$;
>> $\tilde{A} := A + (a - Ae)e^*$;   *update formula*
>> solve $Ax = y$ for x;   *example of a linear equation —*
>>   *to be solved in each cycle*
>> $A := \tilde{A}$;   *prepare for next cycle*
>
> **until** cycles are stopped.

To further simplify matters, we will assume throughout this chapter that the data are given in such a way that the matrix A is always nonsingular: the principal concern here is not singularity but numerical instability. The linear equation $Ax = y$ to be solved for x is only an example to indicate that some decomposition of A is needed at each step. Some problems need a solving of more than one equation and may also involve the transpose A^* of A.

The two important questions to ask in connection with update methods such as (16.1.3) are:

1) How can we cheaply implement algorithm (16.1.3) so that linear equations such as $\tilde{A}x = y$ are solved in each cycle?
2) How stable is this method numerically, i.e. what errors do we have to expect in the solution x after an arbitrary number of cycles have been performed?

16.2 Updates Using the Sherman-Morrison Formula

The simplest implementation of algorithm (16.1.3) makes use of the formula

$$(16.2.1) \quad \tilde{A}^{-1} = A^{-1} - \frac{A^{-1}(Ae - a)e^*A^{-1}}{e^*A^{-1}(Ae - a)} = A^{-1} - \frac{(e - A^{-1}a)e^*A^{-1}}{1 - e^*A^{-1}a},$$

which is usually attributed to Sherman & Morrison (1949) and which is easily checked by multiplying the right-hand side of (16.2.1) with the right-hand side of (16.1.2). If we denote A^{-1} by B, then the implementation takes the form

(16.2.2) Update Via the Sherman-Morrison Formula.   *comment:*

    input A;   *an initial $N \times N$-matrix*

    generate $B := A^{-1}$;

    repeat

        enter $a, e, y \in \mathbf{R}^N$ such that $||e|| = 1$;   *new data*

$$\tilde{B} := B - \frac{(e - Ba)e^*B}{1 - e^*Ba};$$ *update formula*

$$x := \tilde{B}y;$$ *linear equation*

$$B := \tilde{B};$$ *prepare for next cycle*

    until cycles are stopped.

We use this implementation to clarify some ideas which we are going to pursue in this chapter. A **flop** denotes a computer operation which may maximally consist of one addition, one multiplication and some index manipulation, see Golub & Van Loan (1983). This is a typical operation arising in numerical linear algebra.

Initially, $O(N^3)$ flops are used to generate the inverse $B := A^{-1}$. Thereafter, only $O(N^2)$ flops are used per cycle to update B and to calculate the solution x. From the point of view of computational efficiency, the implementation (16.2.2) is the best we can do. However, as was first pointed out by Bartels & Golub (1968–69), this and similar implementations are not numerically stable since the "pivot element" $1 - e^*Ba$ may have a large relative error due to cancellation, without the condition number of $\tilde{A}$ being very large. Let us investigate this question of numerical stability in more detail. We distinguish between two notions of stability:

(16.2.3) Local Stability. Let us assume that B represents the numerically exact inverse of A at the beginning of some cycle. Then, due to round-off errors, at the end of the cycle we actually have that $\tilde{B} = (\tilde{A} + \Delta_{\tilde{A}})^{-1}$ where $\Delta_{\tilde{A}}$ is some (hopefully small) error. This error induces an error in the solution: Instead of $x = \tilde{A}^{-1}y$, we actually calculate $x + \Delta_x = \tilde{B}y = (\tilde{A} + \Delta_{\tilde{A}})^{-1}y$, and a standard argument in numerical linear algebra, see e.g. Golub & Van Loan (1983), leads to the following estimate for the relative error in x:

$$\frac{||\Delta_x||}{||x||} \leq \frac{||\Delta_{\tilde{A}}||}{||\tilde{A}||} \operatorname{cond}(\tilde{A}) + O(||\Delta_{\tilde{A}}||^2).$$

The solving of the equation $\tilde{A}x = y$ is considered to be numerically stable if the relative error in x does not exceed the order of magnitude $\varepsilon_{\text{tol}} \operatorname{cond}(\tilde{A})$ where ε_{tol} is the relative machine error. This means that the relative error in A has to be of the same order of magnitude as ε_{tol}.

However, $1 - e^*Ba$ may have a large relative error ε due to cancellation, and consequently $\tilde{B} = \tilde{A}^{-1} + \Delta_{\tilde{B}}$ may have an error of the magnitude

$$\|\Delta_{\tilde{B}}\| \approx \varepsilon \frac{\|(e - Ba)e^*B\|}{|1 - e^*Ba|}.$$

Since

$$\tilde{A} + \Delta_{\tilde{A}} = \left(\tilde{A}^{-1} + \Delta_{\tilde{B}}\right)^{-1} = \tilde{A} - \tilde{A}\Delta_{\tilde{B}}\tilde{A} + O\left(\|\Delta_{\tilde{B}}\|\right)^2,$$

it is possible that the relative error in $\tilde{A}$ exceeds ε_{tol} by orders of magnitudes. Hence, the method (16.2.2) may be numerically unstable. Similar instabilities can occur in the Gauss decomposition method without row or column pivoting, and this is a well-known effect described in every book on numerical linear algebra, see e.g. Golub & Van Loan (1983). We call the instability described here a **local instability** since it may occur within one single cycle of method (16.2.2).

(16.2.4) Global Stability. On the other hand, we may assume that we begin a cycle already with some error in B:

$$B = (A + \Delta_A)^{-1}.$$

The question now is, how this error is propagated through the current cycle:

$$\tilde{B} = (\tilde{A} + \Delta_{\tilde{A}})^{-1}.$$

In order to simplify the discussion, we neglect local errors (which we consider separately), and assume that all calculations in the current cycle are exact. At the end of the cycle, we hence obtain

$$\tilde{B} = B - \frac{(e - Ba)e^*B}{1 - e^*Ba}$$

$$= \left[(A + \Delta_A) + ((A + \Delta_A)e - a)\right]^{-1}$$

$$= \left[\tilde{A} + \Delta_A(\text{Id} - ee^*)\right]^{-1},$$

and hence we obtain the propagation error

$$\Delta_{\tilde{A}} = \Delta_A(\text{Id} - ee^*).$$

Thus the new error $\Delta_{\tilde{A}}$ is obtained from the old error Δ_A by projecting the rows of Δ_A orthogonally to e, i.e. the error is damped. If for example, the vector e runs through an orthogonal basis of $\mathbf{R}^N$, then after N steps the error is damped to zero. In view of the above discussion, we call an implementation of (16.1.3) **globally stable** if

(16.2.5) $$\|\Delta_{\tilde{A}}\| \le \|\Delta_A\| + O(\|\Delta_A\|^2)$$

holds, and **self-correcting**, see Georg (1982) and Georg & Hettich (1987), if

(16.2.6) $$\Delta_{\tilde{A}} = \Delta_A(\text{Id} - ee^*) + O(\|\Delta_A\|^2)$$

holds. Hence, the implementation (16.2.2) is globally stable and self-correcting.

(16.2.7) Stability. Of course, the numerical stability of an implementation of algorithm (16.1.3) depends on both aspects, i.e. we call an implementation stable if it is globally **and** locally stable. Implementations such as (16.2.2) are very popular since they use the lowest amount of flops possible, are very easy to program, and in view of their global stability, they usually perform well. However, since they may be locally unstable, readers who are tempted to use this implementation must keep in mind that they are not always safe. At least some test for cancellation errors should be employed as a safeguard.

16.3 QR Factorization

Because of the celebrated numerical stability of QR factorizations, see also chapter 4, another very popular implementation of algorithm (16.1.3) consists of updating a QR factorization of A in each cycle. It is well-known, see e.g. Golub & Van Loan (1983), that QR factorizations of a matrix are numerically very stable. As we will see later, it may also be necessary to update a **permutation matrix** P, which consists of some permutation of the columns of the identity matrix Id. We recall that $PP^* = $ Id is a simple consequence.

   Hence, let Q be an orthogonal matrix (i.e. $QQ^* = $ Id), P a permutation matrix and R an upper triangular matrix such that

(16.3.1) $QAP = R$.

It follows that the linear equation $Ax = y$ can be solved by $x := PR^{-1}Qy$, i.e. by a matrix multiplication $z := Qy$, a forward solving of $Rw = z$ and a rearranging $x := Pw$ of the co-ordinates. In order to obtain a QR factorization of the update $\tilde{A}$ in a numerically efficient way, we multiply equation (16.1.2) from the left with Q and from the right with P and use (16.3.1):

(16.3.2) $Q\tilde{A}P = R + (Qa - RP^*e)e^*P$.

By applying some Givens transformations on this situation, we now obtain a QR factorization of the update $\tilde{A}$ in a computationally efficient way. This is best described by means of a pseudo-code as in (4.2.1). We recall that an $N \times N$-matrix has **Hessenberg form** if it has nearly upper triangular form:

$$\begin{pmatrix} x & x & x & x & x & x \\ x & x & x & x & x & x \\ 0 & x & x & x & x & x \\ 0 & 0 & x & x & x & x \\ 0 & 0 & 0 & x & x & x \\ 0 & 0 & 0 & 0 & x & x \end{pmatrix}.$$

(16.3.3) General QR Step.   *comment:*

$u := Qa - RP^*e; \quad v^* := e^*P;$ *initialization*

$\tilde{Q} := Q; \quad \tilde{R} := R; \quad \tilde{P} := P;$

for $i = N - 1$ **downto** 1 **do**

    begin

      $(s_1, s_2) := \big(u[i], u[i+1]\big);$ *calculate Givens rotation*

      if $s_2 \neq 0$ **then**   *else: no rotation is necessary*

        begin

        $s := \sqrt{s_1^2 + s_2^2}; \quad (s_1, s_2) := s^{-1}(s_1, s_2);$

$$\begin{pmatrix} e_i^*\tilde{R} \\ e_{i+1}^*\tilde{R} \end{pmatrix} := \begin{pmatrix} s_1 & s_2 \\ -s_2 & s_1 \end{pmatrix} \begin{pmatrix} e_i^*\tilde{R} \\ e_{i+1}^*\tilde{R} \end{pmatrix};$$ *rows i and $i+1$ are rotated*

$$\begin{pmatrix} e_i^*\tilde{Q} \\ e_{i+1}^*\tilde{Q} \end{pmatrix} := \begin{pmatrix} s_1 & s_2 \\ -s_2 & s_1 \end{pmatrix} \begin{pmatrix} e_i^*\tilde{Q} \\ e_{i+1}^*\tilde{Q} \end{pmatrix};$$

$$\begin{pmatrix} u[i] \\ u[i+1] \end{pmatrix} := \begin{pmatrix} s_1 & s_2 \\ -s_2 & s_1 \end{pmatrix} \begin{pmatrix} u[i] \\ u[i+1] \end{pmatrix};$$

        end;

    end;

$e_1^*\tilde{R} := e_1^*\tilde{R} + u[1]v^*;$ *now $\tilde{Q}\tilde{A}\tilde{P} = \tilde{R}$ has Hessenberg form*

for $i = 1$ **to** $N - 1$ **do**

    begin

      $(s_1, s_2) := \big(\tilde{R}[i, i], \tilde{R}[i+1, i]\big);$ *calculate Givens rotation*

      if $s_2 \neq 0$ **then**   *else: no rotation is necessary*

        begin

        $s := \sqrt{s_1^2 + s_2^2}; \quad (s_1, s_2) := s^{-1}(s_1, s_2);$

$$\begin{pmatrix} e_i^*\tilde{R} \\ e_{i+1}^*\tilde{R} \end{pmatrix} := \begin{pmatrix} s_1 & s_2 \\ -s_2 & s_1 \end{pmatrix} \begin{pmatrix} e_i^*\tilde{R} \\ e_{i+1}^*\tilde{R} \end{pmatrix};$$ *rows i and $i+1$ are rotated*

$$\begin{pmatrix} e_i^*\tilde{Q} \\ e_{i+1}^*\tilde{Q} \end{pmatrix} := \begin{pmatrix} s_1 & s_2 \\ -s_2 & s_1 \end{pmatrix} \begin{pmatrix} e_i^*\tilde{Q} \\ e_{i+1}^*\tilde{Q} \end{pmatrix};$$

        end;

    end.   *now $\tilde{Q}\tilde{A}\tilde{P} = \tilde{R}$ has upper triangular form*

In the above example, the permutation matrix P was not modified at all and could have been omitted. However, permutations play a crucial role if a special but very important case is considered, namely that the vector e coincides with some unit base vector: $e = e_k$. In this case, $P^*e = e_m$

for some $m \in \{1, 2, \ldots, N\}$, and if we introduce the permutation matrix $P_0 := (e_1, \ldots, e_{m-1}, e_{m+1}, \ldots, e_N, e_m)$, then

$$(16.3.4) \qquad Q\tilde{A}PP_0 = \Big(R + (Qa - RP^*e)\, e^*P \Big) P_0 \,.$$

already has Hessenberg form:

$$\begin{pmatrix} x & x & x & x & x & x \\ 0 & x & x & x & x & x \\ 0 & 0 & x & x & x & x \\ 0 & 0 & x & x & x & x \\ 0 & 0 & 0 & x & x & x \\ 0 & 0 & 0 & 0 & x & x \end{pmatrix}.$$

Therefore, (16.3.3) simplifies to the following

(16.3.5) Special QR Step.

comment:

let $m \in \{1, 2, \ldots, N\}$ be such that $P^*e = e_m$;   *initialization*

$P_0 := (e_1, \ldots, e_{m-1}, e_{m+1}, \ldots, e_N, e_m)$;

$\tilde{P} := PP_0$;   *new permutation matrix*

$\tilde{Q} := Q$;

$\tilde{R} := \Big(R + (Qa - RP^*e)\, e^*P \Big) P_0$; $\tilde{Q}\tilde{A}\tilde{P} = \tilde{R}$ *has Hessenberg form*

for $i = m$ **to** $N - 1$ **do**

  begin

  $(s_1, s_2) := \big(\tilde{R}[i, i], \tilde{R}[i+1, i] \big)$;   *calculate Givens rotation*

  if $s_2 \neq 0$ **then**   *else: no rotation is necessary*

    begin

    $s := \sqrt{s_1^2 + s_2^2}$; $(s_1, s_2) := s^{-1}(s_1, s_2)$;

    $\begin{pmatrix} e_i^* \tilde{R} \\ e_{i+1}^* \tilde{R} \end{pmatrix} := \begin{pmatrix} s_1 & s_2 \\ -s_2 & s_1 \end{pmatrix} \begin{pmatrix} e_i^* \tilde{R} \\ e_{i+1}^* \tilde{R} \end{pmatrix}$;   *rows i and $i+1$ are rotated*

    $\begin{pmatrix} e_i^* \tilde{Q} \\ e_{i+1}^* \tilde{Q} \end{pmatrix} := \begin{pmatrix} s_1 & s_2 \\ -s_2 & s_1 \end{pmatrix} \begin{pmatrix} e_i^* \tilde{Q} \\ e_{i+1}^* \tilde{Q} \end{pmatrix}$;

    end;

  end.   *now $\tilde{Q}\tilde{A}\tilde{P} = \tilde{R}$ has upper triangular form*

By using the techniques (16.3.3) or (16.3.5), an implementation of algorithm (16.1.3) can now be described in the following way. Here Q, $\tilde{Q}$, Q_0 denote orthogonal matrices, P, $\tilde{P}$, P_0 denote permutation matrices, and R, $\tilde{R}$ denote upper triangular matrices.

(16.3.6) QR Update.   *comment:*

    input A;   *an initial $N \times N$-matrix*

    generate Q, P, R such that $QAP = R$;   *initial factorization*

    repeat

        enter $a, e, y \in \mathbf{R}^N$ such that $||e|| = 1$;   *new data*

        generate Q_0, P_0 such that   *update formula*

$$\tilde{R} := Q_0 \Big(R + (Qa - RP^*e)\, e^*P \Big) P_0 \text{ is upper triangular;}$$

$$\tilde{Q} := Q_0 Q; \quad \tilde{P} := PP_0; \qquad \tilde{Q}\tilde{A}\tilde{P} = \tilde{R} \text{ is the new factorization}$$

$$x := \tilde{P}\tilde{R}^{-1}\tilde{Q}y; \qquad\qquad\qquad \text{linear equation}$$

$$Q := \tilde{Q}; \quad P := \tilde{P}; \quad R := \tilde{R}; \qquad \text{prepare for next cycle}$$

    until cycles are stopped.

Again, the computational cost amounts to an initial $O(N^3)$ flops and an additional $O(N^2)$ flops per cycle. Let us now investigate the stability of this implementation. The local stability is evident from the well-known facts about the stability of the QR factorization, see e.g. Golub & Van Loan (1983). The global stability requires some discussion. We assume that we begin a cycle with some error already in the factorization:

$$Q(A + \Delta_A)P = R.$$

Again the question is, how this error is propagated through the current cycle:

$$\tilde{Q}(\tilde{A} + \Delta_{\tilde{A}})\tilde{P} = \tilde{R}.$$

In order to simplify the discussion, we neglect local errors (which we consider separately), and assume that all calculations in the current cycle are exact. Furthermore, we neglect errors in the orthogonality relations such as $QQ^* = $ Id. This is permissible, since the update of orthogonal matrices by means of Givens transformations or similar techniques is known to be numerically very stable. A more thorough account of errors in the orthogonality relations is given by Georg & Hettich (1987). We now calculate

$$Q_0 Q \Big(\tilde{A} + \Delta_{\tilde{A}} \Big) PP_0 - \tilde{R}$$

$$= Q_0 \Big(R + (Qa - RP^*e)\, e^*P \Big) P_0$$

$$= Q_0 \Big(Q(A + \Delta_A)P + Q(a - Q^*RP^*e)\, e^*P \Big) P_0$$

$$= Q_0 \Big(Q(A + \Delta_A)P + Q(a - (A + \Delta_A)e)\, e^*P \Big) P_0$$

$$= Q_0 Q \Big(\tilde{A} + \Delta_A(\text{Id} - ee^*) \Big) PP_0 ,$$

and again obtain the propagation error

(16.3.7) $$\Delta_{\tilde{A}} = \Delta_A(\mathrm{Id} - ee^*).$$

Hence the implementation (16.3.6) is very safe, since it is locally stable and globally stable and self correcting. However, the computational cost is considerably higher than in the implementation (16.2.2). More than half of this cost is paid for keeping an update of the orthogonal matrix Q. Therefore Gill & Murray (1973) proposed not to store Q. Indeed, if $QAP = R$ is a QR factorization of A, then $R^*R = P^*A^*AP$ is a Cholesky factorization, and the equation $Ax = y$ can be solved via $PR^*RP^*x = y$ i.e. via a forward and a backward solving and some renumbering of the co-ordinates. In order to obtain an implementation which does not make explicit use of Q, we note that $QAP = R$ implies $QA^{*\,-1}P = (QAP)^{*\,-1} = R^{*\,-1}$ and hence $Q = R^{*\,-1}P^*A^*$. This last expression is substituted for Q in (16.3.6) to obtain a new implementation:

(16.3.8) QR Update Without Storing Q.   *comment:*
    input A;   *an initial $N \times N$-matrix*
    generate Q, P, R *initial factorization $P^*A^*AP = R^*R$*
      such that $QAP = R$;
    repeat
      enter $a, e, y \in \mathbf{R}^N$ such that $\|e\| = 1$;   *new data*
      generate Q_0, P_0 such that   *update formula*

$$\tilde{R} := Q_0\Big(R + R^{*\,-1}P^*A^*(a - Ae)\,e^*P \Big)P_0 \text{ is upper triangular;}$$

$$\tilde{P} := PP_0; \quad \tilde{A} := A + (a - Ae)e^*; \quad \text{new factorization } \tilde{P}^*\tilde{A}^*\tilde{A}\tilde{P} = \tilde{R}^*\tilde{R}$$

$$x := \tilde{P}\tilde{R}^{-1}\tilde{R}^{*\,-1}P^*A^*y; \qquad\qquad\qquad \text{linear equation}$$

$$P := \tilde{P}; \quad R := \tilde{R}; \quad A := \tilde{A}; \qquad\qquad \text{prepare for next cycle}$$

    until cycles are stopped.

The local stability of this implementation is usually quite satisfactory. Let us assume that $P^*A^*AP = R^*R$ represents the numerically exact Cholesky factorization at the beginning of some cycle. Since orthogonal factorizations are involved, at the end of the cycle we can roughly estimate the local error $\Delta_{\tilde{A}}$ by

$$\frac{\|\Delta_{\tilde{A}}\|}{\|\tilde{A}\|} \leq \varepsilon_{\mathrm{tol}}.$$

We note, however, that we solve the linear equation $A^*Ax = A^*y$, and hence the typical role of the condition number $\mathrm{cond}(A)$ is replaced by its square:

$$\frac{\|\Delta_x\|}{\|x\|} \leq \mathrm{cond}^2(A)\frac{\|\Delta_A\|}{\|A\|}.$$

Hence, the above implementation is not advisable if the condition number of the linear systems is expected to be large.

To investigate the global stability, we again assume that we begin a cycle with some error in the factorization:

$$(16.3.9) \qquad\qquad Q(A + \Delta_A)P = R.$$

The question is, how this error is propagated through the current cycle:

$$Q_0 Q(\tilde{A} + \Delta_{\tilde{A}})PP_0 = \tilde{R}.$$

We neglect local errors and assume that all calculations in the current cycle are exact. Furthermore, we neglect errors in the orthogonality relations such as $QQ^* = \mathrm{Id}$. From (16.3.9) we obtain

$$R^{*\,-1} = Q(A + \Delta_A)^{*\,-1}P$$

and hence

$$R^{*\,-1}P^* = Q(A + \Delta_A)^{*\,-1},$$

which is used in the following calculation:

$$
\begin{aligned}
Q_0 Q(\tilde{A} + \Delta_{\tilde{A}})PP_0 \\
= \tilde{R} \\
= Q_0\Big(R + R^{*\,-1}P^*A^*(a - Ae)\,e^*P \Big)P_0 \\
= Q_0\Big(Q(A + \Delta_A)P + Q(A + \Delta_A)^{*\,-1}A^*(a - Ae)\,e^*P \Big)P_0 \\
= Q_0 Q\Big(A + \Delta_A + (A + \Delta_A)^{*\,-1}A^*(a - Ae)\,e^* \Big)PP_0 .
\end{aligned}
$$

Now we use the estimate

$$(A + \Delta_A)^{*\,-1} = \mathrm{Id} - A^{*\,-1}\Delta_A^* A^{*\,-1} + O(\|\Delta_A\|^2)$$

and obtain

$$\tilde{A} + \Delta_{\tilde{A}} = A + \Delta_A + (a - Ae)\,e^* + A^{*\,-1}\Delta_A^*(a - Ae)\,e^* + O(\|\Delta_A\|^2),$$

which implies

$$(16.3.10) \qquad \Delta_A - \Delta_A + A^{*\,-1}\Delta_A^*(u - Ae)\,e^* + O(\|\Delta_A\|^2).$$

The first and last term of the above propagation error equation are harmless, but the second term may give rise to problems if

$$\|A^{*\,-1}\Delta_A^*(a - Ae)\,e^*\| \geq C\,\|\Delta_A\|$$

262 16. Update Methods and their Numerical Stability

for some $C > 1$. Even if this term does not increase dramatically, an occasional modest factor $C > 1$ in some of the cycles may accumulate so that after a performance of several cycles the global instability of the method may have dramatic dimensions. Such effects are shown in numerical experiments given by Georg & Hettich (1987). It is interesting to note that the factor $(a - Ae)$ may help prevent global instability. In fact, if Newton's method is implemented with a quasi-Newton update such as Broyden's formula (7.1.7), see also theorem (7.1.8), then an update algorithm in the sense of (16.1.3) is performed where $\|(a - Ae)\|$ will be small for most cycles. Let us finally note that it is possible to develop other implementations of algorithm (16.1.3) which use some orthogonal factorization method. However, all these methods are globally unstable if they do not store the orthogonal factor Q.

16.4 LU Factorization

The first efficient and numerically stable implementation of update methods such as (16.1.3) was given by Bartels & Golub (1968–69). They proposed a certain update of LU factorizations. Since at least row pivotings have to be performed in order to make the method locally stable, it turns out that the L-matrix cannot be kept lower triangular and becomes full. We now give a brief account of the main ideas which are somewhat analogous to those of section 16.3. Many variations are possible.

Let L be a $N \times N$-matrix, P a permutation matrix and U an upper triangular matrix such that

(16.4.1) $$LAP = U.$$

It follows that the linear equation $Ax = y$ can be solved by $x := PU^{-1}Ly$, i.e. by a matrix multiplication $z := Ly$, a forward solving of $Uw = z$ and a rearranging $x := Pw$ of the co-ordinates. In order to obtain a similar factorization of the update $\tilde{A}$ in a numerically efficient way, we multiply equation (16.1.2) from the left with L and from the right with P and use (16.4.1):

$$L\tilde{A}P = U + (La - UP^*e)e^*P.$$

Instead of applying some Givens transformations to this situation, we now obtain a new factorization of $\tilde{A}$ in a computationally efficient way by performing some elementary row operations as in the Gauss decomposition method, but we incorporate certain row pivotings to increase the stability. This is best described by means of a pseudo-code:

(16.4.2) General LU Step.   *comment:*

$u := La - UP^*e;\quad v^* := e^*P;$ *initialization*

$\tilde{L} := L;\quad \tilde{U} := U;\quad \tilde{P} := P;$

for $i = N - 1$ **downto** 1 **do**

    if $|u[i]| + |u[i+1]| > 0$ **then**   *else: no operations are necessary*

    begin

        if $|u[i]| < |u[i+1]|$ **then**   *row pivots*

        swap $\left(e_i\tilde{L},\, e_i\tilde{U},\, u[i]\right) \leftrightarrow \left(e_{i+1}\tilde{L},\, e_{i+1}\tilde{U},\, u[i+1]\right);$

        $s := \dfrac{u[i+1]}{u[i]};$ *pivot element*

        $e_{i+1}^*\tilde{R} := e_{i+1}^*\tilde{R} - s\,e_i^*\tilde{R};$ *elementary row operation*

        $e_{i+1}^*\tilde{L} := e_{i+1}^*\tilde{L} - s\,e_i^*\tilde{L};$

        $u[i+1] := 0;$

    end;

$e_1^*\tilde{R} := e_1^*\tilde{R} + u[1]v^*;$ *now $\tilde{L}\tilde{A}\tilde{P} = \tilde{U}$ has Hessenberg form*

for $i = 1$ **to** $N - 1$ **do**

    if $|\tilde{R}[i,i]| + |\tilde{R}[i+1,i]| > 0$ **then**   *else: no operations are necessary*

    begin

        if $|\tilde{R}[i,i]| < |\tilde{R}[i+1,i]|$ **then**   *row pivots*

        swap $\left(e_i\tilde{L},\, e_i\tilde{U}\right) \leftrightarrow \left(e_{i+1}\tilde{L},\, e_{i+1}\tilde{U}\right);$

        $s := \dfrac{\tilde{R}[i+1,i]}{\tilde{R}[i,i]};$ *pivot element*

        $e_{i+1}^*\tilde{R} := e_{i+1}^*\tilde{R} - s\,e_i^*\tilde{R};$ *elementary row operation*

        $e_{i+1}^*\tilde{L} := e_{i+1}^*\tilde{L} - s\,e_i^*\tilde{L};$

    end.   *now $\tilde{L}\tilde{A}\tilde{P} = \tilde{U}$ has upper triangular form*

As in section 16.3, the permutation Γ only plays a crucial role in the special case that the vector e coincides with some unit base vector. Then (16.4.2) simplifies to the following

(16.4.3) Special LU Step.   *comment:*

   let $m \in \{1, 2, \ldots, N\}$ be such that $P^*e = e_m$;   *initialization*

   $P_0 := (e_1, \ldots, e_{m-1}, e_{m+1}, \ldots, e_N, e_m)$;

   $\tilde{P} := PP_0$;   *new permutation matrix*

   $\tilde{L} := L$;

   $\tilde{U} := \left(U + (La - UP^*e)\, e^*P \right) P_0$; $\tilde{L}\tilde{A}\tilde{P} = \tilde{U}$ *has Hessenberg form*

   for $i = m$ **to** $N - 1$ **do**

      if $|\tilde{R}[i, i]| + |\tilde{R}[i + 1, i]| > 0$ **then**   *else: no operations are necessary*

         begin

         if $|\tilde{R}[i, i]| < |\tilde{R}[i + 1, i]|$ **then**   *row pivots*

            swap $\left(e_i\tilde{L},\ e_i\tilde{U} \right) \leftrightarrow \left(e_{i+1}\tilde{L},\ e_{i+1}\tilde{U} \right)$;

         $s := \dfrac{\tilde{R}[i + 1, i]}{\tilde{R}[i, i]}$;   *pivot element*

         $e_{i+1}^*\tilde{R} := e_{i+1}^*\tilde{R} - s\, e_i^*\tilde{R}$;   *elementary row operation*

         $e_{i+1}^*\tilde{L} := e_{i+1}^*\tilde{L} - s\, e_i^*\tilde{L}$;

         end.   now $\tilde{L}\tilde{A}\tilde{P} = \tilde{U}$ *has upper triangular form*

By using the techniques (16.4.2–3), an implementation of algorithm (16.1.3) can now be described in the following way:

(16.4.4) QL Update.   *comment:*

   input A;   *an initial $N \times N$-matrix*

   generate L, P, U such that $LAP = U$;   *initial factorization*

   repeat

      enter $a, e, y \in \mathbf{R}^N$ such that $\|e\| = 1$;   *new data*

      generate L_0, P_0 such that   *update formula*

      $\tilde{U} := L_0\left(U + (La - UP^*e)\, e^*P \right) P_0$ is upper triangular;

      $\tilde{L} := L_0 L$; $\tilde{P} := PP_0$; $\tilde{L}\tilde{A}\tilde{P} = \tilde{U}$ *is the new factorization*

      $x := \tilde{P}\tilde{U}^{-1}\tilde{L}y$;   *linear equation*

      $L := \tilde{L}$; $P := \tilde{P}$; $U := \tilde{U}$;   *prepare for next cycle*

   until cycles are stopped.

Again, the computational cost amounts to an initial $O(N^3)$ flops and an additional $O(N^2)$ flops per cycle. However, we emphasize that the elementary row operations are considerably less expensive than the Givens transformations or some similar technique discussed in section 16.3. Let us now investigate the stability of this implementation. Gauss decomposition techniques employing row pivots are generally accepted as being "sufficiently" stable for practical purposes, see e.g. Golub & Van Loan (1983). For the same reason, we propose to accept the above implementation as being locally "sufficiently" stable, see also Bartels & Golub (1968–69), Bartels (1971), Bartels & Stoer & Zengler (1971), Gill & Murray & Wright (1981), Powell (1977, 1985).

The global stability needs some discussion along the same lines as in section 16.3. We assume that we begin a cycle already with some error in the factorization:

$$L(A + \Delta_A)P = U .$$

Again the question is, how this error is propagated through the current cycle:

$$\tilde{L}(\tilde{A} + \Delta_{\tilde{A}})\tilde{P} = \tilde{U} .$$

We neglect local errors (which we consider separately), and assume that all calculations in the current cycle are exact. We have

$$L_0 L\left(\tilde{A} + \Delta_{\tilde{A}}\right) PP_0 = \tilde{U}$$

$$= L_0 \left(U + (La - UP^*e)\, e^*P\right) P_0$$

$$= L_0 \left(L(A + \Delta_A)P + L(a - L^{-1}UP^*e)\, e^*P\right) P_0$$

$$= L_0 L\left(A + \Delta_A + (a - (A + \Delta_A)e)\, e^*\right) PP_0$$

$$= L_0 L\left(\tilde{A} + \Delta_A(\mathrm{Id} - ee^*)\right) PP_0 ,$$

and again obtain the propagation error

$$(16.4.5) \qquad\qquad \Delta_{\tilde{A}} = \Delta_A(\mathrm{Id} - ee^*) .$$

Hence, the implementation (16.4.5) is locally "sufficiently" stable and globally stable and self correcting. The computational cost is considerably lower than in the implementation (16.3.6).

Program 1. A Simple PC Continuation Method

We present a very simple version of a PC continuation method which traces a path c as characterized by the defining initial value problem (2.1.9). It uses an Euler predictor step, then evaluates the Jacobian at the predicted point and performs Newton steps as a corrector back to the curve. Hence only one calculation and decomposition of the Jacobian is needed per predictor-corrector step. A steplength control strives to remain below a maximal contraction factor in the Newton step and a maximal distance to the curve in the spirit of (6.1.10). The algorithm stops at a point at which the function $f(x) := x[N+1] - 1$ vanishes. This is achieved by switching the steplength control over to Newton steps for the arclength, see (9.2.3). A simple test is incorporated to check whether the condition number of the linear systems becomes too bad.

The following test example is furnished: a homotopy $H : \mathbf{R}^N \times \mathbf{R} \to \mathbf{R}^N$ is defined by $H(z, \lambda) := z - \lambda f(z)$ where the ith coordinate of $f(z)$ is given by

(P1.1)
$$f(z)[i] := \exp\left(\cos\left(i \sum_{k=1}^{N} z[k]\right)\right).$$

The algorithm starts at $(z, \lambda) = (0, 0)$ and stops at $(z, \lambda) = (\bar{z}, 1)$ such that $\bar{z}$ is a fixed point of f. We first sketch the essential features of the algorithm:

Sketch of Program 1.   *comment:*

> **input**
>> **begin**
>> $x \in H^{-1}(0)$;   *initial point*
>> $h > 0$;   *initial steplength*
>> $\mathtt{hmin} > 0$;   *minimal stepsize*
>> $\mathtt{tol} > 0$;   *tolerance for corrector loop*
>> $1 > \mathtt{ctmax} > 0$;   *maximal contraction factor*
>> $\mathtt{dmax} > 0$;   *maximal distance to curve*
>> **end input**;

$A := H'(x);$ a QR decomposition of the Jacobian is performed

$t := t(A);$ tangent

newton := false;

label 93;

repeat

    $u := x + ht;$ predictor step

    $A := H'(u);$ a QR decomposition of A is performed

    repeat for iter $= 1, 2, \ldots$ begin corrector loop

      dist $:= \|A^+ H(u)\|;$ length of corrector step

      $u := u - A^+ H(u);$ corrector step

      if iter > 1 **then** contr $:= \dfrac{\text{dist}}{\text{disto}};$ contraction factor

      disto := dist;   save distance

      if dist $>$ dmax **or** contr $>$ ctmax

      then reduce h and **goto 93;**   PC step not accepted

    until dist $<$ tol;   end corrector loop

    $x := u;$ new point approximately on curve

    $t := t(A);$ new tangent

    if $x[N + 1] > 1$ **then** newton := true;   switch to Newton's steplength

    if newton **then**

      $h := -\dfrac{x[N + 1] - 1}{t[N + 1]}$ Newton's steplength, see (9.2.3)

    else adapt h w.r.t. ctmax, dmax;   according to (6.1.10)

until $|h| <$ hmin.   stopping criterion

The following is the complete FORTRAN program listing.

```
program cont
```
                          continuation method, follows a curve H(u) = 0

                          Euler predictor, Newton-correctors

                          stepsize control by asymptotic estimates

                          Jacobian is evaluated only at predictor point

                          stepsize is monitored by two different values:

                          1. contraction rate in corrector steps

                          2. distance to the curve

                          stops at a point x such that x(n1) = 0

                                        arrays:

```
parameter(n = 10, n1 = n+1)
dimension b(n1,n)
dimension q(n1,n1)
dimension x(n1), u(n1)
dimension t(n1)
dimension y(n)
logical succ, newt
```
                          dimension of the problem

                          transpose of Jacobian

                          orth. matrix for QR dec. of b

                          current points on the curve

                          tangent vector

                          stores values y := H(x)

```
                                                             parameters:
        tol = 1.e-4                        tolerance for corrector iteration
        ctmax = 0.6                   maximal contr. rate in corrector step
        dmax = .4                              maximal distance to curve
        hmax = 1.                                     maximal stepsize
        hmin = 1.e-5                                   minimal stepsize
        maxjac = 1000             maximal number of Jacobian evaluations
        cdmax = 1.e3                    maximum for condition estimate
        fmax = 2.                        maximal factor for acceleration
        h = .03                                       initial stepsize
        eta = .1                     perturbation to avoid cancellation
                                     when calculating the contraction rate


                                                          main program
        open(1, file='cont.dat')                            output file
        call stpnt(x, n1)           user defined starting point, H(x) = 0
        newt = .false.
        mapct = 0                           counts the calls of the map H
        jacct = 0                      counts the calls of the Jacobian H'
        call jacob(b, x, n, n1)          b := transpose of Jacobian at x
        jacct = jacct + 1
        call decomp(b, q, cond, n, n1)      b, q := orthog. decomp. of b
        if (cond .gt. cdmax) then
          write(1,*) ' bad cond. estimate in init. point = ', cond
          write(*,*) ' bad cond. estimate in init. point = ', cond
          stop
          endif
        do 91 k = 1, n1                                          tangent
          t(k) = q(n1, k)
91      continue
        call setor(or, t, n1)                             set orientation

12      continue                                        begin PC loop
        if (abs(h).lt.hmin) then
          write(1,*) ' failure at minimal stepsize'
          write(*,*) ' failure at minimal stepsize'
          stop
          endif
        if (jacct .gt. maxjac) then
          write(*,*) ' maximal number of Jacobian eval. exceeded'
          write(1,*) ' maximal number of Jacobian eval. exceeded'
          stop
          endif
        do 92 k = 1, n1
          u(k) = x(k) + h * or * t(k)                     predictor step
92      continue
        fac = 1./ fmax                 initialize deceleration factor
        call jacob(b, u, n, n1)          b := transpose of Jacobian at u
        jacct = jacct + 1
        call decomp(b, q, cond, n, n1)                      decompose b
        if (cond .gt. cdmax) goto 21
        iter = 0                        counts the corrector iterations
93      iter = iter + 1                         begin corrector loop
          call map(u, y, n, n1)
          mapct = mapct + 1
          call newton(q, b, u, y, dist, n, n1)
          if (dist.gt.dmax) goto 21
```

```
         fac = max(fac, sqrt(dist/dmax)*fmax)
         if (iter.ge.2) then
           contr = dist / (disto + tol*eta)            contraction rate
           if (contr.gt.ctmax) goto 21
           fac = max(fac, sqrt(contr/ctmax)*fmax)
         endif
         if (dist.lt.tol) goto 22                   corrector successful
         disto = dist
         goto 93                                     end corrector loop

21       h = h / fmax                                  PC not accepted
         goto 12

22       continue                                     PC step accepted
         succ = .false.
         if (u(n1).ge.1.) newt = .true.       switch to Newton steplength
         if (newt) then
           h = - (u(n1) - 1.) / q(n1, n1)
           if (abs(h).lt.hmin) succ = .true.        solution point found
         else
           if (fac.gt.fmax) fac = fmax
           h = min(abs(h/fac), hmax)                steplength adaptation
           if (h.gt.hmax) h = hmax
         endif
         do 94 k = 1, n1
           x(k) = u(k)                               new point on curve
           t(k) = q(n1, k)                               new tangent
94       continue
         if (succ) then                         stopping the curve tracing
           write(1,*) ' success with', mapct,' calls of "map" and',
     *     jacct, ' calls of "jacob"'
           write(*,*) ' success with', mapct,' calls of "map" and',
     *     jacct, ' calls of "jacob"'
           write(1,*)
           write(*,*)
           write(1,*) ' solution vector:'
           write(*,*) ' solution vector:'
           write(1,*) ' ================'
           write(*,*) ' ================'
           do 95 k = 1, n
             write(1,*) ' x(', k, ') = ', x(k)
             write(*,*) ' x(', k, ') = ', x(k)
95         continue
           stop
         endif
         goto 12
         end

         subroutine map(x, y, n, n1)                      user defined
                                             input: x output: y = H(x)
                                     H(x) = 0 defines the curve to be traced
         dimension x(n1), y(n)
         s = 0.0
         do 91 i = 1, n
           s = s + x(i)
91       continue
         do 92 i = 1, n
           y(i) = x(i) - x(n1) * exp(cos(i * s))
```

```
92      continue
        return
        end

        subroutine jacob(b, x, n, n1)                        user defined
                                                        input: x output: b
                                          evaluates the transpose b of the Jacobian at x
        dimension b(n1,n), x(n1)
        s = 0.0
        do 91 i = 1, n
          s = s + x(i)
91      continue
        do 92 k = 1, n1
          do 93 i = 1, n
            if (k.eq.n1) then
              b(k, i) = -exp(cos(i * s))
            elseif (i.eq.k) then
                b(k,i)=1.+x(n1)*exp(cos(i*s))*sin(i*s)*i
            else
                b(k,i) = x(n1)*exp(cos(i*s))*sin(i*s)*i
            endif
93        continue
92      continue
        return
        end

        subroutine stpnt(x, n1)                              user defined
                                             output: x = starting point on curve
        dimension x(n1)
        do 91 k = 1, n1
          x(k) = 0.0
91      continue
        return
        end

        subroutine setor(or, t, n1)                          user defined
                                                    input: t output: or(t)
                                 decides in which direction the curve will be traversed
        dimension t(n1)
        if (t(n1).gt.0.) then
          or = 1.0
        else
          or = -1.0
        endif
        return
        end

        subroutine givens(b, q, c1, c2, l1, l2, l3, n, n1)
                                                  input: b, q, c1, c2, l1, l2, l3
                                                     output: b, q, c1, c2
                                           one Givens rotation is performed —
                                                  on rows l1 and l2 of b and q
                               the rotation maps c1, c2 onto sqrt(c1**2+c2**2), 0
        dimension b(n1, n), q(n1, n1)
        if (abs(c1)+abs(c2) .eq. 0.) return
        if (abs(c2) .ge. abs(c1)) then
          sn = sqrt(1. + (c1/c2)**2) * abs(c2)
        else
```

```
      sn = sqrt(1. + (c2/c1)**2) * abs(c1)
   endif
   s1 = c1/sn
   s2 = c2/sn
   do 91 k = 1, n1
     sv1 = q(l1, k)
     sv2 = q(l2, k)
     q(l1, k) =  s1 * sv1 + s2 * sv2
     q(l2, k) = -s2 * sv1 + s1 * sv2
91  continue
   do 92 k = 13, n
     sv1 = b(l1, k)
     sv2 = b(l2, k)
     b(l1, k) =  s1 * sv1 + s2 * sv2
     b(l2, k) = -s2 * sv1 + s1 * sv2
92  continue
   c1 = sn
   c2 = 0.0
   return
   end

   subroutine decomp(b, q, cond, n, n1)
```
input: b, output: b, q, cond
a QR decomposition for b is stored in q, b —
by using Givens rotations on b and q = id —
until b is upper triangular
a very coarse condition estimate cond is provided
```
   dimension b(n1, n), q(n1, n1)                    start with q := id
   do 91 k = 1, n1
     do 92 l = 1, n1
       q(k, l) = 0.0
92    continue
     q(k, k) = 1.0
91  continue
   do 93 m = 1, n                          successive Givens transformations
     do 94 k = m+1, n1
       call givens(b, q, b(m, m), b(k, m), m, k, m+1, n, n1)
94    continue
93  continue
   cond = 0.                               very coarse condition estimate
   do 95 i = 2, n
     do 96 k = 1, i - 1
       cond = max(cond, abs(b(k,i)/b(i,i)))
96    continue
95  continue
   return
   end

   subroutine newton(q, b, u, y, d, n, n1)
```
input q, b, u, y = H(u), n, n1
output u, d
y is changed
a Newton step u := u − A+H(u) is performed —
where A approximates the current Jacobian H'
q, b = QR decomposition of A*
d = length of Newton step
```
   dimension q(n1, n1), b(n1, n), u(n1), y(n)
   do 91 k = 1, n
```

```
          do 92 l = 1, k-1
            y(k) = y(k) - b(l, k) * y(l)
92        continue
          y(k) = y(k) / b(k, k)
91      continue
        d = 0.
        do 93 k = 1, n1
          s = 0.0
          do 94 l = 1, n
            s = s + q(l, k) * y(l)
94        continue
          u(k) = u(k) - s
          d = d + s**2
93      continue
        d = sqrt(d)
        return
        end
```

A run of the above program gave the following results:

```
success with 900 calls of 'map' and 280 calls of 'jacob'

solution vector:
================
x( 1) =   1.492
x( 2) =   .5067
x( 3) =   .3890
x( 4) =   .9273
x( 5) =   2.420
x( 6) =   2.187
x( 7) =   .7729
x( 8) =   .3721
x( 9) =   .5866
x( 10) =  1.754
```

Program 2. A PL Homotopy Method

We now present a PL homotopy method in the sense of Eaves & Saigal, see section 13.4. The algorithm is applied to the problem $\min_{x \in \mathbf{R}^2} \varphi(x)$, where

(P2.1) $\qquad \varphi : \mathbf{R}^2 \to \mathbf{R}, \quad \varphi(x_1, x_2) := 100\left(x_2 - x_1^3\right)^2 - \left(1 - x_1\right)^2,$

which we take from the paper of Himmelblau (1972). The problem obviously has the unique solution $x_1 = 1$, $x_2 = 1$. The interesting point of this example is that the standard optimization algorithms perform slowly since the graph of the function displays a steep valley which decreases only gradually. Essentially the same behavior is also shown by the PL method applied to the map

(P2.2) $\qquad H : \mathbf{R}^2 \times \mathbf{R} \to \mathbf{R}^2, \quad H(x, \lambda) = \begin{cases} x - x_0 & \text{for } \lambda \leq 0, \\ \nabla \varphi(x) & \text{for } \lambda > 0. \end{cases}$

Occasionally Newton steps are tempted as described in section 13.5, and automatic pivots in the sense of section 13.6 are performed to save LP steps. It is interesting to see how the algorithm runs up and down the bisection levels until it comes so near to the solution that Newton iterations are finally successful, see figure P2.a.

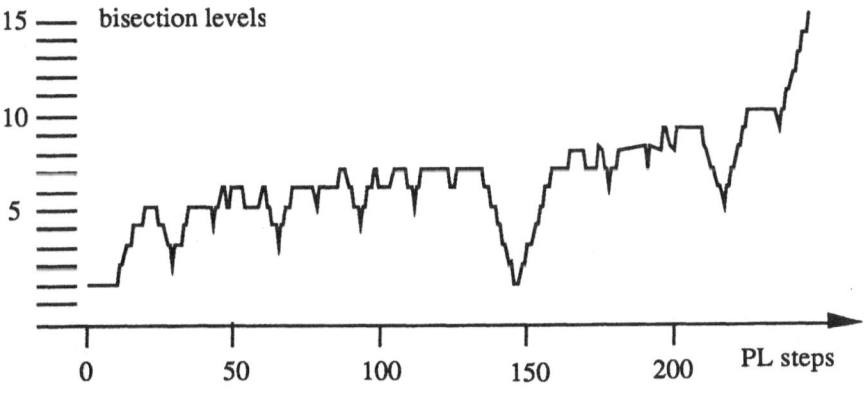

Figure P2.a Running up and down the bisection levels

It should be noted that single precision (real*4) is barely capable of handling the bad conditioning of the problem. We first sketch the essential features of the algorithm:

Sketch of Program 2.   *comment:*

> **input**
>> **begin**
>>> $v_1, \ldots, v_{N+1} \in \mathbf{R}^N$;   *vertices of a starting simplex*
>>>
>>> $x_0 \in \mathbf{R}^N$;   *starting point*
>>>
>>> bis $\in \mathbf{N}$ *maximal bisection level allowed*
>>
>> **end input**;
>
> define an affine image of the triangulation J_3 of $\mathbf{R}^N \times \mathbf{R}$
>> which uses the above starting simplex   *triangulation*
>>
>> and which uses x_0 as a barycenter;
>
> $k_2 = 0$ *pivoting index*
>
> **repeat**   *begin of PL loop*
>> find a new k_2;   *door-in-door-out step*
>>
>> pivot k_2;   *pivoting step*
>>
>> **if** a new bisection level is encountered
>>> **then** try some Newton steps;
>>
>> **if** Newton steps were successful **then** stop;
>
> **until** level = bis is encountered.

The following is the complete FORTRAN program listing.

```
program plhom              Piecewise linear homotopy method
                      in the sense of Eaves and Saigal, see section 13.4
                    with automatic pivoting steps and tentative Newton steps
                        the condition of the labeling matrix is tested

integer bis, i1, i2, k2, count, maxct, k, n, n1, level
real stol, kappa, cdmax, newtl, ferr
parameter(n = 2, n1 = n+1)          dimension of the problem
parameter(ferr = 1.0e-6)          tolerance, used for stopping
parameter(stol = 1.e-4)              ≈ sqrt(machine tolerance)
parameter(cdmax = 1./stol)        maximum for condition estimate
parameter(bis = 18)                maximal number of bisections
parameter(kappa = 0.5)             contr. factor for Newton steps
parameter(maxct = 400)              maximal number of steps
real d(0:n1)                       level of vertices (stepsize)
real v(n, 0:n1)                          vertices of simplex
integer l(0:n1), r(0:n1)           permutations for vertices
integer a(0:n1)                    axis from v(.,i) to v(.,i+1)
real z(n)                          center of virtual simplex
real x(0:n), w(0:n)                points , x(0) = 2.**(-level)
```

```
        real x0(n)                                    starting point
        real y(n)                            current value y= f(x)
        real c(0:n1)                                         column
        real q(0:n1, 0:n1)                        orthogonal matrix
        real b(0:n1, 0:n)                   upper triangular matrix
        logical newl                            new level traversed ?
        logical succ                     success of Newton iterations ?

        i1 = 0                                 first higher level index
        i2 = 0                                  last higher level index
        k2 = 0                           index of vertex being pivoted
        count = 0                 counts number of function evaluations
        newtl = 0.5                      last level for Newton steps
        open(1, file='plhom.dat')                        output file
        call load(v,x0,x,y,z,a,d,r,l,b,q,cdmax,count,n,n1)

1       continue                                   start of PL loop
        call index(k2,stol,q,n,n1)            find new pivoting index k2
        call pivot(k2,newl,d,v,l,r,a,z,x,i1,i2,n,n1)
        level = nint(-alog(abs(d(i1)))/alog(2.))
        write(1,'(i6,4x,"level=",i3)') count, level
        write(*,'(i6,4x,"level=",i3)') count, level
        if (newl .and. (newtl .gt. x(0))) then
          newtl = x(0)                             tentative Newton steps
          call newton(x,x0,y,w,c,cdmax,ferr,kappa,count,
     *    v,k2,q,b,n,n1,succ)
          if (succ) then
            write(1,'(6x,a)') 'Newton iterations succeeded'
            write(*,'(6x,a)') 'Newton iterations succeeded'
            goto 35
          else
            write(1,'(6x,a)') 'Newton iterations did not succeed'
            write(*,'(6x,a)') 'Newton iterations did not succeed'
          endif
          do 91 k = 1, n
            x(k) = v(k, k2)
91        continue
        endif
        if (level .gt. bis) then
          write(1,'(6x,a)') 'maximal bisection level exceeded'
          write(*,'(6x,a)') 'maximal bisection level exceeded'
          goto 35
        endif
        call label(x, x0, y, n)
        count = count + 1
        if (count .gt. maxct) then
          write(1,'(6x,a)') 'maximal number of PL steps exceeded'
          write(*,'(6x,a)') 'maximal number of PL steps exceeded'
          goto 35
        endif
        call update(b,q,y,w,cdmax,k2,n,n1)
        goto 1                                        end of PL loop

35      continue                                    best solution found
        write(1,'(6x,a,i6/)') 'number of label evaluations:',count
        write(*,'(6x,a,i6/)') 'number of label evaluations:',count
        write(1,'(6x,a)') 'approximate solution found:'
        write(*,'(6x,a)') 'approximate solution found:'
```

```
write(1,'(6x,a,i2,a,e16.8)')  ('x(', k, ')=', x(k), k=1,n)
write(*,'(6x,a,i2,a,e16.8)')  ('x(', k, ')=', x(k), k=1,n)
end

subroutine stpnt(x0, n)                                    user defined
                     output: x = starting point for homotopy method

real x0(n)
integer n
x0(1) = -1.2
x0(2) = 1.0
return
end

subroutine stsim(v, n, n1)                                 user defined
                                    output: v = starting simplex

real v(n, 0:n1)
integer n, n1, k, m
do 91 k = 1, n
  v(k, 1) = 1.0
91    continue
do 92 m = 2, n1
  do 93 k = 1, n
    v(k, m) = v(k, m - 1)
93      continue
  v(m - 1, m) = 0.0
92    continue
return
end

subroutine label(x, x0, y, n)                              user defined
                            input: x output: y = label of x

real x(0:n), x0(n), y(n), x12, x13
integer n, k, level
level = nint(-alog(x(0))/alog(2.))
if (level .gt. 0) then                     label = f (interesting level)
  x12 = x(1) * x(1)
  x13 = x12 * x(1)
  y(1) = -600.0 * (x(2) - x13) * x12 - 2.0 * (1.0 - x(1))
  y(2) = 200.0 * (x(2) - x13)
else                                    label on the trivial level
  do 91 k = 1, n
  y(k) = x(k) - x0(k)
91    continue
endif
return
end

subroutine givens(b, q, c1, c2, l1, l2, l3, n, n1)
                              input: b, q, c1, c2, l1, l2, l3
                              output: b, q, c1, c2
                      one Givens rotation is performed —
                          on rows l1 and l2 of b and q
               the rotation maps c1, c2 onto sqrt(c1**2+c2**2), 0

real b(0:n1, 0:n),q(0:n1,0:n1),sn,s1,s2,c1,c2,sv1,sv2
```

```
      integer l1, l2, l3, n, n1, k
      if (abs(c1)+abs(c2) .eq. 0.) return
      if (abs(c2) .ge. abs(c1)) then
        sn = sqrt(1. + (c1/c2)**2) * abs(c2)
      else
        sn = sqrt(1. + (c2/c1)**2) * abs(c1)
      endif
      s1 = c1/sn
      s2 = c2/sn
      do 91 k = 0, n1
        sv1 = q(l1, k)
        sv2 = q(l2, k)
        q(l1, k) =  s1 * sv1 + s2 * sv2
        q(l2, k) = -s2 * sv1 + s1 * sv2
91    continue
      do 92 k = l3, n
        sv1 = b(l1, k)
        sv2 = b(l2, k)
        b(l1, k) =  s1 * sv1 + s2 * sv2
        b(l2, k) = -s2 * sv1 + s1 * sv2
92    continue
      c1 = sn
      c2 = 0.0
      return
      end

      subroutine testcd(b, cdmax, n, n1)
```
test of condition —
a very coarse estimate

```
      real b(0:n1, 0:n), cdmax
      integer n, n1, i, k
      do 91 i = 1, n
        do 92 k = 0, i - 1
          if (abs(b(k,i)) .gt. cdmax*abs(b(i, i))) then
            write(1,'(6x,a)') 'bad cond. estimate'
            write(*,'(6x,a)') 'bad cond. estimate'
            stop
          endif
92      continue
91    continue
      return
      end

      subroutine load(v,x0,x,y,z,a,d,r,l,b,q,cdmax,count,n,n1)
      real v(n,0:n1), x0(n), y(n), z(n), d(0:n1), q(0:n1,0:n1),
     *   b(0:n1,0:n), x(0:n), cdmax

      integer a(0:n1), l(0:n1), r(0:n1), n, n1, k, m, count
      call stsim(v, n, n1)
      call stpnt(x0, n)
      do 81 k = 1, n
        y(k) = 0.0
        v(k,0) = 0.5 * (v(k,1) + v(k,n1))
        do 82 m = 1, n1
          y(k) = y(k) + v(k,m)
82      continue
        y(k) = y(k) / real(n1)
81    continue
```
first new vertex

barycenter of starting simplex

```
      do 83 k = 1, n                              shifting barycenter into x0
        do 91 m = 0, n1
          v(k, m) = v(k, m) - y(k) + x0(k)
91        continue
83    continue
      do 92 k = 1, n
        z(k) = 0.5                                load virtual simplex
        a(k) = k
92    continue
      do 93 k = 1, n1
        d(k) = -1.0
93    continue
      d(0) = 0.5
      do 94 k = 0, n1
        r(k) = k + 1
        l(k) = k - 1
94    continue
      l(0) = n1
      r(n1) = 0
      do 95 m = 0, n1                             loading b and q
        b(m, n) = 1.0
        x(0) = abs(d(m))
          do 96 k = 0, n1
            q(k, m) = 0.0
96        continue
        q(m, m) = 1.0
        do 97 k = 1, n
          x(k) = v(k, m)
97      continue
        call label(x, x0, y, n)
        count = count + 1
        do 98 k = 1, n
          b(m, k - 1) = y(k)
98      continue
95    continue
      do 88 m = 0, n
        do 89 k = m + 1, n1
          call givens(b, q, b(m, m), b(k, m), m, k, m+1, n, n1)
89      continue
88    continue
      call testcd(b, cdmax, n, n1)
      return
      end

      subroutine index(k2,stol,q,n,n1)           find new pivoting index k2

      real q(0:n1, 0:n1), s, stol
      integer k2, n, n1, k
      if (q(n1, k2).gt. 0) then
        do 91 k = 0, n1
          q(n1, k) = -q(n1, k)
91      continue
      endif
      s = 1.e20
      k2 = -1
      do 92 k = 0, n1
        if (q(n1, k) .gt. stol) then
          if (q(n, k) / q(n1, k) .lt. s) then
```

```
            s = q(n, k) / q(n1, k)
            k2 = k
          endif
        endif
92    continue
      if (k2 .eq. -1) then
        write(1,'(6x,a)') 'instability: no index found'
        write(*,'(6x,a)') 'instability: no index found'
        stop
      endif
      return
      end

      subroutine update(b,q,y,w,cdmax,k2,n,n1)
                      the decomposition q transpose(A) = b is updated —
                        for the case that A(k2,1:n) is replaced by y
                                                      see section 16.3

      real w(0:n), y(n), q(0:n1, 0:n1), b(0:n1, 0:n), cdmax
      integer k2, n, n1, k, l
      do 91 k = 1, n
        w(k - 1) = y(k)
91    continue
      w(n) = 1.0
      do 92 k = 0, n
        do 82 l = 0, k
          w(k) = w(k) - b(l, k) * q(l, k2)
82      continue
92    continue
      do 93 k = n, 0, -1
        call givens(b, q, q(k,k2), q(k+1,k2), k, k+1, k, n, n1)
93    continue
      do 94 k = 0, n1                                        correction of q
        q(0, k) = 0.0
94    continue
      q(0, k2) = 1.0
      do 95 k = 0, n
        b(0, k) = b(0, k) + w(k)
95    continue
      do 96 k = 0, n
        call givens(b, q, b(k,k), b(k+1, k), k, k+1, k+1, n, n1)
96    continue
      call testcd(b, cdmax, n, n1)
      return
      end

      function even(a)                          test if the real number a —
                                                    is near an even number

      real a, b
      logical even
      b = abs(a) / 2.0
      if (b - aint(b) .lt. 0.25) then
        even = .true.
      else
        even = .false.
      endif
      return
      end
```

```
      subroutine pivot(k2,newl,d,v,l,r,a,z,x,i1,i2,n,n1)
                                    performs all necessary updates —
                          for pivoting vertex k2 of the current simplex —
                                     with respect to the triangulation J³
                                 indicates whether a new level is traversed
                                 performs automatic pivots, see section 13.6 —
                             in case that the traversed level has height ≤ 0.25

      real d(0:n1), v(n, 0:n1), z(n), x(0:n), s
      integer l(0:n1), r(0:n1), a(0:n1), k2, i1, i2, n, n1, k1,
    * k3, k, i0
      integer pivcase                 cases for pivoting, ordering as in (13.3.7)

      logical newl
      newl = .false.
77    continue                            entry point for automatic pivot
      k1 = l(k2)
      k3 = r(k2)
      if (k2 .eq. i1) then
        if (i1 .ne. i2) then
          pivcase = 2
        else
          pivcase = 5
        endif
      else
        if (k2 .eq. i2) then
          if (even((z(a(k1)) + d(k1)) / d(k3))) then
            pivcase = 7
          else
            pivcase = 8
          endif
        else
          if (k1 .eq. i2) then
            if (k3 .ne. i1) then
              pivcase = 3
            else
              pivcase = 6
            endif
          else
            if (k3 .eq. i1) then
              pivcase = 4
            else
              pivcase = 1
            endif
          endif
        endif
      endif
      goto (1,2,3,4,5,6,7,8) pivcase

1     k = a(k2)
      a(k2) = a(k1)
      a(k1) = k
      s = d(k2)
      d(k2) = d(k1)
      d(k1) = s
      do 91 k = 1, n
        v(k, k2) = v(k, k1) + v(k, k3) - v(k, k2)
```

```
91      continue
        goto 66

2       z(a(k2)) = z(a(k2)) + 2.0 * d(k2)
        d(k2) = -d(k2)
        do 92 k = 1, n
          v(k, k2) = 2.0 * v(k, k3) - v(k, k2)
92      continue
        goto 66

3       i2 = k2
        d(k1) = d(k2) * 0.5
        d(k2) = d(k1)
        a(k1) = a(k2)
        do 93 k = 1, n
          v(k, k2) = v(k, k1) + 0.5 * (v(k, k3) - v(k, k2))
93      continue
        if ((k3 .eq. l(i1)) .and. (abs(d(k2)) .le. 0.25)) then
          r(k1) = k3                              automatic pivot
          l(k3) = k1
          r(k3) = k2
          l(k2) = k3
          r(k2) = i1
          l(i1) = k2
          d(k2) = d(k3)
          d(k3) = d(i1)
          i2 = k3
          goto 77
        endif
        goto 66

4       a(i2) = a(k1)
        d(i2) = -d(k1) * 0.5
        d(k2) = d(i2)
        do 97 k = 1, n
          v(k, k2) = v(k, i2) + 0.5 * (v(k, k1) - v(k, k2))
97      continue
        i3 = r(i2)
        r(k2) = i3
        l(i3) = k2
        r(i2) = k2
        l(k2) = i2
        r(k1) = k3
        l(k3) = k1
        i2 = k2
        if ((r(k2) .eq. k1) .and.(abs(d(k2)) .le. 0.25)) then
          i2 = l(k2)                              automatic pivot
          r(i2) = k1
          l(k1) = i2
          r(k1) = k2
          l(k2) = k1
          r(k2) = k3
          l(k3) = k2
          i2 = k1
          d(k2) = d(k1)
          d(k1) = d(i1)
          goto 77
        endif
```

```
      goto 66

5     i1 = k3
      i2 = k1
      d(k2) = d(k2) * 4.0
      do 87 k = 1, n
        v(k, k2) = v(k, k1)
87    continue
      i0 = l(i1)
      do 94 k = 0, n1
        if ((k .ne. i2) .and. (k .ne. i0))
     *    z(a(k)) = z(a(k)) - 0.5 * d(k)
94    continue
      if (abs(d(k2)) .le. 0.5) then                        automatic pivot
        i2 = l(k1)
        r(i2) = k2
        l(k2) = i2
        r(k2) = k1
        l(k1) = k2
        r(k1) = k3
        l(k3) = k1
        i2 = k2
        d(k1) = d(k2)
        d(k2) = d(k3)
        goto 77
      endif
      goto 66

6     i1 = k2
      i2 = k2
      d(k2) = d(k2) * 0.25
      newl = .true.
      do 95 k = 1, n
        v(k, k2) = 0.5 * (v(k, k1) + v(k, k3))
95    continue
      i0 = l(i1)
      do 96 k = 0, n1
        if ((k .ne. i2) .and. (k .ne. i0))
     *    z(a(k)) = z(a(k)) + 0.5 * d(k)
96    continue
      goto 66

7     a(k2) = a(k1)
      d(k2) = d(k1) * 2.0
      i2 = k1
      do 98 k = 1, n
        v(k, k2) = v(k, k3) + 2.0 * (v(k, k1) - v(k, k2))
98    continue
      goto 66

8     r(k1) = k3
      l(k3) = k1
      k3 = l(i1)
      do 99 k = 1, n
        v(k, k2) = v(k, k3) + 2.0 * (v(k, k1) - v(k, k2))
99    continue
      r(k3) = k2
      l(k2) = k3
```

```
          r(k2) = i1
          l(i1) = k2
          d(k3) = -d(k1) * 2.0
          d(k2) = d(k3)
          a(k3) = a(k1)
          i2 = k1
          goto 66
```

```
66        continue                                    end of pivoting cases
          do 89 k = 1, n
            x(k) = v(k, k2)
89        continue
          x(0) = abs(d(k2))
          return
          end
```

```
          subroutine newton(x,x0,y,w,c,cdmax,ferr,kappa,count,
     *       v,k2,q,b,n,n1,succ)
```
                                        tentative Newton steps w.r.t. barycentric co-ordinates
                                        see section 13.5

```
          real v(n,0:n1),x(0:n),y(n),q(0:n1,0:n1),b(0:n1,0:n),c(0:n1),
     *       x0(n),w(0:n), cdmax, s, y1, y2, ferr, kappa
          integer count, k2, n, n1, l, k
          logical succ
          succ = .false.
          s = q(n, k2) / q(n1, k2)                      first Newton step
          do 91 l = 0, n1
            c(l) = (q(n, l) - s * q(n1, l)) / b(n, n)
91        continue
          do 92 k = 1, n
            x(k) = 0.0
            do 93 l = 0, n1
              if (l .ne. k2) x(k) = x(k) + c(l) * v(k, l)
93          continue
92        continue
          call label(x, x0, y, n)
          count = count + 1
          y2 = 0.0
          do 94 k = 1, n
            y2 = y2 + abs(y(k))
94        continue
          write(1,'(6x,a,e16.8)') 'norm(f)=', y2
          write(*,'(6x,a,e16.8)') 'norm(f)=', y2
```

```
77        continue                         begin loop of successive Newton steps
          call update(b,q,y,w,cdmax,k2,n,n1)
          y1 = y2
          s = (1.0 - q(n, k2) / b(n, n)) / q(n1, k2)
          do 96 l = 0, n1
            c(l) = (q(n, l) / b(n, n) + s * q(n1, l))
96        continue
          do 97 k = 1, n
            do 98 l = 0, n1
              if (l .ne. k2) x(k) = x(k) + c(l) * v(k, l)
98          continue
97        continue
          call label(x, x0, y, n)
```

```
      count = count + 1
      y2 = 0.0
      do 99 k = 1, n
        y2 = y2 + abs(y(k))
99    continue
      write(1,'(6x,a,e16.8)') 'norm(f)=', y2
      write(*,'(6x,a,e16.8)') 'norm(f)=', y2
      if (y2 .lt. ferr) then
        succ = .true.
        return
      endif
      if (y2 .gt. kappa * y1) then
        return
      else
        goto 77
      endif
      end
```

A run of the above program gave the following result. We print only the last part of the output file:

```
249      level=  9
250      level= 10
   norm(f)=   0.50691070E-02
   norm(f)=   0.30294531E-02
   Newton iterations did not succeed
253      level= 10
254      level= 10
255      level= 10
256      level= 10
257      level= 10
258      level= 10
259      level= 10
260      level= 10
261      level= 10
262      level=  9
263      level= 10
264      level= 10
265      level= 11
    norm(f)=   0.12469822E-03
    norm(f)=   0.36960933E-04
    norm(f)=   0.14268670E-03
    Newton iterations did not succeed
269      level= 11
270      level= 12
    norm(f)=   0.11098384E-03
    norm(f)=   0.48157180E-04
    norm(f)=   0.21099754E-03
    Newton iterations did not succeed
274      level= 12
275      level= 12
276      level= 13
    norm(f)=   0.25033182E-03
    norm(f)=   0.38767040E-03
    Newton iterations did not succeed
279      level= 13
280      level= 13
```

```
281     level= 14
     norm(f)=  0.23651090E-03
     norm(f)=  0.19121162E-03
     Newton iterations did not succeed
284     level= 14
285     level= 15
     norm(f)=  0.23746473E-03
     norm(f)=  0.47683720E-04
     norm(f)=  0.00000000E+00
     Newton iterations succeeded
     number of label evaluations:    288

     approximate solution found:
     x( 1)=  0.10000000E+01
     x( 2)=  0.10000000E+01
```

The PL homotopy algorithms depend very much on the choice of the affine image of the triangulation J_3. The affine map is automatically given by the user defined starting point x_0 and the user defined starting simplex in $\mathbf{R}^N$. This is particularly interesting to see for the example (P1.1) which in this context leads to the following homotopy

$$\textbf{(P2.3)} \qquad H : \mathbf{R}^N \times \mathbf{R} \to \mathbf{R}^N, \quad H(x,\lambda) = \begin{cases} x & \text{for } \lambda \le 0, \\ x - f(x) & \text{for } \lambda > 0. \end{cases}$$

The program changes at the following places:

`parameter(n = 6, n1 = n+1)`	*dimension of the problem*
`parameter(ferr = 1.0e-4)`	*tol. norm(f), used for stopping*
`parameter(stol = 1.e-3)`	$\approx$ *sqrt(machine tolerance)*
`parameter(cdmax = 1./stol)`	*maximum for condition estimate*
`parameter(bis = 10)`	*maximal number of bisections*
`parameter(kappa = 0.5)`	*contr. factor for Newton steps*
`parameter(maxct = 2000)`	*maximal number of steps*

```
**************************************************************
     subroutine label(x, x0, y, n)                user defined
                                          input: x output: y = label of x

     real x(0:n), x0(n), y(n), s
     integer n, k
     level = nint(-alog(x(0))/alog(2.))
     if (level .gt. 0) then                 label = f (interesting level)
       s = 0.0
       do 91 k = 1, n
         s = s + x(k)
91     continue
       do 92 k = 1, n
         y(k) = x(k) - exp(cos(k * s))
92     continue
     else                                   label on the trivial level
       do 93 k = 1, n
         y(k) = x(k) - x0(k)
93     continue
     endif
     return
```

```
          end
**********************************************************************
          subroutine stpnt(x0, n)                              user defined
                                        output: x = starting point for homotopy method

          real x0(n)
          integer n, i
          do 91 i = 1, n
            x0(i) = 0.0
91        continue
          return
          end
```

With the starting simplex

```
          subroutine stsim(v, n, n1)                           user defined
                                                output: v = starting simplex

          real v(n, 0:n1)
          integer n, n1, k, m
          do 91 k = 1, n
            v(k, 1) = 10.0
91        continue
          do 92 m = 2, n1
            do 93 k = 1, n
              v(k, m) = v(k, m - 1)
93          continue
            v(m - 1, m) = 0.0
92        continue
          return
          end
```

we obtain the following output (only the last lines are given):

```
   453      level=  8
   454      level=  8
   455      level=  9
     norm(f)=  0.20663770E-01
     norm(f)=  0.43382050E-02
     norm(f)=  0.98764900E-03
     norm(f)=  0.21252040E-03
     norm(f)=  0.79303980E-04
     Newton iterations succeeded
     number of label evaluations:    460

     approximate solution found:
     x( 1)=  0.24149892E+01
     x( 2)=  0.17415463E+01
     x( 3)=  0.11014134E+01
     x( 4)=  0.68083550E+00
     x( 5)=  0.46093451E+00
     x( 6)=  0.37482151E+00
```

However, if we change the starting simplex into a different geometrical form

```
      subroutine stsim(v, n, n1)                      user defined
                                              output: v = starting simplex

      real v(n, 0:n1)
      integer n, n1, k, m
      do 91 k = 1, n
        v(k, 1) = 0.0
91    continue
      do 92 m = 2, n1
        do 93 k = 1, n
          if (k+1 .eq. m) then
            v(k, m) = 10.0
          else
            v(k, m) = 0.0
          endif
93      continue
92    continue
      return
      end
```

then the performance of the algorithm changes drastically as can be seen by the following output:

```
64      level=  3
65      level=  4
  norm(f)=  0.16743760E+01
  norm(f)=  0.26990700E+00
  norm(f)=  0.54594490E-01
  norm(f)=  0.12029620E-01
  norm(f)=  0.27115050E-02
  norm(f)=  0.66077710E-03
  norm(f)=  0.15437602E-03
  norm(f)=  0.33348800E-04
  Newton iterations succeeded
  number of label evaluations:    73

  approximate solution found:
  x( 1)=  0.13212870E+01
  x( 2)=  0.42966222E+00
  x( 3)=  0.47269183E+00
  x( 4)=  0.15330100E+01
  x( 5)=  0.26841723E+01
  x( 6)=  0.11308190E+01
```

Not only is a different solution found, but the algorithm performs much faster.

Program 3. A Simple Euler-Newton Update Method

The following is a modification of program 1 in which the Jacobian $A \approx H'$ of the homotopy is calculated only once at the starting point (via difference approximations), and all subsequent approximations $A \approx H'$ along the curve are obtained by using Broyden's update formulas, see (7.2.13). A steplength adaptation insures an acceptable performance of the Newton corrector step. A simple modification is incorporated to keep the condition number of the linear systems under control. The same numerical example as in program 1 is used so that the two algorithms can be directly compared.

Sketch of Program 3.   *comment:*

    input

      begin

      $x \in H^{-1}(0)$;   *initial point*

      $h > 0$;   *initial steplength*

      $\texttt{hmin} > 0$;   *minimal stepsize*

      end input;

    $\texttt{newton} := \texttt{false}$;

    $A :\approx H'(x)$;   *difference approximation —*

                         a QR decomposition of A is performed

    $t := t(A)$;   *tangent*

    repeat

    label 12;

      $u := x + ht$;   *predictor step*

      update A on x, u;   *a QR decomposition is updated*

      if angle test is negative **then** reduce h and **goto 12**;

      generate a perturbation vector pv;

      $v := u - A^{+}(H(u) - \text{pv})$;   *corrector step*

      update A on u, v;   *a QR decomposition is updated*

      if residual or contraction test is negative **then** reduce h and **goto 12**;

$x := v;$ *new point approximately on curve*

$t := t(A);$ *new tangent*

if $x[N+1] > 1$ **then** newton := true;   *switch to Newton's steplength*

if newton then

$$h := -\frac{x[N+1]-1}{t[N+1]}$$ *Newton's steplength, see (9.2.3)*

else increase h;

until $|h| <$ hmin.   *stopping criterion*

The following is the complete FORTRAN program listing.

```
program contup
```
                          continuation method
                          follows a curve H(u) = 0
           one Euler predictor, one Newton-corrector
       Broyden update after each step, see chapter 7
           stops at a point x such that x(n1) = 0

                                   arrays:

```
parameter(n = 10,  n1 = n+1)
```
             dimension of the problem
```
parameter(pi = 3.1415926535898)
dimension b(n1,n)
```
             transpose of Jacobian
```
dimension q(n1,n1)
```
       orth. matrix for QR dec. of b
```
dimension x(n1),  u(n1),  v(n1)
```
         current points on the curve
```
dimension t(n1)
```
                 tangent vector
```
dimension y(n),w(n),p(n),pv(n),r(n)
```
           values of the map H
```
logical test, succ, newton
```

                               parameters:

```
ctmax = .8
```
      maximal contr. rate in corrector step
```
dmax = .2
```
               maximal norm for H
```
dmin = .001
```
               minimal norm for H
```
pert = .00001
```
              perturbation of H
```
hmax = 1.28
```
              maximal stepsize
```
hmin = .000001
```
              minimal stepsize
```
hmn = .00001
```
      minimal Newton step size
```
h = .32
```
               initial stepsize
```
cdmax = 1000.
```
      maximum for condition estimate
```
angmax = pi/3.
```
              maximal angle
```
maxstp = 9000
```
    maximal number of evaluations of H
```
acfac = 2.
```
   acceleration factor for steplength control

                             main program

```
open(1, file='contup.dat')
```
                   output file
```
call stpnt(x, n1)
```
    user defined starting point, H(x) = 0
```
newton = .false.
mapct = 0
```
          counts the calls of the map H
```
call jac(b, x, y, h, n, n1)
```
                 $b = H'(x)^*$
```
mapct = mapct + 1 + n1
call decomp(b, q, cond, n, n1)
```
       b, q := orthog. decomp. of b
```
if (cond .gt. cdmax) then
  write(1,*) ' bad cond. estimate in init. point = ', cond
  write(*,*) ' bad cond. estimate in init. point = ', cond
  stop
endif
```

```
         do 90 k = 1, n1                                    tangent saved
           t(k) = q(n1, k)
90       continue
         call setor(or, t, n1)                              set orientation

12       continue                                           begin PC loop
         if (abs(h).lt.hmin) then
           write(1,*) ' failure at minimal stepsize'
           write(*,*) ' failure at minimal stepsize'
           stop
         endif
         if (mapct .gt. maxstp) then
           write(*,*) ' maximal number of function eval. exceeded'
           write(1,*) ' maximal number of function eval. exceeded'
           stop
         endif
         do 83 k = 1, n1                                    tangent saved
           t(k) = q(n1, k)
83       continue
         do 92 k = 1, n1
           u(k) = x(k) + h * or * t(k)                      predictor step
92       continue
         call map(u, w, n, n1)
         mapct = mapct + 1
         call upd(q,b,x,u,y,w,t,h,angmax,test,n,n1)         predictor update
         if (test .eq. .false.) goto 21                     angle test is neg.
         call newt(q,b,u,v,w,p,pv,r,pert,dmax,dmin,
     *   ctmax,cdmax,test,n,n1)                  Newton corrector and update
         mapct = mapct + 1
         if (test.eq..false.) goto 21           residual or contr. test is neg.
         goto 22

21       h = h / acfac                                      PC not accepted
         goto 12

22       continue                                           PC step accepted
         succ = .false.
         if (v(n1).ge.1.) newton = .true.        switch to Newton steplength
         if (newton) then
           h = - (v(n1) - 1.) / q(n1, n1)
           if (abs(h).lt.hmn) succ = .true.                 solution point found
         else
           h = abs(h) * acfac                               steplength adaptation
           if (h.gt.hmax) h = hmax
         endif
         do 94 k = 1, n1
           x(k) = v(k)                                      new point on curve
94       continue
         do 95 k = 1, n
           y(k) = r(k)                                      y = H(x)
95       continue
         if (succ) then                           stopping the curve tracing
           write(1,*) ' success with', mapct,' calls of "map"'
           write(*,*) ' success with', mapct,' calls of "map"'
           write(1,*)
           write(*,*)
           write(1,*) ' solution vector:'
           write(*,*) ' solution vector:'
```

```
        write(1,*) ' ================'
        write(*,*) ' ================'
        do 96 k = 1, n
          write(1,*) ' x(', k, ') = ', x(k)
          write(*,*) ' x(', k, ') = ', x(k)
96      continue
        stop
      endif
      goto 12
      end

      subroutine map(x, y, n, n1)                        user defined
                                              input: x output: y = H(x)
                                      H(x) = 0 defines the curve to be traced
      dimension x(n1), y(n)
      s = 0.
      do 91 i = 1, n
        s = s + x(i)
91    continue
      do 92 i = 1, n
        y(i) = x(i) - x(n1) * exp(cos(i * s))
92    continue
      return
      end

      subroutine jac(b, x, y, h, n, n1)            input: x output: b
                                      evaluates the transpose b of the Jacobian at x
                                              by using forward differences
      dimension b(n1,n), x(n1), y(n)
      do 91 i = 1, n1
        x(i) = x(i) + h
        call map(x, y, n, n1)
        x(i) = x(i) - h
        do 92 k = 1, n
          b(i,k) = y(k)
92      continue
91    continue
      call map(x, y, n, n1)
      do 93 i = 1, n1
        do 94 k = 1, n
          b(i,k) = (b(i,k) - y(k)) / h
94      continue
93    continue
      return
      end

      subroutine stpnt(x, n1)                            user defined
                                        output: x = starting point on curve
      dimension x(n1)
      do 91 k = 1, n1
        x(k) = 0.
91    continue
      return
      end

      subroutine setor(or, t, n1)                        user defined
                                              input: t output: or(t)
                              decides in which direction the curve will be traversed
```

```
      dimension t(n1)
      if (t(n1).gt.0.) then
        or = 1.0
      else
        or = -1.0
      endif
      return
      end

      subroutine givens(b, q, c1, c2, l1, l2, l3, n, n1)
```
input: b, q, c1, c2, l1, l2, l3
output: b, q, c1, c2
one Givens rotation is performed —
on rows l1 and l2 of b and q
the rotation maps c1, c2 onto sqrt(c1**2+c2**2), 0
```
      dimension b(n1, n), q(n1, n1)
      if (abs(c1)+abs(c2) .eq. 0.) return
      if (abs(c2) .ge. abs(c1)) then
        sn = sqrt(1. + (c1/c2)**2) * abs(c2)
      else
        sn = sqrt(1. + (c2/c1)**2) * abs(c1)
      endif
      s1 = c1/sn
      s2 = c2/sn
      do 91 k = 1, n1
        sv1 = q(l1, k)
        sv2 = q(l2, k)
        q(l1, k) =  s1 * sv1 + s2 * sv2
        q(l2, k) = -s2 * sv1 + s1 * sv2
91    continue
      do 92 k = l3, n
        sv1 = b(l1, k)
        sv2 = b(l2, k)
        b(l1, k) =  s1 * sv1 + s2 * sv2
        b(l2, k) = -s2 * sv1 + s1 * sv2
92    continue
      c1 = sn
      c2 = 0.
      return
      end

      subroutine decomp(b, q, cond, n, n1)
```
input: b output: b, q, cond
a QR decomposition for b is stored in q, b —
by using Givens rotations on b and q = id —
until b is upper triangular
a very coarse condition estimate cond is provided
```
      dimension b(n1, n), q(n1, n1)
      do 91 k = 1, n1                          start with q := id
        do 92 l = 1, n1
          q(k, l) = 0.
92      continue
        q(k, k) = 1.0
91    continue
      do 93 m = 1, n                  successive Givens transformations
        do 94 k = m+1, n1
          call givens(b, q, b(m, m), b(k, m), m, k, m+1, n, n1)
94      continue
```

```
93    continue
      cond = 0.                                    very coarse condition estimate
      do 95 i = 2, n
        do 96 k = 1, i - 1
          cond = max(cond, abs(b(k,i)/b(i,i)))
96      continue
95    continue
      return
      end

      subroutine newt(q,b,u,v,w,p,pv,r,pert,dmax,dmin,
     *  ctmax,cdmax,test,n,n1)
```

$$\text{input } q, \, b, \, u, \, w = H(u)$$
$$\text{output } v, \, test, \, r = H(v)$$
$$w \text{ is changed}$$
$$\text{one Newton step } v := u - A^+ \, w \text{ is performed}$$
$$\text{where } A \approx H'$$
$$q, \, b = QR \text{ decomposition of } A^*$$
$$q, \, b \text{ are updated}$$
$$\text{perturbations are used for stabilization}$$
$$\text{residual and contraction tests are performed}$$

```
      dimension q(n1,n1),b(n1,n),u(n1),v(n1),w(n),pv(n),p(n),r(n)
      logical test
      test = .true.
      do 81 k = 1, n                                        perturbation
        if (abs(w(k)) .gt. pert) then
          pv(k) = 0.
        else if (w(k) .gt. 0.) then
          pv(k) = w(k) - pert
        else
          pv(k) = w(k) + pert
        endif
        w(k) = w(k) - pv(k)
81    continue
      d1 = ynorm(w, n)
      if (d1 .gt. dmax) then
        test = .false.
        return
      endif
      do 91 k = 1, n
        do 92 l = 1, k-1
          w(k) = w(k) - b(l, k) * w(l)
92      continue
        w(k) = w(k) / b(k, k)
91    continue
      d2 = ynorm(w, n)
      do 93 k = 1, n1
        s = 0.
        do 94 l = 1, n
          s = s + q(l, k) * w(l)
94      continue
        v(k) = u(k) - s
93    continue
      call map(v, r, n, n1)
      do 74 k = 1, n
        p(k) = r(k) - pv(k)
74    continue
      d3 = ynorm(p, n)
```

```
         contr = d3 / (d1 + dmin)
         if (contr .gt. ctmax) test = .false.
         do 95 k = n-1, 1, -1
           call givens(b, q, w(k), w(k+1), k, k+1, k, n, n1)
95       continue
         do 96 k = 1, n
           b(1,k) = b(1,k) - p(k) / d2
96       continue
         do 97 k = 1, n-1
           call givens(b, q, b(k,k), b(k+1,k), k, k+1, k, n, n1)
97       continue
         if (b(n,n) .lt. 0.) then
           test = .false.
           b(n,n) = - b(n,n)
           do 82 k = 1, n1
             q(n,k) = - q(n,k)
             q(n1,k) = - q(n1,k)
82         continue
         endif
         do 85 i = 2, n                    perturbation of upper triangular matrix
           do 86 k = 1, i - 1
             if (abs(b(k,i)) .gt. cdmax * abs(b(i,i))) then
               if (b(i,i) .gt. 0.) then
                 b(i,i) = abs(b(k,i)) / cdmax
               else
                 b(i,i) = - abs(b(k,i)) / cdmax
               endif
             endif
86         continue
85       continue
         do 87 k = 1, n-1
           b(k+1,k) = 0.
87       continue
         return
         end

         subroutine upd(q,b,x,u,y,w,t,h,angmax,test,n,n1)
                         input q, b, x, u = predictor, y = H(x), w = H(u)
                         q, b = QR decomposition of transpose(H')
                         q, b are updated
                         perturbations are used for stabilization
                         an angle test is performed
         dimension q(n1,n1),b(n1,n),x(n1),u(n1),t(n1),y(n),w(n)
         logical test
         test = .true.
         pi = 3.14159265358979323846
         do 91 k = 1, n
           b(n1,k) = (w(k) - y(k)) / h
91       continue
         do 92 k = 1, n                                              update
           call givens(b, q, b(k,k), b(n1,k), k, n1, k, n, n1)
92       continue
         ang = 0.
         do 93 k = 1, n1                                             angle
           ang = ang + t(k) * q(n1, k)
93       continue
         if (ang .gt. 1.0) ang = 1.
         if (ang .lt. -1.0) ang = -1.
```

```
      ang = acos(ang)
      if (ang .gt. angmax) test = .false.
      return
      end

      function ynorm(y,n)
      dimension y(n)
      s = 0.
      do 13 k = 1, n
        s = s + y(k)**2
13    continue
      ynorm = sqrt(s)
      return
      end
```

A run of the above program gave the following results:

```
success with 2912 calls of 'map'

solution vector:
================
x( 1) =   1.492
x( 2) =   .5067
x( 3) =   .3891
x( 4) =   .9273
x( 5) =   2.420
x( 6) =   2.187
x( 7) =   .7729
x( 8) =   .3721
x( 9) =   .5866
x( 10) =   1.754
```

Program 4. A Continuation Algorithm for Handling Bifurcation

We modify an Euler-Newton method e.g. Program 1 in such a way that it has the interactive capability to detect simple bifurcation points and trace a bifurcating branch via perturbations if desired. The implementation will be similar to (8.2.4). In order to detect simple bifurcations, it is crucial that the Jacobian is reevaluated at the predictor point. We first sketch the essential features of the algorithm:

Sketch of Program 4. *comment:*

    input

        begin

        $x \in \mathbf{R}^{N+1}$;   *starting point, $H(x) \approx 0$*

        or $\in \{+1, -1\}$;   *direction in which the curve is traced*

        pert $\in \{\text{true}, \text{false}\}$;   *perturbation is on or off*

        $h > 0$;   *initial stepsize*

        arc > 0;   *arclength for countdown*

        pv $\in \mathbf{R}^{N}$;   *perturbation vector*

        dmax;   *maximal distance to curve*

        end;

    normalize: $\|\text{pv}\| = \frac{1}{2}\text{dmax}$;

    label 75;   *interactive driver*

    interactive driver monitors arc, pert, or, h and stopping;

    corrector equation is

$$\begin{cases} H(x) = 0 & \text{for } \text{pert} = \text{false}, \\ H(x) = \text{pv} & \text{for } \text{pert} = \text{true}; \end{cases}$$

    label 12;   *begin of PC loop*

        if arc < 0 **then goto 75**;   *enter new arc for countdown*

        if h is too small **then goto 75**;   *failure?*

        Newton corrector iteration on x;

        if corrector not successful **then goto 75**;   *failure?*

$u = x + \mathbf{or} \cdot h \cdot t(H'(u));$ *predictor step*

Newton corrector iteration on u;

if corrector not successful **then goto 21**;

if angle test negative **then goto 21**;

label 22;   *PC step accepted*

     $t_1 := t(H'(x)); \quad t_2 := t(H'(u));$

     $\mathbf{arc} := \mathbf{arc} - h;$ *arclength countdown*

     modify h;   *steplength control*

     $x := u;$ *new point on curve*

     if $t_1^* t_2 < 0$ **then goto 75**   *orientation changed, bifurcation?*

     else goto 12;

label 21;   *PC step not accepted*

$h := \frac{1}{2}h; \quad$ **goto 12**.

We illustrate the performance of the above algorithm by generating a bifurcation diagram of the periodic solutions of a differential delay equation

(P4.1) $$\dot{x}(t+1) = -\lambda f\big(x(t)\big).$$

The standard assumptions concerning f are:

(P4.2) Assumptions.
     (i) $f : \mathbf{R} \to \mathbf{R}$ is sufficiently smooth,
     (ii) $f(x)x > 0$ for $x \in \mathbf{R}$, $x \neq 0$,
     (iii) $f'(0) = 1$,
     (iv) $\inf_{x \in \mathbf{R}} f(x) > -\infty$.

Differential delay equations are often used for modeling population growth in natural sciences. The periodic solutions are of particular interest. Their bifurcation diagrams have been studied by Jürgens & Saupe (1979), Jürgens & Peitgen & Saupe (1980), Hadeler (1980), Peitgen (1982), Peitgen & Prüfer (1979), Saupe (1982), Saupe (1983), see also Program 6 of this appendix. We discretize the problem by using a standard numerical integration method in conjunction with interpolation via cubic splines, see Georg (1982). In program 6, a Galerkin method in conjunction with Fast Fourier Transforms is used to discretize a differential delay equation.

     Let us first show how our problem can be formulated as an operator equation in Banach space with one degree of freedom. For a given initial value function $x : [0,1] \to \mathbf{R}$, we consider the uniquely defined extension

$$x_\lambda : [0,\infty) \to \mathbf{R}$$

of x satisfying the delay equation (P4.1). We introduce the Banach space $\mathcal{X}$ of continuous functions $x : [0, 1] \to [0, 1]$ with the sup-norm

$$||x|| = sup_{0 \le t \le 1}|x(t)|.$$

For a given a period $p \ge 0$ and a given eigenvalue parameter λ we define the operator

(P4.3) $T_\lambda^p : \mathcal{X} \to \mathcal{X}, \qquad x(t) \mapsto x_\lambda(t + p), \quad 0 \le t \le 1.$

Clearly, x is a fixed point of T_λ^p if and only if x_λ is a solution of (P4.1) with period p. However, we have an undesirable degree of freedom in this formulation of the problem, since for all $a \ge 0$ the shifted function $t \mapsto x_\lambda(t + a)$ is also a solution with period p. We therefore make the additional normalization $x(0) = 0$. The above discussion shows that the problem of finding the periodic solutions of (P4.1) can be formulated as a zero problem for the following map

(P4.4)
$$\begin{aligned} H : \mathcal{X} \times \mathbf{R} \times (0, \infty) \ &\longrightarrow\ \mathcal{X} \times \mathbf{R}, \\ (x, \lambda, p)\ &\longmapsto\ \left(x - T_\lambda^p x,\ x(0)\right). \end{aligned}$$

This problem has the trivial solution $x = 0$ for all λ and p. Nussbaum (1975) showed that a nontrivial solution branch bifurcates off from this branch of trivial solutions at $\lambda = \frac{\pi}{2}$. Furthermore, a linearization of (P4.1) leads to the equation

(P4.5) $\dot{x}(t + 1) = -\lambda x(t)$

which has the solution

(P4.6) $x_0(t) := \alpha \sin\left(\frac{\pi}{2} t\right)$

where α is an arbitrary constant. This provides the tangent of the bifurcating branch at the bifurcation point, hence for small α, the function (P4.6) together with $p = 4$ and $\lambda = \frac{\pi}{2}$ furnishes a good approximation to a solution of the nonlinear equation $H(x, \lambda, p) = 0$.

To illustrate the method, we use the example

(P4.7) $f(x) = x\,\dfrac{1 + x^2}{1 + x^4}$

which has been studied by Hadeler (1980). To reduce the numerical effort, we only investigate periodic solutions which are odd with respect to half of their period:

(P4.8) $x_\lambda\left(t + \dfrac{p}{2}\right) = -x_\lambda(t).$

This leads to a slight change in the homotopy:

(P4.9)

$$H : \mathcal{X} \times \mathbf{R} \times (0, \infty) \longrightarrow \mathcal{X} \times \mathbf{R},$$

$$(x, \lambda, p) \longmapsto \left(x + T_\lambda^{\frac{p}{2}} x, \, x(0)\right).$$

We discretize $x \in \mathcal{X}$ via nine equidistant points $t_i := \frac{i}{8}$, $i = 0, \ldots, 8$ and approximate the extension x_λ by integrating (P4.1) via Simpson's Rule on the grid t_i, $i = 9, \ldots, J$ for some sufficiently large integer $J > 0$. Then we interpolate the result with free cubic splines and use this interpolation to approximate $x_\lambda(t + \frac{p}{2})$ on the grid t_i, $i = 0, \ldots, 8$. Thus we are led to a discretized problem

(P4.10) $H(x, \lambda, p) = 0,$ where $H : \mathbf{R}^9 \times \mathbf{R} \times \mathbf{R} \longrightarrow \mathbf{R}^9 \times \mathbf{R}.$

Figure P4.a is a bifurcation diagram based on the data of the run below. It is evident that at some $\lambda \in [4.71, 4.83]$ there is a secondary bifurcation point. The two branches of solutions differ by the value of the period p: on the primary branch the solution has constant period $p = 4$, on the secondary branch the period varies continuously.

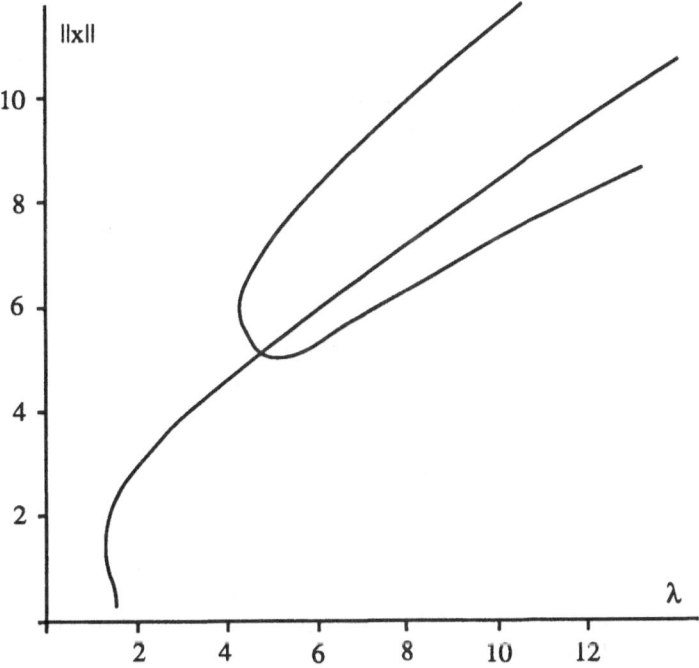

Figure P4.a   Bifurcation diagram

Usually, an additional difficulty for detecting secondary bifurcation points of discretized problems stems from the fact that a discretization often has an effect similar to a perturbation: namely, bifurcation points are destroyed in the sense of Sard's theorem (11.2.3), see the discussion of section 8.2 and in particular figure 8.2.a. Hence, the question arises why the secondary bifurcation is not destroyed in the present case. The answer is surprisingly tricky, see Georg (1982), and will be given here since it provides some insight into the numerical handling of secondary bifurcation points.

Let us consider periodic solutions x_λ of (P4.1) which in addition have the following properties:

1) x_λ has period $p = 4$,
2) x_λ is odd in the sense of (P4.8),
3) $x_\lambda(1 + t) = x_\lambda(1 - t)$ for $t \in [0, 1]$.

It is not difficult to show that the initial value $x(t) = x_\lambda(t)$ for $t \in [0, 1]$ of such a solution can also be characterized by the integral equation

$$(P4.11) \qquad x(t) = \lambda \int_{1-t}^{1} f\big(x(\xi)\big)\, d\xi \,.$$

This integral equation is of the form

$$(P4.12) \qquad x = \lambda F(x)$$

where $F : \mathcal{X} \to \mathcal{X}$ is a nonlinear compact operator with the additional property $F(0) = 0$. It has been shown that a simple eigenvalue of the linearized equation

$$(P4.13) \qquad x = \lambda F'(0)x$$

leads to a bifurcation point of the nonlinear equation, see Krasnosel'skiĭ (1964), Rabinowitz (1971). It can be shown for the nonlinearity (P4.2) that $\lambda = \frac{\pi}{2}$ is a simple eigenvalue of (P4.13). Hence, (P4.12) has a solution branch bifurcating off from the trivial solution $x = 0$ at the eigenvalue $\lambda = \frac{\pi}{2}$. A reasonable discretization method

$$(P4.14) \qquad x_h = \lambda F_h(x_h)$$

should mimic this situation, i.e. $F_h : \mathbf{R}^N \to \mathbf{R}^N$ is a nonlinear operator with the property $F_h(0) = 0$, and the Frechet derivative $F_h'(0)$ has a simple eigenvalue near $\lambda = \frac{\pi}{2}$. Since the integral equation (P4.12) admits only solutions of period $p = 4$, it does not permit the secondary branch of solutions as indicated in figure P4.a. Hence, a reasonable discretization method has a continuous branch of solutions of period $p = 4$ similar to the primary branch in figure P4.a.

Now let us consider a discretization method for the more general equation $H(x, \lambda, p) = 0$ with H as in (P4.8). For the particular case $p = 4$, this method may or may not be equivalent to a reasonable discretization method for (P4.12). In the first case, the secondary bifurcation point is not destroyed by this discretization, and in the second case the secondary bifurcation point is destroyed *with probability one*. It is easy to check that the discretization which we have described above is equivalent to a standard discretization of (P4.12) via Simpson's Rule with mesh size $h = \frac{1}{8}$, and hence the secondary bifurcation is not destroyed.

The algorithm is started by using a small α in formula (P4.6). The following is the complete FORTRAN program listing.

```
      program bif                          continuation method, follows a curve H(u) = 0
                                                    Euler predictor, Newton-correctors
                                                    stepsize control by asymptotic estimates
                                                    Jacobian is evaluated at each point
                                                    an interactive driver monitors:
                                                    orientation, perturbation, arclength countdown
                                                    in order to trace bifurcating curves
                                                    a protocol is written on bif.dat

                                                                              arrays:
      parameter(n = 10, n1 = n+1)                            dimension of the problem
      parameter(pi = 3.1415926535898)
      dimension b(n1,n)                                     transpose of Jacobian
      dimension q(n1,n1)                                orth. matrix for QR dec. of b
      dimension x(n1), u(n1)                             current points on the curve
      dimension t(n1)                                            tangent vector
      dimension y(n), pv(n)                             stores values y := H(x)
      logical pert, corr, succ

                                                                         parameters:
      tol = .0001                                   tolerance for corrector iteration
      ctmax = 0.3                             maximal contr. rate in corrector step
      dmax = .05                                       maximal distance to curve
      amax = pi/180.*30.                                        maximal angle
      hmax = 1.                                              maximal stepsize
      hmin = .0001                                           minimal stepsize
      cdmax = 1000.                            maximum for condition estimate
      fmax = 2.                                 maximal factor for acceleration
      h = .03                                                initial stepsize
      or = -1.                                            initial orientation
      pert = .false.                                     initial perturbation
      arc = 0.                                initial arclength countdown

                                                                     main program
      open(1, file='bif.dat')                                    output file
      call stpnt(x, n1)                     user defined starting point, H(x) = 0
      call setper(pv, n)                                set perturbation vectors
      save = dmax/2./xnorm(pv, n)
        do 23 k = 1, n                                  adapt perturbation to dmax
          pv(k) = pv(k)*save
23      continue
75    call driver(arc,or,pert,h)
      corr = .true.

12    continue                                               begin PC loop
      if (abs(h).lt.hmin) then
        write(*,*) 'minimal stepsize'
```

```
            write(1,*) 'minimal stepsize'
            goto 75
          endif
          if (arc.le.0.) then
            write(*,*) 'enter new arclength for countdown'
            write(1,*) 'enter new arclength for countdown'
            goto 75
          endif
          if (corr) then                              initial corrector necessary
            do 82 k = 1, n1
              u(k) = x(k)
82          continue
            call crloop(u,y,pv,b,q,fac,
     *       tol,fmax,ctmax,dmax,cdmax,pert,succ,n,n1)
            if (succ) then                            corrector loop successful?
              corr = .false.
              do 18 k = 1, n1
                x(k) = u(k)                                            new point
                t(k) = q(n1,k)                                       new tangent
18            continue
              write(*,*) '||x||=',xnorm(x,n-1),' lambda=',x(n),
     *           ' period=', x(n1), ' h=',h
              write(1,*) '||x||=',xnorm(x,n-1),' lambda=',x(n),
     *           ' period=', x(n1), ' h=',h
            else
              write(*,*) 'initial corrector loop not successful'
              write(1,*) 'initial corrector loop not successful'
              goto 75
            endif
          endif
          do 97 k = 1, n1
            u(k) = x(k) + h * or * t(k)                           predictor step
97        continue
          call crloop(u,y,pv,b,q,fac,
     *     tol,fmax,ctmax,dmax,cdmax,pert,succ,n,n1)
          if (.not.succ) goto 21
          angle = 0.
          do 24 k = 1, n1
            angle = angle + t(k)*q(n1,k)
24        continue
          sangle = sign(1.,angle)
          angle = sangle*angle
          if (angle.gt.1.) angle = 1.
          angle = acos(angle)
          if ((pert).and.(sangle.lt.0.)) goto 21
          if (angle.gt.amax) goto 21                                  angle test
          fac = max(fac, angle/amax*fmax)
          goto 22

21        h = h / fmax                                            PC not accepted
          goto 12

22        continue                                               PC step accepted
          arc = arc - abs(h)                                 arclength countdown
          if (fac.gt.fmax) fac = fmax
          h = min(abs(h/fac), hmax)                       steplength adaptation
          if (h.gt.hmax) h = hmax
          do 94 k = 1, n1
```

```
         x(k) = u(k)                                 new point on curve
         t(k) = q(n1, k)                                new tangent
94       continue
         write(*,*) '||x||=',xnorm(x,n-1),' lambda=',x(n),
     *            ' period=', x(n1), ' h=',h
         write(1,*) '||x||=',xnorm(x,n-1),' lambda=',x(n),
     *            ' period=', x(n1), ' h=',h
         if ((.not.pert).and.(sangle.lt.0.)) then
           write(*,*) 'orientation changed'
           write(1,*) 'orientation changed'
           goto 75
         endif
         goto 12
         end


         subroutine driver(arc,or,pert,h)              interactive driver
         logical pert
         realn = 0.
         write(1,*) ' '
         write(1,*) 'interactive driver'
77       continue
         write(*,*) ' 1) stop  2) go  3) arc=',arc,'  4) or=',or,
     *   ' 5) pert=',pert,'  6) h=',h
         write(*,*) 'enter integer (option) and realnumber (value)'
         read(*,*) intgr, realn
         write(1,*) 'integer=',intgr, ' real number=', realn
         if (intgr .eq. 1) stop
         if (intgr .eq. 2) goto 78
         if (intgr .eq. 3) arc = realn
         if (intgr .eq. 4) or = -or
         if (intgr .eq. 5) pert = .not.pert
         if (intgr .eq. 6) h = realn
         goto 77
78       write(1,*) 'arc=', arc, ' or=', or, ' pert=', pert, ' h=', h
         write(1,*) ' '
         end


         subroutine map(x, y, n, n1)                        user defined
                                                  input: x output: y = H(x)
                                          H(x) = 0 defines the curve to be traced
         dimension x(n1), y(n), w(200), d(200), xm(200),
     *   p(200), q(200)
         nm3 = n-3
         nm2 = n-2
         nm1 = n-1
         lim = 4*nm2 + n
         h = 1./ float(nm2)
         xind = x(n1)/ 2./ h
         ind = xind
         if ((ind.lt.1).or.(ind+n.gt.lim)) goto 2
         t = (xind - ind)*h
         r1 = -x(n)*h*5./12.
         r2 = -x(n)*h*8./12.
         r3 = x(n)*h/12.
         q1 = -x(n)*h/3.
         q2 = -x(n)*h*4./3.
         q3 = q1
         do 1 k = 1, lim
```

```
          if (k.le.nm1) then
            w(k) = x(k)
          elseif (mod(k,2).eq.0) then
            w(k) = w(k-1) + r1* xf(w(k-nm1)) + r2* xf(w(k-nm2))
   *              + r3* xf(w(k-nm3))
          else
            w(k) = w(k-2) + q1* xf(w(k-n)) + q2* xf(w(k-nm1))
   *              + q3* xf(w(k-nm2))
          endif
1         continue
          do 3 k = 2, lim - 1
            d(k) = 3./h**2 * ( w(k+1)-2.*w(k)+w(k-1) )
3         continue
          p(2) = sqrt(2.)
          d(2) = d(2)/ p(2)
          do 4 k = 3, lim - 1
            q(k-1) = .5/ p(k-1)
            p(k) = sqrt(2. - q(k-1)**2)
            d(k) = (d(k) - q(k-1)*d(k-1))/ p(k)
4         continue
          xm(lim) = 0.
          xm(lim-1) = d(lim-1)/ p(lim-1)
          do 5 k = lim - 2, 2, -1
            xm(k) = (d(k) - q(k)*xm(k+1))/ p(k)
5         continue
          xm(1) = 0.
          do 6 k = ind + 1, ind + nm1
            a7 = w(k)
            c7 = xm(k)/2.
            b7 = (w(k+1)-w(k))/h - h/6.*(2.*xm(k)+xm(k+1))
            d7 = (xm(k+1)-xm(k))/ (6.*h)
            y(k-ind) = x(k-ind) + (((d7*t)+c7)*t+b7)*t+a7
6         continue
          y(n) = x(1)
          return
2         write(*,*) 'failure in map'
          write(1,*) 'failure in map'
          stop
          end


          function xf(t)                      auxiliary function for above map
          xf = t*(1. + t**2) / (1. + t**4)
          end


          subroutine jacob(b, x, n, n1)                        user defined
                                              input: x output: b
                              evaluates the transpose b of the Jacobian at x
          dimension b(n1,n), x(n1), y(30), w(30)
          h1 = 1024.
          h = 1./h1
          call map(x, y, n, n1)
          do 1 k = 1, n1
            x(k) = x(k) + h
            call map(x, w, n, n1)
            x(k) = x(k) - h
            do 2 l = 1, n
              b(k,l) = h1*(w(l)-y(l))
2           continue
```

```
1       continue
        end

        subroutine stpnt (x, n1)                            user defined
                                              output: x = starting point on curve
        parameter(pi = 3.1415926535898)
        dimension x(n1)
        h = 1./float(n1-3)
        do 1 k=1, n1-2
          tk = (k-1)*h*pi/2.
          x(k) = .1* sin(tk)
1       continue
        x(n1-1) = pi/2.
        x(n1) = 4.
        end

        subroutine setper (pv, n)                           user defined
                                              defines the perturbation vector pv
        dimension pv(n)
        do 1 k = 1, n-1
          pv(k) = 1.
1       continue
        pv(n) = 0.
        end

        subroutine func (x, y, pv, pert, n, n1)
                                              perturbed function evaluation
        dimension x(n1), y(n), pv(n)
        logical pert
        call map(x, y, n, n1)
        if (pert) then
          do 31 k = 1, n
            y(k) = y(k) - pv(k)
31        continue
        endif
        end

        subroutine givens (b, q, c1, c2, l1, l2, l3, n, n1)
                                         input: b, q, c1, c2, l1, l2, l3
                                         output: b, q, c1, c2
                                      one Givens rotation is performed —
                                         on rows l1 and l2 of b and q
                         the rotation maps c1, c2 onto sqrt(c1**2+c2**2), 0
        dimension b(n1, n), q(n1, n1)
        if (abs(c1)+abs(c2) .eq. 0.) return
        if (abs(c2) .ge. abs(c1)) then
          sn = sqrt(1. + (c1/c2)**2) * abs(c2)
        else
          sn = sqrt(1. + (c2/c1)**2) * abs(c1)
        endif
        s1 = c1/sn
        s2 = c2/sn
        do 91 k = 1, n1
          sv1 = q(l1, k)
          sv2 = q(l2, k)
          q(l1, k) =  s1 * sv1 + s2 * sv2
          q(l2, k) = -s2 * sv1 + s1 * sv2
91      continue
```

```
        do 92 k = 13, n
          sv1 = b(11, k)
          sv2 = b(12, k)
          b(11, k) =  s1 * sv1 + s2 * sv2
          b(12, k) = -s2 * sv1 + s1 * sv2
92      continue
        c1 = sn
        c2 = 0.0
        end


        subroutine decomp(b, q, cond, n, n1)
```
                                    input: b, output: b, q, cond
                          a QR decomposition for b is stored in q, b —
                          by using Givens rotations on b and q = id —
                                         until b is upper triangular
                       a very coarse condition estimate cond is provided
```
        dimension b(n1, n), q(n1, n1)
        do 91 k = 1, n1                              start with q := id
          do 92 l = 1, n1
            q(k, l) = 0.0
92        continue
          q(k, k) = 1.0
91      continue
        do 93 m = 1, n                  successive Givens transformations
          do 94 k = m+1, n1
            call givens(b, q, b(m, m), b(k, m), m, k, m+1, n, n1)
94        continue
93      continue
        cond = 0.                       very coarse condition estimate
        do 95 i = 2, n
          do 96 k = 1, i - 1
            cond = max(cond, abs(b(k,i)/b(i,i)))
96        continue
95      continue
        end


        subroutine newton(q, b, u, y, n, n1)
```
                                    input q, b, u, y = H(u), n, n1
                                                       output u
                                                   y is changed
                       a Newton step u := u − A⁺(H(u) − pv) is performed —
                       where A approximates the current Jacobian H'
                                      q, b = QR decomposition of A*
```
        dimension q(n1, n1), b(n1, n), u(n1), y(n)
        do 91 k = 1, n
          do 92 l = 1, k-1
            y(k) = y(k) - b(l, k) * y(l)
92        continue
          y(k) = y(k) / b(k, k)
91      continue
        do 93 k = 1, n1
          s = 0.0
          do 94 l = 1, n
            s = s + q(l, k) * y(l)
94        continue
          u(k) = u(k) - s
93      continue
        end
```

```
       subroutine crloop(x,y,pv,b,q,fac,
     *   tol,fmax,ctmax,dmax,cdmax,pert,succ,n,n1)
```
corrector loop
input x,y
output x,y,b,q,fac,succ

```
       dimension x(n1),y(n),pv(n),b(n1,n),q(n1,n1)
       logical succ, pert
       succ = .false.
       fac = 1./fmax
       call func(x, y, pv, pert, n, n1)
35     continue
       call jacob(b, x, n, n1)
       call decomp(b, q, cond, n, n1)
       if (cond .gt. cdmax) return
       dist1 = xnorm(y,n)
       fac = max(fac, sqrt(dist1/dmax)*fmax)
       if (dist1.lt.tol) goto 34
       if (dist1.gt.dmax) return
       call newton(q, b, x, y, n, n1)
       call func(x, y, pv, pert, n, n1)
       dist2 = xnorm(y,n)
       contr = dist2 / (dist1 + tol)
       fac = max(fac, sqrt(contr/ctmax)*fmax)
       if (contr.gt.ctmax) return
       dist1 = dist2
       goto 35
34     succ = .true.
       end
```

success of corrector loop

begin loop
b := transpose of Jacobian at x
decompose b
bad conditioning

corrector successful

contraction rate

end loop
corrector successful

```
       function xnorm(y, n)
       dimension y(n)
       x = 0.
       do 1 k = 1, n
       x = x + y(k)**2
1      continue
       xnorm = sqrt(x)
       end
```
calculates euclidean norm of y

The bifurcation diagram in figure P4.a was generated by the following protocol of an interactive session:

```
interactive driver
integer= 3 real number= 20.00
integer= 2 real number= 20.00
arc= 20.00 or= -1.000 pert= F h= 3.0000E-02

||x||= .2103 lambda= 1.560 period= 4.000 h= 3.0000E-02
||x||= .2401 lambda= 1.556 period= 4.000 h= 6.0000E-02
||x||= .2995 lambda= 1.548 period= 4.000 h= .1200
||x||= .4177 lambda= 1.528 period= 4.000 h= .2400
||x||= .6516 lambda= 1.476 period= 4.000 h= .3316
||x||= .9733 lambda= 1.395 period= 4.000 h= .6633
||x||= 1.298 lambda= 1.335 period= 4.000 h= .2056
||x||= 1.502 lambda= 1.322 period= 4.000 h= .1434
||x||= 1.645 lambda= 1.327 period= 4.000 h= .1259
||x||= 1.769 lambda= 1.343 period= 4.000 h= .1188
```

```
||x||= 1.885 lambda= 1.368 period= 4.000 h= .1163
||x||= 1.996 lambda= 1.400 period= 4.000 h= .1158
||x||= 2.104 lambda= 1.441 period= 4.000 h= .1177
||x||= 2.210 lambda= 1.489 period= 4.000 h= .1217
||x||= 2.317 lambda= 1.546 period= 4.000 h= .1268
||x||= 2.425 lambda= 1.611 period= 4.000 h= .1341
||x||= 2.535 lambda= 1.686 period= 4.000 h= .1422
||x||= 2.647 lambda= 1.771 period= 4.000 h= .1522
||x||= 2.764 lambda= 1.866 period= 4.000 h= .1646
||x||= 2.886 lambda= 1.974 period= 4.000 h= .1780
||x||= 3.014 lambda= 2.096 period= 4.000 h= .1929
||x||= 3.148 lambda= 2.232 period= 4.000 h= .2121
||x||= 3.292 lambda= 2.385 period= 4.000 h= .2334
||x||= 3.445 lambda= 2.558 period= 4.000 h= .2555
||x||= 3.609 lambda= 2.751 period= 4.000 h= .2813
||x||= 3.785 lambda= 2.967 period= 4.000 h= .3175
||x||= 3.979 lambda= 3.214 period= 4.000 h= .3517
||x||= 4.190 lambda= 3.492 period= 4.000 h= .3915
||x||= 4.420 lambda= 3.805 period= 4.000 h= .4345
||x||= 4.671 lambda= 4.155 period= 4.000 h= .4903
||x||= 4.949 lambda= 4.554 period= 4.000 h= .5361
||x||= 5.248 lambda= 4.994 period= 4.000 h= .5827
orientation changed

interactive driver
integer= 4 real number= .0000
integer= 3 real number= 10.00
integer= 2 real number= 10.00
arc= 10.00 or= 1.000 pert= F h= .5827

||x||= 5.248 lambda= 4.994 period= 4.000 h= .5827
||x||= 5.569 lambda= 5.476 period= 4.000 h= .6815
||x||= 5.940 lambda= 6.043 period= 4.000 h= .7354
||x||= 6.334 lambda= 6.658 period= 4.000 h= .8031
||x||= 6.760 lambda= 7.333 period= 4.000 h= .8929
||x||= 7.228 lambda= 8.088 period= 4.000 h= .9730
||x||= 7.733 lambda= 8.914 period= 4.000 h= .9752
||x||= 8.233 lambda= 9.746 period= 4.000 h= 1.000
||x||= 8.742 lambda= 10.60 period= 4.000 h= 1.000
||x||= 9.246 lambda= 11.46 period= 4.000 h= 1.000
||x||= 9.746 lambda= 12.32 period= 4.000 h= 1.000
||x||= 10.24 lambda= 13.19 period= 4.000 h= 1.000
||x||= 10.73 lambda= 14.05 period= 4.000 h= 1.000
enter new arclength for countdown

interactive driver
integer= 4 real number= .0000
integer= 5 real number= .0000
integer= 3 real number= 20.00
integer= 2 real number= 20.00
arc= 20.00 or= -1.000 pert= T h= 1.000

||x||= 10.73 lambda= 14.06 period= 3.999 h= 1.000
||x||= 10.23 lambda= 13.19 period= 3.999 h= 1.000
||x||= 9.739 lambda= 12.32 period= 3.999 h= 1.000
||x||= 9.239 lambda= 11.46 period= 3.999 h= 1.000
||x||= 8.736 lambda= 10.60 period= 3.999 h= 1.000
||x||= 8.228 lambda= 9.747 period= 3.999 h= 1.000
```

```
||x||= 7.715 lambda= 8.894 period= 3.999 h= .9816
||x||= 7.206 lambda= 8.060 period= 3.999 h= .9325
||x||= 6.716 lambda= 7.272 period= 3.999 h= .9478
||x||= 6.213 lambda= 6.475 period= 3.998 h= .8602
||x||= 5.748 lambda= 5.758 period= 3.997 h= .7875
||x||= 5.313 lambda= 5.108 period= 3.993 h= .6959
||x||= 5.212 lambda= 4.967 period= 3.989 h= .3480
||x||= 5.157 lambda= 4.901 period= 3.983 h= .1423
||x||= 5.133 lambda= 4.875 period= 3.978 h= 5.0860E-02
||x||= 5.114 lambda= 4.864 period= 3.970 h= 2.0477E-02
||x||= 5.097 lambda= 4.858 period= 3.962 h= 1.5860E-02
||x||= 5.086 lambda= 4.858 period= 3.954 h= 1.6296E-02
||x||= 5.075 lambda= 4.861 period= 3.945 h= 2.8311E-02
||x||= 5.059 lambda= 4.871 period= 3.930 h= 3.9659E-02
||x||= 5.040 lambda= 4.890 period= 3.909 h= 7.9318E-02
||x||= 5.010 lambda= 4.933 period= 3.867 h= .1586
||x||= 4.968 lambda= 5.034 period= 3.791 h= .1868
||x||= 4.950 lambda= 5.174 period= 3.713 h= .1757
||x||= 4.957 lambda= 5.317 period= 3.645 h= .2092
||x||= 4.992 lambda= 5.488 period= 3.568 h= .2303
||x||= 5.071 lambda= 5.677 period= 3.495 h= .2260
||x||= 5.172 lambda= 5.866 period= 3.445 h= .2231
||x||= 5.278 lambda= 6.055 period= 3.408 h= .2538
||x||= 5.401 lambda= 6.271 period= 3.375 h= .2763
||x||= 5.535 lambda= 6.508 period= 3.346 h= .3006
||x||= 5.679 lambda= 6.768 period= 3.321 h= .3110
||x||= 5.826 lambda= 7.039 period= 3.299 h= .3277
||x||= 5.979 lambda= 7.326 period= 3.279 h= .3416
||x||= 6.135 lambda= 7.628 period= 3.262 h= .3527
||x||= 6.294 lambda= 7.942 period= 3.246 h= .3784
||x||= 6.461 lambda= 8.280 period= 3.231 h= .3916
||x||= 6.631 lambda= 8.632 period= 3.217 h= .4118
||x||= 6.807 lambda= 9.004 period= 3.204 h= .4508
||x||= 6.996 lambda= 9.412 period= 3.191 h= .4487
||x||= 7.181 lambda= 9.820 period= 3.180 h= .4846
||x||= 7.378 lambda= 10.26 period= 3.169 h= .5155
||x||= 7.583 lambda= 10.73 period= 3.158 h= .5065
||x||= 7.781 lambda= 11.20 period= 3.148 h= .5411
||x||= 7.989 lambda= 11.70 period= 3.138 h= .5458
||x||= 8.196 lambda= 12.20 period= 3.130 h= .5750
||x||= 8.409 lambda= 12.74 period= 3.121 h= .5674
||x||= 8.617 lambda= 13.26 period= 3.114 h= .6170
enter new arclength for countdown

interactive driver
integer= 4 real number= .0000
integer= 5 real number= .0000
integer= 3 real number= 20.00
integer= 2 real number= 20.00
arc= 20.00 or= 1.000 pert= F h= .6170

||x||= 8.617 lambda= 13.26 period= 3.114 h= .6170
||x||= 8.392 lambda= 12.69 period= 3.122 h= .6635
||x||= 8.144 lambda= 12.08 period= 3.132 h= .6422
||x||= 7.900 lambda= 11.48 period= 3.143 h= .5723
||x||= 7.678 lambda= 10.96 period= 3.153 h= .5684
||x||= 7.453 lambda= 10.43 period= 3.165 h= .5405
||x||= 7.234 lambda= 9.941 period= 3.177 h= .4969
```

```
||x||= 7.030 lambda= 9.489 period= 3.190 h= .4922
||x||= 6.824 lambda= 9.043 period= 3.203 h= .4687
||x||= 6.624 lambda= 8.620 period= 3.218 h= .4469
||x||= 6.429 lambda= 8.219 period= 3.234 h= .4173
||x||= 6.244 lambda= 7.846 period= 3.251 h= .4061
||x||= 6.060 lambda= 7.486 period= 3.271 h= .3766
||x||= 5.885 lambda= 7.155 period= 3.292 h= .3528
||x||= 5.719 lambda= 6.847 period= 3.316 h= .3402
||x||= 5.557 lambda= 6.552 period= 3.344 h= .3266
||x||= 5.399 lambda= 6.272 period= 3.378 h= .3123
||x||= 5.248 lambda= 6.006 period= 3.421 h= .3017
||x||= 5.109 lambda= 5.752 period= 3.479 h= .2765
||x||= 5.007 lambda= 5.523 period= 3.561 h= .2559
||x||= 4.964 lambda= 5.313 períod= 3.655 h= .2399
||x||= 4.964 lambda= 5.121 period= 3.748 h= .2362
||x||= 5.008 lambda= 4.955 period= 3.856 h= .2020
||x||= 5.074 lambda= 4.829 period= 3.958 h= .1946
||x||= 5.145 lambda= 4.718 period= 4.055 h= .1752
orientation changed

interactive driver
integer= 4 real number= .0000
integer= 3 real number= 10.00
integer= 2 real number= 10.00
arc= 10.00 or= -1.000 pert= F h= .1752

||x||= 5.145 lambda= 4.718 period= 4.055 h= .1752
||x||= 5.233 lambda= 4.638 period= 4.147 h= .1851
||x||= 5.343 lambda= 4.567 period= 4.243 h= .1644
||x||= 5.443 lambda= 4.503 period= 4.327 h= .1518
||x||= 5.536 lambda= 4.444 period= 4.405 h= .1497
||x||= 5.634 lambda= 4.391 period= 4.483 h= .1561
||x||= 5.743 lambda= 4.346 period= 4.566 h= .1691
||x||= 5.873 lambda= 4.317 period= 4.657 h= .1863
||x||= 6.025 lambda= 4.313 period= 4.754 h= .2156
||x||= 6.205 lambda= 4.343 period= 4.856 h= .2268
||x||= 6.395 lambda= 4.414 period= 4.943 h= .2339
||x||= 6.588 lambda= 4.513 period= 5.019 h= .2850
||x||= 6.814 lambda= 4.656 period= 5.099 h= .2798
||x||= 7.030 lambda= 4.815 period= 5.161 h= .3074
||x||= 7.261 lambda= 5.004 period= 5.215 h= .3407
||x||= 7.510 lambda= 5.225 period= 5.262 h= .3711
||x||= 7.775 lambda= 5.476 period= 5.305 h= .3810
||x||= 8.040 lambda= 5.742 period= 5.340 h= .3885
||x||= 8.304 lambda= 6.021 period= 5.369 h= .4179
||x||= 8.583 lambda= 6.328 period= 5.395 h= .3933
||x||= 8.840 lambda= 6.621 period= 5.415 h= .4174
||x||= 9.107 lambda= 6.937 period= 5.434 h= .4307
||x||= 9.379 lambda= 7.268 period= 5.450 h= .4477
||x||= 9.656 lambda= 7.615 period= 5.465 h= .5119
||x||= 9.968 lambda= 8.016 period= 5.480 h= .5533
||x||= 10.30 lambda= 8.455 period= 5.495 h= .4764
||x||= 10.58 lambda= 8.835 period= 5.506 h= .4794
||x||= 10.86 lambda= 9.221 period= 5.517 h= .5125
||x||= 11.15 lambda= 9.635 period= 5.528 h= .5377
||x||= 11.46 lambda= 10.07 period= 5.538 h= .6606
||x||= 11.83 lambda= 10.61 period= 5.550 h= .5104
enter new arclength for countdown
```

```
interactive driver
integer= 1 real number= .0000
```

Program 5. A PL Surface Generator

by Stefan Gnutzmann

The following program represents a simplified implementation of a PL algorithm for approximating an implicitly defined surface. It is based upon the algorithms discussed in section 15.4. A more general and sophisticated PASCAL version is given in Gnutzmann (1988). The example used here is the surface of the torus described by the equation

$$H(x_1, x_2, x_3) := (x_1^2 + x_2^2 + x_3^2 + R^2 - r^2)^2 - 4R^2(x_2^2 + x_3^2) = 0$$

with $R = 1.0$ and $r = 0.2$. Figure P5.a below is a wire figure plot of a run of the program. The program utilizes the Freudenthal triangulation with the meshsize $\delta > 0$ translated to an initial point x given in the output below. After a transverse edge of a simplex is determined, a zero point of H on the edge is calculated via a modified Newton iteration, or if necessary, via bisection. For each transverse simplex all three or four corresponding zero points are output, thus describing a triangular or quadrilateral piece of the PL approximation. The program stops when all transverse simplices within the user determined bounding domain have been found.

Sketch of Program 5.   *comment:*

    input

      begin

      $x \in H^{-1}(0)$;   *initial point*

      $\delta > 0$;   *mesh size*

      `boundary`;   *data for bounding cell*

      end input;

    calculate a starting simplex σ_1 using x and δ;

    `simplexlist` := $\{\sigma_1\}$;

    index := 1;

    repeat   *determine all transverse edges of σ_{index}*

      find transverse edge $[\text{Vertex}_0, \text{Vertex}_1]$;

      initialize index vector $\text{clface}(j) := j, j = 0, 1, \ldots$;

```
start := clface(N + 1);
transverse.facets := {clface(N + 1), clface(N + 2)};
calculate and write zero point;
```

repeat   *pivoting step*

    if $\text{sign}\left(\text{Vertex}_{\text{clface}(0)}\right) = \text{sign}\left(\text{Vertex}_{\text{clface}(N+1)}\right)$

    then begin leave := clface(0);

       clface(0) := clface(N + 1) **end**;

    else begin leave := clface(1);

       clface(1) := clface(N + 1) **end**;

    clface(N + 1) := clface(N + 2);

    clface(N + 2) := leave;

    transverse.facets := transverse.facets $\cup$ {clface(N + 2)};

    calculate and write zero point;

until start = clface(N + 2);

for $i \in$ transverse.facets **do**

    if the facet corresponding to i is contained in the boundary

    then transverse.facets := transverse.facets $\setminus \{i\}$;

    for $i \in$ transverse.facets **do begin**   *determine all neighbors*

    pivot σ_{index} across facet i obtaining σ_j;

    simplexlist := simplexlist $\cup \{\sigma_j\}$ **end**;

    index := index + 1;

until index $>$ cardinality(simplexlist).

In the interest of conserving space, a number of devices from the version in Gnutzmann (1988) which would contribute to improved efficiency have been omitted. Among these we mention

- Elimination of those simplices from the simplex list which are no longer needed (to save storage).
- Instead of a sequential search through the simplex list, a binary search or Hash method may be used.
- The length of the simplex list can be reduced by a judicious choice pivoting.

The FORTRAN program given below includes a bisection iteartive improvement. A number of steps are possible for improving the surface mesh. Some of these have been described in Allgower & Gnutzmann (1989).

    The following is a complete FORTRAN program listing for the PL surface approximation algorithm.

```
      program surapp                         Piecewise linear approximation of an
                                          implicitly defined surface, see section 15.4

      parameter (n=1, k=2, nplusk=n+k)                  dimension of the problem
      parameter (k1=k+1, n3=n+3)
      parameter (lisdim = 1000)                dimension of the simplex list
      double precision eps
      parameter (eps = 1d-10)                             machine tolerance

      double precision delta, origin(1:nplusk)
                                 mesh size and origin of the triangulation
      double precision lbound(1:nplusk), ubound(1:nplusk)
                                  lower and upper bounds of the problem
      double precision simx(0:nplusk,1:nplusk)
                                           vertices of current simplex
      double precision simf(0:nplusk,1:n)
                                          function values of vertices
      double precision error                  error of the approximation
      double precision u(1:nplusk), v(1:nplusk), x(1:nplusk),
     *                 fx(1:n)                          auxiliary arrays

      integer slist (1:lisdim,1:nplusk)                list of simplices
                           a simplex is characterized by barycenter * (nplusk+1)
      integer inds                       current member of simplex list
      integer maxs                           last entry of simplex list
      integer numver (k1:n3)                             counts pl pieces
      integer pi(1:nplusk), z(1:nplusk)
                                   pi and z values of the current simplex
      integer clface (0:nplusk), i              auxiliary variables
      logical facets (0:nplusk)

                                  transverse facets of the current simplex
                            facets(i) = .true. means that facet i is transverse

      open(1, file='surapp.dat')                            output file
      inds = 1                          starting values (initialization)
      maxs = 1
      error = 0.0
      do 10 i = k1, n3
        numver (i) = 0
10    continue

      write (*,'(/1x,a,a)') 'pl approximation of an implicitly',
     *                      ' defined surface'
      write (1,'(/1x,a,a)') 'pl approximation of an implicitly',
     *                      ' defined surface'

      call start (delta, origin, lbound, ubound, simx, simf,
     *            slist, nplusk, n, lisdim, pi, z, x, v, fx, u)
                                   compute starting simplex, mesh size, origin, lower and
                                                              upper bounds

20    continue                                       begin of pl loop
      call appsim (inds, simx, simf, numver, nplusk, n, k1, n3,
     *             slist, lisdim, maxs, lbound, ubound, eps,
     *             error, pi, z, u, v, x, fx, facets, clface )
                                              process current simplex
      inds = inds+1
      if ((inds .le. maxs) .and. (inds.le.lisdim) ) then
```

```
                                   not all simplices are processed
          call getver (inds, simx, simf, slist, nplusk, lisdim,
     *                  n, pi, z, x, fx, u, origin, delta)
                                   simx and simf of next simplex are computed
          goto 20
        end if                                               end of loop

                                              statistics of program
       write (*,'(//" total number of transverse simplices",
     *        14x,i8)')  maxs
       write (*,'(5x,"pl pieces containing",i2,a,15x,i8)')
     *        (i,' vertices',numver(i),i=k1,n3)
       write (*,'(/" maximum of all function values",16x,d12.6)')
     *        error
       write (1,'(//" total number of transverse simplices",
     *        14x,i8)')  maxs
       write (1,'(5x,"pl pieces containing",i2,a,15x,i8)')
     *        (i,' vertices',numver(i),i=k1,n3)
       write (1,'(/" maximum of all function values",16x,d12.6)')
     *        error

       stop
       end


       subroutine appsim (inds, simx, simf, numver, nplusk, n, k1,
     *             n3, slist, lisdim, maxs, lbound, ubound,
     *             eps, error, pi, newcen, u, v, x, fx,
     *             facets, clface )
                        input: inds, simx, simf, numver, slist, maxs, lbound, ubound
                                   output: numver, slist, maxs

                        this subprogram computes all cl faces of the current simplex,
                        all neighbors of the current simplex which share a common
                                   transverse facet are put on the simplex list

       double precision simx (0:nplusk,1:nplusk),
     *             simf (0:nplusk,1:n),
     *             lbound (1:nplusk), ubound(1:nplusk),
     *             eps, error,
     *             u(1:nplusk), v(1:nplusk),
     *             x(1:nplusk), fx(1:n)
       integer i, j, numver (k1:n3), lisdim, k1, n3, maxs, inds,
     *         slist (1:lisdim,1:nplusk), pi(1:nplusk)
       logical facets (0:nplusk)
                        for an explanation of these variables see the main program

       integer clface (0:nplusk)               indices of cl face (0..n) and of
                                               vertices to be pivoted (n+1..n+k)
       integer start                           first vertex to pivot in cl face
       integer numcl                           counts cl faces
       integer newcen (1:nplusk)        barycenter * (nplusk+1) of a neighbor
       logical bound                     function which checks the bounds

                                              search of a transverse edge
                                              works only if nplusk=3
       i = 1
110    continue
       if ((simf(0,1).le.0.0) .eqv. (simf(i,1).le.0.0)) then
```

```
            i = i+1
            if (i .le. nplusk) goto 110
         end if
      if (i .gt. nplusk) return

      do 120 j = 0, nplusk                    starting values (initialization)
         clface(j) = j
120   continue

      numcl = 1
      if (i.ne.1) then
         clface(1) = i
         clface(i) = 1
      end if

      do 130 i = 0, nplusk
                              facets clface(n+1) and clface(n+2) are transverse
         facets (clface(i)) = i .gt. n
130   continue

      start = clface (n+1)
      call output (simx, simf, clface, u, v, x, fx, inds,
     *             nplusk, n, eps, error)
                                      compute zero point of the cl face
                                      and write it on a file or screen

140      continue                                    begin cl face loop
         call pivot (clface, simf, nplusk, n)
                                          compute next cl face
         call output (simx, simf, clface, u, v, x, fx, inds,
     *               nplusk, n, eps, error)
                                      compute zero point of the cl face
                                      and write it on a file or screen
         numcl = numcl+1
         facets (clface(nplusk)) = .true.
                                      facet clface(nplusk) is transverse
         if (clface(n+2) .ne. start) goto 140
                                      stop test works correctly if k=2

      numver(numcl) = numver(numcl)+1                counts pl pieces

      do 160 i = 0, nplusk
                         loop which checks the bounds of transverse facets
                         a facet is outside the boundary if all vertices of the
                                                      facet are outside
         if (facets(i)) then
            do 150 j = 0, nplusk
               if ((j .ne. i) .and. .not.
     *             bound (simx, j, lbound, ubound, nplusk, eps))
     *             goto 160
150         continue
            facets (i) = .false.
         end if
160   continue

      do 170 i = 0, nplusk
                    loop which computes all neighbors of the current simplex
                            if they share a common transverse facet
```

```
      if (facets(i)) then
         call reflec(slist,inds,i,newcen,nplusk,lisdim,pi)
```
                    compute the barycenter of the neighbor and
                    put it on the simplex list
```
         call newsim (newcen, slist, maxs, lisdim, nplusk)
      end if
170   continue
      return
      end

      logical function bound(simx,ind,lbound,ubound,nplusk,eps)
```
                    input: simx, ind, lbound, ubound
                    output: bound

                    this function checks the bounds of the vertex simx(ind,.)

```
      double precision simx (0:nplusk,1:nplusk),
     *                 lbound(1:nplusk), ubound(1:nplusk),
     *                 eps
      integer nplusk,  ind
```
                    for an explanation of these variables see the main program

```
      integer i
```                                              *auxiliary variables*
```
      logical r

      i = 0
210   i = i+1
      r = ((lbound(i)-simx(ind,i)) .ge. eps) .or.
     *     ((simx(ind,i)-ubound(i)) .ge. eps)
      if ((i .lt. nplusk) .and. .not. r) goto 210
      bound = r
      return
      end

      subroutine fvalue (x, f, nplusk, n)
```                                              *input: x*
                    output: f (function value of x)

                    user defined function e.g. an equation of a torus

```
      double precision x(1:nplusk), f(1:n), help
      integer nplusk, n

      help = x(2)**2 + x(3)**2
      f(1) = (x(1)**2 + help + 0.96d0)**2 - 4.0d0*help
      return
      end

      subroutine getver(inds,simx,simf,slist,nplusk,lisdim,n,
     *                  pi,z,x,fx,vertex,origin,delta)
```
                    input: inds, slist, origin, delta
                    output: simx, simf

                    the subroutine computes the vertices of the current simplex
                    and the function values belonging to the vertices
                    see the rules of the Freudenthal triangulation (12.1.10)

```
      double precision simx(0:nplusk,1:nplusk),
     *                 simf(0:nplusk,1:n),
```

```
*                     x(1:nplusk), fx(1:n), delta,
*                     origin(1:nplusk)
     integer lisdim, nplusk, n, slist (1:lisdim,1:nplusk),
*            inds, pi(1:nplusk), z(1:nplusk)
                          for an explanation of these variables see the main program

     integer i, help
     double precision vertex (1:nplusk)                auxiliary variables

     do 410 i = 1, nplusk
                          permutation pi and integer vector z are calculated
                            (only the barycenter * (nplusk+1) was stored)
        z(i) = slist(inds,i) / (nplusk+1)
        help = mod (slist(inds,i),nplusk+1)
        if (help .lt. 0) then
           help = help+nplusk+1
           z(i) = z(i)-1
        end if
        pi (nplusk+1-help) = i
410  continue

     do 420 i = 1, nplusk              starting value for current vertex
        vertex(i) = z(i)
420  continue

     call simtox (vertex, simx, origin, delta, nplusk, 0)
                       calculate coordinates of vertex 0 and put it on simx(0,.)

                     function value of vertex 0 is computed and stored in simf(0,.)
     do 430 i = 1, nplusk
        x(i) = simx(0,i)
430  continue
     call fvalue (x, fx, nplusk, n)
     do 440 i = 1, n
        simf(0,i) = fx(i)
440  continue

                     all other vertices and function values are calculated
     do 470 i = 1, nplusk
        vertex (pi(i)) = vertex (pi(i)) + 1.0d0
                     rules of the Freudenthal triangulation, see (12.1.10)
        call simtox (vertex, simx, origin, delta, nplusk, i)
        do 450 j = 1, nplusk
           x(j) = simx(i,j)
450     continue
        call fvalue (x, fx, nplusk, n)
        do 460 j = 1, n
           simf(i,j) = fx(j)
460     continue
470  continue

     return
     end

     subroutine newsim (center, slist, maxs, lisdim, nplusk)
                          input: center, slist, smax
                          output: slist, smax
```

```
                              the subroutine puts a new barycenter on the simplex list

        integer maxs, lisdim, nplusk, slist (1:lisdim,1:nplusk)
                              for an explanation of these variables see the main program

        integer i, j                                      auxiliary variables
        integer center (1:nplusk)
                              barycenter * (nplusk+1) of a simplex

                         loop compares all simplices of the list with current center
        do 520 i = maxs, 1, -1
           do 510 j = 1, nplusk
              if (slist(i,j) .ne. center(j)) goto 520
510        continue
           return                     the simplex is already a member of the list
520     continue

                                         center belongs to a new simplex
                                       and must be stored in the simplex list
        maxs = maxs+1

                                   check the capacity of the simplex list
        if (maxs .eq. lisdim+1) then
           write (*,'(" simplex list is too small")')
           write (1,'(" simplex list is too small")')
           return
        end if

                                 storing center at the end of the simplex list
        do 530 i = 1, nplusk
           slist (maxs,i) = center(i)
530     continue
        return
        end

        subroutine output (simx, simf, clface, u, v, x, fx, inds,
     *                      nplusk, n, eps, error)
                                    input: simx, simf, clface, inds, error
                                    output: error, zero point on the screen

                                 output calculates the zero point on the edge with
                               a bisection method and writes it to a file or screen
                               subroutine works correctly if nplusk=3

        double precision simx(0:nplusk,1:nplusk),
     *                   simf(0:nplusk,1:n), eps, error
        integer inds, nplusk, n
                              for an explanation of these variables see the main program

        integer clface (0:nplusk)
                              for an explanation of clface see the subroutine appsim

        double precision u(1:nplusk)
                                        first vertex (simx(clface(0),.))
        double precision v(1:nplusk)                u+v = second vertex
        double precision x(1:nplusk)            zero point approximation
        double precision fx(1:n)                    function value of x
        double precision lambda
```

```
                                          barycentric coordinate of the zero point
      double precision lowerb,  upperb
                                      lower and upper bound for the bisection method
      logical neg                                      =.true. iff f(u) is negative
      integer i                                               auxiliary variable

                                                    starting values are calculated
      do 610  i = 1, nplusk
          u(i) = simx (clface(0),i)
          v(i) = simx (clface(1),i) - u(i)
610   continue

      neg = simf(clface(0),1) .lt. 0.0d0
      lowerb = 0.0d0
      upperb = 1.0d0

620       continue                            begin loop of bisection method
          lambda = (lowerb + upperb)/2.0d0
          do 630 i = 1, nplusk
              x(i) = lambda*v(i) + u(i)
630       continue

          call fvalue ( x, fx, nplusk, n)
          if (neg .eqv. (fx(1) .lt. 0.0d0)) then
              lowerb = lambda
          else
              upperb = lambda
          end if
          if (upperb-lowerb .ge. eps) goto 620

                                                      approximation error
      if (error .lt. dabs(fx(1))) error = dabs (fx(1))

      write (*,'(i6,5(3x,d15.8))') inds,(x(i),i=1,nplusk)
      write (1,'(i6,5(3x,d15.8))') inds,(x(i),i=1,nplusk)
      return
      end

      subroutine pivot (clface, simf, nplusk, n)
                                                  input: clface, simf
                                                  output: clface

                                                  pivot of clface(n+1)
                                          pivot works correctly if nplusk = 3

      double precision simf (0:nplusk,1:n)
      integer nplusk, n
                      for an explanation of these variables see the main program

      integer clface (0:nplusk)
                      for an explanation of clface see the subroutine appsim

      integer leave                         index of vertex which leaves the cl face

      if ((simf(clface (n+1),1).le.0.0d0) .eqv.
     *   (simf(clface (0),1).le.0.0d0))  then
                                  sign of clface(0) equal to sign of clface(n+1)
          leave = clface(0)
```

```
      clface(0) = clface (n+1)
else
                              sign of clface(1) equal to sign of clface(n+1)
      leave = clface(1)
      clface(1) = clface (n+1)
end if

clface (n+1) = clface (n+2)
clface (n+2) = leave
return
end

subroutine reflec (slist, inds, facet, newcen, nplusk,
*                   lisdim, pi)
                                        input: slist, inds, facet
                                        output: newcen

                           newcen is obtained by reflecting the vertex facet
                                             of the current simplex
                           see rules (12.1.10) of the Freudenthal triangulation

integer nplusk, slist(1:lisdim,1:nplusk), pi(1:nplusk)
                      for an explanation of these variables see the main program

integer facet                         index of vertex which should be reflected
                                                  from the current simplex
integer newcen(1:nplusk)
                              barycenter * (nplusk+1) of the neighbor
integer i, help                                   auxiliary variables

                                        computing of starting values
do 810 i = 1, nplusk
   newcen(i) = slist (inds,i)
   help = mod (slist(inds,i),nplusk+1)
   if (help .lt. 0) help = help+nplusk+1
   pi (nplusk+1-help) = i
810   continue

                                        reflection (see (12.1.11))
if ((facet.gt.0) .and. (facet.lt.nplusk)) then
   newcen (pi(facet)) = newcen (pi(facet)) - 1
   newcen (pi(facet+1)) = newcen (pi(facet+1)) + 1
else if (facet.eq.0) then
   newcen (pi(1)) = newcen(pi(1)) + 2
   do 820 i = 2, nplusk
      newcen(pi(i)) = newcen(pi(i)) + 1
820   continue
else
   newcen (pi(nplusk)) = newcen (pi(nplusk)) - 2
   do 830 i = 1, nplusk-1
      newcen (pi(i)) = newcen (pi(i)) - 1
830   continue
end if
return
end

subroutine simtox (vertex, simx, origin, delta,
*                   nplusk, ind)
```

```
                                                      input: vertex, origin, delta
                                                      output: simx(ind,.)

                                         transformation of vertex to true co-ordinates

      double precision delta, origin(1:nplusk),
     *                  simx(0:nplusk,1:nplusk)
      integer nplusk
                             for an explanation of these variables see the main program

      double precision vertex(1:nplusk)
                                                  integer coordinates of vertex
      integer ind                                            index for simx
      integer i

      do 910  i = 1, nplusk
         simx(ind,i) = origin(i) + delta*vertex(i)
910   continue
      return
      end

      subroutine start (delta, origin, lbound, ubound, simx,
     *                  simf, slist, nplusk, n, lisdim, pi, z,
     *                  x, v, fx, u)
                           output: delta, origin, lbound, ubound, simx, simf, slist

                                                  user defined subroutine
                                     calculating starting values of the algorithm

      double precision delta, origin(1:nplusk),
     *                  lbound(1:nplusk), ubound(1:nplusk),
     *                  simx(0:nplusk,1:nplusk),
     *                  simf(0:nplusk,1:n), x(1:nplusk),
     *                  v(1:nplusk), fx(1:n), u(1:nplusk)
      integer lisdim, nplusk, n, slist (1:lisdim,1:nplusk),
     *        pi(1:nplusk), z(1:nplusk)
                             for an explanation of these variables see the main program

      integer i, j, step, max                              auxiliary variables


                                                             setting bounds
      lbound(1) = -0.6d0
      lbound(2) = -1.2d0
      lbound(3) = -1.2d0
      ubound(1) = 0.6d0
      ubound(2) = 1.2d0
      ubound(3) = 1.2d0

                           initial point (should be a approximation of a zero point)
      x(1) = 0.1875d0
      x(2) = 1.125d0
      x(3) = 0.0625d0

      delta = 0.25d0                                         mesh size

      write (*,'(/" initial point")')
      write (*,'(5(1x,d10.5))') (x(i),i=1,nplusk)
```

```
      write (1,'(/" initial point")')
      write (1,'(5(1x,d10.5))') (x(i),i=1,nplusk)
```

first simplex: z = 0 and pi = id

```
      do 1010 i = 1, nplusk
         slist(1,i) = nplusk+1-i
1010  continue

      step = 0
      max = 1
```

construction of of a transverse simplex

```
1020  continue
      do 1030 i = 1, nplusk
         origin(i) = x(i) - delta/(nplusk+1)*(nplusk+1-i)
1030  continue

      call getver (1, simx, simf, slist, nplusk, lisdim, n,
     *                pi, z, v, fx, u, origin, delta)
```

Freudenthal triangulation

simx and simf of the starting simplex are calculated

search for a transverse edge
works only if nplusk=3

```
      i = 1
1040  continue
         if ((simf(0,1).le.0.0) .eqv. (simf(i,1).le.0.0)) then
            i = i+1
            if (i .le. nplusk) goto 1040
         end if

      if (i .gt. nplusk) then
         step = step+1
         if (step .lt. max) then
```

simplex is not transverse

reduce mesh size and try it again

```
            delta = delta*step/(step+1.0)
            goto 1020
         else
            stop
         end if
      end if
```

recording simplex and function values

```
      write (*,'(/" start simplex of mesh size ",f10.5)')
     *      delta
      write (1,'(/" start simplex of mesh size ",f10.5)')
     *      delta
      do 1050 j = 1, nplusk
         write (*,'(6(1x,f11.5))') (simx (i,j), i=0,nplusk)
         write (1,'(6(1x,f11.5))') (simx (i,j), i=0,nplusk)
1050  continue
      write (*,'(/" function values")')
      write (1,'(/" function values")')
      do 1060 j = 1, n
         write (*,'(6(1x,d11.5))') (simf (i,j), i=0,nplusk)
         write (1,'(6(1x,d11.5))') (simf (i,j), i=0,nplusk)
1060  continue
      write (*,'(/" simplex numbers and approximate zero ",
     *      "points:")')
```

```
      write (1,'(/" simplex numbers and approximate zero ",
     *        "points:")')
      return
      end
```

A run of the above program gave the following results. We give only the
beginning and the end of the output list.

```
pl approximation of an implicitly defined surface

initial point
.18750D+00 .11250D+01 .62500D-01

start simplex of mesh size     .25000
      .00000     .25000     .25000     .25000
    1.00000    1.00000    1.25000    1.25000
      .00000     .00000     .00000     .25000

function values
-.15840D+00 0.90506D-01 0.43222D+00 0.50926D+00

simplex numbers and approximate zero points:
      1      0.20000000D+00      0.10000000D+01      0.00000000D+00
      1      0.14142136D+00      0.11414214D+01      0.00000000D+00
      1      0.13723591D+00      0.11372359D+01      0.13723591D+00
      2      0.00000000D+00      0.12000000D+01      0.00000000D+00
      2      0.14142136D+00      0.11414214D+01      0.00000000D+00
      2      0.13723591D+00      0.11372359D+01      0.13723591D+00
      3      0.20000000D+00      0.10000000D+01      0.00000000D+00
      3      0.19903578D+00      0.10000000D+01      0.19903578D+00
      3      0.13723591D+00      0.11372359D+01      0.13723591D+00
      4      0.19999610D+00      0.10000000D+01     -0.50003903D-01
      4      0.20000000D+00      0.10000000D+01      0.00000000D+00
      4      0.14142136D+00      0.11414214D+01      0.00000000D+00
      4      0.13868111D+00      0.11386811D+01     -0.11131889D+00
      5      0.00000000D+00      0.12000000D+01      0.00000000D+00
      5      0.00000000D+00      0.11855655D+01      0.18556546D+00
      5      0.13723591D+00      0.11372359D+01      0.13723591D+00
      6      0.00000000D+00      0.11989120D+01     -0.51087986D-01
      6      0.00000000D+00      0.12000000D+01      0.00000000D+00
      6      0.14142136D+00      0.11414214D+01      0.00000000D+00
      6      0.13868111D+00      0.11386811D+01     -0.11131889D+00

                                .
                                .
                                .

    551      0.00000000D+00     -0.12000000D+01      0.00000000D+00
    551      0.20000000D+00     -0.10000000D+01      0.00000000D+00
    551      0.19903578D+00     -0.10000000D+01      0.19903578D+00
    552     -0.14142136D+00     -0.11414214D+01      0.00000000D+00
    552      0.00000000D+00     -0.12000000D+01      0.00000000D+00
    552      0.00000000D+00     -0.11989120D+01      0.51087986D-01
    552     -0.13868111D+00     -0.11386811D+01      0.11131889D+00
    553     -0.13723591D+00     -0.11372359D+01     -0.13723591D+00
    553     -0.20000000D+00     -0.10000000D+01      0.00000000D+00
    553     -0.14142136D+00     -0.11414214D+01      0.00000000D+00
    554     -0.13723591D+00     -0.11372359D+01     -0.13723591D+00
    554      0.00000000D+00     -0.12000000D+01      0.00000000D+00
    554      0.00000000D+00     -0.11855655D+01     -0.18556546D+00
```

```
555    -0.13868111D+00    -0.11386811D+01     0.11131889D+00
555     0.00000000D+00    -0.11736695D+01     0.25000000D+00
555     0.00000000D+00    -0.11989120D+01     0.51087986D-01
556    -0.13723591D+00    -0.11372359D+01    -0.13723591D+00
556     0.00000000D+00    -0.12000000D+01     0.00000000D+00
556    -0.14142136D+00    -0.11414214D+01     0.00000000D+00
```

```
total number of transverse simplices              556
    pl pieces containing 3 vertices                384
    pl pieces containing 4 vertices                172

maximum of all function values              0.357643D-10
```

Figure P5.a is a wire figure plot of the torus mentioned above. This picture was generated by a similar run with meshsize 0.2 and starting point $(0.15 , 1.1 , 0.05)^*$. The figure contains 1440 edges, 960 triangles or quadrilaterals, and 480 vertices. The surface area of the PL approximation was summed up to 7.6599 (the actual surface area of this torus is $0.8\,\pi^2 \approx 7.8957$).

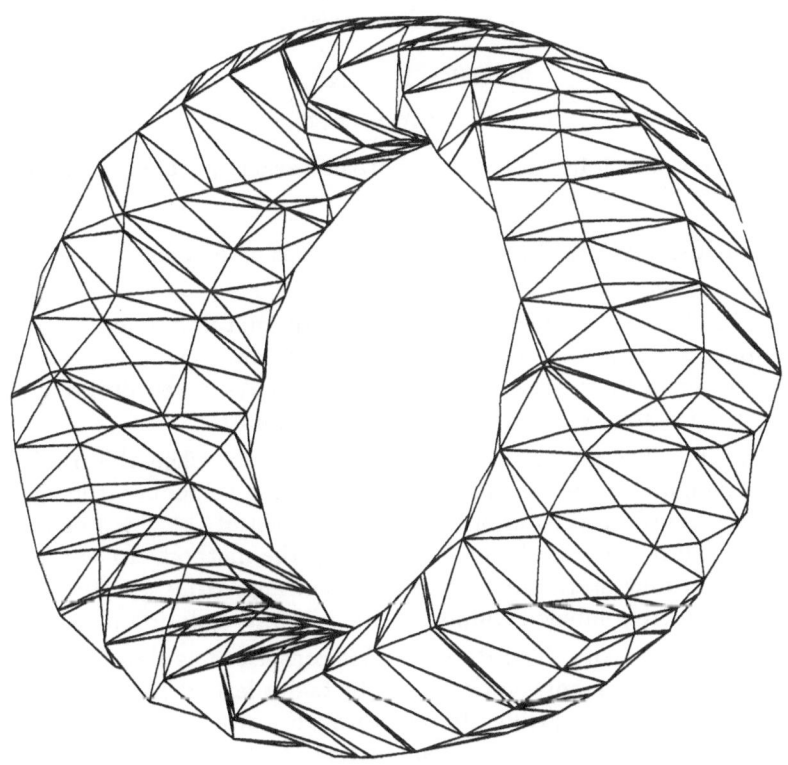

Figure P5.a PL approximation of a torus

Program 6. SCOUT — Simplicial Continuation Utilities

by Hartmut Jürgens and Dietmar Saupe

P6.1 Introduction

SCOUT is a program which implements a series of algorithms based on ideas described in chapter 12 concerning PL continuation methods. The problem to be solved numerically is a nonlinear fixed point or eigenvalue problem, i. e. to find the zeros of

$$F : \mathbf{R}^N \times \mathbf{R} \to \mathbf{R}^N,$$

$$(x, \lambda) \mapsto F(x, \lambda).$$

More precisely, the following is a list of the various problem areas that are handled by SCOUT.

Continuation. SCOUT traces a path of zeros of F using the Generic PL Continuation Algorithm (12.3.9).

Start. For the start of the continuation algorithm a first completely labeled simplex must be provided. Usually an educated guess for a zero of F can be given by the user. Given that information, the program tries to construct a start of the algorithm near the guess.

Refinement. The almost identical problem occurs when the mesh size of the triangulation is changed. One may want to reduce the mesh size to increase the precision of the approximation of the zeros of F, or one might enlarge the mesh size to speed up the method. In any case, the algorithm may start out from the current approximation to generate a new completely labeled simplex of the new triangulation nearby.

Predictor-Corrector. The PL method is considered as slow, as it does not exploit smoothness of solution paths by taking larger steps where possible but rather steadily grinds its way through a triangulation. This shortcoming may be overcome to some extent by superimposing a predictor-corrector scheme onto the basic PL algorithm. For each corrector step an initial guess for a point of the PL approximation of $F^{-1}(0)$ is provided by the predictor step. The corrector step then produces a completely labeled simplex near that point. This is very similar to the generation of a start simplex above or the transition to a new triangulation. In fact, SCOUT uses the identical algorithm for all three tasks.

Perturbation. The basic PL continuation method with a fixed triangulation is very robust in the sense that bifurcation points do not harm the algorithm. At such points due to a change of topological degree the algorithm picks up a bifurcating branch and follows it from then on. Devices are required to enable the detection of the other branches and the continuation of the old path. Also there may be disjoint continua of zeros of F. How can those be found? SCOUT addresses these problems by means of perturbations of the map F. In particular, the program is set up to handle multi-parameter problems with maps

$$F : \mathbf{R}^N \times \mathbf{R}^2 \to \mathbf{R}^N,$$
$$(x, \lambda, \mu) \mapsto F(x, \lambda, \mu).$$

Only one parameter λ or μ is allowed to vary while the other is kept constant. By extending F to a map with several parameters and changing to μ as the variable parameter one may solve the problem of bifurcating branches and disjoint continua. Some a priori knowledge about the solutions may be incorporated by the user as he can specify exactly the extension of $F(x, \lambda)$ to a two parameter map $F(x, \lambda, \mu)$. However, another so called standard perturbation involving a third parameter ρ is always provided by SCOUT. Thus the user does not always have to program his own special perturbations.

Precision improvement. The PL approximation is not as good as an approximation gained by e. g. Newton's method. This is due to the triangulation with a fixed mesh size. This is not a serious limitation since the overall goal of the PL algorithm is to unravel the *global* structure of the zero set of F. Locally, other methods may be used to improve the precision. Since an approximation of a local derivative is contained in the data structures of the PL algorithm, SCOUT uses it to improve the precision of the approximation by a modified Newton's method.

Interaction. From the above it follows that the program should be primarily interactive. The decisions when to use what perturbation techniques and how long to pursue the tracing of a particular solution branch cannot be left to the machine. Therefore a large portion of the efforts in SCOUT went into the design of interactive techniques.

A few remarks on the history of SCOUT are in order. The first version was written in 1979 as part of the authors' diploma thesis Jürgens & Saupe (1979). It contained only the basic PL algorithm and a few perturbation techniques. About a year later the code had matured and included mesh refinement techniques and dynamic adjustments of the triangulation for greater speed, see

Jürgens & Peitgen & Saupe (1980). A portable FORTRAN66 version was implemented at several universities. Subsequently, a completely new version was created, written in structured FORTRAN using the SFTRAN preprocessor, see Saupe (1982). It contains the predictor-corrector scheme and revised techniques for perturbations, mesh refinement and so on. The code has been used heavily in the study of nonlinear boundary value problems, periodic solutions to differential delay equations and other problems, see Saupe (1982–83), Caras & Janata & Saupe & Schmitt (1985).

This section is a condensed excerpt from the report Jürgens & Saupe (1990), which can be obtained from the authors along with the code for both versions of SCOUT. The address is Institut für Dynamische Systeme, Fachbereich Mathematik und Informatik, Universität Bremen, 2800 Bremen 33, West Germany.

P6.2 Computational Algorithms

(P6.2.1) The Predictor-Corrector Method and the Corrector Step

The skeleton of a PC method for tracing a component C of the zeros of F is as follows.

Initial data $x_0 \in C$, $k = 1$.

Step 1 (predictor) Choose $y_k \in \mathbf{R}^{N+1}$ as an approximation of a *next* point of C. Define a codimension 1-manifold $H_k \subset \mathbf{R}^{N+1}$ (e. g. hyperplane) as the kernel of some functional $\gamma_k : \mathbf{R}^{N+1} \to \mathbf{R}$, such that $y_k \in H_k$ and H_k is *sufficiently transverse* to C.

Step 2 (corrector) Solve the system of equations

$$
\begin{aligned}
F(x) &= 0\,, \\
\gamma_k(x) &= 0\,.
\end{aligned}
\tag{P6.2.2}
$$

Step 3 Either stop or increase k by 1 and go to step 1.

(In this subsection we do not use a special symbol for the parameter λ, thus $x \in \mathbf{R}^{N+1}$.) In our hybrid algorithm we use the PL method to solve the system (P6.2.2) in each corrector step. The result of such a corrector step will be a completely labeled simplex with respect to the map F. Thus the PC method contained in SCOUT generates a subsequence of a chain of completely labeled simplexes.

For the discussion of the corrector step, let us drop the index k. We define a suitable path in $H = \gamma^{-1}(0)$ which connects the given estimate $y \in H$ with the solution $x \in H$ of (P6.2.2). For this purpose define

$$F_\gamma : \mathbf{R}^{N+1} \to \mathbf{R}^{N+1},$$

$$x \longmapsto \begin{pmatrix} F(x) \\ \gamma(x) \end{pmatrix}$$

and consider the homotopy

$$G : \mathbf{R}^{N+1} \times \mathbf{R} \longrightarrow \mathbf{R}^{N+1},$$

$$(x, \mu) \longmapsto F_\gamma(x) - \mu F_\gamma(y) = \begin{pmatrix} F(x) - \mu F(y) \\ \gamma(x) \end{pmatrix}.$$

We have that $G(y, 1) = G(x, 0) = 0$ and we assume that $G^{-1}(0)$ contains a path that connects $(y, 1)$ with $(x, 0)$. This assumption can be verified, e. g. in the case where G satisfies some differentiability properties and y is close enough to x.

Instead of applying a continuation method to follow the zeros of G in $\mathbf{R}^{N+2}$ directly we can set up a topological perturbation $\tilde{F}$ of F in the spirit of Jürgens & Peitgen & Saupe (1980) Peitgen (1982), which reduces the dimension by one, thus effectively eliminating the artificial parameter μ. In terms of the PL algorithm $\tilde{F}$ is evaluated only at the vertexes of the triangulation as follows.

(P6.2.3) $$\tilde{F}_T(v) = \begin{cases} F(v) & \text{if } \gamma(v) \geq 0, \\ -\tilde{d} & \text{otherwise,} \end{cases}$$

where $v \in \mathbf{R}^{N+1}$ is any vertex of T. Let $\sigma = [v_1, \ldots, v_{N+1}] \in T$ be completely labeled with respect to $\tilde{F}_T$. Then either σ has exactly one vertex v_j such that $\gamma(v_j) < 0$ and σ carries an $x \in \sigma$ with $F_T(x) - \mu \tilde{d} = 0$ for some $\mu \geq 0$ or σ is completely labeled with respect to F and σ carries an $x \in F_T^{-1}(0)$.

We thus conclude that either σ intersects $H = \gamma^{-1}(0)$ and then approximates a zero of G or σ is already a completely labeled simplex for the original problem $F(x) = 0$. Hence, starting the simplicial algorithm for the perturbed problem $\tilde{F}(x) = 0$ close to the estimate y we will get a chain of completely labeled simplexes which first traces $\{x \in \mathbf{R}^{N+1} \mid F_T(x) = \mu \tilde{d}, \mu > 0\}$ close to H and eventually leaves H, thus approximating the zeros of F.

There are two more technical questions to be answered. How can one make sure that the $(N+1)$-simplex that contains the estimate y in fact has also two completely labeled facets? And in which direction should the PL algorithm trace the chain through y? The first question can be settled by slightly modifying $\tilde{d}$ and γ. The other question is solved using the fact that chains of completely labeled simplices carry an orientation. This orientation is reversed, of course, when the direction of the chain is flipped. Thus one has to match the orientation of the chain generated by the corrector step with the orientation of the chain that belongs to the original unperturbed problem. We omit these technical details here, see Peitgen & Prüfer (1979), Saupe (1982). The corrector step fails, if the number ν of pivoting steps in it exceeds a maximal number ν_{max}.

(P6.2.4) Predictor Step and Step Size Control

In SCOUT there are two ways to compute a first estimate for the next point on the curve, the predictor via linear or quadratic extrapolation. We assume that an $(N+1)$–simplex $\tau = [v_1, ..., v_{N+2}] \in \mathcal{T}$ is given such that $\sigma = [v_1, ..., v_{N+1}]$ is completely labeled. Typically, this will be the case after each linear programming step, e. g. at the termination of a corrector step. The linear extrapolation is obtained from $F_{\mathcal{T}}|_{\tau}$. This map has a one-dimensional kernel, and the coefficients of the kernel with respect to $v_1, ..., v_{N+2}$ are

$$\left(c_1, ..., c_{N+2}\right)^T = \begin{pmatrix} -L^{-1}(1, F(v_{N+2}))^T \\ 1 \end{pmatrix}.$$

Here L denotes as usual the labeling matrix of σ. Thus a tangent is given by $x_{tan} = c_1 v_1 + ... + c_{N+2} v_{N+2}$. If $\alpha > 0$ denotes the step length for the predictor step and $x_0 \in \sigma$ with $F_{\mathcal{T}}(x_0) = 0$, we set

$$y = x_0 + \alpha \frac{x_{tan}}{\|x_{tan}\|}.$$

Since this method uses only local information, it is especially appropriate for small predictor steps or for obtaining the first few predictors. After several cycles of the predictor-corrector scheme we have accumulated enough data to perform higher order predictor steps. The quadratic extrapolation e. g. uses three computed points and quadratic polynomials for the predictor, for brevity we omit the formulae.

In addition to x_{k-1} the last corrector step has produced the coefficient matrix L^{-1} relative to the completely labeled N-simplex which contains x_{k-1}. Assuming that the derivative of F exists and satisfies a Lipschitz condition, we have that L^{-1} implies an approximation of the inverse of the Jacobian at x_{k-1}. If the predictor step is not too large, then the approximation carries over to the predictor point y_k. This can be used to perform a modified Newton step for the point y_k. The size of the Newton correction can be estimated without completely computing the correction vector. If this estimate β does not exceed a maximal number β_{max} we may replace the predictor by the corrected vector. Otherwise the predictor is not accepted. This procedure could even be iterated.

The step size control for the predictor step is very simple and heuristic. If the previous corrector step had been accepted, then the step length is increased by a factor $\overline{\alpha} = 1.5$, otherwise it is reduced by the factor $\underline{\alpha} = 0.5$. A minimal step length of α_{min} equal to the mesh size of the triangulation and a maximal step size α_{max} are prescribed. The predictor-corrector scheme fails when the step size becomes too small. In that case the special setup of the hybrid method offers a fallback in the sense that we can easily switch back to the robust fixed triangulation algorithm starting from the result of the last successful corrector step. From there on one may carry out a number of cycles through the basic PL algorithm until the "difficult" part of the zero set of F is overcome and one may return to the more efficient predictor-corrector scheme.

(P6.2.5) Start Simplex and Changing the Mesh Size

Let $F : \mathbf{R}^{N+1} \to \mathbf{R}^N$ be a continuous map. Also let $y \in \mathbf{R}^N \times \mathbf{R}$ be an estimate for a zero of F and let T be a triangulation of $\mathbf{R}^{N+1}$. When changing the mesh size or when generating a start simplex for the PL algorithm one has to find a completely labeled simplex in T close to the estimate y.

This is the problem to be solved in each corrector step, too. Therefore, we naturally employ the corrector step. As initial data we define an $(N+1)$-simplex τ_y via $y \in \tau_y = [v_1, \ldots, v_{N+2}] \in T$ and a transverse hyperplane $\gamma^{-1}(0)$ by setting

$$\gamma(x) = t_0 \, (x - y)^T, \; 0 \neq t_0 \in \text{Kern } DF_T|_{\tau_y}.$$

Here $DF_T|_{\tau_y}$ denotes the linear part of the affine mapping F_T restricted to the $(N+1)$-simplex τ_y. A successful corrector step with these specifications will solve the problem, see Saupe (1982). For the task of finding the start simplex one has to slightly modify the procedure at the beginning of the corrector step which determines the direction of the chain of completely labeled simplexes.

(P6.2.6) Multi-Parameter Problems and the Standard Perturbation

As mentioned above, SCOUT is formulated for a map F with two parameters λ and μ. Often the original map already has two or more parameters. One typical example is the boundary value problem $-\ddot{u} = \lambda f(u)$ with $u(0) = u(\pi) = \mu$ and a nonlinear function $f : \mathbf{R} \to \mathbf{R}$. This differential equation must be discretized so that one obtains a finite dimensional problem, which then has the two parameters as above. Depending on the choice for the nonlinearity $f(u)$ there may be one or more continua of solutions for the Dirichlet problem ($\mu = 0$). But when both parameters come into play these disjoint branches become connected, and the PL algorithm has access to all of them. For examples of this kind see Jürgens & Peitgen & Saupe (1980), Peitgen & Saupe & Schmitt (1981), Peitgen (1982).

The built-in standard perturbation in SCOUT uses a third parameter ρ in addition to λ and μ. For $\rho = 0$ the old values of F are obtained and for

$\rho \neq 0$ let

$$F(x, \lambda, \mu, \rho) = (1 - |\rho|)F(x, \lambda, \mu, 0) + \rho d$$

where $0 \neq d \in \mathbf{R}^N$. Thus the standard perturbation is just a homotopy to a constant vector. The idea behind this is the following: Since $F(x, \lambda, \mu, \pm 1) \equiv \pm d \neq 0$ we have that the zeros of F are bounded in ρ by ± 1. Therefore, if we keep λ and μ fixed and let only ρ vary, then we hope that the solution path which leaves the level $\rho = 0$ will return to it and not go to ∞, thus yielding another solution of the unperturbed problem. With this technique one may attempt to unravel bifurcation points as well as to find new disjoint continua of solutions. It is the experience of the authors that the method works in many cases.

(P6.2.7) Modified Newton Steps for Local Precision

Assume $F : \mathbf{R}^{N+1} \to \mathbf{R}^N$ and $\mathcal{T}$ is a triangulation of $\mathbf{R}^{N+1}$. Let $\sigma \in \mathcal{T}$ be a completely labeled N-simplex. Then σ spans an affine hyperplane $H \subset \mathbf{R}^{N+1}$ of codimension 1. In this section we describe how to find a zero of the restricted map $F|_H : H \to \mathbf{R}^N$. Assume that $F|_H$ is differentiable and let $\hat{F} : H \to \mathbf{R}^N$ be the affine extension of $F_{\mathcal{T}}|_\sigma$ to H. Then we interpret the differential $D\hat{F}$ as an approximation of the differential $D(F|_H)$. Let x_0 be the zero of $F_{\mathcal{T}}$ carried by $\sigma = [v_1, \dots, v_{N+1}]$ and let L be the labeling matrix with respect to $V = (v_1 \ \ v_2 \ \ \cdots \ \ v_{N+1})$. Then the modified Newton iterates are given by

$$x_{k+1} = x_k - V \cdot L^{-1} \cdot \begin{pmatrix} 0 \\ F(x_k) \end{pmatrix}, \quad k = 0, 1, \dots.$$

Of course, the convergence rate cannot be expected to be quadratic as is the case in Newton's method. One might consider updating methods to improve convergence rates. This has not been done in SCOUT, since the design of the package stresses the global study of $F^{-1}(0)$ rather than local accuracy. The modified Newton steps are typically applied at interesting points along the solution continua as selected by the user.

P6.3 Interactive Techniques

In the following we give a short description of the program operation. The user communicates with SCOUT through an instruction set of about 25 commands. Each command consists of a two letter code and up to four numeral parameters. Commands are buffered and sequentially processed. The "HE" command e. g. activates a help facility which reads a data file and then outputs a portion of the file containing the requested help. Up to 30 commands and their parameters may be recorded in a special command buffer. These can be executed by issuing one single command ("EX"). Incorrect commands or parameters are detected and rejected by the command processor.

An important feature of the program is the capability of dumping the essential portions of the SCOUT data base on a file. At a later time, during the same or a different run of SCOUT, the user can restore this data from the memory file and continue the investigation of his problem. A similar procedure applies to the internal storage of parts of the data base which is carried out before each corrector step. The full data base will automatically be restored after a possible breakdown of the predictor-corrector scheme.

The user periodically receives a feedback on his terminal describing the progress of the path following. Moreover, a protocol file containing more information is written simultaneously. As an option the user may request the output of the solutions on a separate file. This file may then later be processed for instance for a graphical representation of the solutions. These solutions may also be read in by SCOUT to serve as starting guesses for the continuation methods.

Special emphasis is put on the provision of plotting of bifurcation diagrams. There are two options for this purpose: 1. The plot data is output to a file for later graphical processing. 2. The plot data is routed directly to a graphical device ("picture system") while SCOUT is running. Of course, both of the options may be chosen. Either option however, requires the user to add certain software to the SCOUT package. First, the plotted data depends very much on the problem. Often, a plot of the maximum norm of the solution versus the parameters λ and μ is not sufficient. Thus the user has to define his plot data in a special output routine. There he can also define extra output for his terminal and his files. Secondly, if he desires to connect a graphical device directly to SCOUT, he must write an interface routine that drives his device.

The main program controls the execution of the simplicial and predictor-corrector algorithms. It delegates most of the work load to various subroutines. There are six basic cases that are handled by the main program:

- Resetting the data base to its starting values.
- Initialization for the start of the simplicial algorithm: Generation of a start simplex and inversion of the labeling matrix.

- Execution of a single corrector step in the case that the start simplex is not completely labeled.
- Execution of a single corrector step for the purpose of changing the mesh size of the triangulation.
- Execution of the simplicial algorithm in a fixed triangulation: repeated linear programming and pivoting steps together with labeling and output.
- Execution of the predictor-corrector algorithm.

The COMMAND routine is the key program unit which enables the user to input his commands to SCOUT. There are five different command modes. In each mode a certain subset of all commands is accepted as a valid input. These subsets are not disjoint, in fact, many commands are available in all five modes. When a prompt appears on the terminal, the command routine awaits a command to be typed in by the user. This prompt consists of a key word which indicates the command mode and a number denoting the total number of cycles through the simplicial core routines. The five modes are:

INIT : Commands setting initial parameters such as mesh size and dimension of the triangulation, initial values for the start of the path following etc. are expected.

FIX : The fixed triangulation algorithm has been started and all commands are accepted except for two commands which are valid in INIT mode.

P/C : The predictor-corrector algorithm is activated and running. The output is restricted to the end of corrector steps.

SH : The corrector step in the start homotopy has not yet finished. Only commands that continue or abort the corrector step are accepted.

PROG : In this mode COMMAND loads the user commands into a special command buffer. Only the commands that initialize this mode or cause execution of the buffer will not be accepted.

The "HE" and "IF" commands for help and information are available in all modes. Certain commands like "GO" return the control back to the main program which then lets the simplicial algorithm work until the next user input is requested.

The SCOUT package consists of a main program and a collection of 30 subroutines (about 2700 lines). In addition six routines from the LINPACK package are needed. Almost all of the terminal, file and plotting output is defined in one of two routines: OUTPUT or OUTP1. The first routine is the standard SCOUT routine which provides certain general output. Special problem dependent output such as plot data has to be defined by the user in the OUTP1 routine. Of course, The user must supply a routine which evaluates the map $F(x, \lambda, \mu)$. A sample is given in the last section of this appendix.

P6.4 Commands

The following alphabetical list briefly describes the commands in SCOUT. The emphasized words in the first line of each item denote parameters to the command. Optional parameters are enclosed in square brackets.

BL [*mesh*]. The triangulation is "blown up", i. e. a new mesh size *mesh* is prescribed. A corrector step in the new triangulation is performed.

CM [*formfeed*]. Use this comment command to insert a line of text (up to 80 characters) into your output list. A non zero value of *formfeed* will additionally result in a form feed.

EX [*start*]. Triggers the execution of commands from the programmed command buffer. The optional parameter *start* denotes the command at which to start the interpretation. Thus the first *start*-1 commands are skipped (default is 1). See also the **PR** command.

FC *ifc1 ifc2 rfc1 rfc2*. This command sets the values of the parameters *ifc1*, *ifc2*, *rfc1* and *rfc2* which are available in the user supplied subroutine FCT() from a FORTRAN common block. Applications are e. g. collecting several functions in the same source or controlling rarely changed parameters of the problem.

GO [*nout* [*special*]]. This command passes control back to the main program. In **INIT** mode the start simplex will be generated and the program checks if it is completely labeled. In mode **FIX** the basic PL algorithm will be continued. The next user input will be expected after *nout* terminal outputs. This number becomes the default for subsequent **GO** commands. The complete solution vectors are included in these outputs when *special* $= 1$, if *special* $= 2$, then these vectors will also be stored in a file.

HE . The commands available to the current mode are displayed (one line per command and parameters). Additional help for an individual command can be obtained by typing the command and a question mark as parameter.

IF [*case*]. This command causes outputs of either a list of the current special command buffer (*case*=1) or a list of the currently open files and their uses (*case*=2). The user can extend these capabilities.

IN [*incr*]. Sets the maximal number of steps between user commands. This is sometimes useful e. g. in "level" output mode or in corrector steps, when the program does not seem to come to a final completely labeled simplex. When the optional parameter is omitted, the current maximal number of steps is output.

LE [*number*][*type*]. This command activates the output mode "level". Outputs are produced not after a certain number of steps, but rather when the completely labeled simplex is contained in a level of the triangulation, i. e. all vertexes of this simplex carry the same parameter value. In general, these parameter values are multiples of the mesh size. To exit the output mode "level", use the commands **LE**-1 or **OU**. The parameter *number* denotes the number of levels to be traversed for each output, e. g. **LE**2 skips every other level. *type* declares that a different component of the vertex vectors should be regarded for the level criterion.

ME . The command produces a dump file containing all the data necessary to later restart the program at the current point (see the **RE** command).

MF [*special*]. This command will cause a display of the solution vector, a point of the computed PL manifold. If *special* = 2, then this vector will be stored on a new file. To store more vectors onto that file, use **MF**1 or *special* = 2 in the **GO** or **TU** command.

NW [*iters*][*delta*][*eps*]. A certain number of quasi-newton iterations is invoked. *iters* is the maximum number of iteration allowed. The iterations are terminated when the changes in the solution decrease below *delta* or when the value of the function at the solution decreases below *eps* .

OU [*steps*]. The output command sets the number of cycles of the PL algorithm between two consecutive outputs to *steps*. Output mode "level" is terminated if active.

PA *parameter* [*lambda*][*mu*]. This command defines the free parameter and in **INIT** mode it optionally sets the initial values of λ and μ. The values 1,2,3 of *parameter* selects λ, μ and ρ (for the standard perturbation) respectively. In **FIX** and **P/C** mode this is possible only when the current simplex is in a level of the triangulation (see the **LE** command).

PC *alpha_max* [*mf_type*] [*beta_max*][*alpha_0*]. The **PC** command initializes and controls execution of the predictor-corrector algorithm. The first parameter *alpha_max* is the maximal step size. The next parameter *mf_Type* defines the codimension 1 manifold used in the corrector steps, the default (0) is given by hyperplanes. Optionally, spheres may be used (*mf_type*≠0). The number *beta_max* is the bound for the corrector step estimate explained in section (P6.2.4). It is measured in mesh sizes (default 5). The last parameter sets the initial step length. To terminate the predictor-corrector scheme the command **PC**-1 can be used.

PL *case* [*mode*]. This command defines plot action to a file or (if programmed) to an online graphics device. There are four different cases. When *case*=1, plotting is initialized and the *mode* parameter selects the output devices (1 = file output, 2 = graphics output, 3 = both). The plot output may be temporarily disabled and then reinvoked by the command **PL**2. **PL**3 terminates the plotting and closes the devices, while **PL**4 enters an interactive display mode on the graphics device (e. g. allowing for rotation of 2-parameter bifurcation diagrams).

PR [*start*]. Begins a recording of subsequent commands into a special command buffer and, thus the **PROG** mode is entered. The commands are

executed only by means of the **EX** command. All commands are allowed
for recording except for **PR** and **EX**. The stop command **ST** terminates
the recording. The maximum buffer size is 30. Command buffering is
useful when the same sequence of commands must be given over and over
again e. g. when plotting a grid over a 3-dimensional bifurcation diagram
(tracing a square grid in the λ-μ space). The optional parameter *start*
denotes the number of the command in the buffer at which to begin the
recording. Thus several such "makros" can be defined.

RE . This command restarts the program at a previously stored point from
a dump file (created by the **ME** command).

SP *i*. Declares that the homotopy to the constant vector in the standard
perturbation technique uses the *i*-th unit vector.

ST [*restart*]. The stop command **ST** terminates SCOUT. To rerun SCOUT
use the command with *restart* = 1. The command is also used in mode
PROG to finish the recording of commands.

SV [*next*]. The start simplex will be centered around zero. If the program is
supposed to search for a start simplex somewhere else, a vector (or several) may be supplied from a file. Such a file might have been generated
by the commands **MF** or **GO** in a previous run of SCOUT. After the
first vector is read from the file and displayed one may obtain the next
vector from the file by typing **SV** 1.

TR *dim* [*mesh*]. The triangulation command **TR** defines the dimension and
the size of the triangulation. *dim* is the dimension in the problem not
counting parameters (initial default is the highest possible dimension).
mesh is the mesh size of the triangulation (initial default is 0.01).

TU [*nout* [*special*]]. This "turn" command is the same as **GO** except that
the orientation of the current solution curve is reversed.

P6.5 Example: Periodic Solutions to a Differential Delay Equation

The structure of continua of periodic solutions of differential delay equations
of the type $\dot{x}(t) = f(x(t), x(t-\tau))$ constitute an area of interest where continuation methods have aided intuition, and provided material for new insights
and conjectures. Such equations are often motivated from the natural sciences. In 1977 Mackey and Glass proposed the delay equation

(P6.5.1) $$\dot{x}(t) = \frac{ax(t-\tau)}{1 + x(t-\tau)^8} - bx(t), \quad a, b, \tau > 0$$

as a model for the dynamics of the production of red blood cells. The growth
rate (the first term) in the equation is assumed to depend on the concentration

of cells at time $t - \tau$. If the delay time τ is sufficiently large, as it is conjectured to be the case in patients with leukemia, the concentration $x(t)$ will oscillate or even behave chaotically, see Mackey & Glass (1977). Yorke (see Nussbaum (1979)) first considered the simplified model

$$\dot{x}(t) = -\lambda \frac{x(t-1)}{1 + |x(t-1)|^p}, \quad \lambda > 0, \ p \geq 1$$

which seems to generate very similar behavior of solutions for $p = 8$ and sufficiently large parameters λ. The program SCOUT has been used by the authors to provide a comprehensive study of periodic solutions to this latter equation, see Saupe (1982–83). One of the methods used in these studies, a Galerkin approximation, also applies to the case of the Mackey-Glass equation (P6.5.1) as explained in the following.

   To cast the problem of periodic solutions into a finite dimensional setting which is useful for computation, we employ a Galerkin method built upon Fourier analysis. First note, that a T-periodic solution to (P6.5.1) is equivalent to a 2π-periodic solution to

(P6.5.2) $$\dot{x}(t) = \frac{\lambda}{\omega} \frac{x(t-\omega)}{1 + x(t-\omega)^8} - \frac{\mu}{\omega} x(t)$$

where we have set

$$\omega = \frac{2\pi\tau}{T}, \quad \lambda = a\tau, \quad \mu = b\tau.$$

Let $C_{2\pi}$ be the space of continuous and real 2π-periodic functions and let $E_m \subset C_{2\pi}$ be the $(2m+1)$-dimensional subspace of $C_{2\pi}$ given by all trigonometric polynomials x_m of the form

$$x_m(t) = \frac{a_0}{2} + \sum_{k=1}^{m} a_k \cos kt + b_k \sin kt$$

with real coefficients $a_0, a_1, ..., a_m$ and $b_1, ..., b_m$. Introduce the operators

$$\mathcal{S}_\omega : C_{2\pi} \to C_{2\pi}, \quad \mathcal{S}_\omega x(t) = x(t - \omega), \quad \omega > 0$$
$$\mathcal{F} : C_{2\pi} \to C_{2\pi}, \quad \mathcal{F}x(t) = f(x(t)), \quad f(x) = x/(1 + x^8)$$

and the projection $\mathcal{P}_m : C_{2\pi} \to C_{2\pi}$ with $\mathcal{P}_m(C_{2\pi}) = E_m$, where the (Fourier) coefficients of $\mathcal{P}_m x$ are given as usual by

$$a_k = \frac{1}{\pi} \int_0^{2\pi} x(t) \cos(kt)dt, \quad k = 0, 1, ..., m,$$

$$b_k = \frac{1}{\pi} \int_0^{2\pi} x(t) \sin(kt)dt, \quad k = 1, 2, ..., m.$$

We call $x_m \in E_m$ a Galerkin approximation of a 2π-periodic solution of order m, if x_m satisfies the equation

(P6.5.3)
$$\dot{x}_m - \frac{\lambda}{\omega}\, \mathcal{P}_m \mathcal{F} \mathcal{S}_\omega x_m + \frac{\mu}{\omega}\, x_m = 0$$

This means that x_m is a Galerkin approximation, if the first Fourier coefficients of
$$\frac{\lambda}{\omega}\, \frac{x_m(t-\omega)}{1 + x_m(t-\omega)^8} - \frac{\mu}{\omega}\, x_m(t)$$

coincide with the coefficients of the derivative of x_m.

But (P6.5.3) alone is not yet sufficient for the computation of periodic solutions of (P6.5.1), since the differential equation is autonomous and the exact frequencies ω are unknown. Moreover, if x_m solves (P6.5.3) then so does $\mathcal{S}_s x_m$ for all $s \in \mathbf{R}$. Thus solutions are not isolated, which poses another problem for numerical methods. These problems are overcome by regarding ω as an unknown variable and by adding an additional equation to the system, which removes the ambiguity of solutions. One such possible "anchor" equation is

(P6.5.4)
$$\dot{x}_m(0) = 0.$$

In the numerical evaluation of the expressions in equations (P6.5.3), (P6.5.4) the only apparent problem is the computation of $\mathcal{S}_\omega x_m$ from the coefficients of x_m and of the Fourier coefficients of $\mathcal{F}\mathcal{S}_\omega x_m$. While $\mathcal{S}_\omega x_m$ can be obtained directly using basic trigonometric identities, one employs the fast Fourier transform methods to compute $\mathcal{P}_m \mathcal{F} \mathcal{S}_\omega x_m$. Two such transforms are necessary, the first to compute 2^k values of $\mathcal{S}_\omega x_m(t)$ at equidistantly sampled times (here we must require $2^k > 2m + 1$). Then the nonlinearity f is applied to obtain $\mathcal{F}\mathcal{S}_\omega x_m(t)$ at discrete times, and the second (inverse) Fourier transform takes these values back into the frequency domain. Only the first $2m+1$ leading coefficients are kept and used for the evaluation of the Galerkin equation. This program is carried out in the computer code listed below. The Fourier transform routine REALFT() is not listed, we have used the code from Press & Flannery & Teukolsky & Vetterling (1986).

```
      SUBROUTINE FCT(F,V,RPARAM,IPRD)
      REAL F(1), V(1), W(64)
      REAL RPARAM(1)
      INTEGER IPRD                              the dimension of V, F
      COMMON /CFCT/ RFC1,RFC2,IFC1,IFC2         parameters (FC command)
C
C     Mackey Glass Equation
C         dx/dt(t) = lambda/w * x(t-w)/(1+x(t-w)**p) - mu/w * x(t)
C
C     On input V contains:
C         V(1) = a(0)                  double constant part of x(t)
C         V(2) = T                     period (w = 2 pi / T)
```

```
C          V(2k+1) = a(k), k=1,2,...   cosine coefficients
C          V(2k+2) = b(k), k=1,2,...   sine coefficients
C
C      The parameters are:
C          RPARAM(1) = lambda            first factor
C          RPARAM(2) = mu                second factor
C            IFC1 = p              exponent
C
C      On output F contains:
C          F(1) :                   constant part of Galerkin equation
C          F(2) = dx/dt (0)   :     the anchor expression
C           F(2k+1), k=1,2,... :    cosine coefficients
C           F(2k+2), k=1,2,... :    sine coefficients
C
C
C      Compute coefficiencts of shifted function x(t-w)
       OMEGA = 6.2831852 / V(2)                    omega = 2*pi / period
       W(1) = V(1)                                        constant part
       W(2) = 0E0                                            last term
       DO 200 K = 1, IPRD/2-1              use trigonmetric recursion
          C = COS (K * OMEGA)                     for all other terms
          S = SIN ·(K * OMEGA)
          K1 = 2 * K + 1
          K2 = K1 + 1
          W(K1) = C * V(K1) - S * V(K2)
          W(K2) = S * V(K1) + C * V(K2)
   200 CONTINUE
       ILEN = 64                       discretization size, number of samples
       DO 300 I =  IPRD+1, ILEN
   300 W(I) = 0E0                               set extra coefficients to zero
C
C      Shifted function x(t-w) at discrete times and nonlinearity
       CALL REALFT (W, ILEN/2, -1)             inverse Fourier transform
       DO 400 I = 1, ILEN                         apply the nonlinearity
   400 W(I) = W(I) / (1E0 + ABS(W(I))**IFC1)
C
C      Do the Fourier transform back into the frequency domain
       CALL REALFT (W, ILEN/2, 1)              forward Fourier transform
       FACTOR = 2E0 / ILEN                      need to scale the result
       DO 500 I = 1, ILEN                     because REALFT returns a
   500 W(I) = FACTOR * W(I)                   multiple of true transform
C
C      Set up the Galerkin expression
       F(1) = (RPARAM(2) * V(1) - RPARAM(1) * W(1)) / OMEGA
       F(2) = 0
       DO 700 K = 1, IPRD/2 - 1
          K1 = 2 * K + 1
          K2 = K1 + 1
          F(2) = F(2) + K * V(K2)           sum up derivative terms at zero
          F(K1) =  K*V(K2)+(RPARAM(2)*V(K1)-RPARAM(1)*W(K1)) / OMEGA
          F(K2) = -K*V(K1)+(RPARAM(2)*V(K2)-RPARAM(1)*W(K2)) / OMEGA
   700 CONTINUE
       RETURN
       END
```

A suitable start for tracing a continuum of solutions can be derived from the linearization of (P6.5.2). Note, that the trivial (constant) solutions are $x(t) \equiv 0$ and $x(t) \equiv \pm c$ where c is the solution to the equation

$$\frac{\lambda c}{1 + c^8} = \mu c.$$

These trivial solutions cannot be used in the above Galerkin scheme because of nonuniqueness (the variable ω can be chosen arbitrarily). Linearizing about 0 yields the conditions

$$\omega = -\lambda \sin \omega \,,$$
$$\mu = \lambda \cos \omega$$

for solutions of the type $x(t) = e^{it}$. Therefore we expect bifurcating branches to (P6.5.2) for $\lambda = -\omega / \sin \omega$ and $\mu = -\omega / \tan \omega$. The parameter ω must be chosen between $\frac{3}{2}\pi$ and 2π up to multiples of 2π. Linearizing about c yields

$$\omega = -\lambda f'(c) \sin \omega \,,$$
$$\mu = \lambda f'(c) \cos \omega \,.$$

Setting $m = \lambda f'(c)$, we derive the conditions

$$\mu = \frac{-\omega}{\tan \omega} \,,$$
$$m = \frac{-\omega}{\sin \omega} \,,$$
$$\lambda = \frac{8\mu^2}{m + 7\mu} \,,$$

for ω between $\frac{\pi}{2}$ and π up to multiples of 2π. As an example, we get

$$\lambda = 1.688, \quad \mu = 1, \quad \omega = 2.029.$$

For these values we have

$$c = 0.9548, \quad T = \frac{2\pi}{\omega} = 3.0979.$$

Thus, the function $x(t) = c + \varepsilon \cos t$ for small $\varepsilon > 0$ should be very close to true solutions of (P6.5.2). In terms of a start vector we may use ($\varepsilon = 0.2$)

$$(0.9548, \ 3.0979, \ 0.2, \ 0.0, \ 0.0, \ \ldots).$$

together with the above values for λ and μ. The following computer output demonstrates how the continuum of solutions is picked up and followed. The typical lines of the form

    511.St-1 Par: 3.28 1.00   S: 1.123   .546   Per:3.030   E: .004 .014

have the following meaning:

511.St-1	total number of PL steps and orientation of curve
Par: 3.28 1.00	λ and μ
S: 1.123 .546	solution x_m, constant part $\dfrac{a_0}{2}$, oscillatory part

$$||x_m(t) - \frac{a_0}{2}||_\infty$$

Per: 3.030	period of solution $x_m(t)$, i. e. $T = \dfrac{2\pi}{\omega}$
E: .004 .014	errors $||F(x, \lambda, \mu)||_\infty$ and

$$||x_m + \frac{\lambda}{\omega}\mathcal{F}S_\omega x_m - \mu x_m||_\infty \text{ (defect)}$$

    Output related to the performance of the predictor-corrector scheme lists among others the current step length, the corrector step estimator, causes of deceleration and number of PL steps per PC cycle.

```
INIT(       0): TR 16,0.05
INIT(       0): FC 8                        ¿
New Function specifications:   IFC1    8      RFC1      .000
                               IFC2    0      RFC2      .000
INIT(       0): SV
Enter filename: Start.vec
   1.9080   3.0970   .2000    .0000    .0000
    .0000    .0000   .0000    .0000    .0000
    .0000    .0000   .0000    .0000    .0000
    .0000
RPA1 :   1.7000     RPA2 :   1.0000
INIT(       0): GO
Initial triangulation values:  mesh size              .050
                                centered  around
   1.9080   3.0970   .2000    .0000    .0000
    .0000    .0000   .0000    .0000    .0000
    .0000    .0000   .0000    .0000    .0000
    .0000
                                image of center
    .0667    .0000   .0184   -.0373   -.0217
   -.0283   -.0125   .0025    .0004   -.0014
   -.0006   -.0005   .0001    .0000    .0000
    .0000

Initial problem values:       dimension           16
                              parameters  PA1     1.700
                                          PA2     1.000
                              variable    PA1

                              function specifications
                                          IFC1    8
                                          IFC2    0
                                          RFC1    .000
                                          RFC2    .000
```

```
SH(    0): GO
   147.St-1 Par: 1.70 1.00  S:  .955  .037  Per:3.082  E: .002 .003
         X:   1.91008   3.08231    .03613   -.00088   -.00045
              .00040    .00001     .00010   -.00004   -.00008
              .00004    .00000     .00000    .00003   -.00002
             -.00001

FIX(   147): OU 100
FIX(   147): GO 3
   247.St-1 Par: 1.76 1.00  S:  .960  .099  Per:3.098  E: .001 .002
   347.St-1 Par: 1.90 1.00  S:  .973  .175  Per:3.098  E: .003 .005
   447.St-1 Par: 2.05 1.00  S:  .988  .236  Per:3.093  E: .004 .010
FIX(   447): PC 1.0
Max step length:   1.0 (hyperplanes)  Max estimator :   5.0
P/C(   447): GO 10
    1. PC-cycle   Step length :  .1000   PL -steps:         15
   462.St-1 Par: 2.15 1.00  S:  .997  .272  Per:3.095  E: .004 .007
    2. PC-cycle   Step length :  .1500   PL -steps:         10
   472.St-1 Par: 2.30 1.00  S: 1.012  .315  Per:3.085  E: .004 .009
    3. PC-cycle   Step length :  .2250   PL -steps:         13
   485.St-1 Par: 2.53 1.00  S: 1.034  .373  Per:3.069  E: .004 .010
    4. PC-cycle   Step length :  .3375   PL -steps:         14
   499.St-1 Par: 2.80 1.00  S: 1.064  .433  Per:3.057  E: .006 .013
    5. PC-cycle   Step length :  .5063   PL -steps:         12
   511.St-1 Par: 3.28 1.00  S: 1.123  .546  Per:3.030  E: .004 .014
    6. PC-cycle   Step length :  .7594   PL -steps:         35
   546.St-1 Par: 4.00 1.00  S: 1.221  .715  Per:3.037  E: .005 .022
    7. PC-cycle   Step length : 1.0000   PL -steps:          5
   551.St-1 Par: 4.95 1.00  S: 1.364  .942  Per:3.085  E: .004 .037
    8. PC-cycle   Step length : 1.0000   PL -steps:         27
   578.St-1 Par: 5.88 1.00  S: 1.505 1.170  Per:3.158  E: .005 .047
Deceleration:   578   Estimator:  8.393  2.315
    9. PC-cycle   Step length :  .5000   PL -steps:         16
   594.St-1 Par: 6.44 1.00  S: 1.589 1.308  Per:3.206  E: .006 .058
   10. PC-cycle   Step length :  .7500   PL -steps:         15
   609.St-1 Par: 7.17 1.00  S: 1.697 1.490  Per:3.264  E: .004 .066
P/C(   609): PC -1
Predictor-corrector scheme is turned off.
FIX(   609): LE
FIX(   609): GO 1,1
   622.St-1 Par: 7.20 1.00  S: 1.702 1.497  Per:3.266  E: .005 .068
         X:   3.40302   3.26634   1.19040    .01697    .17557
             -.16245    .01753    .03253    .05444    .03352
              .03233    .00272    .01426    .00487    .01199
              .00478
FIX(   622): PA 2
New variable parameter  PA2
FIX(   622): PC 2.0
Max step length:   2.0 (hyperplanes)  Max estimator :   5.0
P/C(   622): GO 13
    1. PC-cycle   Step length :  .2000   PL -steps:         92
   714.St-1 Par: 7.20 1.10  S: 1.600 1.358  Per:3.121  E: .005 .071
    2. PC-cycle   Step length :  .3000   PL -steps:        102
   816.St-1 Par: 7.20 1.25  S: 1.454 1.160  Per:2.942  E: .004 .068
Deceleration:   816   Estimator:  5.660   .773
    3. PC-cycle   Step length :  .2250   PL -steps:         40
   856.St-1 Par: 7.20 1.35  S: 1.368 1.042  Per:2.847  E: .004 .073
    4. PC-cycle   Step length :  .3375   PL -steps:         25
```

```
   881.St-1 Par: 7.20 1.50  S: 1.259  .896  Per:2.735  E: .003 .074
Deceleration:    881    Estimator:  8.055  5.301
      5. PC-cycle   Step length :  .2531   PL -steps:      38
   919.St-1 Par: 7.20 1.65  S: 1.175  .819  Per:2.657  E: .004 .063
....
     13. PC-cycle   Step length :  .8109   PL -steps:      32
  1152.St-1 Par: 7.20 5.05  S:  .883  .110  Per:2.369  E: .018 .017
P/C( 1152): MF
   1.7670   2.3690    .0958   -.0347   -.0087
    .0067    .0028    .0061   -.0010   -.0002
    .0001    .0007   -.0002    .0001    .0000
    .0000
P/C( 1152): ST
```

This listing has been shortened and edited a bit to accomodate some restrictions necessary for printing it here. Let us explain some of the operations that were carried out in this short run of SCOUT. Initially, at step 0, the dimension is set to 16 and the mesh size to 0.05. Also the exponent in the nonlinearity (coded as IFC1) is set to 8. Then the start vector is read in from the file 'Start.vec' and the first GO command lets the program set up a simplex around the given vector, determine the labels and so on. The mode changes from INIT to SH. A corrector step is necessary, since the simplex is not completely labeled (next GO command). After 147 PL steps the algorithm finds a completely labeled simplex, outputs the solution vector and changes the mode to FIX. Three outputs are next requested, each one after 100 steps. The variable parameter λ has increased from 1.70 to 2.05 while the solution grew in norm from 0.037 to 0.326. At step 447 the predictor-corrector algorithm is activated with a maximal step length of 1.0 and a maximal estimator of 5 mesh sizes (default). 10 cycles are performed and the current step length increases from 0.1 to 1.0. In the 9-th cycle a deceleration takes place due to a too large estimator 8.393. The second number 2.315 is the estimator for the modified predictor, i. e. after the correction worth 8.393 mesh sizes has been applied. The ratio between these two numbers can also be used additionally to control the step sizes. At step 609 the PC method is turned off, and the standard algorithm in the fixed triangulation advances one level and outputs the solution vector there. Next the variable parameter is changed to μ (it is necessary, that the current simplex is in a level of the triangulation). Then the solution branch is followed in that new direction using the predictor-corrector method. After 13 cycles the norm of the solution vector has decreased almost to 0, thus we are at another bifurcation point for a different value of ω.

We finish this section with a bifurcation diagram, see figure P6.a, which depicts the solutions for five different values of μ as they bifurcate from the trivial constant solution $x(t) \equiv c$. On the vertical axis the norm of the oscillatory part of the solutions x_m is plotted, i. e. $||x_m(t) - \frac{a_0}{2}||_\infty$. Also there is a secondary bifurcation included for the $\mu = 2, 3, 4, 5$ branches. These are period doubling bifurcations which can be found very easily by starting the algorithm on the primary branch using a doubled period and suitably modified coefficients (set a_k and b_k to 0 for odd indices $k > 0$). The continuum will automatically branch off at the secondary bifurcation points. These secondary periodic solutions cannot be continued to $\mu = 1$. The loop of the secondary branch present for $\mu = 2$ will close itself to a point when the parameter μ decreases. The solutions bifurcating from the other trivial constant solution $x(t) \equiv 0$ result in a very similar bifurcation diagram. We remark, that these studies presented here are only the results of a few runs of SCOUT and are far from complete. Moreover, the dimension in the Galerkin approximation is only 16, i. e. the highest frequency terms are only $\sin 7t$ and $\cos 7t$.

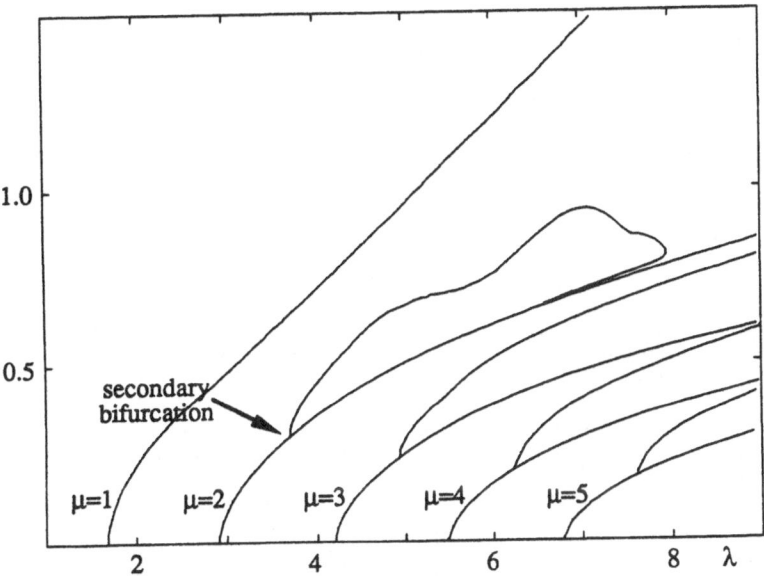

Figure P6.a Differential delay bifurcation diagram

Bibliography

Abbott, J. P. & Brent, R. P. (1978): A note on continuation methods for the solution of nonlinear equations. *J. Austral. Math. Soc.* **20**, 157–164.

Abraham, R. & Robbin, J. (1967): *Transversal mappings and flows*. W. A. Benjamin, New York, Amsterdam.

Acar, R. & Todd, M. J. (1980): A note on optimally dissecting simplices. *Math. Oper. Res.* **5**, 63–66.

Adler, I. & Saigal, R. (1976): Long monotone paths in abstract polytopes. *Math. Oper. Res.* **1**, 89–95.

Alexander, J. C. (1978): The topological theory of an imbedding method. In: *Continuation methods*. H.-J. Wacker editor. Academic Press, New York, London, 37–68.

Alexander, J. C. (1978): Bifurcation of zeros of parametrized functions. *J. Funct. Anal.* **29**, 37–53.

Alexander, J. C. (1979): Numerical continuation methods and bifurcation. In: *Functional differential equations and approximation of fixed points*. H.-O. Peitgen & H.-O. Walther editors. Lecture Notes in Math. **730**. Springer Verlag, Berlin, Heidelberg, New York, 1–15.

Alexander, J. C. (1981): A primer on connectivity. In: *Fixed point theory*. E. Fadell & G. Fournier editors. Lecture Notes in Math. **886**. Springer Verlag, Berlin, Heidelberg, New York, 455–483.

Alexander, J. C. (1987): Average intersection and pivoting densities. *SIAM J. Numer. Anal.* **24**, 129–146.

Alexander, J. C. & Li, T.-Y. & Yorke, J. A. (1983): Piecewise smooth homotopies. In: *Homotopy methods and global convergence*. B. C. Eaves & F. J. Gould & H.-O. Peitgen & M. J. Todd editors. Plenum Press, New York, 1–14.

Alexander, J. C. & Slud, E. V. (1983): Global convergence rates of piecewise-linear continuation methods: a probabilistic approach. In: *Homotopy methods and global convergence*. B. C. Eaves & F. J. Gould & H.-O. Peitgen & M. J. Todd editors. Plenum Press, New York, 15–30.

Alexander, J. C. & Yorke, J. A. (1978): Homotopy continuation method: numerically implementable topological procedures. *Trans. Amer. Math. Soc.* **242**, 271–284.

Alexandroff, P. S. (1956–1960): *Combinatorial topology*, Vol. 1–3. Graylock Press, Rochester, NY.

Allgower, E. L. (1977): Application of a fixed point search algorithm to nonlinear boundary value problems having several solutions. In: *Fixed points: algorithms and applications*. S. Karamardian editor. Academic Press, New York, 87–111.

Allgower, E. L. (1981): A survey of homotopy methods for smooth mappings. In: *Numerical solution of nonlinear equations*. E. L. Allgower & K. Glashoff & H.-O. Peitgen editors. Lecture Notes in Math. **878**. Springer Verlag, Berlin, Heidelberg, New York, 1–29.

Allgower, E. L. (1982): Homotopy methods for calculating specific critical points. In: *Proc. Nordic Symposium on Linear Complementarity Problems and Related Areas*. T. Larsson & P. A. Smeds editors. Linkoping Inst. Tech., 95–98.

Allgower, E. L. (1984): Bifurcations arising in the calculation of critical points via homotopy methods. In: *Numerical methods for bifurcation problems*. T. Küpper & H. D. Mittelmann & H. Weber editors. ISNM **70**. Birkhäuser Verlag, Basel, 15–28.

Allgower, E. L. & Böhmer, K. (1987): Application of the mesh independence principle to mesh refinement strategies. *SIAM J. Numer. Anal.* **24**, 1335–1373.

Allgower, E. L. & Chien, C.-S. (1986): Continuation and local perturbation for multiple bifurcations. *SIAM J. Sci. Statist. Comput.* **7**, 1265–1281.

Allgower, E. L. & Chien, C.-S. (1988): Bifurcations arising from complexification of real homotopies. *Soochow J. Math.* **14**, 1–10.

Allgower, E. L. & Chien, C.-S. & Georg, K. (1989): Large sparse continuation problems. *J. Comput. Appl. Math* **26**, 3–21.

Allgower, E. L. & Georg, K. (1978): Triangulations by reflections with applications to approximation. In: *Numerische Methoden der Approximationstheorie*. L. Collatz & G. Meinardus & H. Werner editors. ISNM **42**. Birkhäuser Verlag, Basel, 10–32.

Allgower, E. L. & Georg, K. (1979): Generation of triangulations by reflections. *Utilitas Math.* **16**, 123–129.

Allgower, E. L. & Georg, K. (1980): Simplicial and continuation methods for approximating fixed points and solutions to systems of equations. *SIAM Rev.* **22**, 28–85.

Allgower, E. L. & Georg, K. (1980): Homotopy methods for approximating several solutions to nonlinear systems of equations. In: *Numerical solution of highly nonlinear problems*. W. Forster editor. North-Holland, Amsterdam, New York, 253–270.

Allgower, E. L. & Georg, K. (1983): Relationships between deflation and global methods in the problem of approximating additional zeros of a system of nonlinear equations. In: *Homotopy methods and global convergence*. B. C. Eaves & F. J. Gould & H.-O. Peitgen & M. J. Todd editors. Plenum Press, New York, 31–42.

Allgower, E. L. & Georg, K. (1983): Predictor-corrector and simplicial methods for approximating fixed points and zero points of nonlinear mappings. In: *Mathematical programming: The state of the art*. A. Bachem & M. Grötschel & B. Korte editors. Springer Verlag, Berlin, Heidelberg, New York, 15–56.

Allgower, E. L. & Georg, K. (1990): Numerically stable homotopy methods without an extra dimension. In: *Computational solution of nonlinear systems of equations*, E. L. Allgower & K. Georg editors, American Mathematical Society, Providence.

Allgower, E. L. & Georg, K. (1989): Estimates for piecewise linear approximations of implicitly defined manifolds. *Appl. Math. Lett.* **1.5**, 1–7.

Allgower, E. L. & Georg, K. (1990), (eds.): *Computational solution of nonlinear systems of equations*. Lectures in Applied Mathematics Series, American Mathematical Society, Providence.

Allgower, E. L. & Glashoff, K. & Peitgen, H.-O. (1981), (eds.): *Numerical solution of nonlinear equations*. Lecture Notes in Math. **878**. Springer Verlag, Berlin, Heidelberg, New York.

Allgower, E. L. & Gnutzmann, S. (1987): An algorithm for piecewise linear approximation of implicitly defined two-dimensional surfaces. *SIAM J. Numer. Anal.* **24**, 452–469.

Allgower, E. L. & Gnutzmann, S. (1989): Polygonal meshes for implicitly defined surfaces. Preprint, March 1989.

Allgower, E. L. & Jeppson, M. M. (1973): The approximation of solutions of nonlinear elliptic boundary value problems having several solutions. In: *Numerische, insbesondere approximationstheoretische Behandlung von Funktionalgleichungen*. R. Ansorge & W. Törnig editors. Lecture Notes in Math. 333. Springer Verlag, Berlin, Heidelberg, New York, 1–20.

Allgower, E. L. & Keller, C. L. (1971): A search routine for a Sperner simplex. *Computing* **8**, 157–165.

Allgower, E. L. & Schmidt, Ph. H. (1983): PL approximation of implicitly defined manifolds. In: *Numerical analysis of parametrized nonlinear equations*. University of Arkansas Seventh Lecture Series in the Mathematical Sciences. Fayetteville, Arkansas, 1–13.

Allgower, E. L. & Schmidt, Ph. H. (1984): Piecewise linear approximation of solution manifolds for nonlinear systems of equations. In: *Selected topics in operations research and
mathematical economics*. G. Hammer & D. Pallaschke editors. Lecture Notes in Economics and Mathematical Systems **226**. Springer Verlag, Berlin, Heidelberg, New York,
339–347.

Allgower, E. L. & Schmidt, Ph. H. (1985): An algorithm for piecewise-linear approximation
of an implicitly defined manifold. *SIAM J. Numer. Anal.* **22**, 322–346.

Allgower, E. L. & Schmidt, Ph. H. (1986): Computing volumes of polyhedra. *Math. Comp.*
46, 171–174.

Alligood, K. (1979): Homological indices and homotopy continuation. Ph. D. Thesis, University of Maryland.

Amann, H. (1974): Lectures on some fixed point theorems. *Monografias de Matemática*.
IMPA, Rio de Janeiro.

Amann, H. (1976): Fixed point equations and nonlinear eigenvalue problems in ordered
Banach spaces. *SIAM Rev.* **18**, 620–709.

Ambrosetti, A. & Hess, P. (1980): Positive solutions of asymptotically linear elliptic eigenvalue problems. *J. Math. Anal. Appl.* **73**, 411–422.

Anselone, P. M. & Moore, R. H. (1966): An extension of the Newton-Kantorovič method
for solving nonlinear equations with an application to elasticity. *J. of Math. Anal. and
Appl.* **13**, 476–501.

Anson, D. K. & Cliffe, K. A. (1988): A numerical investigation of the Schaeffer homotopy
in the problem of Taylor-Couette flows. To appear in: *Proceedings of the Royal Society*.

Asmuth, R. & Eaves, B. C. & Peterson, E. L. (1979): Computing economic equilibria on
affine networks with Lemke's algorithms. *Math. Oper. Res.* **4**, 209–214.

Avila, J.H. (1974): The feasibility of continuation methods for nonlinear equations. *SIAM
J. Numer. Anal.* **11**, 102–122.

Awoniyi, S. A. & Todd, M. J. (1983): An efficient simplicial algorithm for computing a zero
of a convex union of smooth functions. *Math. Programming* **25**, 83–108.

Babuska, I. & Rheinboldt, W. C. (1982): Computational error estimation and adaptive
processes for some nonlinear structural problems. *Comput. Methods Appl. Mech. Engrg.*
32–34, 895–938.

Bachem, A. & Grötschel, M. & Korte, B. (1983): *Mathematical programming: the state of
the art*. Springer Verlag, Berlin, Heidelberg, New York.

Balinski, M. L. & Cottle, R. W. (1978), (eds.): *Complementarity and fixed point problems*.
Math. Programming Study **7**. North-Holland, Amsterdam, New York.

Bank, R. E. & Chan, T. F. (1986): PLTMGC: A multi-grid continuation program for pa-
rameterized nonlinear elliptic systems. *SIAM J. Sci. Statist. Comput.* **7**, 540–559.

Bank, R. E. & Dupont, T. & Yserentant, H. (1987): The hierarchical bases multigrid methods. Preprint. Konrad-Zuse-Zentrum, Berlin.

Bank, R. E. & Mittelmann, H. D. (1986): Continuation and multi-grid for nonlinear elliptic
systems. In: *Multigrid methods II*. W. Hackbusch & U. Trottenberg editors. Lecture Notes
in Math. **1228**. Springer Verlag, Berlin, Heidelberg, New York.

Bank, R. E. & Mittelmann, H. D. (1989): Stepsize selection in continuation procedures and
damped Newton's method. *J. Comput. Appl. Math.* **26**, 67–77.

Bank, R. E. & Rose, D. J. (1980): Parameter selection for Newton-like methods applicable
to nonlinear partial differential equations. *SIAM J. Numer. Anal.* **17**, 806–822.

Bank, R. E. & Rose, D. J. (1981): Global approximate Newton methods. *Numer. Math.*
37, 279–295.

Bárány, I. (1979): Subdivisions and triangulations in fixed point algorithms. International
Research Institute for Management Science, Moscow, USSR.

Bárány, I. (1980): Borsuk's theorem through complementary pivoting. *Math. Programming*
18, 84–88.

Baranov, S. I. (1972): A certain method of computing the complex roots of a system of
nonlinear equations (in Russian). *Ž. Vyčisl. Mat. i Mat. Fiz.* **12**, 199–203.

Barnett, S. (1974): A new look at classical algorithms for polynomials resultant and g.c.d.
calculation. *SIAM Rev.* **16**, 193–206.

Bartels, R. H. (1971): A stabilization of the simplex method. *Numer. Math.* **16**, 414–434.

Bartels, R. H. & Golub, G. H. (1968): Stable numerical methods for obtaining the Chebyshev solution to an overdetermined system of equations. *Comm. ACM* **11**, 401–406.

Bartels, R. H. & Golub, G. H. (1969): The simplex method of linear programming using LU decompositions. *Comm. ACM* **12**, 266–268.

Bartels, R. H. & Stoer, J. & Zengler, C. (1971): A realization of the simplex method based on triangular decompositions. In: *Linear Algebra*. J. H. Wilkinson & C. Reinsch editors. Springer Verlag, Berlin, Heidelberg, New York.

Batoz, J. L. & Dhatt, G. (1979): Incremental displacement algorithms for nonlinear problems. *Internat. J. Numer. Methods Engrg.* **14**, 1262–1267.

Bauer, L. & Keller, H. B. & Reiss, E. L. (1975): Multiple eigenvalues lead to secondary bifurcation. *SIAM J. Appl. Math.* **17**, 101–122.

Becker, K.-H. & Seydel, R. (1981): A Duffing equation with more than 20 branch points. In: *Numerical solution of nonlinear equations*. E. L. Allgower & K. Glashoff & H.-O. Peitgen editors. Lecture Notes in Math. **878**. Springer Verlag, Berlin, Heidelberg, New York, 98–107.

Ben-Israel, A. (1966): A Newton-Raphson method for the solution of systems of equations. *J. Math. Anal. Appl.* **15**, 243–252.

Ben-Israel, A. & Greville, T. N. E. (1974): *Generalized inverses: theory and applications*. John Wiley & Sons, New York.

Berge, C. (1963): *Topological spaces*. MacMillan Comp., New York.

Berger, M. S. (1977): *Nonlinearity and functional analysis*. Academic Press, New York, London.

Bernstein, S. (1910): Sur la généralisation du problème de Dirichlet. *Math. Ann.* **69**, 82–136.

Bertsekas, D. P. (1984): *Constrained optimization and Lagrange multiplier methods*. Academic Press, New York, London.

Beyn, W.-J. (1980): On discretizations of bifurcation problems. In: *Bifurcation problems and their numerical solution*. H. D. Mittelmann & H. Weber editors. ISNM **54**. Birkhäuser Verlag, Basel, 46–73.

Beyn, W.-J. (1981): Lösungszweige nichtlinearer Randwertaufgaben und ihre Approximation mit dem Differenzenverfahren. Habilitationsschrift, Univ. Konstanz.

Beyn, W.-J. (1984): Defining equations for singular solutions and numerical applications. In: *Numerical methods for bifurcation problems*. T. Küpper & H. D. Mittelmann & H. Weber editors. ISNM **70**. Birkhäuser Verlag, Basel, 42–56.

Beyn, W.-J. & Doedel, E. (1981): Stability and multiplicity of solutions to discretizations of nonlinear ordinary differential equations. *SIAM J. Sci. Statist. Comput.* **2**, 107–120.

Bhargava, R. & Hlavacek, V. (1984): Experience with adapting one-parameter imbedding methods toward calculation of countercurrent separation process. *Chem. Eng. Commun.* **28**, 165.

Bissett, E. J. & Cavendish, J. C. (1979): A numerical technique for solving nonlinear equations possessing multiple solutions. Tech. Report GMR-3057, MA-169, General Motors Research Laboratories.

Bloomenthal, J. (1988): Polygonization of implicit surfaces. *Computer Aided Geometric Design* **5**, 341–355.

Boggs, P. T. (1976): The convergence of the Ben-Israel iteration for nonlinear least squares problems. *Math. Comp.* **30**, 512–522.

Bohl, E. (1980): Chord techniques and Newton's methods for discrete bifurcation problems. *Numer. Math.* **34**, 111–124.

Bohl, E. (1981): Applications of continuation techniques in ordinary differential equations. In: *Numerical treatment of nonlinear problems*. C. T. H. Baker & C. Phillips editors, Oxford University Press, 159–170.

Bohl, E. & Bigge, J. (1985): Deformations of the bifurcation diagram due to discretization. *Math. Comp.* **45**, 393–403.

Bohl, P. (1904): Über die Bewegung eines mechanischen Systems in der Nähe einer Gleichgewichtslage. *J. Reine Angew. Math.* **127**, 179–276.

Bolstad, J. H. (1987): A high order multigrid method for elliptic problems with folds. Preprint. Submitted to: *Numer. Math.*.

Bolstad, J. H. & Keller, H. B. (1986): A multigrid continuation method for elliptic problems with folds. *SIAM J. Sci. Statist. Comput.* **7**, 1081–1104.

Bolstad, J. H. & Keller, H. B. (1987): Computation of anomalous modes in the Taylor experiment. *J. Comp. Phys.* **69**, 230–251.

Bosarge, W. E. (1971): Iterative continuation and the solution of nonlinear two-point boundary value problems. *Numer. Math.* **17**, 268–283.

Bourji, S. K. & Walker, H. F. (1987): Least-change secant updates of nonsquare matrices. Research report, Utah State University.

Bowman, C. & Karamardian, S. (1977): Error bounds for approximate fixed points. In: *Fixed points: algorithms and applications.* S. Karamardian editor. Academic Press, New York, London, 181–191.

Branin, F. H., Jr., (1971): Solution of nonlinear DC network problems via differential equations. Memoirs Mexico 1971 IEEE conference on systems, networks, and computers. Oaxtepec, Mexico.

Branin, F. H., Jr. (1972): Widely convergent method for finding multiple solutions of simultaneous nonlinear equations. *IBM J. Res. Develop.* **16**, 504–522.

Branin, F. H., Jr. & Hoo, S. K. (1972): A method for finding multiple extrema of a function of *n* variables. In: *Numerical methods for non-linear optimization.* F. A. Lootsma editor. Academic Press, New York, London, 231–237.

Brent, R. P. (1972): On the Davidenko-Branin method for solving simultaneous nonlinear equations. *IBM J. Res. Develop.* **16**, 434–436.

Brent, R. P. (1973): Some efficient algorithms for solving systems of nonlinear equations. *SIAM J. Numer. Anal.* **10**, 327–344.

Brezzi, F. & Descloux, J. & Rappaz, J. & Zwahlen, B. (1984): On the rotating beam: Some theoretical and numerical results. *Calcolo* **21**, 345–367.

Brezzi, F. & Rappaz, J. & Raviart, P. A. (1980–1981): Finite dimensional approximation of nonlinear problems.Part 1: Branches of nonsingular solutions. *Numer. Math.* **36**,1–25. Part 2: Limit points. *Numer. Math.* **37**, 1–28. Part 3: Simple bifurcation points. *Numer. Math.* **38**, 1–30.

Bristeaux, M. O. & Glowinski, R. & Perriaux, J. & Poirier, G. (1984): Non unique solutions of the transonic equation by arc length continuation techniques and finite element least squares methods. In: *Proceedings of the 5th international conference on finite elements and flow problems,* held at Austin, Texas, Jan. 1984.

Broadie, M. N. (1983): OCTASOLV user's guide. Systems Optimization Laboratory Technical Report 83/8. Department of Oper. Research, Stanford University.

Broadie, M. N. (1983): Subdivisions and antiprisms for PL homotopy algorithms. Systems Optimization Laboratory Technical Report 83/14. Depart. of Oper. Research, Stanford University.

Broadie, M. N. (1985): An introduction to the octahedral algorithm for the computation economic equilibria. In: *Economic equilibrium: model formulation and solution.* A. S. Manne editor. *Math. Programming Study* **23**. North-Holland, Amsterdam, New York, 121–143.

Broadie, M. N. & Eaves, B. C. (1987): A variable rate refining triangulation. *Math. Programming* **38**, 161–202.

Brooks, P. S. (1980): Infinite retrogression in the Eaves-Saigal algorithm. *Math. Programming* **19**, 313–327.

Brouwer, L. E. J. (1911): Beweis der Invarianz der Dimensionenzahl. *Math. Ann.* **70**, 161–165.

Brouwer, L. E. J. (1912): Über Abbildung von Mannigfaltigkeiten. *Math. Ann.* **71**, 97–115.

Browder, F. E. (1960): On continuity of fixed points under deformations of continuous mappings. *Summa Brasilia Mathematica* **4**, 183–191.

Brown, K. M. & Gearhart, W. B. (1971): Deflation techniques for the calculation of further solutions of a nonlinear system. *Numer. Math.* **16**, 334–342.

Broyden, C. G. (1965): A class of methods for solving nonlinear simultaneous equations. *Math. Comp.* **19**, 577–593.

Broyden, C. G. (1969): A new method of solving nonlinear simultaneous equations. *Computer J.* **12**, 94–99.

Broyden, C. G. (1970): The convergence of single-rank quasi-Newton methods. *Math. Comp.* **24**, 365–382.

Broyden, C. G. & Dennis, J. E., Jr. & Moré, J. J. (1973): On the local and superlinear convergence of quasi-Newton methods. *J. Inst. Math. Appl.* **12**, 223–245.

de Bruijn, N. G. (1961): *Asymptotic methods in analysis*. North-Holland, Amsterdam, second edition.

Brunovský, P. & Meravý, P. (1984): Solving systems of polynomial equations by bounded and real homotopy. *Numer. Math.* **43**, 397–418.

Butz, A. (1972): Solution of nonlinear equations with space filling curves. *J. Math. Anal. Appl.* **37**, 351–383.

Byrne, G. D. & Baird, L. A. (1985): Distillation calculations using a locally parametrized continuation method. *Comput. Chem. Eng.* **9**, 593.

Caras, S. & Janata, J. & Saupe, D. & Schmitt, K. (1985): pH-based enzyme potentiometric sensors, part 1, theory. *Analytical Chemistry* **57**, 1917–1920.

Carr, J. & Mallet-Paret, J. (1983): Smooth homotopies for finding zeros of entire functions. In: *Homotopy methods and global convergence*. B. C. Eaves & F. J. Gould & H.-O. Peitgen & M. J. Todd editors. Plenum Press, New York, 43–62.

Castelo, A. & Freitas, S. R. & Tavares dos Santos, G. (1990): PL approximation to manifolds and its application to implicit ODE's. In: *Computational solution of nonlinear systems of equations*, E. L. Allgower & K. Georg editors, American Mathematical Society, Providence.

Cellina, A. & Sartori, C. (1978): The search for fixed points under perturbations. *Rend. Sem. Mat. Univ. Padova* **59**, 199–208.

Chan, T. F. (1984): Techniques for large sparse systems arising from continuation methods. In: *Numerical methods for bifurcation problems*. T. Küpper & H. Mittelmann & H. Weber editors. ISNM **70**. Birkhäuser Verlag, Basel, 116–128.

Chan, T. F. (1984): Newton-like pseudo-arclength methods for computing simple turning points. *SIAM J. Sci. Statist. Comput.* **5**,135–148.

Chan, T. F. (1984): Deflation techniques and block-elimination algorithms for solving bordered singular systems. *SIAM J. Sci. Stat. Comp.* **5**, 121–134.

Chan, T. F. (1984): Efficient numerical methods for large nonlinear systems. Preprint, Yale University.

Chan, T. F. (1985): An approximate Newton method for coupled nonlinear systems. *SIAM J. Numer. Anal.* **22**, 904–913.

Chan, T. F. (1987): Rank revealing QR-factorizations. *Linear Algebra Appl.* **88/89**, 67–82.

Chan, T. F. & Keller, H. B. (1982): Arc-length continuation and multi-grid techniques for nonlinear eigenvalue problems. *SIAM J. Sci. Statist. Comput.* **3**, 173–194.

Chan, T. F. & Saad, Y. (1985): Iterative methods for solving bordered systems with applications to continuation methods. *SIAM J. Sci. Statist. Comput.* **6**, 438–451.

Chao, K. S. & Liu, D. K. & Pan, C. T. (1975): A systematic search method for obtaining multiple solutions of simultaneous nonlinear equations. *IEEE Trans. Circuits and Systems* **22**, 748–753.

Chao, K. S. & Saeks, R. (1977): Continuation methods in circuit analysis. *Proc. IEEE* **65**, 1187–1194.

Charnes, A. (1952): Optimality and degeneracy in linear programming. *Econometrica* **20**, 160–170.

Charnes, A. & Cooper, W. W. & Henderson, A. (1953): *An introduction to linear programming*. John Wiley & Sons, New York.

Charnes, A. & Garcia, C. B. & Lemke, C. E. (1977): Constructice proofs of theorems relating to: $F(x) = y$, with applications. *Math. Programming.* **12**, 328–343.

Chavez, R. & Seader, J. D. & Wayburn, T. L. (1986): Multiple steady state solutions for interlinked separation systems. *Ind. Eng. Chem. Fundam.* **25**, 566.

Chien, C.-S. (1989): Secondary bifurcations in the buckling problem. *J. Comput. Appl. Math* **25**, 277–287.

Chow, S. N. & Hale, J. K. (1982): *Methods of bifurcation theory*. Springer Verlag, Berlin, Heidelberg, New York.

Chow, S. N. & Mallet-Paret, J. & Yorke, J. A. (1978): Finding zeros of maps: homotopy methods that are constructive with probability one. *Math. Comp.* **32**, 887–899.

Chow, S. N. & Mallet-Paret, J. & Yorke, J. A. (1979): A homotopy method for locating all zeros of a system of polynomials. In: *Functional differential equations and approximation of fixed points*. H.-O. Peitgen & H.-O. Walther editors. Lecture Notes in Math. **730**. Springer Verlag, Berlin, Heidelberg, New York, 77–88.

Chu, M. T. (1983): On a numerical treatment for the curve-tracing of the homotopy method. *Numer. Math.* **42**, 323–329.

Chu, M. T. (1984): A simple application of the homotopy method to symmetric eigenvalue problems. *Linear Algebra Appl.* **59**, 85–90.

Chu, M. T. (1986): A continuous approximation to the generalized Schur's decomposition. *Linear Algebra Appl.* **78**, 119–132.

Chu, M. T. (1988): On the continuous realization of iterative processes. *SIAM Review* **30**, 375–387.

Chu, M. T. & Li, T.-Y. & Sauer, T. (1988): Homotopy methods for general λ-matrix problems. *SIAM J. Matrix Anal. Appl.* **9**, 528–536.

Chua, L. O. & Lin, P. M. (1976): *Computer-aided analysis of electronic circuits: algorithms and computational techniques, I.* Prentice-Hall, Englewood Cliffs, NJ.

Chua, L. O. & Ushida, A. (1976): A switching-parameter algorithm for finding multiple solutions of nonlinear resistive circuits. *Circuit Theory and Appl.* **4**, 215–239.

Cliffe, K. A. (1988): Numerical calculations of the primary-flow exchange process in the Taylor problem. *J. Fluid Mech.* **197**, 57–79.

Cliffe, K. A. & Jepson, A. D. & Spence, A. (1985): The numerical solution of bifurcation problems with symmetry with application to the finite Taylor problem. In: *Numerical methods for fluid dynamics II.* K. W. Morton & M. J. Baines editors, Oxford University Press, 155–176.

Cliffe, K. A. & Spence, A. (1985): Numerical calculation of bifurcations in the finite Taylor problem. In: *Numerical methods for fluid dynamics II.* K. W. Morton & M. J. Baines editors, Oxford University Press, 177–197.

Cliffe, K. A. & Winters, K. H. (1986): The use of symmetry in bifurcation calculations and its application to the Bénard problem. *J. of Comp. Physics* **67**, 310–326.

Cohen, A. I. (1972): Rate of convergence of several conjugate gradient algorithms. *SIAM J. Numer. Anal.* **9**, 248–259.

Cohen, D. I. A.(1967): On the Sperner lemma. *J. Combin. Theory* **2**, 585–587.

Cohen, D. I. A.(1980): On the Kakutani fixed point theorem. In: *Numerical solution of highly nonlinear problems.* W. Forster editor. North-Holland, 239–240.

Collins, G. E. (1967): Subresultants and reduced polynomial remainder sequences. *J. Assoc. Comput. Mach.* **14**, 128–142.

Collins, G. E. (1971): The calculation of multivariate polynomial resultants. *J. Assoc. Comput. Mach.* **18**, 515–532.

Collins, G. E. & Loos, R. (1976): Polynomial real root isolation by differentiation. In: Proceedings of the 1976 ACM Symposium on Symbolic and Algebraic Computation, 15–25.

Colville, A. R. (1970): A comparative study of nonlinear programming codes. In: *Proceedings of the Princeton Symposium on Mathematical Programming.* H. W. Kuhn editor. Princeton University Press, Princeton, NJ, 487–501.

Conrad, F. & Herbin, R. & Mittelmann, H. D. (1988): Approximation of obstacle problems by continuation methods. *SIAM J. Numer. Anal.* **25**, 1409–1431.

Cottle, R. W. (1972): Monotone solutions of the parametric linear complementarity problem. *Math. Programming* **3**, 210–224.

Cottle, R. W. (1974): Solution rays for a class of complementarity problems. *Mathematical Programming Study* **1**, 58–70.

Cottle, R. W. & Dantzig, G. B. (1968): Complementary pivot theory of mathematical programming. *Linear Algebra Appl.* **1**, 103–125.

Cottle, R. W. & Dantzig, G. B. (1970): A generalization of the linear complementarity problem. *Journal of Combinatorial Theory* **8**, 295–310.

Cottle, R. W. & Gianessi, F. & Lions, J. L. (1980), (eds.): *Variational inequalities and complentarity problems.* John Wiley & Sons, London.

Cottle, R. W. & Golub, G. H. & Sacher, R. S. (1978): On the solution of large structured linear complementarity problems: The block partitioned case. *Applied Mathematics and Optimization* **4**, 347–363.

Cottle, R. W. & Stone, R. E. (1983): On the uniqueness of solutions to linear complementarity problems. *Math. Programming* **27**, 191–213.

Coughran, W. M., Jr., & Pinto, M. R. & Smith, R. K. (1989): Continuation methods in semiconductor device simulation. *J. Comput. Appl. Math* **26**, 47–65.

Coxeter, H. S. M. (1934): Discrete groups generated by reflections. *Ann. of Math.* **6**, 13–29.

Coxeter, H. S. M. (1973): *Regular polytopes.* Third edition, Dover Publ., New York.

Crandall, M. G. & Rabinowitz, P. H. (1971): Bifurcation from simple eigenvalues. *J. Funct. Anal.* **8**, 321–340.

Crisfield, M. A. (1980): Incremental / iterative solution procedures for nonlinear structural analysis. In: *Numerical methods for nonlinear problems. Vol. 1.* C. Taylor & E. Hinton & D. R. J. Owen editors. Pineridge Press, Swansea, UK.

Crisfield, M. A. (1983): An arc-length method including line searches and accelerations. *Internat. J. Numer. Methods Engrg.* **19**, 1269–1289.

Cromme, L. J. (1980): Sind Fixpunktverfahren effektiv? In: *Konstruktive Methoden der finiten nichtlinearen Optimierung.* L. Collatz & G. Meinardus & W. Wetterling editors. ISNM **55**. Birkhäuser Verlag, Basel, 47–71.

Cromme, L. J. & Diener, I. (1989): Fixed point theorems for discontinuous mappings. To appear in: *Math. Programming.*

Cronin-Scanlon, J. (1964): *Fixed points and topological degree in nonlinear analysis.* Math. Surveys **11**. AMS, Providence, RI.

Crouzeix, M. & Rappaz, J. (1989): *On numerical approximation in bifurcation theory.* To appear in RMA, Masson, Paris.

Dai, Y. & Yamamoto, Y. (1989): The path following algorithm for stationary point problems on polyhedral cones. *J. Op. Res. Soc. of Japan* **32**, 286–309.

Dantzig, G. B. (1963): *Linear programming and extensions.* Princeton Univ. Press, Princeton, NJ.

Dantzig, G. B. & Eaves, B. C. (1974), (eds.): *Studies in optimization.* MAA Studies in Math. **10**. MAA, Washington, DC.

Davidenko, D. (1953): On a new method of numerical solution of systems of nonlinear equations (in Russian). *Dokl. Akad. Nauk USSR* **88**, 601–602.

Davidenko, D. (1953): On approximate solution of systems of nonlinear equations (in Russian). *Ukraïne Mat. Ž.* **5**, 196–206.

Davis, J. (1966): The solution of nonlinear operator equations with critical points. Ph. D. Thesis. Oregon State University.

Decker , D. W. & Keller, H. B. (1980): Multiple limit point bifurcation. *J. Math. Anal. Appl.* **75**, 417–430.

Decker , D. W. & Keller, H. B. (1980): Solution branching – a constructive technique. In: *New approaches to nonlinear problems in dynamics.* P. J. Holmes editor. SIAM, Philadelphia, PA, 53–69.

Decker , D. W. & Keller, H. B. (1981): Path following near bifurcation. *Comm. Pure Appl. Math.* **34**, 149–175.

Deimling, K. (1974): *Nichtlineare Gleichungen und Abbildungsgrade.* Springer Verlag, Berlin, Heidelberg, New York.

Dellnitz, M. & Werner, B. (1989): Computational methods for bifurcation problems with symmetries — with special attention to steady state and Hopf bifurcation points. *J. Comput. Appl. Math* **26**, 97–123.

Den Heijer, C. (1976): Iterative solution of nonlinear equations by imbedding methods. *Math. Centre Amsterdam* **32**.

Den Heijer, C. (1979): The numerical solution of nonlinear operator equations by imbedding methods. Ph. D. Thesis, Amsterdam.

Den Heijer, C. & Rheinboldt, W. C. (1981): On steplength algorithms for a class of continuation methods. *SIAM J. Numer. Anal.* **18**, 925–948.

Dennis, J. E., Jr. & Moré, J. J. (1974): A characterization of superlinear convergence and its application to Quasi-Newton methods. *Math. Comput.* **28**, 549–560.

Dennis, J. E., Jr. & Moré, J. J. (1977): Quasi-Newton methods, motivation and theory. *SIAM Review* **19**, 46–89.

Dennis, J. E., Jr. & Schnabel, R. B.(1983): *Numerical methods for unconstrained optimization and nonlinear equations.* Prentice-Hall, Englewood Cliffs, NJ.

Dennis, J. E., Jr. & Turner, K. (1987): Generalized conjugate directions. *Linear Algebra Appl.* **88/89**, 187–209.

Dennis, J. E., Jr. & Walker, H. F. (1981): Convergence theorems for least change secant update methods. *SIAM J, Numer. Anal.* **18**, 949–987.

Descloux, J. (1984): Two remarks on continuation procedures for solving some nonlinear equations. *Math. Methods Appl. Sci.* **6**, 512–514.

Descloux, J. & Rappaz, J. (1982): Approximation of solution branches of nonlinear equations. *RAIRO Anal. Numér.* **16**, 319–349.

Deuflhard, P. (1979): A step size control for continuation methods and its special application to multiple shooting techniques. *Numer. Math.* **33**, 115–146.

Deuflhard, P. & Fiedler, B. & Kunkel, P. (1987): Efficient numerical pathfollowing beyond critical points. *SIAM J. Numer. Anal.* **24**, 912–927.

Deuflhard, P. & Pesch, H.-J. & Rentrop, P. (1976): A modified continuation method for the numerical solution of nonlinear two-point boundary values problems by shooting techniques. *Numer. Math.* **26**, 327–343.

Diener, I. (1986): Trajectory nets connecting all critical points of a smooth function. *Math. Programming* **36**, 340–352.

Diener, I. (1987): On the global convergence of path-following methods to determine all solutions to a system of nonlinear equations. *Math. Programming* **39**, 181–188.

Doedel, E. J. (1981): AUTO: A program for the automatic bifurcation analysis of autonomous systems. In: *Proceedings of the tenth Manitoba conference on numerical mathematics and computing. Vol. I.* D. S. Meek & H. C. Williams editors. Utilitas Mathematica Publishing, Winnipeg, Man., 265–284.

Doedel, E. J. (1984): The computer-aided bifurcation analysis of predator-prey models. *J. Math. Biol.* **20**,1–14.

Doedel, E. J. (1986): *AUTO: Software for continuation and bifurcation problems in ordinary differential equations.* California Institute of Technology, Pasadena.

Doedel, E. J.& Jepson, A. D. & Keller H. B. (1984): Numerical methods for Hopf bifurcation and continuation of periodic solution paths. In: *Computing methods in applied sciences and engineering, Vol. VI.* R. Glowinski & J.-L. Lions editors. North-Holland, Amsterdam, New York,127–138.

Dontchev, A. L. & Jongen, H. Th. (1986): On the regularity of the Kuhn-Tucker curve. *SIAM J. Control Optim.* **24**, 169-176.

Doup, T. M. (1988): *Simplicial algorithms on the simplotope.* Lecture Notes in Economics and Mathematical Systems. Springer Verlag, Berlin, Heidelberg, New York.

Doup, T. M. & van der Elzen, A. H. & Talman, A. J. J. (1987): Simplicial algorithm for solving the nonlinear complementarity problem on the simplotope. In: *The computation and modelling of economic equilibria*, A. J. J. Talman & G. van der Laan editors, North Holland, Amsterdam, 125–154.

Doup, T. M. & van der Laan, G. & Talman, A. J. J. (1987): The $(2^{n+1} - 2)$-ray algorithm: a new simplicial algorithm to compute economic equilibria. *Math. Programming* **39**, 241–252.

Doup, T. M. & Talman, A. J. J. (1987): The 2-ray algorithm for solving equilibrium problems on the unit simplex. *Methods of Operations Res.* **57**, 269–285.

Doup, T. M. & Talman, A. J. J. (1987): A continous deformation algorithm on the product space of unit simplices. *Math. Oper. Res.* **12**, 485–521.

Doup, T. M. & Talman, A. J. J. (1987): A new simplicial variable dimension algorithm to find equilibria on the product space of unit simplices. *Math. Programming* **37**, 319–355.

Drexler, F.-J. (1977): Eine Methode zur Berechnung sämtlicher Lösungen von Polynomgleichungssystemen. *Numer. Math.* **29**, 45–58.

Drexler, F.-J. (1978): A homotopy method for the calculation of all zeros of zero-dimensional polynomial ideals. In: *Continuation methods.* H.-J. Wacker editor. Academic Press, New York, London, 69–93.

Dunford, N. & Schwartz, J. T. (1963): *Linear operators. Part II: Spectral theory.* Interscience Publ., New York.

Dunyak, J. P. & Junkins, J. L. & Watson, L. T. (1984): Robust nonlinear least squares estimation using the Chow-Yorke homotopy method. *J. Guidance, Control, Dynamics* **7**, 752–755.

Duvallet, J. (1986): Étude de systèmes différentiels du second ordre avec conditions aux deux bornes et résolution par la méthode homotopique simpliciale. Ph.D. Thesis. Académie de Bordeaux.

Duvallet, J. (1990): Computation of solutions of two-point boundary value problems by a simplicial homotopy algorithm. In: *Computational solution of nonlinear systems of equations,* E. L. Allgower & K. Georg editors, American Mathematical Society, Providence.

Eaves, B. C. (1970): An odd theorem. *Proc. Amer. Math. Soc.* **26**, 509–513.

Eaves, B. C. (1971): Computing Kakutani fixed points. *SIAM J. Appl. Math.* **21**, 236–244.

Eaves, B. C. (1971): On the basic theorem of complementarity. *Math. Programming* **1**, 68–75.

Eaves, B. C. (1971): The linear complementarity problem. *Management Sci.* **17**, 612–634.

Eaves, B. C. (1972): Homotopies for the computation of fixed points. *Math. Programming* **3**, 1–22.

Eaves, B. C. (1973): Piecewise linear retractions by reflexion. *Linear Algebra Appl.* **7**, 93–98.

Eaves, B. C. (1974): Solving piecewise linear convex equations. In: *Pivoting and extensions: in honor of A. W. Tucker.* M. L. Balinski editor. *Math. Programming Study* **1**. North-Holland, Amsterdam, New York, 96–119.

Eaves, B. C. (1974): Properly labeled simplexes. In: *Studies in Optimization.* G. B. Dantzig & B. C. Eaves editors. MAA Studies in Math. **10**. MAA, Washington, DC, 71–93.

Eaves, B. C. (1976): A short course in solving equations with PL homotopies. In: *Nonlinear programming.* R. W. Cottle & C. E. Lemke editors. SIAM-AMS Proc. **9**. AMS, Providence, RI, 73–143.

Eaves, B. C. (1977): Complementary pivot theory and markovian decision chains. In: *Fixed points: algorithms and applications.* S. Karamardian editor. Academic Press, New York, London, 59–85.

Eaves, B. C. (1978): Computing stationary points. *Math. Programming Study* **7**, 1–14.

Eaves, B. C. (1978): Computing stationary points, again. In: *Nonlinear Programming* **3**, O. L. Mangasarian & R. R. Meyer & S. M. Robinson editors, Academic Press, N. Y., 391–405.

Eaves, B. C. (1979): A view of complementary pivot theory (or solving equations with homotopies). In: *Functional differential equations and approximation of fixed points.* H.-O. Peitgen & H.-O. Walther editors. Lecture Notes in Math. **730**. Springer Verlag, Berlin, Heidelberg, New York, 89–111.

Eaves, B. C. (1983): Where solving for stationary points by LCP's is mixing Newton iterates. In: *Homotopy methods and global convergence.* B. C. Eaves & F. J. Gould & H.-O. Peitgen & M. J. Todd editors. Plenum Press, New York, 63–78.

Eaves, B. C. (1984): *A course in triangulations for solving equations with deformations.* Lecture Notes in Economics and Mathematical Systems **234**, Springer Verlag, Berlin, Heidelberg, New York.

Eaves, B. C. (1984): Permutation congruent transformations of the Freudenthal triangulation with minimal surface density. *Math. Programming* **29**, 77–99.

Eaves, B. C. (1984): Subdivision from primal and dual cones and polytopes. *Linear Algebra Appl.* **62**, 277–285.

Eaves, B. C. & Freund, R. M. (1982): Optimal scaling of balls and polyhedra. *Math. Programming* **23**, 138–147.

Eaves, B. C. & Gould, F. J. & Peitgen, H.-O. & Todd, M. J. (1983), (eds.): *Homotopy methods and global convergence.* Plenum Press, New York.

Eaves, B. C. & Rothblum, U. G. (1986): Invariant polynomial curves of piecewise linear maps. Preprint.

Eaves, B. C. & Rothblum, U.: Homotopies for solving equations on ordered fields with application to an invariant curve theorem. Preprint.

Eaves, B. C. & Saigal, R. (1972): Homotopies for computation of fixed points on unbounded regions. *Math. Programming* **3**, 225–237.

Eaves, B. C. & Scarf, H. (1976): The solution of systems of piecewise linear equations. *Math. Oper. Res.* **1**, 1–27.

Eaves, B. C. & Yorke, J. A. (1984): Equivalence of surface density and average directional density. *Math. Oper. Res.* **9**, 363–375.

Edelen, D. G. B. (1976): On the construction of differential systems for the solution of nonlinear algebraic and trascendental systems of equations. In: *Numerical methods for differential systems*. L. Lapidus & W. E. Schiesser editors. Academic Press, New York, London, 67–84.

Edelen, D. G. B. (1976): Differential procedures for systems of implicit relations and implicitly coupled nonlinear boundary value problems. In: *Numerical methods for differential systems*. L. Lapidus & W. E. Schiesser editors. Academic Press, New York, London, 85–95.

Van der Elzen, A. H. & van der Laan, G. & Talman, A. J. J. (1985): Adjustment process for finding equilibria on the simplotope. Preprint. Tilburg University.

Engl, H. W. (1976): On the change of parameter in continuation methods using an idea of Anselone-Moore. *SIAM Rev.* 19, 767.

Engles, C. R. (1980): Economic equilibrium under deformation of the economy. In: *Analysis and computation of fixed points*. S. M. Robinson editor. Academic Press, New York, London, 213–410.

Fawcett, J. & Keller, H. B. (1985): 3-dimensional ray tracing and geophysical inversion in layered media. *SIAM J. Appl. Math.* 45, 491–501.

Feilmeier, M. (1972): Numerische Aspekte bei der Einbettung nichtlinearer Probleme. *Computing* 9, 355–364.

Felippa, C. A. (1983): Solution of nonlinear static equations. Preprint. Lockheed Palo Alto Research Laboratory. Revised April 1986.

Felippa, C. A. (1984): Dynamic relaxation under general increment control. In: *Innovative methods for nonlinear problems*. W. K. Liu & T. Belytschko & K. C. Park editors. Pineridge Press, Swansea, UK, 103–133.

Feng Guochen & Cong Luan (1985): On convergence and stability of the global Newton method. *Northeastern Math. J.* 1, 1–4.

Ficken, F. (1951): The continuation method for functional equations. *Comm. Pure Appl. Math.* 4, 435–456.

Filus, L. (1980): Fixed point algorithms and some graph problems. In: *Numerical solution of highly nonlinear problems*. W. Forster editor. North-Holland, Amsterdam, New York, 241–252.

Fink, J. P. & Rheinboldt, W. C. (1983): On the error behavior of the reduced basis technique for nonlinear finite element approximations. *Zeit. Angew. Math. Mech.* 63, 21–28.

Fink, J. P. & Rheinboldt, W. C. (1983): On the discretization error of parametrized nonlinear equations. *SIAM J. Numer. Anal.* 20, 732–746.

Fink, J. P. & Rheinboldt, W. C. (1983): Some analytic techniques for parametrized nonlinear equations and their discretizations. In: *Numerical methods*. V. Pereyra & A. Reinoza editors. Lecture Notes in Math. 1005, Springer Verlag, Berlin, Heidelberg, New York, 108–119.

Fink, J. P. & Rheinboldt, W. C. (1984): Solution manifolds and submanifolds of parametrized equations and their discretization errors. *Numer. Math.* 45, 323–343.

Fink, J. P. & Rheinboldt, W. C. (1984): The role of the tangent mapping in analyzing bifurcation behavior. *Zeit. Angew. Math. Mech.* 64, 407–410.

Fink, J. P. & Rheinboldt, W. C. (1985): Local error estimates for parametrized nonlinear equations. *SIAM J. Numer. Anal.* 22, 729–735.

Fink, J. P. & Rheinboldt, W. C. (1986): Folds on the solution manifold of a parametrized equation. *SIAM J. Numer. Anal.* 23, 693–706.

Fink, J. P. & Rheinboldt, W. C. (1987): A geometric framework for the numerical study of singular points. *SIAM J. Numer. Anal.* 24, 618–633.

Fisher, M. L. & Gould, F. J. (1974): A simplicial algorithm for the nonlinear complementarity problem. *Math. Programming* 6, 281–300.

Fisher, M. L. & Gould, F. J. & Tolle, J. W. (1974): A simplicial algorithm for the mixed nonlinear complementarity problem with applications to convex programming. Preprint.

Fisher, M. L. & Gould, F. J. & Tolle, J. W. (1976): A simplicial approximation algorithm for solving systems of nonlinear equations. *Symposia Mathematica* 19, 73–90.

Fisher, M. L. & Gould, F. J. & Tolle, J. W. (1977): A new simplicial approximation algorithm with restarts: relations between convergence and labeling. In: *Fixed points: algorithms and applications*. S. Karamardian editor. Academic Press, New York, London, 41–58.

Fisher, M. L. & Tolle, J. W. (1977): The nonlinear complementarity problem: existence and determination of solutions. *SIAM J. Control Optim.* 15, 612–624.

Fletcher, R. (1980–1981): *Practical methods of optimization.* Vol. I: Unconstrained optimization. Vol. II: Constrained optimization. John Wiley & Sons, Chichester.

Forster, W. (1980), (ed.): *Numerical solution of highly nonlinear problems.* North-Holland, Amsterdam, New York.

Freudenthal, H. (1942): Simplizialzerlegungen von beschränkter Flachheit. *Ann. of Math.* 43, 580–582.

Freund, R. M. (1984): Variable dimension complexes. Part I: Basic theory. *Math. Oper. Res.* 9, 479–497.

Freund, R. M. (1984): Variable dimension complexes. Part II: A unified approach to some combinatorial lemmas in topology. *Math. Oper. Res.* 9, 498–509.

Freund, R. M. (1986): Combinatorial theorems on the simplotope that generalize results on the simplex and cube. *Math. Oper. Res.* 11, 169–179.

Freund, R. M. (1987): Dual gauge programs, with applications to quadratic programming and the minimum-norm problem. *Math. Programming* 38, 47–67.

Freund, R. M. & Todd, M. J. (1981): A constructive proof of Tucker's combinatorial lemma. *J. Combin. Theory, Ser. A* 30, 321–325.

Fried, I. (1984): Orthogonal trajectory accession to the nonlinear equilibrium curve. *Comput. Methods Appl. Mech. Engrg.* 47, 283–297.

Fujisawa, T. & Kuh, E. (1972): Piecewise linear theory of nonlinear networks. *SIAM J. Appl. Math.* 22, 307–328.

Fukushima, M. (1982): Solving inequality constrained optimization problems by differential homotopy continuation methods. Preprint. Kyoto University, Japan.

Fukushima, M. (1983): A fixed point approach to certain convex programs with applications in stochastic programming. *Math. Oper. Res.* 8, 517–524.

Fulton, W. (1984): *Intersection theory.* Springer Verlag, Berlin, Heidelberg, New York.

Garcia, C. B. (1975): A fixed point theorem including the last theorem of Poincaré. *Math. Programming*, 9, 227–239.

Garcia, C. B. (1975): A global existence theorem for the equation $F(x) = y$. Report #7527. Univ. of Chicago.

Garcia, C. B. (1976): A hybrid algorithm for the computation of fixed points. *Management Sci.* 22, 606–613.

Garcia, C. B. (1977): Continuation methods for simplicial mappings. In: *Fixed points: algorithms and applications*. S. Karamardian editor. Academic Press, New York, London, 149–163.

Garcia, C. B. (1977): Computation of solutions to nonlinear equations under homotopy invariance. *Math. Oper. Res.* 2, 25–29.

Garcia, C. B. & Gould, F. J. (1976): An algorithm based on the equivalence of vector and scalar labels in simplicial approximation. Report #7626, Univ. of Chicago.

Garcia, C. B. & Gould, F. J. (1976): An improved scalar generated homotopy path for solving $f(x) = 0$. Report #7633, Univ. of Chicago.

Garcia, C. B. & Gould, F. J. (1978): A theorem on homotopy paths. *Math. Oper. Res.* 3, 282–289.

Garcia, C. B. & Gould, F. J. (1979): Scalar labelings for homotopy paths. *Math. Programming* 17, 184–197.

Garcia, C. B. & Gould, F. J. (1980): Relations between several path following algorithms and local and global Newton methods. *SIAM Rev.* 22, 263–274.

Garcia, C. B. & Gould, F. J. & Turnbull, T. R. (1984): A PL homotopy method for the linear complementarity problem. In: *Mathematical programming* R. W. Cottle & M. L. Kelmanson & B. Korte editors. North-Holland, Amsterdam, New York, 113–145.

Garcia, C. B. & Li, T.-Y. (1980): On the number of solutions to polynomial systems of equations. *SIAM J. Numer. Anal.* 17, 540–546.

Garcia, C. B. & Li, T.-Y. (1981): On a path following methods for systems of equations. *Bull. Inst. Math. Acad. Sinica* **9**, 249–259.

Garcia, C. B. & Zangwill, W. I. (1979): Determining all solutions to certain systems of nonlinear equations. *Math. Oper. Res.* **4**, 1–14.

Garcia, C. B. & Zangwill, W. I. (1979): Finding all solutions to polynomial systems and other systems of equations. *Math. Programming* **16**, 159–176.

Garcia, C. B. & Zangwill, W. I. (1979): An approach to homotopy and degree theory. *Math. Oper. Res.* **4**, 390–405.

Garcia, C. B. & Zangwill, W. I. (1980): The flex simplicial algorithm. In: *Numerical solution of highly nonlinear problems.* W. Forster editor. North-Holland, Amsterdam, New York, 71–92.

Garcia, C. B. & Zangwill, W. I. (1980): Global continuation methods for finding all solutions to polynomial systems of equations in n variables. In: *Extremal methods and systems analysis.* A. V. Fiacco & K. O. Kortanek editors. Lecture Notes in Econom. and Math. Systems **174**. Springer Verlag, Berlin, Heidelberg, New York, 481–497.

Garcia, C. B. & Zangwill, W. I. (1981): *Pathways to solutions, fixed points, and equilibria.* Prentice-Hall, Englewood Cliffs, NJ.

Gavurin, M. K. (1958): Nonlinear functional equations and continuous analogues of iteration methods (in Russian). *Izv. Vysš. Učebn. Zaved. Matematika* **6**, 18–31.

Gear, C. W. (1971): *Numerical initial value problems in ordinary differential equations.* Prentice-Hall, Englewood Cliffs, NJ.

Georg, K. (1979): Algoritmi simpliciali come realizzazione numerica del grado di Brouwer. In: *A survey on the theoretical and numerical trends in nonlinear analysis.* Gius. Laterza & Figli, Bari, 69–120.

Georg, K. (1979): An application of simplicial algorithms to variational inequalities. In: *Functional differential equations and approximation of fixed points.* H.-O. Peitgen & H.-O. Walther editors. Lecture Notes in Math. **730**. Springer Verlag, Berlin, Heidelberg, New York, 126–135.

Georg, K. (1979): On the convergence of an inverse iteration method for nonlinear elliptic eigenvalue problems. *Numer. Math.* **32**, 69–74.

Georg, K. (1980): A simplicial deformation algorithm with applications to optimization, variational inequalities and boundary value problems. In: *Numerical solution of highly nonlinear problems.* W. Forster editor, North-Holland, Amsterdam, New York, 361–375.

Georg, K. (1981): A numerically stable update for simplicial algorithms. In: *Numerical solution of nonlinear equations.* E. Allgower & K. Glashoff & H.-O. Peitgen editors. Lecture Notes in Math. **878**. Springer Verlag, Berlin, Heidelberg, New York, 117–127.

Georg, K. (1981): Numerical integration of the Davidenko equation. In: *Numerical solution of nonlinear equations.* E. Allgower & K. Glashoff & H.-O. Peitgen editors. Lecture Notes in Math. **878**. Springer Verlag, Berlin, Heidelberg, New York, 128–161.

Georg, K. (1981): On tracing an implicitly defined curve by quasi-Newton steps and calculating bifurcation by local perturbation. *SIAM J. Sci. Statist. Comput.* **2**, 35–50.

Georg, K. (1982): Zur numerischen Realisierung von Kontinuitätsmethoden mit Prädiktor-Korrektor- oder simplizialen Verfahren. Habilitationsschrift, Univ. Bonn.

Georg, K. (1983): A note on stepsize control for numerical curve following. In: *Homotopy methods and global convergence.* B. C. Eaves & F. J. Gould & H.-O. Peitgen & M. J. Todd editors. Plenum Press, New York, 145–154.

Georg, K. (1989): An introduction to PL algorithms. To appear in: *Computational solution of nonlinear systems of equations,* E. L. Allgower & K. Georg editors, American Mathematical Society, Providence.

Georg, K. & Hettich, R. (1987): On the numerical stability of simplex-algorithms. *Optimization* **18**, 361–372.

Georganas, N. D. & Chatterjee, A. (1975): Invariant imbedding and the continuation method: a comparison. *Internat. J. Systems Sci.* **6**, 217–224.

Gfrerer, H. & Guddat, J. & Wacker, H.-J. (1983): A globally convergent algorithm based on imbedding and parametric optimization. *Computing* **30**, 225–252.

Gill, P. E. & Golub, G. H. & Murray, W. & Saunders, M. A. (1974): Methods for modifying matrix factorizations. *Math. Comp.* **28**, 505–535.

Gill, P. E. & Murray, W. (1973): A numerically stable form of the simplex algorithm. *Linear Algebra Appl.* **7**, 99-138.

Gill, P. E. & Murray, W. & Saunders, M. A. & Tomlin, J. A. & Wright, M. H. (1986): On projected Newton barrier methods for linear programming and an equivalence to Karmarkar's projective method. *Math. Programming* **36**, 183-209.

Gill, P. E. & Murray, W. & Wright, M. H. (1981): *Practical optimization.* Academic Press, New York, London.

Glashoff, K. (1974): Eine Einbettungsmethode zur Lösung nichtlinearer restringierter Optimierungsprobleme. In: *Sechste Oberwolfach-Tagung über Operations Research.* H. P. Künzi co-editor. Operations Research Verfahren **18**. Hain, Meisenheim am Glan, 92-102.

Glowinski, R. & Keller, H. B. & Reinhart, L. (1985): Continuation-conjugate gradient methods for the least squares solution of nonlinear boundary value problems. *SIAM J. Sci. Statist. Comput.* **6**, 793-832.

Gnutzmann, S. (1988): *Stückweise lineare Approximation implizit definierter Mannigfaltigkeiten.* Ph.D. thesis, University of Hamburg.

Golub, G. H. & Van Loan, Ch. F. (1983): *Matrix computations.* The Johns Hopkins University Press, Baltimore, MD.

Golubitsky, M. & Schaeffer, D. G. (1985): *Singularities and groups in bifurcation theory.* Vol. I, Springer Verlag, Berlin, Heidelberg, New York.

Golubitsky, M. & Stewart, I. & Schaeffer, D. G. (1988): *Singularities and groups in bifurcation theory.* Vol. II, Springer Verlag, Berlin, Heidelberg, New York.

Gonzaga, C. C. (1987): An algorithm for solving linear programming problems in $O(n^3 L)$ operations. Preprint, University of California, Berkeley.

Górniewicz, L. (1976): Homological methods in fixed point theory of multivalued maps. *Diss. Math.* **129**.

Gould, F. J. (1980): Recent and past developments in the simplicial approximation approach to solving nonlinear equations – a subjective view. In: *Extremal methods and systems analysis.* A. V. Fiacco & K. O. Kortanek editors. Lecture Notes in Econom. and Math. Systems **174**. Springer Verlag, Berlin, Heidelberg, New York, 466-480.

Gould, F. J. & Schmidt, C. P. (1980): An existence result for the global Newton method. In: *Variational inequalities and complementarity problems.* R. Cottle & F. Giannessi & J.-L. Lions editors. John Wiley & Sons, Chichester, 187-194.

Gould, F. J. & Tolle, J. W. (1974): A unified approach to complementarity in optimization. *Discrete Math.* **7**, 225-271.

Gould, F. J. & Tolle, J.. (1976): An existence theorem for solutions to $f(x) = 0$. *Math. Programming* **11**, 252-262.

Gould, F. J. & Tolle, J. W. (1983): *Complementary pivoting on a pseudomanifold structure with applications on the decision sciences.* Sigma Series in Applied Mathematics **2**. Heldermann Verlag, Berlin.

Granas, A. (1959): Sur la notion du degree topologique pour une certaine classe de transformations mutivalentes dans espaces de Banach. *Bull. Acad. Polon. Sci.* **7**,271-275.

Granas, A. (1976): Sur la méthode de continuité de Poincaré. *C. R. Acad. Sci. Paris Sér. A-B* **282**, 983-985.

Griewank, A. & Reddien, G. W. (1984): Characterization and computation of generalized turning points. *SIAM J. Numer. Anal.* **21**, 176-185.

Groetsch, C. W. (1984): *The theory of Tikhonov regularization for Fredholm equations of the first kind.* Pitman, London.

Guddat, J. & Jongen, H. Th. & Kummer, B. & Nozicka, F. (1987), (eds.): *Parametric optimization and related topics.* Akademie Verlag, Berlin.

Hackbusch, W. (1982): Multi-grid solution of continuation problems. In: *Iterative solution of nonlinear systems of equations.* R. Ansorge & Th. Meis & W. Törnig editors. Lecture Notes in Math. **953**. Springer Verlag, Berlin, Heidelberg, New York, 20-45.

Hackl, J. (1978): Solution of optimization problems with nonlinear restrictions via continuation methods. In: *Continuation methods.* H. J. Wacker editor. Academic Press, New York, London, 95-127.

Hackl, J. & Wacker, H. J. & Zulehner, W. (1979): Aufwandsoptimale Schrittweitensteuerung bei Einbettungsmethoden. In: *Constructive methods for nonlinear boundary value problems and nonlinear oscillations.* J. Albrecht & L. Collatz & K. Kirchgässner editors. ISNM **48**. Birkhäuser Verlag, Basel, 48–67.

Hackl, J. & Wacker, H. J. & Zulehner, W. (1980): An efficient step size control for continuation methods. *BIT* **20**, 475–485.

Hadeler, K. P. (1980): Effective computation of periodic orbits and bifurcation diagrams in delay equations. *Numer. Math.* **34**, 457–467.

Hansen, T. (1974): On the approximation of Nash equilibrium points in an N-person noncooperative game. *SIAM J. Appl. Math.* **26**, 622–637.

Harker, P. T. & Pang, J.-S. (1987): Finite-dimensional variational inequality and nonlinear complementarity problems: a survey of theory, algorithms and applications. Preprint, The Wharton School, University of Pennsylvania.

Harray, F. (1969): *Graph theory.* Addison-Wesley.

Hartmann, Ph. (1964): *Ordinary differential equations.* John Wiley and Sons, New York.

Haselgrove, C. B. (1961): The solution of nonlinear equations and of differential equations with two-point boundary conditions. *Comput. J.* **4**, 255–259.

Hlavacek, V. & Van Rompay, P. (1982): Calculation of parametric dependence and finite-difference methods. *AIChE J.* **28**, 1033–1036.

Hlavacek, V. & Seydel, R. (1987): Role of continuation in engineering analysis. *Chemical Eng. Science* **42**, 1281–1295.

Hayes, L. & Wasserstrom, E. (1976): Solution of nonlinear eigenvalue problems by the continuation method. *J. Inst. Math. Appl.* **17**, 5–14.

Healey, T. J. (1988): A group theoretic approach to computational bifurcation problems with symmetry. To appear in: *Comput. Meth. Appl. Mech. & Eng.*.

Heindel, L. E. (1971): Integer arithmetic algorithms for polynomial real zero determination. *J. Assoc. Comput. Mach.* **18**, 533–548.

Henrici, P. (1974-1977): *Applied and computational complex analysis. Vol. 1 and 2.* John Wiley & Sons, New York.

Hestenes, M. R. & Stiefel, E. (1952): Methods of conjugate gradients for solving linear systems. *J. Res. Nat. Bur. Standards* **49**, 409–436.

Hestenes, M. R. (1980): *Conjugate direction methods in optimization.* Springer Verlag, Berlin, Heidelberg, New York.

Hettich, R. (1985): Parametric optimization: applications and computational methods. In: *Methods of Operations Research* **53**, M. J. Beckmann, K.-W. Gaede, K. Ritter, H. Schneeweiss editors, 85–102.

Van der Heyden, L. (1980): A variable dimension algorithm for the linear complementarity problem. *Math. Programming* **19**, 328–346.

Van der Heyden, L. (1982): A refinement procedure for computing fixed points using Scarf's primitive sets. *Math. Oper. Res.* **7**, 295–313.

Himmelblau, D. M. (1972):A uniform evaluation of unconstrained optimization techniques. In: *Numerical methods for nonlinear optimization.* F. Lootsma editor, Academic Press, 69–97.

Himmelblau, D. M. (1972): *Applied nonlinear programming.* McGraw-Hill.

Hirsch, M. W. (1963): A proof of the nonretractibility of a cell onto its boundary. *Proc. Amer. Math. Soc.* **14**, 364–365.

Hirsch, M. W. (1976): *Differential topology.* Springer Verlag, Berlin, Heidelberg, New York.

Hirsch, M. W. & Smale, S. (1974): *Differential equations, dynamical systems, and linear algebra.* Academic Press, New York, London.

Hirsch, M. W. & Smale, S. (1979): On algorithms for solving $F(x) = 0$. *Comm. Pure Appl. Math.* **32**, 281–312.

Hlavacek, V. & Rompay, P. V. (1985): Simulation of counter-current separation processes via global approach. *Comput. Chem. Eng.* **9**, 343.

von Hohenbalken, B. (1975): A finite algorithm to maximize certain pseudoconcave functions of polytopes. *Math. Programming* **9**, 189–206.

von Hohenbalken, B. (1977): Simplicial decomposition in nonlinear programming algorithms. *Math. Programming* **13**, 49–68.

von Hohenbalken, B. (1981): Finding simplicial subdivisions of polytopes. *Math. Programming* 21, 233–234.

Holodniok, M. & Kubíček, M. (1984): DERPER — An algorithm for the continuation of periodic solutions in ordinary differential equations. *J. Comp. Phys.* 55, 254–267.

Holodniok, M. & Kubíček, M. (1984): Continuation of periodic solutions in ordinary differential equations — Numerical algorithm and application to Lorenz model. In: *Numerical methods for bifurcation problems*. T. Küpper & H. D. Mittelmann & H. Weber editors. ISNM 70. Birkhäuser Verlag, Basel, 181–194.

Hoppe, R. H. W. & Mittelmann, H. D. (1988): A multi-grid continuation strategy for parameter-dependent variational inequalities. *J. Comput. Appl. Math* 26, 35–46.

Hornung, U. & Mittelmann, H. D. (1989): A finite element method for capillary surfaces with volume constraints. To appear in: *J. Comput. Phys.*.

Hsu, C. S. & Zhu, W. H. (1984): A simplicial method for locating zeroes of a function. *Quart. of Appl. Math.* 42, 41–59.

Hudson, J. F. P. (1969): *Piecewise linear topology.* W. A. Benjamin, New York, Amsterdam.

Huitfieldt, J. & Ruhe, A. (1988): A new algorithm for numerical path following applied to an example from hydrodynamical flow. Preprint, Chalmers University of Technology, Göteborg..

Ikeno, E. & Ushida, A. (1976): The arc-length method for the computation of characteristic curves. *IEEE Trans.Circuits and Systems*, 181–183.

Irani, K. M. & Ribbens, C. J. & Walker, H. F. & Watson, L. T. & Kamat, M. P. (1989): Preconditioned conjugate gradient algorithms for homotopy curve tracking. Preprint, Virginia Polytechnic Institute.

Itoh, K. (1975): On the numerical solution of optimal control problems in Chebyshev series by the use of a continuation method. *Mem. Fac. Sci. Kyushu Univ. Ser. A* 29, 149–172.

Jakovlev, M. N. (1964): On the solutions of systems of nonlinear equations by differentiation with respect to a parameter (in Russian). *Ž. Vyčisl. Mat. i Mat. Fiz.* 4, 146–149.

Jansen, R. & Louter, R. (1980): An efficient way of programming Eaves' fixed point algorithm. In: *Numerical solution of highly nonlinear problems*. W. Foster editor. North-Holland, Amsterdam, New York, 115–168.

Jarre, F. & Sonnevend, G. & Stoer, J. (1987): An implementation of the method of analytic centers. Technical report, University of Würzburg.

Jeppson, M. M. (1972): A search for the fixed points of a continuous mapping. In: *Mathematical topics in economic theory and computation*. R. H. Day & S. M. Robinson editors. SIAM, Philadelphia, PA, 122–129.

Jepson, A. D. & Keller, H. B. (1984): Steady state and periodic solution paths: their bifurcation and computations. In: *Numerical methods for bifurcation problems*. T. Küpper & H. D. Mittelmann & H. Weber editors. ISNM 70. Birkhäuser Verlag, Basel, 219–246.

Jepson, A. D. & Spence, A. (1985): Folds in solutions of two parameter systems and their calculation. Part I. *SIAM J. Numer. Anal.* 22,347–368.

Jongen, H. Th. (1990): Parametric optimization: Critical points and local minima. In: *Computational solution of nonlinear systems of equations*, E. L. Allgower & K. Georg editors, American Mathematical Society, Providence.

Jongen, H. Th. & Jonker, P. & Twilt, F. (1983): *Nonlinear optimization in R^N. I. Morse theory, Chebyshev approximation.* Peter Lang Verlag.

Jongen, H. Th. & Jonker, P. & Twilt, F. (1986): *Nonlinear optimization in R^N. II. Transversality, flows, parametric aspects.* Peter Lang Verlag.

Jongen, H. Th. & Jonker, P. & Twilt, F. (1982): On one-parameter families of sets defined by (in)equality constraints. *Nieuw. Arch. Wisk.* 30, 307–322.

Jongen, H. Th. & Jonker, P. & Twilt, F. (1986): One-parameter families of optimization problems: equality constraints. *J. Optimization Theory and Applications* 48, 141–161.

Jongen, H. Th. & Jonker, P. & Twilt, F. (1986): Critical sets in parametric nonlinear complementarity problems. *Math. Programming* 34, 333–353.

Jongen, H. Th. & Jonker, P. & Twilt, F. (1987): A note on Branin's method for finding the critical points of smooth functions. In: *Parametric optimization and related topics*. J. Guddat & H. Th. Jongen & B. Kummer & F. Nozicka editors, Akademie Verlag, Berlin, 196–208.

Jongen, H. Th. & Zwier, G. (1985): On the local structure of the feasible set in semi-infinite optimization. In: ISNM **72**. Birkhäuser Verlag, Basel, 185–202.

Jongen, H. Th. & Zwier, G. (1988): On regular minimax optimization. To appear in: *J. Optim. Theory Appl.*.

Jürgens, H. & Peitgen, H.-O. & Saupe, D. (1980): Topological perturbations in the numerical study of nonlinear eigenvalue and bifurcation problems. In: *Analysis and computation of fixed points*. S. M. Robinson editor. Academic Press, New York, London, 139–181.

Jürgens, H. & Saupe, D. (1979): Numerische Behandlung von nichtlinearen Eigenwert- und Verzweigungsproblemen mit Methoden der simplizialen Topologie. Diplomarbeit, Universität Bremen.

Jürgens, H. & Saupe, D. (1990): SCOUT – Simplicial COntinuation UTilities user's manual. Technical report, Universität Bremen.

Kakutani, S. (1941): A generalization of Brouwer's fixed point theorem. *Duke Math. J.* **8**, 457–459.

Kamat, M. P. & Watson, L. T. & Venkayya, V. V. (1983): A quasi-Newton versus a homotopy method for nonlinear structural analysis. *Comput. and Structures* **17**, 579–585.

Kamiya, K. (1987): The decomposition method for systems of nonlinear equations. Core Discussion Paper # 8724, Louvain-la-Neuve, Belgium.

Karamardian, S. (1977), (ed.): *Fixed points: algorithms and applications*. Academic Press, New York, London.

Karmarkar, N. (1984): A new polynomial-time algorithm for linear programming. *Combinatorica* **4**, 373–395.

Karon, J. M. (1978): Computing improved Chebyshev approximations by the continuation method. I. Description of an algorithm. *SIAM J. Numer. Anal.* **15**, 1269–1288.

Katzenelson, J. (1965): An algorithm for solving nonlinear resistive networks. *Bell System Tech. J.* **44**, 1605–1620.

Kearfott, R. B. (1981): A derivative-free arc continuation method and a bifurcation technique. In: *Numerical solution of nonlinear equations*. E. Allgower & K. Glashoff & H.-O. Peitgen editors. Lecture Notes in Math. **878**. Springer Verlag, Berlin, Heidelberg, New York, 182–198.

Kearfott, R. B. (1983): Some general bifurcation techniques. *SIAM J. Sci. Statist. Comput.* **4**, 52–68.

Kearfott, R. B. (1983): Continuation methods and parametrized nonlinear least squares: techniques and experiments. In: *Numerical methods*. V. Pereyra & A. Reinoza editors. Lecture Notes in Math. **1005**. Springer Verlag, Berlin, Heidelberg, New York, 142–151.

Kearfott, R. B. (1983): Second-order predictors and continuation methods: implementation and practice. In: *Numerical analysis of parametrized nonlinear equations*. University of Arkansas Seventh Lecture Series in the Mathematical Sciences. Fayetteville, Arkansas.

Kearfott, R. B. (1984): On a general technique for finding directions proceeding from bifurcation points. In: *Numerical methods for bifurcation problems*. T. Küpper & H. D. Mittelmann & H. Weber editors. ISNM **70**. Birkhäuser Verlag, Basel, 210–218.

Kearfott, R. B.: A note on homotopy methods for solving systems of polynomial equation defined on the $(n-1)$-sphere and other manifolds. Preprint. Univ. of Southwest Louisiana.

Kearfott, R. B. (1987): Abstract generalized bisection and a costbound. *Math. Comput.* **49**, 187–202.

Kearfott, R. B. (1988): The role of homotopy techniques in biomedical modelling: a case study. To appear in the Proceedings of the Twelfth IMACS World Congress on Scientific Computation.

Kearfott, R. B. (1989): An interval step control for continuation methods. To appear in Math. Comp..

Kearfott, R. B. (1990): Interval arithmetic techniques in the computational solution of nonlinear systems of equations: Introduction, examples and comparisons. In: *Computational solution of nonlinear systems of equations*, E. L. Allgower & K. Georg editors, American Mathematical Society, Providence.

Keenan. D. (1981): Further remarks on the global Newton method. *Journal of Mathematical Economics* **8**, 159–166.

Keener, J. P. & Keller, H. B. (1974): Perturbed bifurcation theory. *Arch. Rational Mech. Anal.* **50**, 159–175.

Keller, H. B. (1968): *Numerical methods for two-point boundary value problems.* Blaisdell, Waltham, MA.

Keller, H. B. (1970): Nonlinear bifurcation. *J. Diff. Equ.* **7**, 417–434.

Keller, H. B. (1971): Shooting and imbedding for two-point boundary value problems. *J. Math. Anal. Appl.* **36**, 598–610.

Keller, H. B. (1977): Numerical solution of bifurcation and nonlinear eigenvalue problems. In: *Application of bifurcation theory.* P. H. Rabinowitz editor. Academic Press, New York, London, 359–384.

Keller, H. B. (1978): Global homotopies and Newton methods. In: *Recent advances in numerical analysis.* C. de Boor & G. H. Golub editors. Academic Press, New York, London, 73–94.

Keller, H. B. (1979): Constructive methods for bifurcation and nonlinear eigenvalue problems. In: *Computing methods in applied sciences and engineering. Vol. I.* R. Glowinski & J.-L. Lions editors. Lecture Notes in Math. **704**. Springer Verlag, Berlin, Heidelberg, New York, 241–251.

Keller, H. B. (1980): Isolas and perturbed bifurcation theory. In: *Nonlinear partial differential equations in engineering and applied science.* R. L. Sternberg & A. J. Kalinowski & J. S. Papadakis editors. Marcel Dekker, New York, Basel, 45–50.

Keller, H. B. (1981): Geometrically isolated nonisolated solutions and their approximation. *SIAM J. Numer. Anal.* **18**, 822–838.

Keller, H. B. (1982): Continuation methods in computational fluid dynamics. In: *Numerical and physical aspects of aerodynamic flows.* T. Cebeci editor. Springer Verlag, Berlin, Heidelberg, New York, 3–13.

Keller, H. B. (1982): Practical procedures in path following near limit points. In: *Computing methods in applied sciences and engineering. Vol. V.* R. Glowinski & J.-L. Lions editors. North-Holland, Amsterdam, New York.

Keller, H. B. (1983): The bordering algorithm and path following near singular points of higher nullity. *SIAM J. Sci. Statist. Comput.* **4**, 573–582.

Keller, H. B. (1987): *Lectures on numerical methods in bifurcation problems.* Springer Verlag, Berlin, Heidelberg, New York.

Keller, H. B. (1988): Complex bifurcation. Lecture at a conference on: *Recent trends in nonlinear computational mathematics and mechanics* , University of Pittsburgh.

Keller, H. B. & Lentini, M. (1982): Invariant imbedding, the box scheme and an equivalence between them. *SIAM J. Numer. Anal.* **19**, 942–962.

Keller, H. B. & Perozzi, D. J. (1983): Fast seismic ray tracing. *SIAM J. Appl. Math.* **43**, 981–992.

Keller, H. B. & Schreiber, R. (1983): Spurious solutions in driven cavity calculations. *J. Comp. Phys.* **49**, 165–172.

Keller, H. B. & Schreiber, R. (1983): Driven cavity flows by efficient numerical techniques. *J. Comp. Phys.* **49**, 310–333.

Keller, H. B. & Szeto, R. K.-H. (1980): Calculation of flows between rotating disks. In: *Computing methods in applied sciences and engineering.* R. Glowinski & J. L. Lions editors. North-Holland, Amsterdam, New York, 51–61.

Kellogg, R. B. & Li, T.-Y. & Yorke, J. A. (1976): A constructive proof of the Brouwer fixed point theorem and computational results. *SIAM J. Numer. Anal.* **13**, 473–483.

Kellogg, R. B. & Li, T.-Y. & Yorke, J. A. (1977): A method of continuation of calculating a Brouwer fixed point. In: *Fixed points: algorithms and applications.* S. Karamardian editor. Academic Press, New York, London, 133–147.

Khachiyan, L. G. (1979): A polynomial algorithm in linear programming. *Soviet Math. Doklady* **20**, 191–194.

Kikuchi, F. (1979): Finite element approximations to bifurcation problems of turning point type. *Theor. Appl. Mech.* **27**, 99–114.

Kim, B. B. (1980): On stepsize of parameter continuation method combined with Newton method (in Korean) *Cho-son In-min Kong-hwa-kuk Kwa-hak-won Tong-bo* **4**, 1–9.

Kirszenblat, A. & Chetrit, M. (1974): Least squares nonlinear parameter estimation by the iterative continuation method. *Technical Notes AIAA Journal* **12**, 1751–1752.

Klein, F. (1882–1883): Neue Beiträge zur Riemannschen Funktionentheorie. *Math. Ann.* **21**.

Klopfenstein, R. W. (1961): Zeros of nonlinear functions. *J. Assoc. Comput. Mach.* **8**, 366.

Knaster, B. & Kuratowski, C. & Mazurkiewicz, S. (1929): Ein Beweis des Fixpunktsatzes für n-dimensionale Simplexe. *Fund. Math.* **14**, 132–137.

Köberl, D. (1980): The solution of nonlinear equations by the computation of fixed points with a modification of the sandwich method. *Computing* **25**, 175–179.

Kojima, M. (1974): Computational methods for solving the nonlinear complementarity problem. *Keio Engrg. Rep.* **27**, 1–41.

Kojima, M. (1975): A unification of the existence theorems of the nonlinear complementarity problem. *Math. Programming* **9**, 257–277.

Kojima, M. (1978): On the homotopic approach to systems of equations with separable mappings. In: *Complementarity and fixed point problems.* M. L. Balinski & R. W. Cottle editors. *Math. Programming Study* **7**. North-Holland, Amsterdam, New York, 170–184.

Kojima, M. (1978): A modification of Todd's triangulation J_3. *Math. Programming* **15**, 223–227.

Kojima, M. (1978): Studies on piecewese-linear approximations of piecewise-C^1 mappings in fixed points and complementarity theory. *Math. Oper. Res.* **3**, 17–36.

Kojima, M. (1979): A complementarity pivoting approach to parametric programming. *Math. Oper. Res.* **4**, 464–477.

Kojima, M. (1980): Strongly stable stationary solutions in nonlinear programs. In: *Analysis and computation of fixed points.* S. M. Robinson editor. Academic Press, New York, London, 93–138.

Kojima, M. (1980): A note on "A new algorithm for computing fixed points" by van der Laan and Talman. In: *Numerical solution of highly nonlinear problems.* W. Forster editor. North-Holland, Amsterdam, New York, 37–42.

Kojima, M. (1981): An introduction to variable dimension algorithms for solving systems of equations. In: *Numerical solution of nonlinear equations.* E. L. Allgower & K. Glashoff & H.-O. Peitgen editors. Lecture Notes in Math. **878**. Springer Verlag, Berlin, Heidelberg, New York, 199–237.

Kojima, M. (1981): Recent advances in mathematical programming. X. Computation of fixed points by continuation methods. *Systems and Control* **25**, 421–430.

Kojima, M. & Hirabayashi, R. (1984): Continuous deformation of nonlinear programs. In: *Sensitivity, stability and parametric analysis.* A. V. Fiacco editor. *Math. Programming Study* **21**. North-Holland, Amsterdam, New York, 150–198.

Kojima, M. & Megiddo, N. (1977): On the existence and uniqueness of solutions in nonlinear complementarity theory. *Math. Programming* **12**, 110–130.

Kojima, M. & Mizuno, S. (1983): Computation of all solutions to a system of polynomial equations. *Math. Programming* **25**, 131–157.

Kojima, M. & Mizuno, S. & Noma, T. (1989): A new continuation method for complementarity problems with uniform P-functions. *Math. Programming* **43**, 107–113.

Kojima, M. & Mizuno, S. & Noma, T. (1988): Limiting behavior of trajectories generated by a continuation method for monotone complementarity problems. Preprint B-199, Tokyo Institute of Technology.

Kojima, M. & Mizuno, S. & Yoshise, A. (1987): A primal-dual interior point algorithm for linear programming. To appear in: *Research issues in linear programming.* N. Meggido editor, Springer-Verlag.

Kojima, M. & Mizuno, S. & Yoshise, A. (1989): A polynomial-time algorithm for a class of linear complementarity problems. *Math. Programming* **44**, 1–26.

Kojima, M. & Mizuno, S. & Yoshise, A. (1988): An $O(\sqrt{n}L)$ iteration potential reduction algorithm for linear complementarity problems. Preprint, Tokyo Institute of Technology.

Kojima, M. & Nishino, H. & Arima, N. (1979): A PL homotopy for finding all the roots of a polynomial. *Math. Programming* **16**, 37–62.

Kojima, M. & Nishino, H. & Sekine, T. (1976): An extension of Lemke's method to the piecewise linear complementarity problem. *SIAM J. Appl. Math.* **31**, 600–613.

Kojima, M. & Oishi, S. & Sumi, Y. & Horiuchi, K. (1985): A PL homotopy continuation method with the use of an odd map for the artificial level. *Math. Programming* **31**, 235–244.

Kojima, M. & Yamamoto, Y. (1982): Variable dimension algorithms: basic theory, interpretation, and extensions of some existing methods. *Math. Programming* **24**, 177–215.

Kojima, M. & Yamamoto, Y. (1984): A unified approach to the implementation of several restart fixed point algorithms and a new variable dimension algorithm. *Math. Programming* **28**, 288–328.

Kovach, J. W. & Seider, W. D. (1987): Heterogeneous azeotropic distillation: Experimental and simulation results. *A. I. Ch. E. J.* **33** 1300.

Krasnosel'skiĭ, M. A. (1964): *Topological methods in the theory of nonlinear integral equations*. Pergamon Press, New York, NY.

Kubíček, M. (1976): Algorithm 502. Dependence of solutions of nonlinear systems on a parameter. *ACM Trans. Math. Software* **2**, 98–107.

Kubíček, M. & Hlaváček, V. (1978): One-parameter imbedding techniques for the solution of nonlinear boundary-value problems. *Appl. Math. Comput.* **4**, 317–357.

Kubíček, M. & Holodniok, M. & Hlaváček, V. (1979): Solution of nonlinear boundary value problems XI. One-parameter imbedding methods. *Chem. Engrg. Sci.* **34**, 645–650.

Kubíček, M. & Holodniok, M. & Marek, I. (1981): Numerical solution of nonlinear equations by one-parameter imbedding methods. *Numer. Funct. Anal. Optim.* **3**, 223–264.

Kubíček, M. & Marek, I. (1983): *Computational methods in bifurcation theory and dissipative structures*. Springer Verlag, Berlin, Heidelberg, New York.

Kuhn, H. W. (1960): Some combinatorial lemmas in topology. *IBM J. Res. Develop.* **4**. 518–524.

Kuhn, H. W. (1968): Simplicial approximation of fixed points. *Proc. Nat. Acad. Sci. U.S.A.* **61**, 1238–1242.

Kuhn, H. W. (1969): Approximate search for fixed points. In: *Computing methods in optimization problems 2*. L. A. Zadek & L. W. Neustat & A. V. Balakrishnan editors. Academic Press, New York, London, 199–211.

Kuhn, H. W. (1974): A new proof of the fundamental theorem of algebra. In: *Pivoting and extensions: in honor of A. W. Tucker*. M. L. Balinski editor. *Math. Programming Study* **1**. North-Holland, Amsterdam, New York, 148–158.

Kuhn, H. W. (1977): Finding roots of polynomials by pivoting. In: *Fixed points: algorithms and applications*. S. Karamardian editor. Academic Press, New York, London, 11–39.

Kuhn, H. W. (1980): On the Shapley-Sperner lemma. In: *Numerical solution of highly nonlinear problems*. W. Forster editor. North-Holland, Amsterdam, New York, 233–238.

Kuhn, H. W. & MacKinnon, J. G. (1975): Sandwich method for finding fixed points. *J. Optim. Theory Appl.* **17**, 189–204.

Kuhn, H. W. & Wang, Z. K. & Xu, S. L. (1984): On the cost of computing roots of polynomials. *Math. Programming* **28**, 156–163.

Kuno, M. & Seader, J. D. (1988): Computing all real solutions to a system of nonlinear equations with a global fixed-point homotopy. *Ind. Eng. Chem. Res.* **27**, 1320–1329.

Küpper, T. & Mittelmann, H. D. & Weber, H. (1984), (eds.): *Numerical methods for bifurcation problems*. ISNM **70**. Birkhäuser Verlag, Basel.

Küpper, T. & Seydel, R. & Troger, H. (1987), (eds.): *Bifurcation: analysis, algorithms, applications*. ISNM **79**. Birkhäuser Verlag, Basel.

van der Laan, G. (1980): Simplicial fixed point algorithms. Ph.D. Thesis. Math. Centre, Amsterdam.

van der Laan, G. (1984): On the existence and approximation of zeros. *Math. Programming* **28**, 1–24.

van der Laan, G. (1985): The computation of general equilibrium in economies with a block diagonal pattern. *Econometrica* **53**, 659-665.

van der Laan, G. & Seelen, L. P. (1984). Efficiency and implementation of simplicial zero point algorithms. *Math. Programming* **30**, 196–217.

van der Laan, G. & Talman, A. J. J. (1978): A new algorithm for computing fixed points. Preprint.

van der Laan, G. & Talman, A. J. J. (1979): A restart algorithm for computing fixed points without an extra dimension. *Math. Programming* **17**, 74–84.

van der Laan, G. & Talman, A. J. J. (1979): A restart algorithm without an artificial level for computing fixed points on unbounded regions. In: *Functional differential equations and approximation of fixed points*. H.-O. Peitgen & H.-O. Walther editors. Lecture Notes in Math. **730**. Springer Verlag, Berlin, Heidelberg, New York, 247–256.

van der Laan, G. & Talman, A. J. J. (1980): An improvement of fixed point algorithms by using a good triangulation. *Math. Programming* **18**, 274–285.

van der Laan, G. & Talman, A. J. J. (1980): A new subdivision for computing fixed points with a homotopy algorithm. *Math. Programming* **19**, 78–91.

van der Laan, G. & Talman, A. J. J. (1980): Convergence and properties of recent variable dimension algorithms. In: *Numerical solution of highly nonlinear problems*. W. Forster editor. North-Holland, Amsterdam, New York, 3–36.

van der Laan, G. & Talman, A. J. J. (1981): Labelling rules and orientation: on Sperner's lemma and Brouwer degree. In: *Numerical solution of nonlinear equations*. E. L. Allgower & K. Glashoff & H.-O. Peitgen editors. Lecture Notes in Math. **878**. Springer Verlag, Berlin, Heidelberg, New York, 238–257.

van der Laan, G. & Talman, A. J. J. (1981): A class of simplicial restart fixed point algorithms without an extra dimension. *Math. Programming* **20**, 33–48.

van der Laan, G. & Talman, A. J. J. (1982): On the computation of fixed points in the product space of unit simplices and an application to noncooperative N person games. *Math. Oper. Res.* **7**, 1–13.

van der Laan, G. & Talman, A. J. J. (1983): Note on the path following approach of equilibrium programming. *Math. Programming* **25**, 363–367.

van der Laan, G. & Talman, A. J. J. (1987): Adjustment processes for finding economic equilibria. *Economics Letters* **23**, 119–123.

van der Laan, G. & Talman, A. J. J. (1987): Adjustment processes for finding economic equilibria. In: *The computation and modelling of economic equilibria*, A. J. J. Talman & G. van der Laan editors, North Holland, Amsterdam, 85–124.

van der Laan, G. & Talman, A. J. J. (1986): Simplicial algorithms for finding stationary points, a unifying description. *Journal of Optimiziation Theory and Applications* **50**, 262–281.

van der Laan, G. & Talman, A. J. J., (1987) (eds.): *The computation and modelling of economic equilibria*. North-Holland, Amsterdam.

van der Laan, G. & Talman, A. J. J., (1987): Computing economic equilibria by variable dimension algorithms: State of the art. Preprint FEW 270, Tilburg University.

van der Laan, G. & Talman, A. J. J. (1988): An algorithm for the linear complementarity problem with upper and lower bounds. To appear in: *Journal of Optimization Theory and Applications*.

van der Laan, G. & Talman, A. J. J. & van der Heyden, L. (1987): Simplicial variable dimension algorithms for solving the nonlinear complementarity problem on a product of unit simplices using a general labelling. *Math. Oper. Res.* **12**, 377–397.

van der Laan, G. & Talman, A. J. J. (1984) & van der Heyden, L. (1988): Shortest paths for simplicial algorithms. To appear in: *Math. Programming*.

Laasonen, P. (1970): An imbedding method of iteration with global convergence. *Computing* **5**, 253–258.

Langford, W. F. (1977): Numerical solution of bifurcation problems for ordinary diffenrential equations. *Numer. Math.* **28**, 171–190.

Lahaye, E. (1934): Une méthode de resolution d'une categorie d'equations transcendantes. *C. R. Acad. Sci. Paris* **198**, 1840–1842.

Lahaye, E. (1948): Sur la résolution des systèmes d'équations trascendantes. *Acad. Roy. Belg. Bull. Cl. Sci. (5)* **34**, 809–827.

Lazard, D. (1981): Résolution des systèmes d'équations algébriques. *Theoret. Comput. Sci.* **15** 77–110.

Leder, D. (1970): Automatische Schrittweitensteuerung bei global konvergenten Einbettungsmethoden. *Z. Angew. Math. Mech.* **54**, 319–324.

Leder, D. (1974): Zur Lösung nichtlinearer Gleichungssysteme mittels diskreter Einbettungsmethoden. Dissertation. Tech. Univ. Dresden.

Lemke, C. E. (1965): Bimatrix equilibrium points and mathematical programming. *Management Sci.* **11**, 681–689.

Lemke, C. E. (1968): On complementary pivot theory. In: *Mathematics of the decision sciences. Part I*. G. B. Dantzig & A. F. Veinott, Jr. editors. Lectures in Appl. Math. **11**. AMS, Providence, RI, 95–114.

Lemke, C. E. (1980): A survey of complementarity theory. In: *Variational inequalities and complentarity problems*. R. W. Cottle & F. Gianessi & J. L. Lions editors. John Wiley & Sons, London.

Lemke, C. E. & Grotzinger, S. J. (1976): On generalizing Shapley's index theory to labelled pseudo manifolds. *Math. Programming* 10, 245–262.

Lemke, C. E. & Howson, J. T. (1964): Equilibrium points of bimatrix games. *SIAM J. Appl. Math.* 12, 413–423.

Lentini, M. & Reinoza, A. (1983): Piecewise nonlinear homotopies. In: *Numerical methods*. V. Pereyra & A. Reinoza editors. Lecture Notes in Math. 1005. Springer Verlag, Berlin, Heidelberg, New York, 162–169.

Leray, J. & Schauder, J. (1934): Topologie et équations fonctionelles. *Ann. Sci. École Norm. Sup.* 51, 45–78.

Li, S. B. (1983): The funcion-factor method: a new method of passing the singularities which arise in the continuation methods for solving systems of nonlinear equations. *Math. Numer. Sinica* 5, 162–175.

Li, T.-Y. (1976): Computing the Brouwer fixed point by following the continuation curve. In: *Fixed point theory and its applications*. Academic Press, New York, London, 131–135.

Li, T.-Y. (1983): On Chow, Mallet-Paret and Yorke homotopy for solving systems of polynomials. *Bull. Inst. Math. Acad. Sinica* 11, 433–437.

Li, T.-Y. (1987): Solving polynomial systems. *Math. Intelligencer* 9, 33–39.

Li, T.-Y. & Rhee, N. H. (1989): Homotopy algorithm for symmetric eigenvalue problems. *Numer. Math.* 55, 265–280.

Li, T.-Y. & Mallet-Paret, J. & Yorke, J. A. (1985): Regularity results for real analytic homotopies. *Numer. Math.* 46, 43–50.

Li, T.-Y. & Sauer, T. (1987): Regularity results for solving systems of polynomials by homotopy method. *Numer. Math.* 50, 283–289.

Li, T.-Y. & Sauer, T. (1987): Homotopy method for generalized eigenvalue problems $Ax = \lambda Bx$. *Linear Algebra Appl.* 91, 65–74.

Li, T.-Y. & Sauer, T. & Yorke, J. A. (1987): Numerical solution of a class of deficient polynomial systems. *SIAM J. Numer. Anal.* 24, 435–451.

Li, T.-Y. & Sauer, T. & Yorke, J. A. (1987): The random product homotopy and deficient polynomial systems. *Numer. Math.* 51, 481–500.

Li, T.-Y. & Sauer, T. & Yorke, J. A. (1988): Numerically determining solutions of systems of polynomial equations. *Bull. AMS* 18, 173–177.

Li, T.-Y. & Sauer, T. & Yorke, J. A. (1989): The cheater's homotopy: An efficient procedure for solving systems of polynomial equations. To appear in: *SIAM J. Numer. Anal.* 26, 1241–1251.

Li, T.-Y. & Sun, H. Z. & Sun, X.-H. (1989): Parallel homotopy algorithm for symmetric tridiagonal eigenvalue problem. Preprint, Michigan State University.

Li, T.-Y. & Wang, X. (1989): A more efficient homotopy for solving deficient polynomial systems. Preprint, Michigan State University.

Li, T.-Y. & Yorke, J. A. (1979): Path following approach for solving nonlinear equations: homotopy, continuous Newton and projection. In: *Functional differential equations and approximation of fixed points*. H.-O. Peitgen & H.-O. Walther editors. Lecture Notes in Math. 730. Springer Verlag, Berlin, Heidelberg, New York, 257–264.

Li, T.-Y. & Yorke, J. A. (1980): A simple reliable numerical algorithm for following homotopy paths. In: *Analysis and computation of fixed points*, S. M. Robinson editor. Academic Press, New York, London, 73–91.

Lin, W. J. & Seader, J. D. & Wayburn, T. L. (1987): Computing multiple solutions to systems of interlinked separation columns. *A. I. Ch. E. J.* 33 886.

Lindfield, G. R. & Simpson, D. C. (1979): Modifications of the continuation method for the solution of systems of nonlinear equations. *Internat. J. Math. Math. Sci.* 2, 299–308.

Lozi, R. (1975): A computing method for bifurcation boughs of nonlinear eigenvalue problems. *Bull. Amer. Math. Soc.* 81, 1127–1129.

Lundberg, B. N. & Poore, A. B. (1989): Variable order Adams-Bashforth predictors with error-stepsize control for continuation methods. Preprint, Colorado State University, Ft. Collins.

Lundberg, B. N. & Poore, A. B. & Yang, B. (1990): Smooth penalty functions and continuation methods for constrained optimization. In: *Computational solution of nonlinear systems of equations*, E. L. Allgower & K. Georg editors, American Mathematical Society, Providence.

Lüthi, H.-J. (1975): A simplicial approximation of a solution for the nonlinear complementarity problem. *Math. Programming* 9, 278–293.

Lüthi, H.-J. (1976): *Komplementaritäts- und Fixpunktalgorithmen in der mathematischen Programmierung, Spieltheorie und Ökonomie*. Lecture Notes in Economics and Mathematical Systems 129. Springer Verlag, Berlin, Heidelberg, New York.

Mackens, W. (1989): Numerical differentiation of implicitly defined space curves. *Computing* 41, 237–260.

Mackey, M. C. & Glass, L. (1977): Oscillations and chaos in physiological control systems. *Science* 197, 287–289.

MacKinnon, J. G. (1977): Solving economic general equilibrium models by the sandwich method. In: *Fixed points: algorithms and applications*. S. Karamardian editor. Academic Press, New York, London, 367–402.

MacKinnon, J. G. (1980): Solving urban general equilibrium problems by fixed point methods. In: *Analysis and computation of fixed points*. S. M. Robinson editor. Academic Press, New York, London, 197–212.

Magnanti, T. L. & Orlin, J. B. (1988): Parametric linear programming and anti-cycling pivoting rules. *Math. Progr.* 41, 317–325.

Mangasarian, O. L. (1969): *Nonlinear programming*. McGraw-Hill, New York.

Mansfield, L. (1981): Finite element methods for nonlinear shell analysis. *Numer. Math.* 37, 121–131.

Manteuffel, T. A. (1979): Shifted incomplete Cholesky factorization. In: *Sparse Matrix Proceedings 1978*. I. S. Duff & G. W. Stewart editors. SIAM, Philadelphia, PA, 41–61.

Marden, M. (1949): *The geometry of the zeros of a polynomial in a complex variable*. Mathematical Surveys 3. AMS, New York, NY.

McCormick, G. P. (1983): *Nonlinear programming. Theory, algorithms and applications*. John Wiley & Sons, New York.

McCormick, G. P. & Ritter, K. (1974): Alternate proofs of the convergence properties of the conjugate-gradient method. *J. Optim. Theory Appl.* 13, 497–518.

Megiddo, N. (1978): On the parametric nonlinear complementarity problem. In: *Complementarity and fixed point problems*. M. L. Balinski & R. W. Cottle editors. *Math. Programming Study* 7. North-Holland, Amsterdam, New York, 142–150.

Megiddo, N. (1986): Pathways to the optimal set in linear programming. In: Proceedings of the 6th Mathematical Programming Syposium of Japan, Nagoya, 1–35.

Megiddo, N. & Kojima, M, (1977): On the existence and uniqueness of solutions in nonlinear complementarity theory. *Math. Programming* 12, 110–130.

Mehlem, R. G. & Rheinboldt, W. C. (1982): A comparison of methods for determining turning points of nonlinear equations. *Computing* 29, 201–226.

Mehra, R. K. & Washburn, R. B. Jr. (1982): Application of continuation methods in stability and optimization problems of engineering. In: *Mathematical programming with data perturbations*, I. Lecture Notes in Pure and Appl. Math. 73, Dekker, New York, 169–203.

Meintjes, K. & Morgan, A. P. (1987): A methodology for solving chemical equilibrum systems. *Appl. Math. Comput.* 22, 333–361.

Mejia, R. (1986): CONKUB: A conversational path-follower for systems of nonlinear equations. *J. Comp. Phys.* 63, 67–84.

Mejia, R. (1990): Interactive program for continuation of solutions of large systems of nonlinear equations. In: *Computational solution of nonlinear systems of equations*, E. L. Allgower & K. Georg editors, American Mathematical Society, Providence.

Menzel, R. (1980): Ein implementierbarer Algorithmus zur Lösung nichtlinearer Gleichungssysteme bei schwach singulärer Einbettung. *Beiträge Numer. Math.* 8, 99–111.

Menzel, R. & Schwetlick, H. (1976): Über einen Ordnungsbegriff bei Einbettungsalgorithmen zur Lösung nichtlinearer Gleichungen. *Computing* 16, 187–199.

Menzel, R. & Schwetlick, H. (1978): Zur Lösung parameterabhängiger nichtlinearer Gleichungen mit singulëren Jacobi-Matrizen. *Numer. Math.* 30, 65–79.

Menzel, R. & Schwetlick, H. (1985): Parametrization via secant length and application to path following. *Numer. Math.* **47**, 401–412.

Merrill, O. (1972): A summary of techniques for computing fixed points of continuous mappings. In: *Mathematical topics in economic theory and computation*. R. Day & S. Robinson editors. SIAM, Philadelphia, PA, 130–149.

Merrill, O. (1972): Applications and extensions of an algorithm that computes fixed points of a certain upper semi-continuous point to set mapping. Ph. D. Thesis. Univ. of Michigan, Ann Arbor, MI.

Meyer, G. H. (1968): On solving nonlinear equations with a one-parameter operator imbedding. *SIAM J. Numer. Anal.* **5**, 739–752.

Meyerson, M. D. & Wright, A. H. (1979): A new and constructive proof of the Borsuk-Ulam theorem. *Proc. Amer. Math. Soc.* **73**, 134–136.

Meyer-Spasche, R. & Keller, H. B. (1980): Computation of the axisymmetric flow between rotating cylinders. *J. Comput. Phys.* **35**, 100–109.

Meyer-Spasche, R. & Keller, H. B. (1985): Some bifurcation diagrams for Taylor vortex flows. *Phys. Fluids* **28**, 1248–1252.

Miersemann, E. & Mittelmann, H. D. (1989): On the continuation for variational inequalities depending on an eigenvalue parameter. *Math. Methods Appl. Sci.* **11**, 95–104.

Miersemann, E. & Mittelmann, H. D. (1989): Continuation for parametrized nonlinear variational inequalities. *J. Comput. Appl. Math* **26**, 23-34.

Miersemann, E. & Mittelmann, H. D. (1989): Extension of Beckert's continuation method to variational inequalities. To appear in: *Math. Nachr.*.

Milnor, J. W. (1968): *Singular points of complex hypersurfaces*. Princeton University Press and the University of Tokyo Press, Princeton, New Jersey.

Milnor, J. W. (1969): *Topology from the differentiable viewpoint*. Univ. Press of Virginia, Charlottesville, VA.

Mittelmann, H. D. (1982): A bibliography on numerical methods for bifurcation problems. Preprint. Universität Dortmund.

Mittelmann, H. D. (1982): A fast solver for nonlinear eigenvalue problems. In: *Iterative solution of nonlinear systems of equations*. R. Ansorge & Th. Meis & W. Törnig editors. Lecture Notes in Math. **953**. Springer Verlag, Berlin, Heidelberg, New York, 46–67.

Mittelmann, H. D. (1984): Continuation near symmetry-breaking bifurcation points. In: *Numerical methods for bifurcation problems*. T. Küpper & H. Mittelmann & H. Weber editors. ISNM **70**. Birkhäuser Verlag, Basel, 319–334.

Mittelmann, H. D. (1986): Multilevel continuation techniques for nonlinear boundary value problems with parameter-dependence. *Appl. Math. Comput.* **19**, 265–282.

Mittelmann, H. D. (1986): A pseudo-arclength continuation method for nonlinear eigenvalue problems. *SIAM J. Numer. Anal.* **23**, 1007–1916.

Mittelmann, H. D. (1988): Multi-grid continuation and spurious solutions for nonlinear boundary value problems. *Rocky Mountain J. Math.* **18**, 387–401.

Mittelmann, H. D. (1987): On continuation for variational inequalities. *SIAM J. Numer. Anal.* **24**, 1374–1381.

Mittelmann, H. D. (1987): Continuation methods for parameter-dependent boundary value problems. To appear in: *Lectures in Appl. Math.* AMS, Providence, RI.

Mittelmann, H. D. (1990): Nonlinear parametrized equations: New results for variational problems and inequalities. In: *Computational solution of nonlinear systems of equations*, E. L. Allgower & K. Georg editors, American Mathematical Society, Providence.

Mittelmann, H. D. & Roose, D. (1989) (eds.): *Continuation techniques and bifurcation problems*. Special volume: J. Comput. Appl. Math **26**.

Mittelmann, H. D. & Weber, H. (1980): Numerical methods for bifurcation problems – a survey and classification. In: *Bifurcation problems and their numerical solution*. H. D. Mittelmann & H. Weber editors. ISNM **54**. Birkhäuser Verlag, Basel, 1–45.

Mittelmann, H. D. & Weber, H. (1980), (eds.): *Bifurcation problems and their numerical solution*. ISNM **54**. Birkhäuser Verlag, Basel.

Mittelmann, H. D. & Weber, H. (1985): Multi-grid solution of bifurcation problems. *SIAM J. Sci. Stat. Comput.* **6**, 49–60.

Mizuno, S. (1984): An analysis of the solution set to a homotopy equation between polynomials with real coefficients. *Math. Programming* **28**, 329–336.

Mizuno, S. & Yoshise, A. & Kikuchi, T. (1988): Practical polynomial time algorithms for linear complementarity problems. Technical Report #13 , Tokyo Institute of Technology.

Moore, G. & Spence, A. (1980): The calculation of turning points of nonlinear equations. *SIAM J. Numer. Anal.* **17**, 567–576.

Moré, J. J. (1974): Coercivity conditions in nonlinear complementarity problems. *SIAM Rev.* **16**, 1–16.

Moré, J. J. (1974): Classes of functions and feasibility conditions in nonlinear complementarity problems. *Math. Programming* **6**, 327–338.

Morgan, A. P. (1983): A method for computing all solutions to systems of polynomial equations. *ACM Trans. Math. Software* **9**, 1–17.

Morgan, A. P. (1986): A transformation to avoid solutions at infinity for polynomial systems. *Appl. Math. Comput.* **18**, 77–86.

Morgan, A. P. (1986): A homotopy for solving polynomial systems. *Appl. Math. Comput.* **18**, 87–92.

Morgan, A. P. (1987): *Solving polynomial systems using continuation for engineering and scientific problems.* Prentice-Hall, Englewood Cliffs, NJ.

Morgan, A. P. & Sommese, A. J. (1987): A homotopy for solving general polynomial systems that respects m-homogeneous structures. *Appl. Math. Comput.* **24**, 101–113.

Morgan, A. P. & Sommese, A. J. (1987): Computing all solutions to polynomial systems using homotopy continuation. *Appl. Math. Comput.* **24**, 115–138.

Morgan, A. P. & Sommese, A. J. (1989): Coefficient parameter polynomial continuation. *Appl. Math. Comput* **29**, 123–160.

Morgan, A. P. & Sommese, A. J. (1990): Generically nonsingular polynomial continuation. In: *Computational solution of nonlinear systems of equations*, E. L. Allgower & K. Georg editors, American Mathematical Society, Providence.

Morgan, A. P. & Sommese, A. J. & Wampler, C. W. (1990): Polynomial continuation for mechanism design problems. In: *Computational solution of nonlinear systems of equations*, E. L. Allgower & K. Georg editors, American Mathematical Society, Providence.

Morgan, A. P. & Sommese, A. J. & Wampler, C. W. (1989): Computing singular solutions to nonlinear analytic systems. Preprint, University of Notre Dame.

Morgan, A. P. & Sommese, A. J. & Watson, L. T. (1988): Finding all solutions to polynomial systems using HOMPACK. Res. Pub. GMR-6109, G. M. Research Labs.

Muller, R. E. (1982): Numerical solution of multiparameter eigenvalue problems. *Z. Angew. Math. Mech.* **62**, 681–686.

Murota,K. (1982): Global convergence of a modified Newton iteration for algebraic equations. *SIAM J. Numer. Anal.* **19**, 793–799.

Murty, K. G. (1978): Computational complexity of complementary pivot methods. In: *Complementarity and fixed point problems.* M. L. Balinski & R. W. Cottle editors. *Math. Programming Study* **7**. North-Holland, Amsterdam, New York, 61–73.

Neta, B. & Victory, H. D., Jr. (1982): A higher order method for determining nonisolated solutions of a system of nonlinear equations. *Computing* **32**, 163–166.

Netravali, A. N. & Saigal, R. (1976): Optimum quantizer design using a fixed-point algorithm. *Bell System Tech. J.* **55**, 1423–1435.

Nirenberg, L. (1974): *Topics in nonlinear functional analysis.* Courant Institute, New York.

Nussbaum, R. D. (1975): A global bifurcation theorem with applications to functional differential equations. *J. Func. Anal.* **19**, 319–338.

Nussbaum, R. D. (1979): Uniqueness and nonuniqueness for periodic solutions of $x'(t) = -g(x(t-1))$. *J. Diff. Equ.* **34**, 25–54.

Ogneva, V. A. & Chernyshenko, V. M. (1978): An analogue of the method for the continuation of the solution with respect to the parameter for nonlinear operator equations. *Mat. Zametki* **23**, 601–606.

Ojika, T. (1982): Sequential substitution method for the resultant of a system of nonlinear algebraic equations. *Mem. Osaka Kyoiku Univ. III Natur. Sci. Appl. Sci.* **31**, 63–69.

Ortega, J. M. & Rheinboldt, W. C. (1970): *Iterative solution of nonlinear equations in several variables.* Academic Press, New York, London.

Padovan, J. & Tovichakchaikul, S. (1982): Self-adaptive predictor-corrector algorithms for static nonlinear structural analysis. *Comput. & Structures* **15**, 365–377.

Pan, C. T. & Chao, K. S. (1978): A computer-aided root-locus method. *IEEE Trans. Automatic Control* **23**, 856–860.

Pan, C. T. & Chao, K. S. (1980): Multiple solutions of nonlinear equations: roots of polynomials. *IEEE Trans. Circuits and Systems* **27**, 825–832.

van der Panne, C. (1974): A complementary variant of Lemke's method for the linear complementarity problem. *Math. Programming* **7**, 283–310.

Peitgen, H.-O. (1982): Topologische Perturbationen beim globalen numerischen Studium nichtlinearer Eigenwert- und Verzweigungsprobleme. *Jahresbericht des Deutschen Mathematischen Vereins* **84**, 107–162.

Peitgen, H.-O. & Prüfer, M. (1979): The Leray-Schauder continuation method is a constructive element in the numerical study of nonlinear eigenvalue and bifurcation problems. In: *Functional differential equations and approximation of fixed points*. H.-O. Peitgen & H.-O. Walther editors. Lecture Notes in Math. **730**. Springer Verlag, Berlin, Heidelberg, New York, 326–409.

Peitgen, H.-O. & Saupe, D. & Schmitt, K. (1981): Nonlinear elliptic boundary value problems versus their finite difference approximations: Numerically irrelevant solutions. *J. Reine Angew. Math.* **322**, 74–117.

Peitgen, H.-O. & Schmitt, K. (1981): Positive and spurious solutions of nonlinear eigenvalue problems. In: *Numerical solutions of nonlinear equations*. E. L. Allgower & K. Glashoff & H.-O. Peitgen editors. Lecture Notes in Math. **878**. Springer Verlag, Berlin, Heidelberg, New York, 275–324.

Peitgen, H.-O. & Schmitt, K. (1983): Global topological perturbations in the study of nonlinear eigenvalue problems. *Math. Methods Appl. Sci.* **5**, 376–388.

Peitgen, H.-O. & Siegberg, H. W. (1981): An $\bar{\varepsilon}$-perturbation of Brouwer's definition of degree. In: *Fixed point theory*. E. Fadell & G. Fournier editors. Lecture Notes in Math. **886**. Springer Verlag, Berlin, Heidelberg, New York, 331–366.

Peitgen, H.-O. & Walther, H.-O. (1979), (eds.): *Functional differential equations and approximation of fixed points*. Lecture Notes in Math. **730**. Springer Verlag, Berlin, Heidelberg, New York.

Percell, P. (1980): Note on a global homotopy. *Numer. Funct. Anal. Optim.* **2**, 99–106.

Perozzi, D. J. (1980): Analysis of optimal stepsize selection in homotopy and continuation methods. Ph.D. Thesis, Part II, California Institute of Technology.

Peters, G. & Wilkinson, J. H. (1979): Inverse iteration, ill-conditioned equations and Newton's method. *SIAM Review* **21**, 339–360.

Plastock, R.: Global aspects of the continuation method..

Poincaré, H. (1881–1886): *Sur les courbes defini par une équation differentielle. I–IV*. In: Oeuvres I. Gauthier-Villars, Paris.

Polak, E. (1971): *Computational methods in optimization: a unified approach*. Academic Press, New York, London.

Polak, E. & Ribière, G. (1969): Note sur la convergence de méthodes de directions conjugées. *Rev. Française Informat. Recherche Opérationelle* **3**, 35–43.

Pönisch, G. (1985): Empfindlichkeitsanalyse von Rückkehrpunkten implizit definierter Raumkurven. *Wiss. Z. Techn. Univers. Dresden* **34**, 79–82.

Pönisch, G. & Schwetlick, H. (1981): Computing turning points of curves implicitly defined by nonlinear equations depending on a parameter. *Computing* **26**, 107–121.

Pönisch, G. & Schwetlick, H. (1982): Ein lokal überlinear konvergentes Verfahren zur Bestimmung von Rückkehrpunkten implizit definierter Raumkurven. *Numer. Math.* **38**, 455–466.

Poore, A. B. (1986): The expanded Lagrangian system for constrained optimization problems. Preprint. Colorado State University.

Poore, A. B. (1986): Continuation algorithms for linear programming. Preprint. Colorado State University.

Poore, A. B. & Al-Hassan, Q. (1988): The expanded Lagrangian system for constrained optimization problems. *SIAM J. Control and Optimization* **26**, 417–427.

Poore, A. B. & Tiahrt, C. A. (1987): Bifurcation problems in nonlinear parametric programming. *Math. Programming* **39**, 189–205.

Powell, M. J. D. (1970): A Fortran subroutine for solving nonlinear algebraic equations. In: *Numerical Methods For Nonlinear Algebraic Equations*. Ph. Rabinowitz editor, Gordon and Breach, New York, 115–161.

Powell, M. J. D. (1977): Restart procedures for the conjugate gradient method. *Math. Programming* **12**, 241–254.

Powell, M. J. D. (1981): An upper triangular matrix method for quadratic programming. In: *Nonlinear Programming* **4**, O. L. Mangasarian et al. editors, Academic Press, New York, 1–24.

Powell, M. J. D. (1985): On error growth in the Bartels-Golub and Fletcher-Matthews algorithms for updating matrix factorization. University of Cambridge, England, preprint DAMTP NA 10.

Press, W. H. & Flannery, B. P. & Teukolsky, S. A. & Vetterling, W. T. (1986): *Numerical recipes — The art of scientific computing.*Cambridge University Press, Cambridge.

Prüfer, M. (1978): Calculating global bifurcation. In: *Continuation methods.* H.-J. Wacker editor. Academic Press, New York, London, 187–213.

Prüfer, M. & Siegberg, H. W. (1979): On computational aspects of topological degree in R^n. In: *Functional differential equations and approximation of fixed points.* H.-O. Peitgen & H.-O. Walther editors. Lecture Notes in Math. **730**. Springer Verlag, Berlin, Heidelberg, New York, 410–433.

Prüfer, M. & Siegberg, H. W. (1981): Complementary pivoting and the Hopf degree theorem. *J. Math. Anal. Appl.* **84**, 133–149.

Rabinowitz, P. H. (1971): Some global results for nonlinear eigenvalue problems. *J. Funct. Anal.* **7**, 487–513.

Ramm, E. (1982): The Riks/Wempner approach – An extension of the displacement control method in nonlinear analysis. In: *Recent advances in non-linear computational mechanics.* E. Hinton & D. R. J. Owen & C. Taylor editors. Pineridge Press, Swansea, UK, 63–86.

Rao, C. R. & Mitra, S. K. (1971): *Generalized inverse of matrices and its applications.* John Wiley & Sons, New York.

Rappaz, J. (1983): Estimations d'erreur dans différentes normes pour l'approximation de problèmes de bifurcation. *C. R. Acad. Sc. Paris, Série I* **296**, 179–182.

Reinhart, L. (1980): Sur la résolution numérique de problèmes aux limites non linéaires par des méthodes de continuation. Thèse de 3-ème cycle. Univ. Paris VI.

Reinhart, L. (1982): On the numerical analysis of the von Karman equations: Mixed finite element approximation and continuation techniques. *Numer. Math.* **39**, 371–404.

Reinoza, A. (1985): Solving generalized equations via homotopy. *Math. Programming* **31**, 307–320.

Reiser, P. M. (1981): A modified integer labelling for complementarity algorithms. *Math. Oper. Res.* **6**, 129–139.

Renegar, J. (1984): Piecewise linear algorithms and integral geometry. Preprint. Colorado State University.

Renegar, J. (1985): On the complexity of a piecewise linear algorithm for approximating roots of complex polynomials. *Math. Programming* **32**, 301–318.

Renegar, J. (1985): On the cost of approximating all roots of a complex polynomial. *Math. Programming* **32**, 319–336.

Renegar, J. (1986): On the efficiency of a piecewise linear homotopy algorithm in approximating all zeros of a system of complex polynomials. To appear in: *Math. Oper. Res.*.

Renegar, J. (1988): A polynomial-time algorithm, based on Newton's method, for linear programming. *Math. Programming* **40**, 59–93.

Renegar, J. (1988): Rudiments of an average case complexity theory for piecewise-linear path following algorithms. *Math. Progr.* **40**, 113–164.

Rex, G. (1989): E-Verfahren und parameterabhängige nichtlineare Gleichungssysteme. Preprint, Karl-Marx-Universität Leibzig.

Rheinboldt, W. C. (1977): Numerical continuation methods for finite element applications. In: *Formulations and computational algorithms in finite element analysis.* K. - J. Bathe & J. T. Oden & W. Wunderlich editors. MIT Press, Cambridge, MA, 599–631.

Rheinboldt, W. C. (1978): An adaptive continuation process for solving systems of non-linear equations. In: *Mathematical models and numerical methods*. A. M. Tikhonov & F. Kuhnert & N. N. Kuznecov & K. Moszyński & A. Wakulícz editors. Banach Center Publ. **3**, 129–142.

Rheinboldt, W. C. (1978): Numerical methods for a class of finite dimensional bifurcation problems. *SIAM J. Numer. Anal.* **15**, 1–11.

Rheinboldt, W. C. (1980): Solution fields of nonlinear equations and continuation methods. *SIAM J. Numer. Anal.* **17**, 221–237.

Rheinboldt, W. C. (1981): Numerical analysis of continuation methods for nonlinear structural problems. *Comput. & Structures* **13**, 103–113.

Rheinboldt, W. C. (1982): Computation of critical boundaries on equilibrium manifolds. *SIAM J. Numer. Anal.* **19**, 653–669.

Rheinboldt, W. C. (1984): Differential-algebraic systems as differential equations on manifolds. *Math. Comput.* **43**, 473–482.

Rheinboldt, W. C. (1984): On some methods for the computational analysis of manifolds. In: *Numerical methods for bifurcation problems*. T. Küpper & H. D. Mittelmann & H. Weber editors. ISNM **70**. Birkhäuser Verlag, Basel, 401–425.

Rheinboldt, W. C. (1984): Error estimation for nonlinear parametrized equations. In: *Innovative methods for nonlinear problems*. W. K. Liu & T. Belytschko & K. C. Park editors. Pineridge Press, Swansea, UK, 295–312.

Rheinboldt, W. C. (1985): Folds on the solution manifolds of a parametrized equation. *SIAM J. Numer. Anal.* **22**, 729–735.

Rheinboldt, W. C. (1986): *Numerical analysis of parametrized nonlinear equations*. John Wiley & Sons, New York.

Rheinboldt, W. C. (1986): Error estimates and adaptive techniques for nonlinear parametrized equations. In: *Accuracy estimates and adaptivity for finite elements*. I.-Babuska & E. A. Oliveira & O. C. Zienkiewicz editors. Wiley-Interscience Publ., New York, 163–180.

Rheinboldt, W. C. (1987): On a moving-frame algorithm and the triangulation of equilibrium manifolds. In: *Bifurcation: analysis, algorithms, applications*. T. Küpper & R. Seydel & H. Troger editors. ISNM **79**. Birkhäuser Verlag, Basel, 256–267.

Rheinboldt, W. C. (1988): On the computation of multi-dimensional solution manifolds of parametrized equations. *Numer. Math.* **53**, 165–182.

Rheinboldt, W. C. (1988): Error questions in the computation of solution manifolds of parametrized equations. To appear in: *The role of interval methods in scientific computing*. R. E. Moore editor. Academic Press, New York, London.

Rheinboldt, W. C. & Burkardt, J. V. (1983): A locally parametrized continuation process. *ACM Trans. Math. Software* **9**, 215–235.

Rheinboldt, W. C. & Burkardt, J. V. (1983): Algorithm 596: a program for a locally parametrized continuation process. *ACM Trans. Math. Software* **9**, 236–241.

Ribarič, M. & Seliškar, M. (1974): On optimization of stepsize in the continuation method. *Math. Balkanica* **4**, 517–521.

Richter, S, L. & DeCarlo, R. A. (1983): Continuation methods: theory and applications. *IEEE Trans. Systems Man Cybernet.* **13**, 459–464. *IEEE Trans. Circuits and Systems* **30**, 347–352. *IEEE Trans. Automat. Control* **28**, 660–665.

Richter, S. & DeCarlo, R. A. (1984): A homotopy method for eigenvalue assignement using decentralized state feedback. *IEEE Trans. Automat. Control* **29**,148–158.

Riks, E. (1979): An incremental approach to the solution of snapping and buckling problems. *Internat. J. Solids and Structures* **15**, 529–551.

Riley, D. S. & Winters, K. H. (1989): Modal exchange mechanisms in Lapwood convection. To appear in: *J. Fluid Mech.*.

Riley, D. S. & Winters, K. H. (1989): A numerical bifurcation study of natural convection in a tilted two-dimensional porous cavity. Preprint TP.1319, Harwell Laboratory, England.

Roberts, S. M. & Shipman, J. S. (1967): Continuation in shooting methods for two-point boundary value problems. *J. Math. Anal. Appl.* **18**, 45–58.

Roberts, S. M. & Shipman, J. S. (1968): Justification for the continuation methods in two-point boundary value problems. *J. Math. Anal. Appl.* **21**, 23–30.

Roberts, S. M. & Shipman, J. S. (1974): The extended continuation method and invariant imbedding. *J. Math. Anal. Appl.* **45**, 32–42.

Roberts, S. M. & Shipman, J. S. & Roth, C. V. (1968): Continuation in quasilinearization. *J. Optim. Theory Appl.* **2**, 164–178.

Robinson, S. M. (1980), (ed.): *Analysis and computation of fixed points.* Academic Press, New York, London.

Rockafellar, R. T. (1970): *Convex analysis.* Princeton University Press, Princeton, NJ.

Ronto, V. A. (1980): Determination of the initial values of the solutions of nonlinear boundary value problems by the method of continuation of the solution with respect to the parameter. *Ukrain. Mat. Zh.* **32**, 128–133, 144.

Rosenmüller, J. (1971): On a generalization of the Lemke-Howson algorithm to noncooperative *N*-person games. *SIAM J. Appl. Math.* **21**, 73–79.

Rourke, C. P. & Sanderson, B. J. (1972): *Introduction to piecewise-linear topology.* Springer Verlag, Berlin, Heidelberg, New York.

Ruhe, A. (1973): Algorithms for the nonlinear eigenvalue problem. *SIAM J. Numer. Anal.* **10**, 674–689.

Rupp, T. (1988): *Kontinuitätsmethoden zur Lösung einparametrischer semi-infiniter Optimierungsprobleme.* Ph.D.-thesis, Universität Trier.

Ruys, P. H. M. & van der Laan, G. (1987): Computation of an industrial equilibrium. In: *The computation and modelling of economic equilibria,* A. J. J. Talman & G. van der Laan editors, North Holland, Amsterdam, 205–230.

Saad, Y. & Schultz, M. (1986): GMRES: a generalized minimal residual method for solving nonsymmetric linear systems. *SIAM J. Sci. Stat. Comp.* **7**, 856–869.

Saari, D. G. & Saigal, R. (1980): Some generic properties of paths generated by fixed point algorithms. In: *Analysis and computation of fixed points.* S. M. Robinson editor. Academic Press, New York, London, 57–72.

Sagara, N. & Fukushima, M. (1984): A continuation method for solving separable nonlinear least squares problems. *J. Comput. Appl. Math.* **10**, 157–161.

Saigal, R. (1971): Lemke's algorithm and a special linear complementarity problem. *Oper. Res.* **8**, 201–208.

Saigal, R. (1972): On the class of complementary cones and Lemke's algorithm. *SIAM J. Appl. Math.* **23**, 46–60.

Saigal, R. (1976): Extension of the generalized complementarity problem. *Math. Oper. Res.* **1**, 260–266.

Saigal, R. (1976): On paths generated by fixed point algorithms. *Math. Oper. Res.* **1**, 359–380.

Saigal, R. (1977): Investigations into the efficiency of fixed point algorithms. In: *Fixed points: algorithms and applications.* S. Karamardian editor. Academic Press, New York, London, 203–223.

Saigal, R. (1977): On the convergence rate of algorithms for solving equations that are based on methods of complementary pivoting. *Math. Oper. Res.* **2**, 108–124.

Saigal, R. (1979): On piecewise linear approximations to smooth mappings. *Math. Oper. Res.* **4**, 153–161.

Saigal, R. (1979): Fixed point computing methods. In: *Encyclopedia of computer science and technology.* Vol. **8**. Marcel Dekker Inc., New York, 545–566.

Saigal, R. (1979): The fixed point approach to nonlinear programming. In: *Point-to-set mappings and mathematical programming.* P. Huard editor. *Math. Programming Study* **10**. North-Holland, Amsterdam, New York, 142–157.

Saigal, R. (1983): A homotopy for solving large, sparse and structural fixed point problems. *Math. Oper. Res.* **8**, 557–578.

Saigal, R. (1983): An efficient procedure for traversing large pieces in fixed point algorithms. In: *Homotopy methods and global convergence.* B. C. Eaves & F. J. Gould & H.-O. Peitgen & M. J. Todd editors. Plenum Press, New York, 239–248.

Saigal, R. (1984): Computational complexity of a piecewise linear homotopy algorithm. *Math. Programming* **28**, 164–173.

Saigal, R. & Shin, Y. S. (1979): Perturbations in fixed point algorithms. In: *Functional differential equations and approximation of fixed points.* H.-O. Peitgen & H.-O. Walther editors. Lecture Notes in Math. **730**. Springer Verlag, Berlin, Heidelberg, New York, 434–441.

Saigal, R. & Simon, C. (1973): Generic properties of the complementarity problem. *Math. Programming* **4**, 324–335.

Saigal, R. & Solow, D. & Wolsey, L. A. (1975): A comparative study of two algorithms to compute fixed points over unbounded regions. In: *Proceedings 7th mathem. programming symposium*. Held in Stanford.

Saigal, R. & Todd, M. J. (1978): Efficient acceleration techniques for fixed point algorithms. *SIAM J. Numer. Anal.* **15**, 997–1007.

Salgovic, A. & Hlavacek, V. & Ilavsky, J. (1981): Global simulation of countercurrent separation processes via one-parameter imbedding techniques. *Chem. Eng. Sci.* **36**, 1599.

Santaló, L. A. (1976): Integral geometry and geometric probability. In: *Encyclopedia of Mathematics and its Applications, # 1*. Addison-Wesley, Reading, MA.

Sard, A. (1942): The measure of the critical values of differentiable maps. *Bull. Amer. Math. Soc.* **48**, 883–890.

Saupe, D. (1982): On accelerating PL continuation algorithms by predictor-corrector methods. *Math. Programming* **23**, 87–110.

Saupe, D. (1982): Topologische Perturbationen zum Wechsel der Triangulierung in PL-Algorithmen. *Z. Angew. Math. Mech.* **62**, 350–351.

Saupe, D. (1982): Beschleunigte PL-Kontinuitätsverfahren und periodische Lösungen parametrisierter Differentialgleichungen mit Zeitverzögerung. Ph.D. Thesis., Univ. of Bremen.

Saupe, D. (1982): Characterization of periodic solutions of special differential delay equations. In: *Proceedings Equadiff Würzburg 1982*, Springer Lecture Notes.

Saupe, D. (1982): Topologische Perturbationen zum Wechsel der Triangulierung in PL Algorithmen. *ZAMM* **62**, T350–T351.

Saupe, D. (1983): Global bifurcation of periodic solutions to some autonomous differential delay equations. *Appl. Math. Comput.* **13**, 185–211.

Scarf, H. E. (1967): The approximation of fixed points of a continuous mapping. *SIAM J. Appl. Math.* **15**, 1328–1343.

Scarf, H. E. (1967): The core of an N person game. *Econometrica* **35**, 50–69.

Scarf, H. E. (1984): The computation of equilibrium prices. In: *Applied general equilibrium analysis*. H. Scarf & J. Shoven editors. Cambridge University Press, Cambridge, MA, 207–230.

Scarf, H. E. & Hansen, T. (1973): *The computation of economic equilibria*. Yale University Press, New Haven, CT.

Schellhorn, J.-P. (1987): Fitting data through homotopy methods. In: *Statistical data analysis based on the L_1-norm and related methods*. Y. Dodge editor, Elsevier Science Publishers, 131–137.

Schilling, K. (1986): *Simpliziale Algorithmen zur Berechnung von Fixpunkten mengenwertiger Operatoren*. WVT Wissenschaftlicher Verlag, Trier, West Germany.

Schilling, K. (1988): Constructive proofs of fixed point theorems for set valued operators by simplicial algorithms. *Nonlinear Analysis, Theory, Methods and Applications* **12**, 565–580.

Schmidt, J. W. (1978): Selected contributions to imbedding methods for finite dimensional problems. In: *Continuation methods*. H.-J. Wacker editor. Academic Press, New York, London, 215–247.

Schmidt, Ph. H. (1990): PL methods for constructing a numerical implicit function. In: *Computational solution of nonlinear systems of equations*, E. L. Allgower & K. Georg editors, American Mathematical Society, Providence.

Schmidt, W. F. (1078): Adaptive step size selection for use with the continuation method. *Internat. J. Numer. Methods Engrg.* **12**, 677–694.

Schmitt, K. (1982): *A study of eigenvalue and bifurcation problems for nonlinear elliptic partial differential equations via topological continuation methods*. Lecture Notes, Institut de Mathématique Pure et Appliquée, Université Catholique de Louvain.

Schrauf, G. (1983): Lösungen der Navier-Stokes Gleichungen für stationäre Strömungen im Kugelspalt. Ph.D. Thesis. Preprint #611, SFB 72, Univ. Bonn.

Schrauf, G. (1983): Numerical investigation of Taylor-vortex flows in a spherical gap. *ZAMM* **63**, T282–T286.

Schreiber, R. & Keller, H. B. (1983): Driven cavity by efficient numerical techniques. *J. Comput. Phys.* **49**, 310–333.

Schwartz, J. T. (1969): *Nonlinear functional analysis*. Gordon and Breach Science Publishers, New York.

Schwetlick, H. (1975–1976): Ein neues Prinzip zur Konstruktion implementierbarer, global konvergenter Einbettungsalgorithmen. *Beiträge Numer. Math.* **4**, 215–228 and *Beiträge Numer. Math.* **5**, 201–206.

Schwetlick, H. (1979): *Numerische Lösung nichtlinearer Gleichungen*. VEB Deutscher Verlag der Wissenschaft, Berlin.

Schwetlick, H. (1982): *Zur numerischen Behandlung nichtlinearer parameterabhängiger Gleichungen*. Deutsche Akademie der Naturforscher LEOPOLDINA.

Schwetlick, H. (1984): Effective methods for computing turning points of curves implicitly defined by nonlinear equations. In: *Computational mathematics*. A. Wakulícz editor. Banach Center Publ. **13**, 623–645.

Schwetlick, H.(1984): On the choice of steplength in path following methods. *Z. Angew. Math. Mech.* **64**, 391–396.

Schwetlick, H. & Cleve, J. (1987): Higher order predictors and adaptive steplength control in path following algorithms. *SIAM J. Numer. Anal.* **14**, 1382–1393.

Seydel, R. (1979): Numerical computation of branch points in ordinary differential equations. *Numer. Math.* **32**, 51–68.

Seydel, R. (1979): Numerical computation of branch points of nonlinear equations. *Numer. Math.* **33**, 339–352.

Seydel, R. (1981): Numerical computation of periodic orbits that bifurcate from stationary solutions of ordinary differential equations. *Appl. Math. Comput.* **9**, 257–271.

Seydel, R. (1983): Branch switching in bifurcation problems for ordinary differential equations. *Numer. Math.* **41**, 93–116.

Seydel, R. (1984): A continuation algorithm with step control. In: *Numerical methods for bifurcation problems*. T. Küpper & H. D. Mittelmann & H. Weber editors. ISNM **70**. Birkhäuser Verlag, Basel, 480–494.

Seydel, R. (1988): BIFPACK: A program package for continuation, bifurcation and stability analysis, version 2.3. University of Würzburg.

Seydel, R. (1988): *From equilibrium to chaos. Practical bifurcation and stability analysis*. Elsevier, New York.

Shaidurov, V. V. (1973): Continuation with respect to the parameter in the method of regularization. In: *Numerical methods of linear algebra (in Russian)*. Vychisl. Centr Akad. Nauk SSSR Sibirsk. Otdel., Novosibirsk, 77–85.

Shamir, S. (1979): Fixed-point computation methods – some new high performance triangulations and algorithms. Ph.D. Thesis. Engineering-Economic Systems. Stanford University.

Shamir, S. (1980): Two new triangulations for homotopy fixed point algorithms with an arbitrary grid refinement. In: *Analysis and computation of fixed points*. S. M. Robinson editor. Academic Press, New York, London, 25–56.

Shampine, L. F. & Gordon, M. K. (1975): *Computer solutions of ordinary differential equations. The initial value problem*. W. H. Freeman and Co., San Francisco, CA.

Shapley, L. S. (1973): On balanced games without side payments. In: *Mathematical programming*. T. C. Hu & S. M. Robinson editors. Academic Press, New York, London, 261–290.

Shapley, L. S. (1974): A note on the Lemke-Howson algorithm. In: *Pivoting and extensions: in honor of A. W. Tucker*. M. L. Balinski editor. *Math. Programming Study* 1. North-Holland, Amsterdam, New York, 175–189.

Sherman, J. & Morrison, W. J. (1949): Adjustement of an inverse matrix corresponding to changes in the elements of a given column or a given row of the original matrix. *Ann. Math. Statist.* **20**, 621.

Shin, Y. S. & Haftka, R. T. & Watson, L. T. & Plaut, R. H. (1988): Tracing structural optima as a function of available resources by a homotopy method. *Computer Methods in Applied Mechanics and Engineering* **70**, 151–164.

Shoven, J. B. (1977): Applying fixed point algorithms to the analysis of tax policies. In: *Fixed points: algorithms and applications*. S. Karamardian editor. Academic Press, New York, London, 403–434.

Shub, M. & Smale, S. (1985–1986): Computational complexity: on the geometry of polynomials and a theory of cost. Part I: *Ann. Scient. École Norm. Sup.* **4** série t **18**, 107–142. Part II: *SIAM J. Computing* **15**, 145–161.

Sidorov, N. A. (1977): The method of continuation with respect to the parameter in the neighborhood of a branch point. In: *Questions in applied mathematics (in Russian)*. Sibirsk. Energet. Inst., Akad. Nauk SSSR Sibirsk. Otdel., Irkutsk, 109–113.

Siegberg, H. W. (1981): Some historical remarks concerning degree theory. *Amer. Math. Monthly* **88**, 125–139.

Siegberg, H. W. & Skordev, G. (1982): Fixed point index and chain approximations. *Pacific J. of Math.* **102**, 455–486.

Smale, S. (1976): A convergent process of price adjustement and global Newton methods. *J. Math. Econom.* **3**, 1–14.

Smale, S. (1981): The fundamental theorem of algebra and complexity theory. *Bull. Amer. Math. Soc.* **4**, 1–36.

Smale, S. (1985): On the efficiency of algorithms of analysis. *Bull. Amer. Math. Soc.* **13**, 87–121.

Smale, S. (1986): Algorithms for solving equations. Intern. Congress of Math..

Solow, D. (1981): Homeomorphisms of triangulations with applications to computing fixed points. *Math. Programming* **20**, 213–224.

Sonnevend, Gy. & Stoer, J. (1988): Global ellipsoidal approximations and homotopy methods for solving convex analytic programs. Preprint, University of Würzburg.

Spanier, E. H. (1966): *Algebraic topology*. McGraw-Hill, New York, Toronto, London.

Spence, A & Jepson, A. D. (1984): The numerical calculation of cusps, bifurcation points and isola formation points in two parameter problems. In: *Numerical methods for bifurcation problems*. T. Küpper & H. D. Mittelmann & H. Weber editors. ISNM **70**. Birkhäuser Verlag, Basel, 502–514.

Sperner, E. (1928): Neuer Beweis über die Invarianz der Dimensionszahl und des Gebietes. *Abh. Math. Sem. Univ. Hamburg* **6**, 265–272.

Stakgold, I. (1971): Branching of solutions of nonlinear equations. *SIAM Rev.* **13**, 289–332.

Stoer, J. (1983): Solution of large linear systems of equations by conjugate gradient type methods. In: *Mathematical programming: The state of the art*. A. Bachem & M. Grötschel & B. Korte editors. Springer Verlag, Berlin, Heidelberg, New York, 540–565.

Stoer, J. & Bulirsch, R. (1980): *Introduction to numerical analysis*. Springer Verlag, Berlin, Heidelberg, New York.

Stoer, J. & Witzgall, C. (1970): *Convexity and optimization in finite dimensions*. *I*. Springer Verlag, Berlin, Heidelberg, New York.

Stummel, F. & Hainer, K. (1982): *Praktische Mathematik*. Second edition, B. G. Teubner, Stuttgart.

Talman, A. J. J. (1980): Variable dimension fixed point algorithms and triangulations. Ph.D. Thesis. Math. Center Tracts **128**, Amsterdam.

Talman, A. J. J. & Van der Heyden, L. (1983): Algorithms for the linear complementarity problem which allow an arbitrary starting point. In: *Homotopy methods and global convergence*. B. C. Eaves & F. J. Gould & H.-O. Peitgen & M. J. Todd editors. Plenum Press, New York, 267–285.

Talman, A. J. J. & Yamamoto, Y. (1989): A simplicial algorithm for stationary point problems on polytopes. *Math. of Operations Research* **14**, 383–399.

Tanabe, K. (1979): Continuous Newton-Raphson method for solving an underdetermined system of nonlinear equations. *Nonlinear Anal.* **3**, 495–503.

Tanabe, K. (1987): Complementarity-enforcing centered Newton method for linear programming: Global method. In *New methods for linear programming*. Tokyo.

Thurston, G. A. (1969): Continuation of Newton's method through bifurcation points. *J. Appl. Mech. Tech. Phys.* **36**, 425–430.

Tikhonov, A. N. & Arsenin, V. Y. (1977): *Solution of ill-posed problems*. V. H. Winston & Sons, Washington, D. C..

Tillerton, J. R. & Stricklin, J. A. & Haisler, W. E. (1972): Numerical methods for the solution of nonlinear problems in structural analysis. In: *Numerical solution of nonlinear problems*. R. F. Hartung editor. AMD Vol. 6. ASME, New York.

Todd, M. J. (1974): A generalized complementary pivoting algorithm. *Math. Programming* **6**, 243–263.

Todd, M. J. (1976): Extensions of Lemke's algorithm for the linear complementarity problem. *J. Optim. Theory Appl.* **20**, 397–416.

Todd, M. J. (1976): *The computation of fixed points and applications*. Lecture Notes in Economics and Mathematical Systems **124**. Springer Verlag, Berlin, Heidelberg, New York.

Todd, M. J. (1976): On triangulations for computing fixed points. *Math. Programming* **10**, 322–346.

Todd, M. J. (1976): Orientation in complementary pivot algorithms. *Math. Oper. Res.* **1**, 54–66.

Todd, M. J. (1977): Union Jack triangulations. In: *Fixed points: algorithms and applications*. S. Karamardian editor. Academic Press, New York, London, 315–336.

Todd, M. J. (1978): Bimatrix games – an addendum. *Math. Programming* **14**, 112–115.

Todd, M. J. (1978): Fixed-point algorithms that allow restarting without extra dimension. Preprint. Cornell University. Ithaca, NY.

Todd, M. J. (1978): Improving the convergence of fixed-point algorithms. In: *Complementarity and fixed point problems*. M. L. Balinski & R. W. Cottle editors. *Math. Programming Study* **7**. North-Holland, Amsterdam, New York, 151–169.

Todd, M. J. (1978): Optimal dissection of simplices. *SIAM J. Appl. Math.* **34**, 792–803.

Todd, M. J. (1978): On the Jacobian of a function at a zero computed by a fixed point algorithm. *Math. Oper. Res.* **3**, 126–132.

Todd, M. J. (1979): Hamiltonian triangulations of R^n. In: *Functional differential equations and approximation of fixed points*. H.-O. Peitgen & H.-O. Walther editors. Lecture Notes in Math. **730**. Springer Verlag, Berlin, Heidelberg, New York, 470–483.

Todd, M. J. (1979): Piecewise linear paths to minimize convex functions may not be monotonic. *Math. Programming* **17**, 106–108.

Todd, M. J. (1980): Global and local convergence and monotonicity results for a recent variable-dimension simplicial algorithms. In: *Numerical solution of highly nonlinear problems*. W. Forster editor. North-Holland, Amsterdam, New York, 43–69.

Todd, M. J. (1980): A quadratically convergent fixed point algorithm for economic equilibria and linearly constrained optimization. *Math. Programming* **18**, 111–126.

Todd, M. J. (1980): Exploiting structure in piecewise-linear homotopy algorithms for solving equations. *Math. Programming* **18**, 233–247.

Todd, M. J. (1980): Traversing large pieces of linearity in algorithms that solve equations by following piecewise-linear paths. *Math. Oper. Res.* **5**, 242–257.

Todd, M. J. (1980): Numerical stability and sparsity in piecewise-linear algorithms. In: *Analysis and computation of fixed points*. S. M. Robinson editor. Academic Press, New York, London, 1–24.

Todd, M. J. (1981): Approximate labelling for simplicial algorithms and two classes of special subsets of the sphere. *Math. Oper. Res.* **6**, 579–592.

Todd, M. J. (1981): PLALGO: a FORTRAN implementation of a piecewise-linear homotopy algorithm for solving systems of nonlinear equations. Tech. Rep. No. 452. School of Operations Research and Industrial Engineering. Cornell University, Ithaca, NY.

Todd, M. J. (1982): An implementation of the simplex method for linear programming problems with variable upper bounds. *Math. Programming* **23**, 34–49.

Todd, M. J. (1982): An introduction to piecewise-linear homotopy algorithms for solving systems of equations. In: *Topics in numerical analysis*. P. R. Turner editor.Lecture Notes in Math. **965**. Springer Verlag, Berlin, Heidelberg, New York, 149–202.

Todd, M. J. (1982): On the computational complexity of piecewise-linear homotopy algorithms. *Math. Programming* **24**, 216–224.

Todd, M. J. (1982): Fixed-point methods for linear constraints. In: *Nonlinear Optimization*. M. J. D. Powell editor. Academic Press, New York, London, 147–154.

Todd, M. J. (1983): Piecewise-linear homotopy algorithms for sparse systems of nonlinear equations. *SIAM J. Control Optim.* **21**, 204–214.

Todd, M. J. (1983): Computing fixed points with applications to economic equilibrium models. In: *Discrete and system models*. W. F. Lucas & F. S. Roberts & R. M. Thrall editors. Springer Verlag, Berlin, Heidelberg, New York, 279–314.

Todd, M. J. (1984): Efficient methods of computing economic equilibria. In: *Applied general equilibrium analysis.* H. Scarf & J. Shoven editors. Cambridge University Press, Cambridge, MA, 51–68.

Todd, M. J. (1984): J': a new triangulation of $\mathbf{R}^n$. *SIAM J. Algebraic Discrete Methods* **5**, 244–254.

Todd, M. J. (1984): Complementarity in oriented matroids. *SIAM J. Algebr. Discr. Methods* **5**, 467–485.

Todd, M. J. (1985): "Fat" triangulations, or solving certain nonconvex matrix optimization problems. *Math. Programming* **31**, 123–136.

Todd, M. J. (1986): Polynomial expected behavior of a pivoting algorithm for linear complementarity and linear programming problems. *Math. Programming* **35**, 173–192.

Todd, M. J. (1987): Reformulation of economic equilibrium problems for solution by quasi-Newton and simplicial algorithms. In: *The computation and modelling of economic equilibria.* G. van der Laan & A. J. J. Talman editors. North-Holland, Amsterdam, 19–37.

Todd, M. J. & Acar, R. (1980): A note on optimally dissecting simplices. *Math. Oper. Res.* **5**, 63–66.

Todd, M. J. & Wright, A. H. (1980): A variable-dimension simplicial algorithm for antipodal fixed-point theorems. *Numer. Funct. Anal. Optim.* **2**, 155–186.

Toint, Ph. (1979): On the superlinear convergence of an algorithm for solving a sparse minimization problem. *SIAM J. Numer. Anal.* **16**, 1036–1045.

Troger, H. (1975): Ein Beitrag zum Durchschlagen einfacher Strukturen. *Acta Mech.* **23**, 179–191.

Tsai, L.-W. & Morgan, A. P. (1985): Solving the kinematics of the most general six- and five-degree-of-freedom manipulators by continuation methods. *ASME J. of Mechanisms, Transmissions and Automation in Design* **107**, 48–57.

Tuy, H. (1979): Pivotal methods for computing equilibrium points: unified approach and new restart algorithm. *Math. Programming* **16**, 210–227.

Tuy, H. (1980): Solving equations $0 \in f(x)$ under general boundary conditions. In: *Numerical solution of highly nonlinear problems.* W. Forster editor. North-Holland, Amsterdam, New York, 271–296.

Tuy, H. (1980): Three improved versions of Scarf's method using conventional subsimplices and allowing restart and continuation procedures. *Math. Operationsforsch. Statist. Ser. Optim.* **11**, 347–365.

Tuy, H. & v. Thoai, N. & d. Muu, L. (1978): A modification of Scarf's algorithm allowing restarting. *Math. Operationsforsch. Statist. Ser. Optim.* **9**, 357–372.

Ushida, A. & Chua, L. O. (1984): Tracing solution curves of nonlinear equations with sharp turning points. *Internat. J. Circuit Theory Appl.* **12**, 1–21.

Vasudevan, G. & Watson, L. T. & Lutze, F. H. (1988): A homotopy approach for solving constrained optimization problems. Preprint, Virginia Polytechnic Institute.

Vertgeïm, B. A. (1970): The approximate determination of fixed points of continuous mappings. *Soviet. Math. Dokl.* **11**, 295–298.

Vickery, D. J. & Taylor, R. (1986):Path-following approaches to the solution of multicomponent, multistage separation process problems. *A. I. Ch. E. J.* **32**, 547.

de Villiers, N. & Glasser, D. (1981): A continuation method for nonlinear regression. *SIAM J. Numer. Anal.* **18**, 1139–1154.

Wacker, H.-J. (1977): Minimierung des Rechenaufwandes bei Globalisierung spezieller Iterationsverfahren vom Typ Minimales Residuums. *Computing* **18**, 209–224.

Wacker, H.-J. (1978), (ed.): *Continuation methods* Academic Press, New York, London.

Wacker, H.-J. (1978): A summary of the development on imbedding methods. In: *Continuation methods.* H.-J. Wacker editor. Academic Press, New York, London, 1–35.

Wacker, H.-J. & Engl, H. W. & Zarzer, E. (1977): Bemerkungen zur Aufwandsminimierung bei Stetigkeitsmethoden sowie Alternativen bei der Behandlung der singulären Situation. In: *Numerik und Anwendungen von Eigenwertaufgaben und Verzweigungsproblemen.* E. Bohl & L. Collatz & K. P. Hadeler editors. ISNM 38. Birkhäuser Verlag, Basel, 175–193.

Wacker, H.-J. & Zarzer, E. & Zulehner, W. (1978): Optimal stepsize control for the globalized Newton method. In: *Continuation methods.* H.-J. Wacker editor. Academic Press, New York, London, 249–276.

van der Waerden, B. L. (1953): Modern algebra. Volumes I and II. Ungar, New York.

Walker, H. F. (1988): Implementation of the GMRES method using Householder transformations. *SIAM J. Sci. Stat. Comput.* **9**, 152–163.

Walker, H. F. (1990): Newton-like methods for underdetermined systems. In: *Computational solution of nonlinear systems of equations*, E. L. Allgower & K. Georg editors, American Mathematical Society, Providence.

Walker, H. F. & Watson, L. T. (1988): Least change secant update methods for underdetermined systems. Preprint, Utah State University.

Wampler, C. W. & Morgan, A. P. & Sommese, A. J. (1988): Numerical continuation methods for solving polynomial systems arising in kinematics. Research Publication GMR-6372, General Motors Research Laboratories, Warren, MI.

Wang, C. Y. & Watson, L. T. (1979): Squeezing of a viscous fluid between elliptic plates. *Appl. Sci. Res.* **35**, 195–207.

Wang, C. Y. & Watson, L. T. (1979): Viscous flow between rotating discs with injection on the porous disc. *Z. Angew. Math. Phys.* **30**, 773–787.

Wang, C. Y. & Watson, L. T. (1980): On the large deformations of C-shaped springs. *Intern. J. Mech. Sci.* **22**, 395–400.

Wasserstrom, E. (1971): Solving boundary-value problems by imbedding. *J. Assoc. Comput. Mach.* **18**, 594–602.

Wasserstrom, E. (1973): Numerical solutions by the continuation method. *SIAM Rev.* **15**, 89–119.

Watson, L. T. (1979): An algorithm that is globally convergent with probability one for a class of nonlinear two-point boundary value problems. *SIAM J. Numer. Anal.* **16**, 394–401.

Watson, L. T. (1979): Fixed points of C^2 maps. *J. Comput. Appl. Math.* **5**, 131–140.

Watson, L. T. (1979): A globally convergent algorithm for computing fixed points of C^2 maps. *Appl. Math. Comput.* **5**, 297–311.

Watson, L. T. (1979): Solving the nonlinear complementarity problem by a homotopy method. *SIAM J. Control Optim.* **17**, 36–46.

Watson, L. T. (1980): Computational experience with the Chow-Yorke algorithm. *Math. Programming* **19**, 92–101.

Watson, L. T. (1980): Solving finite difference approximations to nonlinear two-point boundary value problems by a homotopy method. *SIAM J. Sci. Statist. Comput.* **1**, 467–480.

Watson, L. T. (1981): Numerical study of porous channel flow in a rotating system by a homotopy method. *J. Comput. Appl. Math.* **7**, 21–26.

Watson, L. T. (1981): Engineering application of the Chow-Yorke algorithm. *Appl. Math. Comput.* **9**, 111–133.

Watson, L. T. (1983): Quadratic convergence of Crisfield's method. *Comput. & Structures* **17**, 69–72.

Watson, L. T. (1983): Engineering applications of the Chow-Yorke algorithm. In: *Homotopy methods and global convergence*. B. C. Eaves & F. J. Gould & H.-O. Peitgen & M. J. Todd editors. Plenum Press, New York, 287–308.

Watson, L. T. (1986): Numerical linear algebra aspects of globally convergent homotopy methods. *SIAM Rew.* **28**, 529–545.

Watson, L. T. (1989): Globally convergent homotopy methods: A tutorial..

Watson, L. T. & Billups, S. C. & Morgan, A. P. (1987): Algorithm 652. Hompack: A suite of codes for globally convergent homotopy algorithms. *ACM Transactions on Mathematical Software* **13**, 281–310.

Watson, L. T. & Bixler, J. P. & Poore, A. B. (1987): Continuous homotopies for the linear complementarity problem. Preprint, Virginia Polytechnic Institute.

Watson, L. T. & Fenner, D. (1980): Chow-Yorke algorithm for fixed points or zeros of C^2 maps. *ACM Trans. Math. Software* **6**, 252–260.

Watson, L. T. & Haftka, R. T. (1988): Modern homotopy methods in optimization. Preprint, Virginia Polytechnic Institute.

Watson, L. T. & Holzer, S. M. & Hansen, M. C. (1983): Tracking nonlinear equilibrium paths by a homotopy method. *Nonlinear Anal.* **7**, 1271–1282.

Watson, L. T. & Li, T.-Y. & Wang, C. Y. (1978): Fluid dynamics of the elliptic porous slider. *J. Appl. Mech.* **45**, 435–436.

Watson, L. T. & Scott, L. R. (1987): Solving Galerkin approximations to nonlinear two-point boundary value problems by a globally convergent homotopy method. *SIAM J. Sci. Stat. Comput.* **8**, 768–789.

Watson, L. T. & Scott, M. R. (1987): Solving spline-collocation approximations to nonlinear two-point boundary value problems by a homotopy method. *Appl. Math. Comput.* **24**, 333–357.

Watson, L. T. & Wang, C. Y. (1979): Deceleration of a rotating disc in a viscous fluid. *Phys. Fluids* **22**, 2267–2269.

Watson, L. T. & Wang, C. Y. (1981): A homotopy method applied to elastica problems. *Internat. J. Solids and Structures* **17**, 29–37.

Watson, L. T. & Yang, W. H. (1980): Optimal design by a homotopy method. *Applicable Anal.* **10**, 275–284.

Watson, L. T. & Yang, W. H. (1981): Methods for optimal engineering design problems based on globally convergent methods. *Comput. & Structures* **13**, 115–119.

Wayburn, T. L. & Seader, J. D. (1987): Homotopy continuation methods for computer-aided process design. *Comput. Chem. Eng.* **11**, 7.

Weber, H. (1979): Numerische Behandlung von Verzweigungsproblemen bei gewöhnlichen Differentialgleichungen. *Numer. Math.* **32**, 17–29.

Weber, H. (1980): Numerical solution of Hopf bifurcation problems. *Math. Methods Appl. Sci.* **2**, 178–190.

Weber, H. (1981): On the numerical solution of some finite-dimensional bifurcation problems. *Numer. Funct. Anal. Optim.* **3**, 341–366.

Weber, H. (1982): Zur Verzweigung bei einfachen Eigenwerten. *Manuscripta Math.* **38**, 77–86.

Weber, H. (1982): Numerical solution of a class of nonlinear boundary value problems for analytic functions. *Z. Angew. Math. Phys.* **33**, 301–314.

Weber, H. (1984): An efficient technique for the computation of stable bifurcation branches. *SIAM J. Sci. Statist. Comput.* **5**, 332–348.

Weber, H. (1985): Multigrid bifurcation iteration. *SIAM J. Numer. Anal.* **22**, 262–279.

Weber, H. & Werner, W. (1981): On the accurate determination of nonisolated solutions of nonlinear equations. *Computing* **26**, 315–326.

Wendland, W. L. (1978): On the imbedding method for semilinear first order elliptic systems and related finite element methods. In: *Continuation methods.* H.-J. Wacker editor. Academic Press, New York, London, 277–336.

Werner, B. & Spence, A. (1984): The computation of symmetry-breaking bifurcation points. *SIAM J. Numer. Anal.* **21**, 388–399.

Whalley, J. (1977): Fiscal harmonization in the EEC; some preliminary findings of fixed point calculations. In: *Fixed points: algorithms and applications.* S. Karamardian editor. Academic Press, New York, London, 435–472.

Whalley, J. & Piggott, J. (1980): General equilibrium analysis of taxation policy. In: *Analysis and computation of fixed points.* S. M. Robinson editor. Academic Press, New York, London, 183–195.

Wilmuth, R. (1977): A computational comparison of fixed point algorithms which use complementary pivoting. In: *Fixed points: algorithms and applications.* S. Karamardian editor. Academic Press, New York, London, 249–280.

Wilson, R. B. (1963): A simplicial algorithm for concave programming. Ph.D. Thesis. Harvard University.

Winters, K. H. (1988): A bifurcation analysis of three-dimensional Bénard convection. Preprint TP.1293, Harwell Laboratory, England.

Winkler, R. (1985): Path-following for two-point boundary value problems. Seminarbericht Nr. 78. Humboldt-Univerität, Berlin.

Wolsey, L. A. (1974): Convergence, simplicial paths and acceleration methods for simplicial approximation algorithms for finding a zero of a system of nonlinear equations. CORE Discussion Paper #7427. Univ. Cath. de Louvain, Belgium.

Wright, A. H. (1981): The octahedral algorithm, a new simplicial fixed point algorithm. *Math. Programming* **21**, 47–69.

Wright, A. H. (1985): Finding all solutions to a system of polynomial equations. *Math. Comp.* **44**, 125–133.

Yamamoto, Y. (1981): Subdivisions and triangulations induced by a pair of subdivided manifolds. Preprint.

Yamamoto, Y. (1982): The 2-ray method: a new variable dimension fixed point algorithm with integer labelling. Discussion Paper Series 154. Univ. of Tsukuba, Japan.

Yamamoto, Y. (1983): A new variable dimension algorithm for the fixed point problem. *Math. Programming* 25, 329–342.

Yamamoto, Y. (1984): A unifying model on retraction for fixed point algorithms. *Math. Programming* 28, 192–197.

Yamamoto, Y. (1984): A variable dimension fixed point algorithm and the orientation of simplices. *Math. Programming* 30, 301–312.

Yamamoto, Y. (1987): A path following algorithm for stationary point problems. *J. Oper. Res. Soc. Japan* 30, 181–198.

Yamamoto, Y. (1987): Stationary point problems and a path following algorithm. In: *Proceedings of the 8th Mathematical Programming Symposium*, Hiroshima, Japan, 153–170.

Yamamoto, Y. (1988): Orientability of a pseudomanifold and generalization of Sperner's lemma. *Journal of the Operations Research Society of Japan* 31, 19–42.

Yamamoto, Y. (1989): Fixed point algorithms for stationary point problems. In: *Mathematical Programming*, M. Iri & K. Tanabe editors, KTK Scientific Publishers, Tokyo, 283–307.

Yamamoto, Y. & Kaneko, M. (1986): The existence and computation of competitive equilibria in markets with an indivisible commodity. *J. Econom. Theory* 38, 118–136.

Yamamura, K. & Horiuchi, K. (1988): Solving nonlinear resistive networks by a homotopy method using a rectangular subdivision. To appear in: *IEEE Trans. on Circuits and Systems*.

Yamashita, H. (1979): A continuous path method of optimization and its application global optimization. In: *Survey of mathematical programming. Vol. I*. A. Prékopa editor. North-Holland, Amsterdam, New York, 539–546.

Yang, Z.-H. (1982): Continuation Newton method for boundary value problems of nonlinear elliptic differential equations. *Numer. Math. J. Chinese Univ.* 4, 28–37.

Yang, Z.-H. & Keller, H. B. (1986): A direct method for computing higher order folds. *SIAM J. Sci. Statist. Comput.* 7, 351–361.

Yomdin, Y. (1990): Sard's theorem and its improved versions in numerical analysis. In: *Computational solution of nonlinear systems of equations*, E. L. Allgower & K. Georg editors, American Mathematical Society, Providence.

Ypma, T. J. (1982): Following paths through turning points. *BIT* 22, 368–383.

Zangwill, W. I. (1977): An eccentric barycentric fixed point algorithm. *Math. Oper. Res.* 2, 343–359.

Zangwill, W. I. & Garcia, C. B. (1981): Equilibrium programming: the path-following approach and dynamics. *Math. Programming* 21, 262–289.

Židkov, E. P. & Puzynin, I. V. (1967): A method of introducing a parameter in the solution of boundary value problems for second order nonlinear ordinary differential equations (in Russian). *Ž. Vyčisl. Mat. i Mat. Fiz.* 7, 1086–1095.

Zienkiewicz, O. C. (1976): *The finite element method in engineering science*. 3rd edition. McGraw-Hill, London.

Zhou, Y. F. & Ruhe, A. (1985): Numerical path following and eigenvalue criteria for branch following. Preprint. Univ. of Göteborg, Sve..

Zulehner, W. (1988): A simple homotopy method for determining all isolated solutions to polynomial systems. *Math. Comp.* 50, 167–177.

Zulehner, W. (1988): On the solutions to polynomial systems obtained by homotopy methods. *Numer. Math.* 54, 303–317.

Index and Notation

$\|\,.\,\|$: usually denotes a norm, *see* norm

$|\,.\,|$: usually stands for the absolute value of a number, but occasionally may indicate a union, namely if $\mathcal{M}$ is a system of sets such as a PL manifold, then $|\mathcal{M}| = \cup_{\sigma \in \mathcal{M}} \sigma$ is the set subdivided by $\mathcal{M}$

$.\,[.]$ usually denotes a co-ordinate, e.g. $x[i]$ for $x \in \mathbf{R}^N$ and $i \in \{1, 2, \dots, N\}$ denotes the $i^{\,\mathrm{th}}$ co-ordinate of x

$[.\,,\dots,.]$, *see* simplex

$(.)^*$: denotes transposition, e.g. x^*y for $x, y \in \mathbf{R}^N$ is the scalar product of x and y

$\hat{\ }$: denotes the deletion of the element beneath

$(.)'$, *see* derivative

∇, *see* derivative

∂ : symbol for a partial derivative, or also for a subgradient, or also for the boundary of a set

$(.)^+$: Moore-Penrose inverse, *cf.* (3.2.2)

$(.)_+$, *cf.* (11.7.2)

$\perp$: sign for orthogonality, *cf. preceding* (3.2.5)

$(.)_-$, *cf.* (11.7.2)

$\#$ usually indicates the cardinality of a set, e.g. $\#\{2, 3, \dots, k\} = k - 1$

$(.)^\#$, *see* set valued hull, *cf.* (13.1.3)

$\setminus$: usually indicates the set-theoretical difference i.e. $A \setminus B = \{x \in A \mid x \notin B\}$

adjacent simplices, *cf.* (12.1.7)

aff(.) symbol for affine hull, *cf.* (14.1.3)

affinely independent, *cf.* (12.1.1)

angle, measure of curvature, *cf.* (6.1.7)

arclength, *cf. preceding* (1.7), *following* (2.1.4), (5.2.1), *preceding* (6.3.2), *preceding* (9.2.1)

arg min, *see* minimization

asymptotically linear map, *cf.* (13.1.2)

augmented Jacobian, *cf.* (2.1.5)

automatic pivot, *cf. section* 13.6

band structure, *cf.* (10.3.17)

barycenter, *cf.* (12.1.5)

barycentric co-ordinates, *cf.* (12.1.4)

Bezout's Theorem, *cf.* (11.6.4)

bifurcation equation, approximation of, *cf.* (8.1.7), (8.3.7)–(8.3.9)

bifurcation point, *cf.* (8.1.1)

bifurcation point, detection and approximation, *cf.* (8.3.1), *end of section* 9.2

bifurcation point, simple, *cf.* (8.1.11)

boundary of a set is usually denoted by the symbol ∂

SPRINGER SERIES IN COMPUTATIONAL MATHEMATICS

Springer-Verlag Berlin
Heidelberg New York London
Paris Tokyo Hong Kong

Springer

SPRINGER SERIES IN COMPUTATIONAL MATHEMATICS

Springer-Verlag
Berlin Heidelberg New York London Paris Tokyo Hong Kong